Protein Secretion in Bacteria

Protein Secretion in Bacteria

Edited by

Maria Sandkvist
Department of Microbiology and Immunology
University of Michigan Medical School
Ann Arbor, Michigan

Eric Cascales
Mediterranean Institute of Microbiology
CNRS Aix-Marseille Université
Marseille, France

Peter J. Christie
Department of Microbiology and Molecular Genetics
McGovern Medical School
Houston, Texas

ASM PRESS

Washington, DC

Library of Congress Cataloging-in-Publication Data

Names: Sandkvist, Maria, editor. | Cascales, Eric, editor. | Christie, Peter James, 1956- editor.
Title: Protein secretion in bacteria / edited by Maria Sandkvist, Eric Cascales, Peter J. Christie.
Description: Washington, DC : ASM Press, 2019. | Includes bibliographical references and index.
  | Summary: "Protein transport into and across membranes is a fundamental process in bacteria that touches upon and unites many areas of microbiology, including bacterial cell physiology, adhesion and motility, nutrient scavenging, intrabacterial signaling and social behavior, toxin deployment, interbacterial antagonism and collaboration, host invasion and disruption, and immune evasion. A broad repertoire of mechanisms and macromolecular machines are required to deliver protein substrates across bacterial cell membranes for intended effects. Some machines are common to most, if not all bacteria, whereas others are specific to Gram-negative or Gram-positive species or species with unique cell envelope properties such as members of Actinobacteria and Spirochetes. Protein Secretion in Bacteria, authored and edited by an international team of experts, draws together the many distinct functions and mechanisms involved in protein translocation in one concise tome. This comprehensive book presents updated information on all aspects of bacterial protein secretion encompassing: Individual secretory systems-Sec, Tat, and T1SS through the newly discovered T9SS Mechanisms, structures, and functions of bacterial secretion systems Lipoprotein sorting pathways, outer membrane vesicles, and the sortase system Structures and roles of surface organelles, including flagella, pili, and curli Emerging technologies and translational implications Protein Secretion in Bacteria serves as both an introductory guide for students and postdocs, and a ready reference for seasoned researchers whose work touches on protein export and secretion. This volume synthesizes the diversity of mechanisms of bacterial secretion across the microbial world into a digestible resource to stimulate new research, inspire continued identification and characterization of novel systems, and bring about new ways to manipulate these systems for biotechnological, preventative, and therapeutic applications"-- Provided by publisher.
Identifiers: LCCN 2019030619 | ISBN 9781683670278 (hardcover) | ISBN
  9781683670285 (ebook)
Subjects: LCSH: Bacterial proteins. | Secretion.
Classification: LCC QR92.P75 P757 2019 | DDC 612.4--dc23
LC record available at https://lccn.loc.gov/2019030619

10 9 8 7 6 5 4 3 2 1

Address editorial correspondence to
ASM Press, 1752 N St., N.W.,
Washington, DC 20036-2904, USA

Send orders to ASM Press, P.O. Box 605, Herndon, VA 20172, USA
Phone: 800-546-2416; 703-661-1593
Fax: 703-661-1501
E-mail: books@asmusa.org
Online: http://www.asmscience.org

Cover: Schematic illustration of the suggested model of the assembled SpoIIIA-IIQ channel. Courtesy of Natalie Zeytuni and Natalie C.J. Strynadka (*see* chapter 31). Cover design by Debra Naylor (Naylor Design, Inc).

Contents

Foreword

As a model organism, *Escherichia coli* K-12 allowed us to unravel the basics of molecular genetics and metabolism. However, about half a century of research devoted to *E. coli* K-12 had led in the 1970s to the dogma that Gram-negative bacteria do not secrete proteins. Although it was well known that some proteins were inserted into the outer membrane of *E. coli* K-12, the question of how these proteins could become inserted into this membrane did not haunt molecular biologists in the least. In the same vein, the flagellum was recognized as a sophisticated organelle extending from the bacterial body, but how this machine could assemble outside of the bacterial body remained a mystery. Last, but not least, conjugation was extensively studied, but all the attention was focused on DNA and the pilus. That conjugation also constituted transmembrane protein trafficking and thus "protein secretion" remained below the radar.

The study of uropathogenic *E. coli* strains by adventurous microbiologists interested in microbial pathogenesis led to the first evidence that a protein, hemolysin in this case, could be released from the periplasm across the outer membrane and hence was secreted (1). However, the former dogma did not fall all at once! Indeed, in the early 1990s, it was still very difficult for the scientific community to accept the fact that *Yersinia* organisms do indeed secrete proteins called Yops (*Yersinia* outer proteins), especially since some of them were poorly soluble in the culture medium (2). Moreover, it appeared next that this unorthodox secretion required many more genes than expected from what was known of the Sec system, raising a great perplexity. However, the genes involved in Yop secretion revealed a totally unexpected commonality with the *Salmonella* flagellar genes studied by the McNab school! This discovery solved two enigmas at once: first, the missing protein export apparatus of the flagellum was identified; second, the Yops secretion apparatus was shown to be related to the flagellum, which was then definitively demonstrated by an electron micrograph of the *Salmonella* needle complex (3). Last, but not least, the needle structure was consistent with the intracellular injection of Yops by extracellular *Yersinia* initially proposed by the Wolf-Watz group (4).

During the same period, the Hultgren-Waksman school, studying the pilus of uropathogenic *E. coli*, a structure that had been initially identified by simple adhesion tests and transmission electron microscopy pictures, unravelled the chaperone-usher pathway, one of the most elegant assembly pathways in bacterial biology (5).

These discoveries from the late 1990s opened a new and extraordinarily fruitful era of research on protein transport: many other export systems and surface nanomachines, as fascinating as type III secretion (T3S) and the chaperone-usher pathway, were discovered, unravelling an extraordinary diversity of structures and functions and opening a whole new domain of microbial biology that one could call "interorganism interaction biology." The outer membrane is no longer seen only as a permeability barrier but is also recognized as an assembly platform for an enormous variety of protein complexes ensuring vital functions such as motility, adherence, attack, capture of new genetic information, etc. But the list is not complete yet: the more recent interest in commensal bacteria of the *Bacteroidetes* phylum draws attention to the central role of an array of large feeding complexes of a new type, made of surface-exposed lipoproteins, and how these abundant lipoproteins reach the surface is not yet understood! Gram-positive bacteria have not been left out from this new wave of discoveries: unlike what has been long thought, they do have pili whose assembly mechanism, based on sortases (6), is as elegant as it is original. Finally, one of the most spectacular discoveries in the field of interorganism interactions is that of type VI secretion (T6S): the domestication of bacteriophages by bacteria to fight one another...or is it T6S that allowed some DNA to become viral? Like the understanding of T3S was greatly helped by the pioneering work on the flagellum, the rapid progress in the understanding of T6S owes much to the phage research carried out by giants in the 1960s and 1970s and even to the concept of bacteriocins, discovered by Andre Gratia...in 1925! All this illustrates once more, if needed, how science progresses by random encounters.

If you are among those who are, like me, fascinated by complex molecular machines and by the way bacteria interact with their environment and in particular with their pairs and predators, I strongly recommend the reading of this book!

Guy R. Cornelis
University of Namur
Namur, Belgium

References

1. **Springer W, Goebel W.** 1980. Synthesis and secretion of hemolysin by *Escherichia coli*. *J Bacteriol* **144:**53–59.
2. **Michiels T, Wattiau P, Brasseur R, Ruysschaert JM, Cornelis G.** 1990. Secretion of Yop proteins by yersiniae. *Infect Immun* **58:**2840–2849.
3. **Kubori T, Matsushima Y, Nakamura D, Uralil J, Lara-Tejero M, Sukhan A, Galán JE, Aizawa SI.** 1998. Supramolecular structure of the *Salmonella typhimurium* type III protein secretion system. *Science* **280:**602–605.
4. **Rosqvist R, Magnusson KE, Wolf-Watz H.** 1994. Target cell contact triggers expression and polarized transfer of Yersinia YopE cytotoxin into mammalian cells. *EMBO J* **13:**964–972.
5. **Sauer FG, Fütterer K, Pinkner JS, Dodson KW, Hultgren SJ, Waksman G.** 1999. Structural basis of chaperone function and pilus biogenesis. *Science* **285:**1058–1061.
6. **Ton-That H, Schneewind O.** 2003. Assembly of pili on the surface of *Corynebacterium diphtheriae*. *Mol Microbiol* **50:**1429–1438.

Preface

The bacterial cell envelope consists of one or two relatively impermeable lipid membranes, which play important roles in confining cellular contents and guarding the cell interior against environmental assault. However, protective, hydrophobic membranes also pose a barrier to the movement of small molecules, (e.g., ions and metals) and macromolecules (e.g., proteins and DNA) between the cytosol, extracytoplasmic compartments, and milieu. Bacteria have solved the challenge of membrane transport by deploying various translocation systems, each dedicated to the export or import of specific repertoires of small or large molecules. Approximately 30% of all bacterial proteins are transported into or across membranes where they contribute to the interplay between bacteria and their environment and engage in a wide range of functions including nutrient acquisition, adhesion and motility, exchange of genetic material, killing of neighboring bacteria, toxin deployment, invasion and/or disruption of host cells, and suppression of the host immune system. The role of protein secretion cannot be overstated in the battle between pathogen and host, a process that is generally multifaceted and involves virulence factors acting extracellularly or within the host cell.

Bacteria have evolved an exceptionally diverse array of mechanisms for inserting proteins into membranes or delivering them to or beyond the cell surface. Most mechanisms require the assembly of highly sophisticated molecular machines comprised of multiple proteins that span one or both membranes. Some secretion systems, such as the general secretory machine (Sec) and the twin-arginine translocation (Tat) system are highly conserved among most or all bacterial species. Other systems are more specialized and found only in Gram-negative or -positive species or among specific taxonomic groups. Proteins have many intrinsic properties that dictate whether they are translocated and by which system, including the presence of signal sequences, hydrophobicity, propensity to fold, cofactor binding requirements, and posttranslational modifications, e.g., lipid modification. Other factors influencing the mode of transport pertain to the final destination of a protein within the bacterial cell, on the bacterial cell surface, in the extracellular environment, or in a host cell. Questions surrounding how the many proteins destined for translocation are recruited to and delivered through dedicated protein translocation "nanomachines" are fascinating, although highly challenging topics for scientific inquiry. Through a combination of molecular, genetic, biochemical, and biophysical approaches, together with recently-developed cutting-edge ultrastructural and single-cell imaging technologies, we are making rapid progress in defining underlying mechanisms, dynamic actions, and structures at near atomic-resolution of bacterial protein translocation machines and processes.

This comprehensive volume encapsulates the latest information about protein transport systems in Gram-positive and -negative species. Topics cover mechanisms and dynamics of secretion, the intricate structures of secretion machines, fundamental roles of surface organelles and effectors in pathogenesis, and translational implications, all with an emphasis on recent insights gained through the use of state-of-the-art technologies. We hope that this comprehensive—yet succinct—set of chapters written by leading scientists in their respective fields will attract a wide readership of established researchers, postdocs, and students in academia and industry. A cover-to-cover reading will thoroughly inform the reader of underlying mechanistic themes as well as the exquisite diversity of bacterial protein translocation systems, inclusive of the Sec and Tat systems, the types I-IX secretion systems, lipoprotein sorting pathways, outer membrane vesicle trafficking systems, organellar (flagella, pilus) assembly systems, the Gram-positive sortase system, bacterial sporulation, and the colicin and filamentous phage uptake systems. By assembling this treatise, our overarching goals are to stimulate new research that will further define known secretion systems, lead to discovery of new secretion systems, and create new strategies aimed at manipulating these systems for novel biotechnological, preventative, and therapeutic applications.

We thank all the authors and the many anonymous reviewers for their outstanding contributions.

Maria Sandkvist
Eric Cascales
Peter J. Christie

Contributors

Marie-Stephanie Aschtgen
Department of Microbiology, Tumor and Cell Biology
Department of Clinical Microbiology
Karolinska University Hospital
Stockholm, Sweden

Tobias Beer
Institute of Biochemistry
Heinrich Heine University Düsseldorf
Düsseldorf, Germany

Barbara A. Bensing
Department of Medicine
San Francisco Veterans Affairs Medical Center and the University of California
San Francisco, California

Harris D. Bernstein
Genetics and Biochemistry Branch
National Institute of Diabetes and Digestive and Kidney Diseases
National Institutes of Health
Bethesda, Maryland

Sujeet Bhoite
Department of Molecular, Cellular, and Developmental Biology
University of Michigan
Ann Arbor, Michigan

Miriam Braunstein
Department of Microbiology and Immunology
University of North Carolina-Chapel Hill
Chapel Hill, North Carolina

Roland Brosch
Institut Pasteur
Unit for Integrated Mycobacterial Pathogenomics
UMR3525 CNRS
Paris, France

Lori L. Burrows
Department of Biochemistry and Biomedical Sciences
Michael G. DeGroote Institute for Infectious Disease Research
McMaster University
Hamilton, Ontario, Canada

Eric Cascales
Laboratoire d'Ingénierie des Systèmes Macromoléculaires (LISM)
Institut de Microbiologie de la Méditerranée (IMM)
Aix-Marseille Université, CNRS
UMR 7255
Marseille, France

Jessica D. Cecil
Oral Health Cooperative Research Centre
Melbourne Dental School
Bio21 Institute
The University of Melbourne
Melbourne, Victoria, Australia

Yunjie Chang
Department of Microbial Pathogenesis
Yale University School of Medicine
New Haven, Connecticut
Microbial Sciences Institute
Yale University
West Haven, Connecticut

Matthew R. Chapman
Department of Molecular, Cellular, and Developmental Biology
University of Michigan
Ann Arbor, Michigan

Yassine Cherrak
Laboratoire d'Ingénierie des Systèmes Macromoléculaires (LISM)
Institut de Microbiologie de la Méditerranée (IMM)
Aix-Marseille Université, CNRS
UMR 7255
Marseille, France

Peter J. Christie
Department of Microbiology and Molecular Genetics
McGovern Medical School
Houston, Texas

Laurie E. Comstock
Division of Infectious Diseases
Brigham and Women's Hospital
Harvard Medical School
Boston, Massachusetts

Peggy A. Cotter
Department of Microbiology and Immunology
University of North Carolina—Chapel Hill
Chapel Hill, North Carolina

Michael J. Coyne
Division of Infectious Diseases
Brigham and Women's Hospital
Harvard Medical School
Boston, Massachusetts

Ross E. Dalbey
Department of Chemistry and Biochemistry
The Ohio State University
Columbus, Ohio

Arnold J. M. Driessen
Molecular Microbiology
Groningen Biomolecular Sciences and Biotechnology Institute
Zernike Institute of Advanced Materials
University of Groningen
Groningen, The Netherlands

Denis Duché
Laboratoire d'Ingénierie des Systèmes Macromoléculaires, UMR7255
Institut de Microbiologie de la Méditerranée, Aix-Marseille Université—CNRS
Marseille, France

Eric Durand
Laboratoire d'Ingénierie des Systèmes Macromoléculaires (LISM)
Institut de Microbiologie de la Méditerranée (IMM)
Aix-Marseille Université, CNRS
UMR 7255
Marseille, France

Bevin C. English
Department of Medical Microbiology & Immunology
University of California, Davis
Davis, California

Isabelle N. Erenburg
Institute of Biochemistry
Heinrich Heine University Düsseldorf
Düsseldorf, Germany

Alyssa C. Fasciano
Program in Immunology
Sackler School of Graduate Biomedical Sciences
Tufts University School of Medicine
Boston, Massachusetts

Nicolas Flaugnatti
Laboratoire d'Ingénierie des Systèmes Macromoléculaires (LISM)
Institut de Microbiologie de la Méditerranée (IMM)
Aix-Marseille Université, CNRS
UMR 7255
Marseille, France
Laboratory of Molecular Microbiology
Global Health Institute, School of Life Sciences
Ecole Polytechnique Fédérale de Lausanne (EPFL)
Lausanne, Switzerland

Kelly M. Frain
The School of Biosciences
University of Kent
Canterbury, United Kingdom

Jorge E. Galán
Department of Microbial Pathogenesis
Yale University School of Medicine
New Haven, Connecticut

Marcin Grabowicz
Emory Antibiotic Resistance Center
Department of Microbiology & Immunology
Division of Infectious Diseases
Department of Medicine
Emory University School of Medicine
Atlanta, Georgia

Birgitta Henriques-Normark
Department of Microbiology, Tumor and Cell Biology
Department of Clinical Microbiology
Karolinska University Hospital
Stockholm, Sweden

I. Barry Holland
Institute of Genetics and Microbiology
University of Paris-Sud
Orsay, France

Manuela K. Hospenthal
Institute of Structural and Molecular Biology
University College London and Birkbeck
London, United Kingdom
Institute of Molecular Biology and Biophysics
ETH Zürich
Zürich, Switzerland

Laetitia Houot
Laboratoire d'Ingénierie des Systèmes Macromoléculaires, UMR7255
Institut de Microbiologie de la Méditerranée, Aix-Marseille Université—CNRS
Marseille, France

P. Lynne Howell
Department of Biochemistry
University of Toronto
Program in Molecular Medicine
Peter Gilgan Centre for Research and Learning
The Hospital for Sick Children,
Toronto, Ontario, Canada

Bo Hu
Department of Microbiology and Molecular Genetics
McGovern Medical School
Houston, Texas

Grant J. Jensen
Department of Biology and Biological Engineering
California Institute of Technology
Howard Hughes Medical Institute
Pasadena, California

Laure Journet
Laboratoire d'Ingénierie des Systèmes Macromoléculaires (LISM)
Institut de Microbiologie de la Méditerranée (IMM)
Aix-Marseille Université, CNRS
UMR 7255
Marseille, France

Kerstin Kanonenberg
Institute of Biochemistry
Heinrich Heine University Düsseldorf
Düsseldorf, Germany

Amalina Ghaisani Komarudin
Molecular Microbiology
Groningen Biomolecular Sciences and Biotechnology Institute
Zernike Institute of Advanced Materials
University of Groningen
Groningen, The Netherlands

Konstantin V. Korotkov
Department of Molecular and Cellular Biochemistry
University of Kentucky
Lexington, Kentucky

Anne Marie Krachler
Department of Microbiology and Molecular Genetics
McGovern Medical School
The University of Texas Health Science Center at Houston
Houston, Texas

Maria Lara-Tejero
Department of Microbial Pathogenesis
Yale University School of Medicine
New Haven, Connecticut

Yang Grace Li
Department of Microbiology and Molecular Genetics
McGovern Medical School
Houston, Texas

Trevor Lithgow
Infection & Immunity Program
Biomedicine Discovery Institute
Department of Microbiology
Monash University
Clayton, Australia

Jun Liu
Department of Microbial Pathogenesis
Yale University School of Medicine
New Haven, Connecticut
Microbial Sciences Institute
Yale University
West Haven, Connecticut

Mark J. Mcbride
Department of Biological Sciences
University of Wisconsin-Milwaukee
Milwaukee, Wisconsin

Matthew Mccallum
Department of Biochemistry
University of Toronto
Program in Molecular Medicine
Peter Gilgan Centre for Research and Learning
The Hospital for Sick Children,
Toronto, Ontario, Canada

Joan Mecsas
Department of Molecular Biology and Microbiology
Tufts University School of Medicine
Boston, Massachusetts

Dominique Missiakas
Department of Microbiology
University of Chicago
Chicago, Illinois

Sandra Muschiol
Department of Microbiology, Tumor and Cell Biology
Department of Clinical Microbiology
Karolinska University Hospital
Stockholm, Sweden

Priyanka Nannapaneni
Department of Microbiology, Tumor and Cell Biology
Department of Clinical Microbiology
Karolinska University Hospital
Stockholm, Sweden

Zachary M. Nash
Department of Microbiology and Immunology
University of North Carolina—Chapel Hill
Chapel Hill, North Carolina

Neil M. O'Brien-Simpson
Oral Health Cooperative Research Centre
Melbourne Dental School
Bio21 Institute
The University of Melbourne
Melbourne, Victoria, Australia

Catherine M. Oikonomou
Department of Biology and Biological Engineering
California Institute of Technology
Pasadena, California

John J. Psonis
Department of Molecular Genetics and Microbiology
School of Medicine
Center for Infectious Diseases
Stony Brook University
Stony Brook, New York

Han Remaut
Structural Biology Brussels
Vrije Universiteit Brussel
Structural and Molecular Microbiology
Structural Biology Research Center, VIB
Brussels, Belgium

Dante P. Ricci
Department of Early Research
Achaogen, Inc.
South San Francisco, California

Colin Robinson
The School of Biosciences
University of Kent
Canterbury, United Kingdom

Maria Sandkvist
Department of Microbiology and Immunology
University of Michigan Medical School
Ann Arbor, Michigan

Lutz Schmitt
Institute of Biochemistry
Heinrich Heine University Düsseldorf
Düsseldorf, Germany

Olaf Schneewind
Department of Microbiology
University of Chicago
Chicago, Illinois

Lamyaa Shaban
Program in Molecular Microbiology
Sackler School of Graduate Biomedical Sciences
Tufts University School of Medicine
Boston, Massachusetts

Sri Karthika Shanmugam
Department of Chemistry and Biochemistry
The Ohio State University
Columbus, Ohio

Thomas J. Silhavy
Department of Molecular Biology
Princeton University
Princeton, New Jersey

Natalie Sirisaengtaksin
Department of Microbiology and Molecular Genetics
McGovern Medical School
The University of Texas Health Science Center at Houston
Houston, Texas

Olivia Spitz
Institute of Biochemistry
Heinrich Heine University Düsseldorf
Düsseldorf, Germany

Natalie C.J. Strynadka
Department of Biochemistry and Molecular Biology and the Center
for Blood Research
University of British Columbia
Vancouver, British Columbia, Canada

Christopher J. Stubenrauch
Infection & Immunity Program
Biomedicine Discovery Institute
Department of Microbiology
Monash University
Clayton, Australia

Paul M. Sullam
Department of Medicine
San Francisco Veterans Affairs Medical Center and the University of California
San Francisco, California

David G. Thanassi
Department of Molecular Genetics and Microbiology
School of Medicine
Center for Infectious Diseases
Stony Brook University
Stony Brook, New York

April Y. Tsai
Department of Medical Microbiology & Immunology
University of California, Davis
Davis, California

Renée M. Tsolis
Department of Medical Microbiology & Immunology
University of California, Davis
Davis, California

Jan Maarten Van Dijl
University of Groningen
University Medical Center Groningen
Department of Medical Microbiology
Groningen, The Netherlands

Nani Van Gerven
Structural Biology Brussels
Vrije Universiteit Brussel
Structural and Molecular Microbiology
Structural Biology Research Center, VIB
Brussels, Belgium

Farzam Vaziri
Institut Pasteur
Unit for Integrated Mycobacterial Pathogenomics
UMR3525 CNRS
Paris, France
Department of Mycobacteriology and Pulmonary Research
Microbiology Research Center
Pasteur Institute of Iran
Tehran, Iran

Gabriel Waksman
Institute of Structural and Molecular Biology
University College London and Birkbeck
London, United Kingdom

Natalie Zeytuni
Department of Biochemistry and Molecular Biology and the Center
for Blood Research
University of British Columbia
Vancouver, British Columbia, Canada

Wolfram R. Zückert
Department of Microbiology, Molecular Genetics and Immunology
University of Kansas School of Medicine
Kansas City, Kansas

Electron Cryotomography of Bacterial Secretion Systems

1

CATHERINE M. OIKONOMOU[1] and GRANT J. JENSEN[1,2]

INTRODUCTION

The envelope of bacterial cells consists of at least one—and often two—membranes, a cell wall, and possibly a surface layer. This envelope allows cells to differentiate themselves from their environment and handle the resulting osmotic pressure, but it also presents a significant obstacle. Anything a cell wishes to export, from a motility appendage to a plasmid, needs to be ushered across this barrier. To accomplish this, bacteria have evolved a battery of secretion systems. Secretion systems are often constructed from dozens of protein building blocks embedded in the cell's envelope. The size, complexity, and location of these machines make them a particular challenge for structural characterization. High-resolution structure determination techniques such as X-ray crystallography and transmission electron microscopy (TEM)-based single-particle reconstruction (SPR) require objects to be purified from their cellular environment. This is problematic for membrane-associated proteins, which embed in the lipid bilayer by means of exposed hydrophobic patches. These patches can be protected during purification by adding detergents to the solvent, but often structural alterations still occur. Secretion systems are also

[1]Department of Biology and Biological Engineering, California Institute of Technology, Pasadena, CA 91125
[2]Howard Hughes Medical Institute, Pasadena, CA 91125

Protein Secretion in Bacteria
Edited by Maria Sandkvist, Eric Cascales, and Peter J. Christie
© 2019 American Society for Microbiology, Washington, DC
doi:10.1128/microbiolspec.PSIB-0019-2018

unusually large targets and frequently lose peripheral or loosely-associated components during purification. In addition, they often cross two membranes and are avidly linked to the cell wall.

Before the advent of electron cryotomography (ECT), we knew an impressive amount about what secretion systems are made of; genetics revealed components, biochemistry revealed their properties, and X-ray crystallography and SPR revealed the structures of purified pieces. But we knew much less about how the pieces fit together into a functional whole. For that, we turned to a complementary technique: ECT. While other techniques reveal structural details of purified components, ECT can be applied directly to intact cells. Briefly, a small volume of bacterial culture is applied to an EM support grid. This grid is then plunged into a cryogen that freezes the sample so quickly that the water cannot crystallize, instead remaining amorphous. The cells are thereby immobilized in a fully hydrated, native state. This frozen sample is then transferred to the TEM for cryo-EM imaging. A three-dimensional reconstruction of a cell is built up from two-dimensional projection images taken from different angles, accomplished by rotating the sample between images. This reconstruction, or tomogram, shows a snapshot of the contents of the cell without fixatives or stains, typically with a mid-range resolution (~5 nm) sufficient to see the shapes of large macromolecular complexes like secretion systems and their abundance and distribution in the cell. This can be particularly powerful for machines that exist in multiple states *in vivo*, only one of which may persist through purification. Images of different copies of the same structure, in the same cell or from many cells, can be merged by subtomogram averaging to generate a higher signal-to-noise view of the structure's inflexible regions (flexible or variable regions get washed out in the average, along with density from nearby but unassociated proteins) (1). While the resolution of these averages is still too low to unambiguously orient atomic

structures, in ideal cases, we can combine the information ECT provides about the form (s) of a structure *in situ* with high-resolution knowledge about its isolated components from complementary structural biology techniques.

ECT has served the study of bacterial secretion systems in two main ways: first, by simply showing what structures look like inside cells, and second, by revealing the relative arrangement of building blocks within those structures. This has provided some spectacular glimpses into how form is transformed into function, which in some cases have dramatically advanced fields of research. Here we briefly highlight some examples of these insights into bacterial secretion systems provided by ECT; the biology of the systems is detailed in other chapters. For recent, in-depth reviews of the subject, including both ECT and SPR, we recommend references 2 and 3. In addition, reference 4 provides an excellent overview of the diversity of bacterial secretion systems.

TYPE II SECRETION SYSTEMS

Capturing snapshots of cellular machinery in different states can often yield insights into how it works. For instance, ECT of the type IVa pilus (T4aP)—a type II secretion system (T2SS)-related machine that extends and retracts a long extracellular pilus for motility—showed that the structure comprises a series of stacked rings spanning the periplasm (Fig. 1A) (5). Imaging structures with and without pili assembled revealed how these rings undergo an ~3-nm rearrangement to ungate the channel and allow the assembling filament to extend out of the cell (5, 6).

For the T4aP, the high-resolution structures of many components had been solved, and biochemical interactions between many components had been worked out (7–28). Still, how the pieces fit together was a puzzle. To solve it, we applied ECT to a series of mutant strains in which T4aP components were

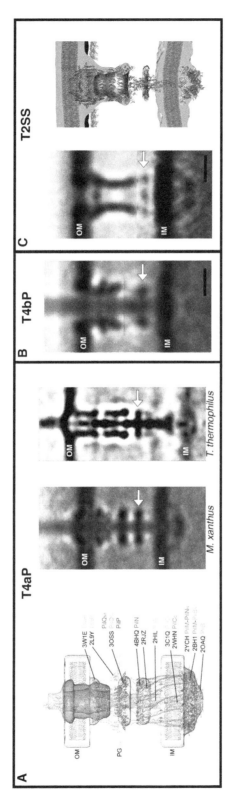

FIGURE 1 T2SS. (A) ECT revealed the structures of the T4aP in *Myxococcus xanthus* and *Thermus thermophilus*, enabling high-resolution structures to be placed into the map to produce a hypothetical model of the machine. Further studies revealed structural convergence of the related T4bP in *Vibrio cholerae* (B) and T2SS in *Legionella pneumophila* (C), with a similar periplasmic ring (arrows) formed by nonhomologous proteins in each system. Note in this and subsequent figures how bacterial secretion systems often locally distort the cell envelope *in vivo*. Images are reprinted from the following with permission: panel A left and middle, reference 6; panel A right, reference 5; panel B, reference 29; and panel C, reference 31.

deleted or tagged with additional protein density. By looking for corresponding absences or additions in the structure, we deduced the approximate locations of the components. Guided by clues from biochemical studies, we could thus place each high-resolution three-dimensional puzzle piece in its proper location (Fig. 1A) (6). Knowing where different pieces lie in the finished form often gives clues about how they might function. For instance, we saw that the ATPase in the cytoplasm was leveraged by a clamp connected to the integrated system of rings. This likely allows the assembly ATPase to rotate an adapter protein in the inner membrane (IM) to scoop pilin subunits from the membrane into the assembling pilus. Switching to a retraction mode, the clamp recruits a different ATPase to release subunits back into the IM. The signal for this switch from extension to retraction is likely sensed by a ring in the periplasm and transmitted through coiled-coil domains to the ATPase clamp (6).

Often, knowing the map of one structure has a domino effect, triggering insights into other, related machines. ECT of the type IVb pilus (T4bP) revealed a structure strikingly similar to that of the T4aP despite the fact that only half its components have homologs in the T4aP (Fig. 1B) (29). For instance, a ring in the periplasm thought to act as a sensor in the T4aP is also present in the T4bP, but it is formed by an unrelated protein. Another member of the family, the pathogenic T2SS, does not assemble a long extracellular pilus filament but rather pumps out effector proteins (30). Again, ECT mapping found the same ring, again formed by a nonconserved protein (Fig. 1C). In this case, the ring was flexible with respect to the rest of the machine, suggesting a possible role in loading cargo (31).

TYPE III SECRETION SYSTEMS

The first secretion system studied by ECT was the type III secretion system (T3SS)-based flagellar motor, responsible for assembling,

and subsequently rotating, the long flagellar filaments that propel the motility of many bacterial cells (32). The flagellar motor had been a subject of fascination for decades, and elements of its mechanism (33) and components (34) were well understood. Numerous EM studies, including SPR, of purified motors had revealed its core structure (for examples, see references 35 to 40). Still, key details remained unresolved, including the structure of the export complex in the cytoplasm responsible for assembling the machine and the interaction between the rotating (rotor) and stationary (stator) portions of the motor. ECT studies were able to reveal the interactions between these components, as well as the structure and location of peripheral components lost upon purification, such as the export apparatus (32, 41–43).

One of the most striking observations that came from studying flagellar motors *in situ* was how their form is adapted to the lifestyles of different species (Fig. 2A) (44). For instance, homologous components come together in different stoichiometries to form rings of different diameters, allowing the motor to gear up or down and generate different amounts of torque (32, 45–47). Some species (particularly pathogens) have elaborated their motors with additional components, perhaps to stabilize them against the increased load of more viscous environments (48) or to keep their associated filaments sheathed in membrane (49).

Pathogenic bacteria like *Yersinia* and *Salmonella* use a different T3SS to assemble, instead of a motility apparatus, a needle-like injectisome to deliver effector proteins not only across the bacterium's envelope but also further past the membrane of a target host cell (50). Injectisomes and flagellar motors are constructed from homologous components but serve strikingly different functions (Fig. 2B) (51). In an example of how seeing the shapes of machines *in situ* can provide clues about how they work, ECT revealed how the export apparatus—also known as the sorting complex—differs from that of the flagellar motor, with subunits forming no

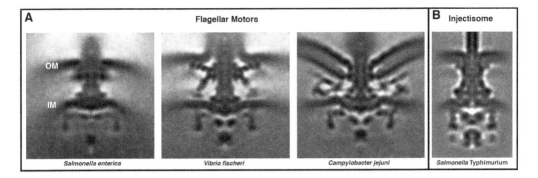

FIGURE 2 **T3SS. ECT and subtomogram averaging revealed the** *in situ* **structures of diverse T3SSs, including the flagellar motors of many bacteria (three examples, from** *Salmonella enterica, Vibrio fischeri,* **and** *Campylobacter jejuni***) (A) and the injectisome from** *Salmonella* **Typhimurium (B). Structures shown from left to right: EMD-3154, EMD-3155, and EBD-3150 from reference 45 and EMD-2667 from reference 52.**

longer a wide rotary ring but rather radial spokes better suited to recognize and feed diverse effector proteins to the central export channel (52). Comparing ECT images of injectisomes *in vivo* with and without the sorting complex showed how the radial symmetry induces a dramatic rearrangement of a ring at the base of the needle (53). Even inside the cell, the sorting complex is fragile; deletion of individual components often disrupts the whole complex (53, 54). These pieces can, however, be functionally tagged with additional protein density, which allowed them to be mapped by ECT. Knowing where components are located, and in what orientation, often provides important clues about how the system works. For instance, seeing that the substrate/chaperone-binding domain of the ATPase sits adjacent to the export gate suggests that the cage of the sorting complex forms an antechamber in which substrates are prepared for secretion (53).

Recently, ECT was used to capture the interaction of *Salmonella enterica* serovar Typhimurium injectisomes with host cells, revealing the translocon pore at the tip of the complex in the host membrane and supporting a model of direct injection of substrates across the host membrane (55). The translocon was much smaller (~13.5-nm outer diameter) than a homologous structure from enteropathogenic *Escherichia coli* (EPEC) previously reconstituted in liposomes (~55 to 65 nm) (56), underscoring the importance of studying secretion systems *in situ*.

TYPE IV SECRETION SYSTEMS

Attempting to purify structures out of the cell often leads to confounding alterations. For instance, SPR structures of purified type IV secretion systems (T4SSs) showed a chamber adjacent to the outer membrane (OM), connected to the IM by a solid stalk, suggesting that substrates enter the outer chamber from the periplasm (57, 58). It was therefore a surprise when ECT structures of the Dot/Icm T4SS *in situ* revealed instead an open channel extending through the periplasm (Fig. 3A) (59, 60). This feature, together with observations of complex assemblies of multiple ATPases in the cytoplasm and flexible "wings" in the periplasm (59–61), suggests that there might be two paths by which substrates could pass through the Dot/Icm T4SS: one through the central channel from the cytoplasm and one laterally into the outer chamber from the periplasm (60). This could help explain how the system can export such a spectacular number (~300) of discrete effectors (62).

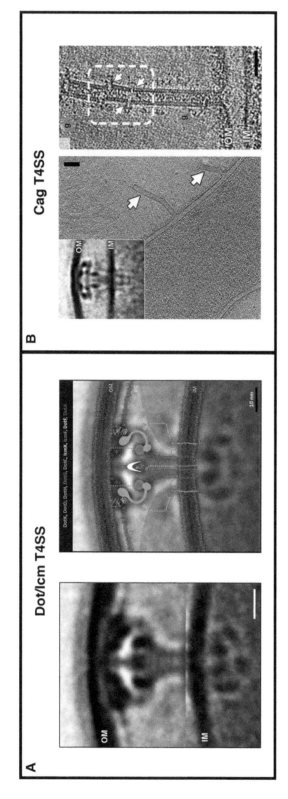

FIGURE 3 T4SS. (A) ECT revealed the *in situ* structure of the *Legionella pneumophila* Dot/Icm T4SS, including a central channel extending from the inner membrane, and enabled components to be placed in the map. (B) It also revealed the similar structure of the *Helicobacter pylori cag* T4SS (inset), as well as novel, related OM tubes. Images are reprinted from the following with permission: panel A, reference 60 (but see also reference 59); panel B, reference 61.

Putting structures in context inside the cell can answer many questions, but it often raises more. ECT images of the *cag* T4SS in *Helicobacter pylori* showed periplasmic machines structurally very similar to the Dot/Icm T4SS. Next to them, however, we sometimes saw wide (~40-nm) OM tubes, studded with portals, extending outward up to 500 nm from the cell (Fig. 3B) (61). The two structures appear to be related, but how, and what the resulting mechanism of effector secretion might be, remains puzzling.

TYPE V SECRETION SYSTEMS

Sometimes, even seeing something as simple as a stick can be illuminating. Many bacteria kill nearby cells through the process of contact-dependent inhibition (CDI). In *E. coli* this process is mediated by a type V secretion system (T5SS) comprising two proteins: (i) CdiB, a β-barrel in the OM that secretes (ii) CdiA, a large multidomain protein containing the toxin (63). It was known that toxin translocation is mediated by interaction of CdiA with a receptor, BamA, on the target cell (64), but the mechanism was unknown. Using ECT, we saw that CdiA forms rods extending out from the surface of the cell (Fig. 4A), but the length of the rod was only half that predicted for the protein. This explained why the receptor recognition domain is located in the middle of the protein (65) and helped elucidate a novel mechanism whereby CdiB initially secretes only the N-terminal half of CdiA, which forms a structured rod. The rest of the protein, including the C-terminal toxin,

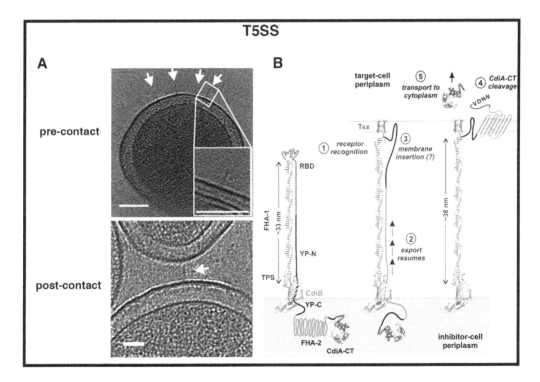

FIGURE 4 T5SS. ECT revealed the stick-like form of the *Escherichia coli* CDI T5SS (A) (arrows) and helped elucidate its mechanism, in which half of CdiA extends out from the cell surface (B). Upon target binding, the remaining half of CdiA is exported to deliver the toxin to the target cell. Images reprinted with permission from reference 66.

remains sequestered in the periplasm; target binding at the end of the rod stimulates secretion to continue, bringing the toxin-laden tip out of the periplasm and, like a tetherball, into the target cell (Fig. 4B) (66).

TYPE VI SECRETION SYSTEMS

In 2005, our group observed tubular structures in cryotomograms of *Vibrio cholerae* (Fig. 5A) but did not know what they were until, with John Mekalanos, we recognized them as type VI secretion systems (T6SS), which allow some bacteria to kill others from different species (67). Our cryotomograms immediately revealed the basic mechanism: two nested tubes, a sheath and a rod, function as a nanoscale dart gun (68;

also recognized in reference 69). Firing entails a constriction in the outer sheath, providing the energy to propel the inner rod, loaded with toxin, through the machine's baseplate in the cell's own envelope and then through the membrane of a neighboring target (68). Subsequent ECT studies revealed additional details, including the structure of the baseplate (Fig. 5B) (70, 71) and higher-resolution details of the sheath that showed how its contraction may expose recycling domains, allowing it to be disassembled into the building blocks of another T6SS (70). We showed that T6SS assemble extensive batteries in some species (Fig. 5C) and that other cells lyse to release micrometer-scale porcupine arrays of T6SS-related structures (Fig. 5D) (72). Some details remain mysterious: could the extra-

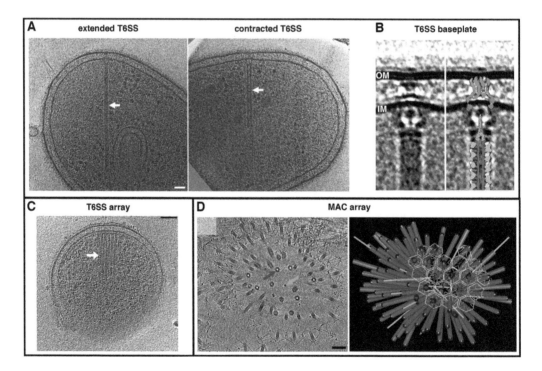

FIGURE 5 T6SS. ECT revealed the contractile mechanism of the T6SS (A) and its structure *in situ* in *Myxococcus xanthus* (B). It also revealed higher-order arrays of T6SS in *Amoebophilus asiaticus* (C) and a T6SS-related metamorphosis-associated contractile structure (MAC) in *Pseudoalteromonas luteoviolacea* (D). Images are reprinted from the following with permission: panels A and B, reference 70; panel D, reference 72.

cellular fibers associated with T6SS *in vivo* (70) mediate target recognition?

OUTLOOK

Much work lies ahead. As described above, the machines we have managed to glimpse inside cells, or even map in detail, still hold many mysteries. In addition, systems studied in one species need to be compared across others; as the T3SS flagellar motor showed, they can be strikingly diverse. Some machines, including the type I, VII, VIII, and IX secretion systems, have yet to be seen at all *in vivo*. Others, including the T2SS and T4SS, have been observed only in an inactive state, and it is now of great interest to see their active forms. In some cases, this will require capturing pathogens interacting with their hosts, samples too thick for direct imaging by ECT. The developing application of focused ion beam (FIB) milling to thin such biological samples should soon open up even these targets for observation (73).

The study of bacterial secretion systems has always been intricately linked to the development of ECT technology. ECT requires highly specialized, expensive equipment and considerable expertise. As a result, the studies we have described here come from only a handful of labs. Fortunately, secretion systems have proven to be wonderful test beds for pioneering new ECT technology. Our subtomogram average of the flagellar motor was one of the first subtomogram averages (32). Our subsequent comparison of flagellar motors in different species was the result of establishing a pipeline to collect many (~40) datasets per day (44). Our dissection of the T4aP by imaging strains with green fluorescent protein tags was the first to demonstrate that approach (6). The Pilhofer lab's cryo-FIB milling of bacterial cells to study the T6SS was one of the first examples of that application (74). No doubt more discoveries about these fascinating nanomachines will come from continuing development of ECT.

ACKNOWLEDGMENTS

We apologize to our colleagues whose work we could not discuss due to limited space, and we thank members of the Jensen lab for helpful discussions.

Work on bacterial secretion systems in the lab is supported by the NIH (grant R01 AI127401 to G.J.J.).

CITATION

Oikonomou CM, Jensen GJ. 2019. Electron cryotomography of bacterial secretion systems. Microbiol Spectrum 7(2):PSIB-0019-2018.

REFERENCES

1. **Oikonomou CM, Jensen GJ.** 2017. Cellular electron cryotomography: toward structural biology in situ. *Annu Rev Biochem* **86:**873–896.
2. **Rapisarda C, Tassinari M, Gubellini F, Fronzes R.** 2018. Using cryo-EM to investigate bacterial secretion systems. *Annu Rev Microbiol* **72:**231–254.
3. **Kooger R, Szwedziak P, Böck D, Pilhofer M.** 2018. CryoEM of bacterial secretion systems. *Curr Opin Struct Biol* **52:**64–70.
4. **Costa TR, Felisberto-Rodrigues C, Meir A, Prevost MS, Redzej A, Trokter M, Waksman G.** 2015. Secretion systems in Gram-negative bacteria: structural and mechanistic insights. *Nat Rev Microbiol* **13:**343–359.
5. **Gold VA, Salzer R, Averhoff B, Kühlbrandt W.** 2015. Structure of a type IV pilus machinery in the open and closed state. *eLife* **4:**e07380.
6. **Chang YW, Rettberg LA, Treuner-Lange A, Iwasa J, Søgaard-Andersen L, Jensen GJ.** 2016. Architecture of the type IVa pilus machine. *Science* **351:**aad2001.
7. **Abendroth J, Murphy P, Sandkvist M, Bagdasarian M, Hol WG.** 2005. The X-ray structure of the type II secretion system complex formed by the N-terminal domain of EpsE and the cytoplasmic domain of EpsL of *Vibrio cholerae. J Mol Biol* **348:**845–855.
8. **Craig L, Volkmann N, Arvai AS, Pique ME, Yeager M, Egelman EH, Tainer JA.** 2006. Type IV pilus structure by cryo-electron microscopy and crystallography: implications for pilus assembly and functions. *Mol Cell* **23:**651–662.

9. Balasingham SV, Collins RF, Assalkhou R, Homberset H, Frye SA, Derrick JP, Tønjum T. 2007. Interactions between the lipoprotein PilP and the secretin PilQ in *Neisseria meningitidis*. *J Bacteriol* **189**:5716–5727.

10. Yamagata A, Tainer JA. 2007. Hexameric structures of the archaeal secretion ATPase GspE and implications for a universal secretion mechanism. *EMBO J* **26**:878–890.

11. Satyshur KA, Worzalla GA, Meyer LS, Heiniger EK, Aukema KG, Misic AM, Forest KT. 2007. Crystal structures of the pilus retraction motor PilT suggest large domain movements and subunit cooperation drive motility. *Structure* **15**:363–376.

12. Ayers M, Sampaleanu LM, Tammam S, Koo J, Harvey H, Howell PL, Burrows LL. 2009. PilM/N/O/P proteins form an inner membrane complex that affects the stability of the *Pseudomonas aeruginosa* type IV pilus secretin. *J Mol Biol* **394**:128–142.

13. Abendroth J, Mitchell DD, Korotkov KV, Johnson TL, Kreger A, Sandkvist M, Hol WG. 2009. The three-dimensional structure of the cytoplasmic domains of EpsF from the type 2 secretion system of *Vibrio cholerae*. *J Struct Biol* **166**:303–315.

14. Bulyha I, Schmidt C, Lenz P, Jakovljevic V, Höne A, Maier B, Hoppert M, Søgaard-Andersen L. 2009. Regulation of the type IV pili molecular machine by dynamic localization of two motor proteins. *Mol Microbiol* **74**:691–706.

15. Sampaleanu LM, Bonanno JB, Ayers M, Koo J, Tammam S, Burley SK, Almo SC, Burrows LL, Howell PL. 2009. Periplasmic domains of *Pseudomonas aeruginosa* PilN and PilO form a stable heterodimeric complex. *J Mol Biol* **394**:143–159.

16. Misic AM, Satyshur KA, Forest KT. 2010. *P. aeruginosa* PilT structures with and without nucleotide reveal a dynamic type IV pilus retraction motor. *J Mol Biol* **400**:1011–1021.

17. Karuppiah V, Hassan D, Saleem M, Derrick JP. 2010. Structure and oligomerization of the PilC type IV pilus biogenesis protein from *Thermus thermophilus*. *Proteins* **78**:2049–2057.

18. Karuppiah V, Derrick JP. 2011. Structure of the PilM-PilN inner membrane type IV pilus biogenesis complex from *Thermus thermophilus*. *J Biol Chem* **286**:24434–24442.

19. Korotkov KV, Johnson TL, Jobling MG, Pruneda J, Pardon E, Héroux A, Turley S, Steyaert J, Holmes RK, Sandkvist M, Hol WG. 2011. Structural and functional studies on the interaction of GspC and GspD in the type II secretion system. *PLoS Pathog* **7**:e1002228.

20. Tammam S, Sampaleanu LM, Koo J, Sundaram P, Ayers M, Chong PA, Forman-Kay JD, Burrows LL, Howell PL. 2011. Characterization of the PilN, PilO and PilP type IVa pilus subcomplex. *Mol Microbiol* **82**:1496–1514.

21. Gu S, Kelly G, Wang X, Frenkiel T, Shevchik VE, Pickersgill RW. 2012. Solution structure of homology region (HR) domain of type II secretion system. *J Biol Chem* **287**:9072–9080.

22. Georgiadou M, Castagnini M, Karimova G, Ladant D, Pelicic V. 2012. Large-scale study of the interactions between proteins involved in type IV pilus biology in *Neisseria meningitidis*: characterization of a subcomplex involved in pilus assembly. *Mol Microbiol* **84**:857–873.

23. Karuppiah V, Collins RF, Thistlethwaite A, Gao Y, Derrick JP. 2013. Structure and assembly of an inner membrane platform for initiation of type IV pilus biogenesis. *Proc Natl Acad Sci U S A* **110**:E4638–E4647.

24. Li C, Wallace RA, Black WP, Li YZ, Yang Z. 2013. Type IV pilus proteins form an integrated structure extending from the cytoplasm to the outer membrane. *PLoS One* **8**:e70144.

25. Lu C, Turley S, Marionni ST, Park YJ, Lee KK, Patrick M, Shah R, Sandkvist M, Bush MF, Hol WG. 2013. Hexamers of the type II secretion ATPase GspE from *Vibrio cholerae* with increased ATPase activity. *Structure* **21**:1707–1717.

26. Tammam S, Sampaleanu LM, Koo J, Manoharan K, Daubaras M, Burrows LL, Howell PL. 2013. PilMNOPQ from the *Pseudomonas aeruginosa* type IV pilus system form a transenvelope protein interaction network that interacts with PilA. *J Bacteriol* **195**:2126–2135.

27. Friedrich C, Bulyha I, Søgaard-Andersen L. 2014. Outside-in assembly pathway of the type IV pilus system in *Myxococcus xanthus*. *J Bacteriol* **196**:378–390.

28. Siewering K, Jain S, Friedrich C, Webber-Birungi MT, Semchonok DA, Binzen I, Wagner A, Huntley S, Kahnt J, Klingl A, Boekema EJ, Søgaard-Andersen L, van der Does C. 2014. Peptidoglycan-binding protein TsaP functions in surface assembly of type IV pili. *Proc Natl Acad Sci U S A* **111**:E953–E961.

29. Chang YW, Kjær A, Ortega DR, Kovacikova G, Sutherland JA, Rettberg LA, Taylor RK, Jensen GJ. 2017. Architecture of the *Vibrio cholerae* toxin-coregulated pilus machine revealed by electron cryotomography. *Nat Microbiol* **2**:16269.

30. Thomassin JL, Santos Moreno J, Guilvout I, Tran Van Nhieu G, Francetic O. 2017. The

trans-envelope architecture and function of the type 2 secretion system: new insights raising new questions. *Mol Microbiol* **105**:211–226.

31. **Ghosal D, Kim KW, Zheng H, Kaplan M, Vogel JP, Cianciotto NP, Jensen GJ.** 2019. In vivo structure of the *Legionella* type II secretion system by electron cryotomography. *bioRxiv.*

32. **Murphy GE, Leadbetter JR, Jensen GJ.** 2006. In situ structure of the complete *Treponema primitia* flagellar motor. *Nature* **442**:1062–1064.

33. **Berg HC.** 2003. The rotary motor of bacterial flagella. *Annu Rev Biochem* **72**:19–54.

34. **Macnab RM.** 2003. How bacteria assemble flagella. *Annu Rev Microbiol* **57**:77–100.

35. **DePamphilis ML, Adler J.** 1971. Fine structure and isolation of the hook-basal body complex of flagella from *Escherichia coli* and *Bacillus subtilis. J Bacteriol* **105**:384–395.

36. **Stallmeyer MJ, Aizawa S, Macnab RM, DeRosier DJ.** 1989. Image reconstruction of the flagellar basal body of *Salmonella typhimurium. J Mol Biol* **205**:519–528.

37. **Francis NR, Sosinsky GE, Thomas D, DeRosier DJ.** 1994. Isolation, characterization and structure of bacterial flagellar motors containing the switch complex. *J Mol Biol* **235**:1261–1270.

38. **Thomas DR, Morgan DG, DeRosier DJ.** 1999. Rotational symmetry of the C ring and a mechanism for the flagellar rotary motor. *Proc Natl Acad Sci U S A* **96**:10134–10139.

39. **Suzuki H, Yonekura K, Namba K.** 2004. Structure of the rotor of the bacterial flagellar motor revealed by electron cryomicroscopy and single-particle image analysis. *J Mol Biol* **337**:105–113.

40. **Thomas DR, Francis NR, Xu C, DeRosier DJ.** 2006. The three-dimensional structure of the flagellar rotor from a clockwise-locked mutant of *Salmonella enterica* serovar Typhimurium. *J Bacteriol* **188**:7039–7048.

41. **Liu J, Lin T, Botkin DJ, McCrum E, Winkler H, Norris SJ.** 2009. Intact flagellar motor of *Borrelia burgdorferi* revealed by cryo-electron tomography: evidence for stator ring curvature and rotor/C-ring assembly flexion. *J Bacteriol* **191**:5026–5036.

42. **Kudryashev M, Cyrklaff M, Wallich R, Baumeister W, Frischknecht F.** 2010. Distinct in situ structures of the *Borrelia* flagellar motor. *J Struct Biol* **169**:54–61.

43. **Abrusci P, Vergara-Irigaray M, Johnson S, Beeby MD, Hendrixson DR, Roversi P, Friede ME, Deane JE, Jensen GJ, Tang CM, Lea SM.** 2013. Architecture of the major

component of the type III secretion system export apparatus. *Nat Struct Mol Biol* **20**:99–104.

44. **Chen S, Beeby M, Murphy GE, Leadbetter JR, Hendrixson DR, Briegel A, Li Z, Shi J, Tocheva EI, Müller A, Dobro MJ, Jensen GJ.** 2011. Structural diversity of bacterial flagellar motors. *EMBO J* **30**:2972–2981.

45. **Beeby M, Ribardo DA, Brennan CA, Ruby EG, Jensen GJ, Hendrixson DR.** 2016. Diverse high-torque bacterial flagellar motors assemble wider stator rings using a conserved protein scaffold. *Proc Natl Acad Sci U S A* **113**:E1917–E1926.

46. **Qin Z, Lin WT, Zhu S, Franco AT, Liu J.** 2017. Imaging the motility and chemotaxis machineries in *Helicobacter pylori* by cryo-electron tomography. *J Bacteriol* **199**:e00695-16.

47. **Chaban B, Coleman I, Beeby M.** 2018. Evolution of higher torque in *Campylobacter*-type bacterial flagellar motors. *Sci Rep* **8**:97.

48. **Kaplan M, Ghosal D, Subramanian P, Oikonomou CM, Kjaer A, Pirbadian S, Ortega DR, Briegel A, El-Naggar MY, Jensen GJ.** 2019. The presence and absence of periplasmic rings in bacterial flagellar motors correlates with stator type. *eLife* **8**:e43487.

49. **Zhu S, Nishikino T, Hu B, Kojima S, Homma M, Liu J.** 2017. Molecular architecture of the sheathed polar flagellum in *Vibrio alginolyticus. Proc Natl Acad Sci U S A* **114**:10966–10971.

50. **Galán JE, Lara-Tejero M, Marlovits TC, Wagner S.** 2014. Bacterial type III secretion systems: specialized nanomachines for protein delivery into target cells. *Annu Rev Microbiol* **68**:415–438.

51. **Diepold A, Armitage JP.** 2015. Type III secretion systems: the bacterial flagellum and the injectisome. *Philos Trans R Soc Lond B Biol Sci* **370**:20150020.

52. **Hu B, Morado DR, Margolin W, Rohde JR, Arizmendi O, Picking WL, Picking WD, Liu J.** 2015. Visualization of the type III secretion sorting platform of *Shigella flexneri. Proc Natl Acad Sci U S A* **112**:1047–1052.

53. **Hu B, Lara-Tejero M, Kong Q, Galan JE, Liu J.** 2017. In situ molecular architecture of the *Salmonella* type III secretion machine. *Cell* **168**:1065–1074.e1010.

54. **Wagner S, Königsmaier L, Lara-Tejero M, Lefebre M, Marlovits TC, Galán JE.** 2010. Organization and coordinated assembly of the type III secretion export apparatus. *Proc Natl Acad Sci U S A* **107**:17745–17750.

55. **Park D, Lara-Tejero M, Waxham MN, Li W, Hu B, Galán JE, Liu J.** 2018. Visualization of

the type III secretion mediated *Salmonella*-host cell interface using cryo-electron tomography. *eLife* 7:e39514.

56. **Ide T, Laarmann S, Greune L, Schillers H, Oberleithner H, Schmidt MA. 2001.** Characterization of translocation pores inserted into plasma membranes by type III-secreted Esp proteins of enteropathogenic *Escherichia coli*. *Cell Microbiol* 3:669–679.

57. **Low HH, Gubellini F, Rivera-Calzada A, Braun N, Connery S, Dujeancourt A, Lu F, Redzej A, Fronzes R, Orlova EV, Waksman G. 2014.** Structure of a type IV secretion system. *Nature* 508:550–553.

58. **Frick-Cheng AE, Pyburn TM, Voss BJ, McDonald WH, Ohi MD, Cover TL. 2016.** Molecular and structural analysis of the *Helicobacter pylori cag* type IV secretion system core complex. *mBio* 7:e02001-15.

59. **Chetrit D, Hu B, Christie PJ, Roy CR, Liu J. 2018.** A unique cytoplasmic ATPase complex defines the *Legionella pneumophila* type IV secretion channel. *Nat Microbiol* 3:678–686.

60. **Ghosal D, Chang YW, Jeong KC, Vogel JP, Jensen GJ. 2018.** Molecular architecture of the *Legionella* Dot/Icm type IV secretion system. *bioRxiv*.

61. **Chang YW, Shaffer CL, Rettberg LA, Ghosal D, Jensen GJ. 2018.** In vivo structures of the *Helicobacter pylori cag* type IV secretion system. *Cell Rep* 23:673–681.

62. **Ensminger AW. 2016.** *Legionella pneumophila*, armed to the hilt: justifying the largest arsenal of effectors in the bacterial world. *Curr Opin Microbiol* 29:74–80.

63. **Aoki SK, Pamma R, Hernday AD, Bickham JE, Braaten BA, Low DA. 2005.** Contact-dependent inhibition of growth in *Escherichia coli*. *Science* 309:1245–1248.

64. **Aoki SK, Malinverni JC, Jacoby K, Thomas B, Pamma R, Trinh BN, Remers S, Webb J, Braaten BA, Silhavy TJ, Low DA. 2008.** Contact-dependent growth inhibition requires the essential outer membrane protein BamA (YaeT) as the receptor and the inner membrane transport protein AcrB. *Mol Microbiol* 70:323–340.

65. **Ruhe ZC, Nguyen JY, Xiong J, Koskiniemi S, Beck CM, Perkins BR, Low DA, Hayes CS. 2017.** CdiA effectors use modular receptor-binding domains to recognize target bacteria. *mBio* 8:e00290-17.

66. **Ruhe ZC, Subramanian P, Song K, Nguyen JY, Stevens TA, Low DA, Jensen GJ, Hayes CS. 2018.** Programmed secretion arrest and receptor-triggered toxin export during antibacterial contact-dependent growth inhibition. *Cell* 175:921–933.e14.

67. **MacIntyre DL, Miyata ST, Kitaoka M, Pukatzki S. 2010.** The *Vibrio cholerae* type VI secretion system displays antimicrobial properties. *Proc Natl Acad Sci U S A* 107:19520–19524.

68. **Basler M, Pilhofer M, Henderson GP, Jensen GJ, Mekalanos JJ. 2012.** Type VI secretion requires a dynamic contractile phage tail-like structure. *Nature* 483:182–186.

69. **Bönemann G, Pietrosiuk A, Mogk A. 2010.** Tubules and donuts: a type VI secretion story. *Mol Microbiol* 76:815–821.

70. **Chang YW, Rettberg LA, Ortega DR, Jensen GJ. 2017.** *In vivo* structures of an intact type VI secretion system revealed by electron cryotomography. *EMBO Rep* 18:1090–1099.

71. **Rapisarda C, Cherrak Y, Kooger R, Schmidt V, Pellarin R, Logger L, Cascales E, Pilhofer M, Durand E, Fronzes R. 2018.** In situ and high-resolution cryo-EM structure of the type VI secretion membrane complex. *bioRxiv*.

72. **Shikuma NJ, Pilhofer M, Weiss GL, Hadfield MG, Jensen GJ, Newman DK. 2014.** Marine tubeworm metamorphosis induced by arrays of bacterial phage tail-like structures. *Science* 343:529–533.

73. **Medeiros JM, Böck D, Pilhofer M. 2018.** Imaging bacteria inside their host by cryo-focused ion beam milling and electron cryotomography. *Curr Opin Microbiol* 43:62–68.

74. **Böck D, Medeiros JM, Tsao HF, Penz T, Weiss GL, Aistleitner K, Horn M, Pilhofer M. 2017.** In situ architecture, function, and evolution of a contractile injection system. *Science* 357:713–717.

SecA-Mediated Protein Translocation through the SecYEG Channel

2

AMALINA GHAISANI KOMARUDIN[1] and ARNOLD J. M. DRIESSEN[1]

INTRODUCTION

Protein transport occurs in all domains of life (1). Proteins that function outside the cytosol are translocated across membranes. The general system for protein translocation is formed by the Sec translocase at its core the translocon: SecYEG in bacteria (2), SecYEβ in archaea (3), and Sec61αβγ in the endoplasmic reticulum of eukaryotes (4, 5). The translocon forms a protein conducting channel in the membrane for unfolded preproteins (6) but also mediates cotranslational insertion of nascent membrane proteins into the membrane (Fig. 1).

During posttranslational translocation, preproteins are synthesized at the ribosome with a cleavable N-terminal signal sequence and bound by the molecular chaperone SecB, which stabilizes the preprotein in an unfolded state (7). SecB targets preproteins to the SecYEG-bound SecA (8–10). SecA is an ATPase (11, 12) that directs preproteins in a stepwise manner into the pore (2). The SecDF complex (13) aids this process by facilitating proton motive force (PMF)-dependent translocation (14). In the cotranslational pathway, nascent membrane proteins are guided to the translocon by signal recognition particle

[1]Molecular Microbiology, Groningen Biomolecular Sciences and Biotechnology Institute, and the Zernike Institute of Advanced Materials, University of Groningen, Nijenborgh 7, 9747AG Groningen, The Netherlands

Protein Secretion in Bacteria
Edited by Maria Sandkvist, Eric Cascales, and Peter J. Christie
© 2019 American Society for Microbiology, Washington, DC
doi:10.1128/microbiolspec.PSIB-0028-2019

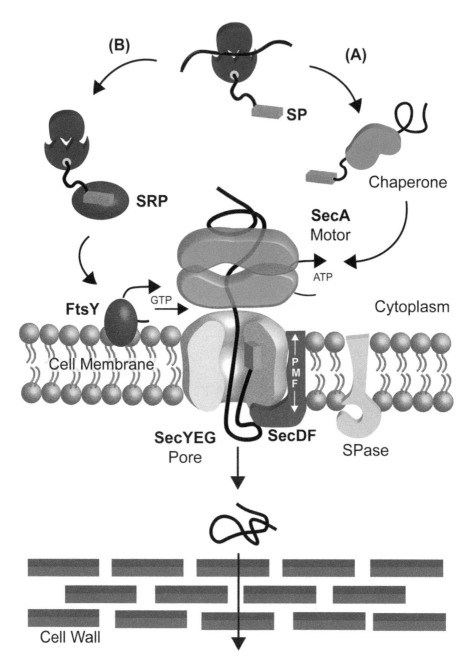

FIGURE 1 The Sec pathway. (A) Posttranslational pathway: after complete synthesis at the ribosome, the unfolded preprotein is recognized by the molecular chaperone SecB (blue) and targeted to SecA (green). SecA guides the preprotein through the SecYEG pore (lime), employing the energy from ATP binding and hydrolysis. The signal peptide is cleaved by the signal peptidase (SPase [yellow]). SecDF (pink) pulls the preprotein across the membrane at the expense of the PMF. (B) Cotranslational pathway: once a hydrophobic transmembrane domain of a nascent membrane protein emerges from the ribosomes, signal recognition particle (SRP) (brown) binds to the ribosome nascent chain (RNC) and guides the complex to the SR receptor FtsY (dark brown) at the membrane. Upon the binding of GTP to the SRP:FtsY heterodimer, the RNC is released from SRP and transferred to the SecYEG channel, where chain elongation at the ribosome is directly coupled to membrane insertion of the nascent membrane protein.

(SRP) to the SRP receptor, FtsY, at the membrane. Subsequently, GTP binding to the SRP-FtsY heterodimer results in release of the nascent chain from SRP to the translocon. Eukaryotes may use this pathway for both translocation and membrane insertion, whereas in bacteria, it is mostly used for insertion (15). This review focuses on bacterial protein translocation.

SecYEG, THE PROTEIN CONDUCTING CHANNEL

Structural analysis of SecYEG provides strong support for its role as a protein conducting channel. SecY, the major subunit, consists of two halves formed by transmembrane segments (TMS) 1 to 5 and 6 to 10 (16) (Fig. 2A). The two halves are connected by a loop of TMS 5/6, resulting in a clamshell-like structure of the translocon (17). SecY is shaped like an hourglass with a funnel-like entrance and a subcentral constriction (Fig. 2D). At the front of SecY, a lateral gate between TMS 2 and 7 can open to the lipid bilayer (16). The exit site of the pore is closed by an α-helical plug (TMS2a) that folds back into the channel (18, 19). The *Methanococcus jannaschii* SecYEβ structure concerns a resting state with a sealed pore where six hydrophobic residues close the constriction ring and the plug closes the exit funnel (16).

SecE surrounds SecY at its back and embraces the SecY clamshell structure with a long transmembrane helix that via a hinge is

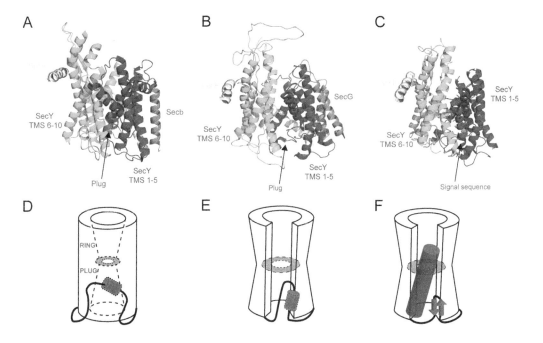

FIGURE 2 **Structural stages of the translocation channel. (A to C) The SecYEG/β crystal structures viewed from the membrane: SecY TMS 1 to 5 (blue), TMS 6 to 10 (green), plug domain (red), SecE (yellow), and SecG/β (orange). (D to F) Cartoon illustration of SecYEG/β. The illustrations depict the opening of the constriction and movement of the plug domain depending on the state of the translocon. (A and D)** *Methanococcus jannaschii* **SecYEβ (PDB entry 1RH5), known as the closed or resting conformation. (B and E)** *Thermotoga maritima* **SecYEG cocrystallized with SecA (not shown) in an Mg-ADP-BeFx-bound transition state (PDB entry 3DIN) as a preopen conformation. (C and F)** *Geobacillus thermodenitrificans* **SecYEG cocrystallized with SecA (not shown) and a signal sequence (magenta) latched into the lateral gate (PDB entry 5EUL), resembling an actively engaged translocation channel.**

connected to a surface-exposed amphipathic helix that contacts loops of SecY (Fig. 2A). The third subunit SecG is peripherally bound to SecY. This subunit is not essential for cell viability (20, 21) but stabilizes the closed channel (22) whereby the cytosolic loop of SecG folds back into the channel at the *cis*-side of the membrane (23).

SecYEG Channel Opening

The lateral gate creates a pathway for the insertion of membrane proteins (24) and provides a binding site for the signal sequence of preproteins (25–27). The signal sequence intercalates into the lateral gate, causing a conformational change in the pore region (26). Three of the six pore ring residues are located on TMS 2 and 7 of the lateral gate, and thus, intercalation of the signal sequence between these TMS is directly coupled to channel opening (26). SecE presumably stabilizes the two halves of SecY when the channel opens and when the plug is displaced from its subcentral position (16, 19). Channel opening is also influenced by SecA, as shown in the *Thermotoga maritima* SecA-SecYEG complex structure (Fig. 2B and E) (28), where a partial opening of lateral gate by ~5 Å provides a gap that may allow an inserting signal sequence to sample the phospholipid bilayer. For translocation, the lateral gate needs to open up to 10 to 12 Å (25, 29).

The structure of *Geobacillus thermodenitrificans* SecYE-SecA complex with a covalently linked signal sequence in the channel further shows large conformational changes in the lateral gate region (30) (Fig. 2C). Now also the plug has shifted to the back of the channel in close proximity to the TMS of SecE, in line with cross-linking studies (19, 31) (Fig. 2C and F). Compared to the *T. maritima* SecA-SecYEG structure (Fig. 2B), TMS 7 of the *G. thermodenitrificans* SecY is tilted 10° relative to the membrane and the periplasmic ends of TMS 3 and 7 are now in close proximity (Fig. 2C). This results in a large opening in the lateral gate that allows

for signal sequence intercalation (30). The plug of *G. thermodenitrificans* SecY adopted a β-strand structure, which differs from the resting α-helical structure (16, 32, 33). The plug domain is poorly conserved (16), and its deletion only reduces the efficiency of translocation (34, 35). It is important for signal sequence recognition (34, 36, 37) and appears to sample the hydrophobicity of the incoming polypeptide (38) to coordinate channel opening. A very large movement of the plug, by ~20 to 27 Å, creates an unobstructed path for protein translocation (19, 31), but this large displacement is not critical for translocation *per se* (39). The plug stabilizes the closed state of SecY (34) and acts as a periplasmic seal to prevent ion leakage.

Pore Constriction and Width

Protein localization (Prl) mutations in *sec* genes allow the translocation of preproteins with a defective or even missing signal sequence. The most dominant *prl* variants are found in SecY (40–43), and these destabilize the closed state of the channel (22, 36, 40), possibly mimicking the function of the signal sequence, SecA, or the ribosome (16). The PrlA4 mutant (44) exhibits a tighter binding of SecA (45), allowing more efficient translocation (45–47) and a lower PMF dependence (48). Overall, this can be understood as a reduced proofreading activity (49–52), as signal sequence recognition is less stringent in PrlA mutants, likely because the channel is already in a partially opened state.

Many of the PrlA mutants cluster around the pore constriction and cause increased ion leakage (53). The hydrophobic constriction ring functions as a gasket around the translocating preprotein to seal the pore (16, 18). The pore exhibits a high plasticity and even supports translocation of preproteins with an internal disulfide bridge, a stable fold induced by chemical cross-linking (54, 55) or bulky fluorophores (56). Structures with a cross section of up to ~22 Å can be translocated (57).

Oligomeric State of SecYEG

The oligomeric state of the SecYEG translocon remains a topic of debate. SecYEG can be purified as a monomer (16, 28) but may also form dimers and higher oligomers (58–60). Crystallography and cross-linking experiments have suggested that SecYEG is dimeric (58, 59, 61–63), but the functional role of the dimer has remained obscure, as only one channel is used to translocate proteins (58, 64). Single SecYEG complexes reconstituted into nanodiscs show that the monomer is sufficient for translocation, as well as for ribosome nascent chain (RNC) binding (47, 64, 65). Further, the cryo-electron microscopic structure of the RNC-SecYEG complex (28, 30) defines the monomer as the minimal functional unit.

SecA, AN ATP-DEPENDENT MOTOR PROTEIN

SecA is a molecular motor that drives protein translocation at the expense of ATP hydrolysis (66). SecA associates with SecYEG but also binds to the phospholipid bilayer and to ribosomes (67, 68).

SecA Structure

SecA is a relatively large protein with a subunit mass of about 102 kDa. It consists of functional and structural subdomains (Fig. 3). The nucleotide binding domain (NBD), comprising NBD1 and NBD2 (also termed intramolecular regulator of ATPase 2 [IRA2]), is essential for ATP binding and hydrolysis (69). The ATPase activity occurs at the interface of NBD1 and NBD2 (70). These NBDs form the so-called DEAD motor, which is also found in DNA/RNA helicases, and contain the highly conserved Walker A and B motifs (71).

Preprotein can cross-link to the PPXD domain (72), which plays an important role in the activation of the ATPase activity (73, 74). The C-terminal domain of SecA can be divided into four subdomains: the α-helical scaffold domain (HSD), which interconnects all other SecA domains (75); the α-helical wing domain (HWD); the two-helix finger (2HF), which is part of the intramolecular regulator of ATP hydrolysis 1 (IRA1) (76); and the C-terminal linker domain (CTL). In *Escherichia coli*, the CTL harbors a zinc finger, which plays a role in the interaction with SecB (7, 9) and phospholipids (77). SecA exhibits a low basal ATPase activity (78),

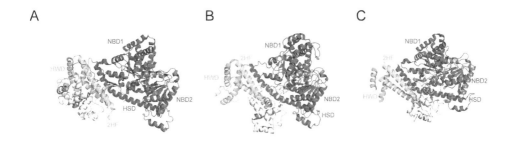

FIGURE 3 **Conformational states of SecA. Structures of SecA from *Bacillus subtilis* (PDB entry 1M6N) (A), Mg-ADP-BeFx-bound SecA cocrystallized with SecYEG (not shown) from *T. maritima* (PDB 3DIN) (B), and Mg-ADP-BeFx-bound SecA from *B. subtilis* engaged with the *G. thermodenitrificans* SecYEG and a signal sequence (not shown) (PDB entry 5EUL) (C). The locations of the PPXD domain (yellow), NBD1 (red), NBD2 (blue), HWD (green), HSD (purple), and 2HF (cyan) are indicated. A large movement of the PPXD domain (yellow) suggests a closed (A) or open (B and C) conformation of SecA.**

which is allosterically stimulated by binding of SecB, SecYEG, and preprotein and by anionic phospholipids (79–84). The ATPase activity of cytosolic SecA is inhibited by IRA1, or 2HF, which forms a helix-loop-helix structure of the HSD that contacts both NBD2 and PPXD (85).

Oligomeric State of SecA

The functional oligomeric state of SecA is a major topic of controversy. SecA purified from cells is mainly dimeric (86). Although SecA appears to function as a dimer (87–90), the monomer-dimer equilibrium is affected by ligands of SecA (91, 92). SecA is highly thermolabile in the presence of phospholipids, but inactivation is prevented by preproteins (93).

Only a few studies have addressed the oligomeric state of SecA while bound to SecYEG. SecA remains dimeric during translocation (90, 94) and is active as a dimer (88, 89, 94, 95), likely as a discrete anti-parallel dimer (96). Dimeric SecA binds the SecYEG with high affinity, where one of the protomers binds tightly to SecYEG and the other protomer is bound to the SecYEG-bound SecA (88). Mutation-induced monomerization abolishes SecA activity (97), but this defect can be overcome by high concentrations of SecA that restore the dimer (88, 98). SecB interacts with the SecYEG-bound dimeric SecA (99).

Binding Partners of SecA

The N-terminal signal sequence of preproteins functions as a targeting signal (100, 101) and induces channel opening; also, the targeting function of signal sequences has been challenged (102–104). Prior to translocation, the preprotein is stabilized in an unfolded state by SecB. SecYEG-bound SecA binds SecB, and this interaction results in a transfer of the preprotein from SecB to SecA (7). SecB is a homotetramer arranged as a dimer of dimers (105). SecB contains two peptide binding grooves that run along either side of the tetramer (106, 107), where preproteins are bound in their folding core (108). During the ATP-dependent initiation of translocation, SecB is released into the cytosol to bind another preprotein.

SecA binds to SecYEG via a phospholipid-bound intermediate (93, 109–111) that involves its amphipathic positively charged N terminus (93). This region serves to tether SecA to the membrane (109), but the membrane interaction also enforces a conformational change which primes SecA for high-affinity SecYEG binding (93). Phospholipid-bound SecA likely functions as a membrane queue of SecA-preprotein complexes before they are delivered to SecYEG for translocation.

Structural Mechanism of SecA Function

SecA undergoes a multitude of ATP-dependent conformation changes during translocation (28, 70, 112) (Fig. 3). In the *T. maritima* SecA-SecYEG structure, where SecA is in an open transition state stabilized by ADP beryllium fluoride, the PPXD domain has moved towards the NBD2 and away from the HWD (28, 113), while the NBDs are in close proximity with the PPXD domain (Fig. 3B). In contrast, the *Bacillus subtilis* SecA structure (Fig. 3A) represents a closed state in which the PPXD domain is located near the HWD (70). The formation of the nucleotide-binding pocket between NBD1 and NBD2 allows the preprotein to bind in a groove between NBD2 and PPXD. This opening stabilizes the preprotein-SecA interaction, which allows for an increased rate of nucleotide exchange, resulting in activation of the ATPase activity (114, 115). The structure of the *G. thermodenitrificans* SecYE engaged with *B. subtilis* SecA (30) and a signal peptide suggests that SecA does not undergo further dramatic conformational changes compared to the *T. maritima* SecA-SecYEG complex (Fig. 3C). It has been proposed that SecA in its ATP-bound state prevents the two halves of SecY from moving further apart.

The 2HF of SecA is inserted into the cytoplasmic opening of the SecY channel (Fig. 4), where it is in close proximity to the translocating preprotein (28). The 2HF makes contact with C4 loop of SecY, and the insertion may result in the opening of lateral gate by a rigid body movement of the two halves of SecY (29). The tip of the loop of the 2HF contains a highly conserved tyrosine residue which is crucial for translocation (76). It has been suggested that the 2HF associates with the unfolded preprotein through hydrophobic side chain interactions, but this model does not explain how SecA can mediate the translocation of stretches of glycine residues (116), which would only allow for main chain interactions. Alternatively, the 2HF acts by opening the channel through its interaction with the SecY C4 loop. Strikingly, chemical cross-linking of the 2HF with the C4 loop did not interfere with translocation (117), suggesting that the 2HF does not function as an ATP-dependent lever to push preproteins through the channel but rather serves to push the two halves of SecY

apart. Additionally, the 2HF of SecA may act as a template by inserting the hairpin formed by the signal peptide and the early mature region of the preprotein (118).

TRANSLOCATION MODELS

Various models for SecA-mediated translocation have been proposed as outlined below (Fig. 5).

Power Stroke Model

A large class of ATPases contains a RecA-like structural domain and uses the energy of ATP binding and hydrolysis to move proteins or nucleic acids (71). SecA has a DEAD box typically found in helicase, and therefore a DNA helicase molecular mechanism has been proposed (119). In this power stoke mechanism, SecA acts as a mechanical device that pushes preproteins into the pore (82). The 2HF may function as an ATP-dependent lever to support such a power stroke mechanism

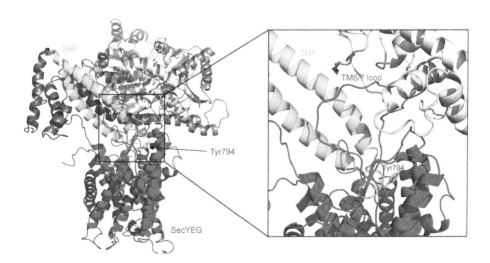

FIGURE 4 Structure of *T. maritima* SecA-SecYEG complex. SecA penetrates into the SecYEG channel (red) via the so-called two-helix finger (2HF [light blue]). The SecA PPXD domain (yellow) also binds to TMS6/7 loop of SecYEG. The conserved tyrosine 794 is depicted in green.

(A) Power Stroke

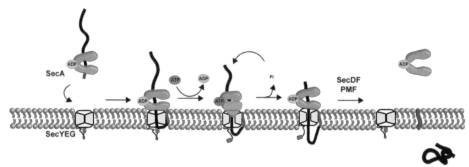

(B) Brownian Ratchet

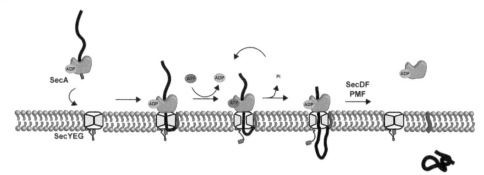

(C) Push and Slide

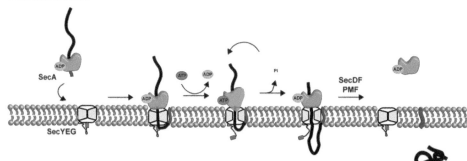

(D) Reciprocating Piston

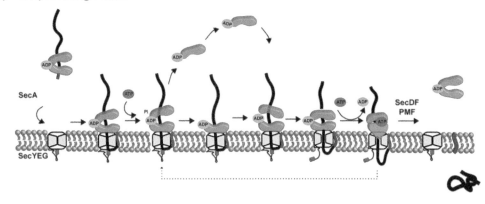

causing stepwise translocation (120). To apply the DNA helicase principle, SecA is required to multimerize in order to have multiple substrate binding sites since monomeric SecA appears to have only one substrate binding site (81, 121). One SecA protomer may act as a clamp and move the preprotein into the channel, while the other SecA protomer traps the chains to prevent retrograde transport, as SecYEG is not able to make a stable anchor for preproteins (122). This implies that a high degree of cooperativity is needed between the two protomers of SecA to ensure that the preprotein is bound to one of the protomers at any given time. Currently, it is unclear how a small movement of the 2HF can drive translocation of polypeptide segments ~20 amino acids long.

Brownian Ratchet Model

In the Brownian ratchet model, SecA acts as the regulator for channel opening of SecYEG (123), while translocation occurs by Brownian movement of the unfolded preprotein through the channel. Because of the contact of the 2HF of SecA with SecYEG (28), movement of the 2HF could potentially result in an opening of the channel. Backsliding of the preprotein would be prevented by the SecA association and provide directionality to the process, which might be further facilitated by folding of the polypeptide at the *cis*-side of the membrane and/or binding

by SecDF (124). This model explains the promiscuity of the system for diverse preprotein substrates (125) but does not explain stepwise translocation (11, 82, 122).

Push-and-Slide Model

The "push-and-slide" mechanism (109) combines the power stroke and Brownian ratchet models and explains earlier observations that SecA-mediated translocation occurs stepwise, whereas in the absence of SecA association, the preprotein may slide within the pore (122). Again, this model depends on the 2HF for the power stroke (76, 109). Once ATP is hydrolyzed, the 2HF would return to its pretranslocation position and dissociate from the preprotein to allow passive sliding of the protein into the channel. This model, however, does not explain that a complex of SecA-SecYEG wherein the SecA 2HF is crosslinked to the C4 loop of SecY is functional in translocation (117). Alternatively, stepwise translocation may arise from binding and release of SecA to and from SecYEG (122, 126).

Reciprocating Piston Model

SecA exists as a dimer during translocation (89, 90, 94, 127), but monomeric states have also been reported (28, 97, 127, 128). The reciprocating piston model combines the power stroke model with the SecA monomer-dimer transition (29). Translocation is initiated by

---

FIGURE 5 Proposed models of SecA-mediated protein translocation. (A) Power stroke: ATP binding and hydrolysis induce conformational changes of SecA that result in a mechanical force on the preprotein, pushing it through the SecYEG channel. In this model, oligomerization of SecA is required to prevent backsliding of the preprotein. (B) Brownian ratchet: SecA regulates the SecYEG channel opening via the 2HF of SecA, allowing the protein translocation via diffusion. Trapping and release of the translocating preprotein at the *cis*-side result in translocation, while SecA may fulfill an additional function by opening the translocation channel. The oligomeric state of SecA is not explicitly shown in this model. (C) Push and slide: this model uses both SecA-dependent pushing and Brownian motion. The oligomeric state of SecA is not explicitly shown in this model. (D) Reciprocating piston: this model is a combination of a power stroke mechanism with the conversion of dimeric-monomeric SecA. Repeated cycles of SecA monomerization-rebinding and ATP binding-hydrolysis yield a stepwise translocation process. In none of these models is the exact role of the PMF and SecDF included, but they contribute to efficient translocation.

binding of dimeric SecA to SecYEG. Next, ATP hydrolysis induces SecA monomerization where one of the SecA monomers remains anchored to SecYEG to prevent backsliding of the partially translocated preprotein, while the other monomer is released from the membrane. Rebinding of another SecA monomer to SecYEG-SecA-preprotein complex then promotes ATP-independent translocation of a preprotein segment, while subsequent binding of ATP drives the translocation by a power stroke. These steps are repeated until the preprotein is fully translocated. This model explains the two consecutive translocation stages observed biochemically, i.e., translocation driven by SecA binding to the preprotein and by ATP binding (82, 122). Complete dissociation of SecA from SecYEG may allow translocation by Brownian diffusion and enable PMF-driven translocation.

ROLE OF THE SecDF COMPLEX

The aforementioned models do not take the role of the PMF into account. Although ATP suffices for translocation *in vitro, in vivo* it strongly depends on the PMF. SecA may mainly serve to initiate and direct translocation by releasing a looped structure of the signal sequence and early mature protein domain into the pore, whereupon translocation is further driven by the PMF (14). Indeed, *in vitro* translocation at low SecA concentrations is highly PMF dependent (129, 130). Late stages of translocation allow large unfolded regions of the preprotein to be translocated without ATP and are SecDF and PMF dependent (131).

SecDF is a subcomplex that associates with the translocon to form the holo-translocon complex (124). The crystal structure of SecDF shows a single polypeptide with 12 TMS, 6 TMS each in both SecD and SecF (132, 133). SecDF also contains 6 periplasmic domains (P1 to P6); P1 and P4 form a periplasmic protruding structure. P1 has been proposed to interact with the polypeptide substrate, and movement of P1 may result in a PMF-dependent pulling action by SecDF at the periplasmic side of the membrane (133, 134).

CONCLUDING REMARKS

Integrating biochemical, biophysical, and structural studies has led to a basic understanding of the molecular mechanism of protein translocation. However, still many mechanistic questions remain unresolved. Although translocation exhibits power stroke- and Brownian diffusion-like mechanistic features, it remains unclear how translocation is linked to the SecA dimer. To unify potentially conflicting results, the process needs to be examined at the single-molecule level to reveal the dynamic interplay between the components and identify their roles at the different stages of the process.

ACKNOWLEDGMENTS

This work is supported by the incentive scheme of the Zernike Institute of Advanced Materials and by the Foundation of Life Sciences of the Netherlands Organisation for Scientific Research (ALW-NWO).

CITATION

Komarudin AG, Driessen AJM. 2019. SecA-mediated protein translocation through the SecYEG channel. Microbiol Spectrum 7(4): PSIB-0028-2019.

REFERENCES

1. **Tsirigotaki A, De Geyter J, Šoštaric N, Economou A, Karamanou S.** 2017. Protein export through the bacterial Sec pathway. *Nat Rev Microbiol* **15:**21–36.
2. **Driessen AJM, Nouwen N.** 2008. Protein translocation across the bacterial cytoplasmic membrane. *Annu Rev Biochem* **77:**643–667.
3. **Bolhuis A.** 2004. The archaeal Sec-dependent protein translocation pathway. *Philos Trans R Soc Lond B Biol Sci* **359:**919–927.

4. **Bondar A-N, del Val C, Freites JA, Tobias DJ, White SH.** 2010. Dynamics of SecY translocons with translocation-defective mutations. *Structure* **18:**847–857.

5. **Pohlschröder M, Prinz WA, Hartmann E, Beckwith J.** 1997. Protein translocation in the three domains of life: variations on a theme. *Cell* **91:**563–566.

6. **Natale P, Brüser T, Driessen AJM.** 2008. Sec- and Tat-mediated protein secretion across the bacterial cytoplasmic membrane—distinct translocases and mechanisms. *Biochim Biophys Acta* **1778:**1735–1756.

7. **Fekkes P, van der Does C, Driessen AJ.** 1997. The molecular chaperone SecB is released from the carboxy-terminus of SecA during initiation of precursor protein translocation. *EMBO J* **16:**6105–6113.

8. **Bechtluft P, Nouwen N, Tans SJ, Driessen AJM.** 2010. SecB—a chaperone dedicated to protein translocation. *Mol Biosyst* **6:**620–627.

9. **Fekkes P, Driessen AJ.** 1999. Protein targeting to the bacterial cytoplasmic membrane. *Microbiol Mol Biol Rev* **63:**161–173.

10. **Cabelli RJ, Chen L, Tai PC, Oliver DB.** 1988. SecA protein is required for secretory protein translocation into *E. coli* membrane vesicles. *Cell* **55:**683–692.

11. **Economou A, Wickner W.** 1994. SecA promotes preprotein translocation by undergoing ATP-driven cycles of membrane insertion and deinsertion. *Cell* **78:**835–843.

12. **du Plessis DJF, Nouwen N, Driessen AJM.** 2011. The Sec translocase. *Biochim Biophys Acta* **1808:**851–865.

13. **Duong F, Wickner W.** 1997. The SecDFyajC domain of preprotein translocase controls preprotein movement by regulating SecA membrane cycling. *EMBO J* **16:**4871–4879.

14. **Tsukazaki T, Nureki O.** 2011. The mechanism of protein export enhancement by the SecDF membrane component. *Biophysics (Nagoya-Shi)* **7:**129–133.

15. **Müller M, Koch HG, Beck K, Schäfer U.** 2001. Protein traffic in bacteria: multiple routes from the ribosome to and across the membrane. *Prog Nucleic Acid Res Mol Biol* **66:**107–157.

16. **Van den Berg B, Clemons WM Jr, Collinson I, Modis Y, Hartmann E, Harrison SC, Rapoport TA.** 2004. X-ray structure of a protein-conducting channel. *Nature* **427:**36–44.

17. **Gumbart J, Schulten K.** 2007. Structural determinants of lateral gate opening in the protein translocon. *Biochemistry* **46:**11147–11157.

18. **Park E, Rapoport TA.** 2011. Preserving the membrane barrier for small molecules during bacterial protein translocation. *Nature* **473:**239–242.

19. **Tam PCK, Maillard AP, Chan KKY, Duong F.** 2005. Investigating the SecY plug movement at the SecYEG translocation channel. *EMBO J* **24:**3380–3388.

20. **Brundage L, Hendrick JP, Schiebel E, Driessen AJ, Wickner W.** 1990. The purified *E. coli* integral membrane protein SecY/E is sufficient for reconstitution of SecA-dependent precursor protein translocation. *Cell* **62:**649–657.

21. **Hanada M, Nishiyama KI, Mizushima S, Tokuda H.** 1994. Reconstitution of an efficient protein translocation machinery comprising SecA and the three membrane proteins, SecY, SecE, and SecG (p12). *J Biol Chem* **269:**23625–23631.

22. **Belin D, Plaia G, Boulfekhar Y, Silva F.** 2015. *Escherichia coli* SecG is required for residual export mediated by mutant signal sequences and for SecY-SecE complex stability. *J Bacteriol* **197:**542–552.

23. **Tanaka Y, Sugano Y, Takemoto M, Mori T, Furukawa A, Kusakizako T, Kumazaki K, Kashima A, Ishitani R, Sugita Y, Nureki O, Tsukazaki T.** 2015. Crystal structures of SecYEG in lipidic cubic phase elucidate a precise resting and a peptide-bound state. *Cell Rep* **13:**1561–1568.

24. **Heinrich SU, Mothes W, Brunner J, Rapoport TA.** 2000. The Sec61p complex mediates the integration of a membrane protein by allowing lipid partitioning of the transmembrane domain. *Cell* **102:**233–244.

25. **du Plessis DJF, Berrelkamp G, Nouwen N, Driessen AJM.** 2009. The lateral gate of SecYEG opens during protein translocation. *J Biol Chem* **284:**15805–15814.

26. **Corey RA, Allen WJ, Komar J, Masiulis S, Menzies S, Robson A, Collinson I.** 2016. Unlocking the bacterial SecY translocon. *Structure* **24:**518–527.

27. **Plath K, Mothes W, Wilkinson BM, Stirling CJ, Rapoport TA.** 1998. Signal sequence recognition in posttranslational protein transport across the yeast ER membrane. *Cell* **94:**795–807.

28. **Zimmer J, Nam Y, Rapoport TA.** 2008. Structure of a complex of the ATPase SecA and the protein-translocation channel. *Nature* **455:**936–943.

29. **Kusters I, Driessen AJM.** 2011. SecA, a remarkable nanomachine. *Cell Mol Life Sci* **68:**2053–2066.

30. **Li L, Park E, Ling J, Ingram J, Ploegh H, Rapoport TA.** 2016. Crystal structure of a substrate-engaged SecY protein-translocation channel. *Nature* **531:**395–399.

31. **Harris CR, Silhavy TJ.** 1999. Mapping an interface of SecY (PrlA) and SecE (PrlG) by

using synthetic phenotypes and in vivo cross-linking. *J Bacteriol* **181**:3438–3444.

32. **Tsukazaki T, Mori H, Fukai S, Ishitani R, Mori T, Dohmae N, Perederina A, Sugita Y, Vassylyev DG, Ito K, Nureki O.** 2008. Conformational transition of Sec machinery inferred from bacterial SecYE structures. *Nature* **455**:988–991.

33. **Egea PF, Stroud RM.** 2010. Lateral opening of a translocon upon entry of protein suggests the mechanism of insertion into membranes. *Proc Natl Acad Sci U S A* **107**:17182–17187.

34. **Maillard AP, Lalani S, Silva F, Belin D, Duong F.** 2007. Deregulation of the SecYEG translocation channel upon removal of the plug domain. *J Biol Chem* **282**:1281–1287.

35. **Junne T, Schwede T, Goder V, Spiess M.** 2006. The plug domain of yeast Sec61p is important for efficient protein translocation, but is not essential for cell viability. *Mol Biol Cell* **17**:4063–4068.

36. **Li W, Schulman S, Boyd D, Erlandson K, Beckwith J, Rapoport TA.** 2007. The plug domain of the SecY protein stabilizes the closed state of the translocation channel and maintains a membrane seal. *Mol Cell* **26**:511–521.

37. **Junne T, Schwede T, Goder V, Spiess M.** 2007. Mutations in the Sec61p channel affecting signal sequence recognition and membrane protein topology. *J Biol Chem* **282**:33201–33209.

38. **Zhang B, Miller TF III.** 2010. Hydrophobically stabilized open state for the lateral gate of the Sec translocon. *Proc Natl Acad Sci U S A* **107**:5399–5404.

39. **Lycklama A, Nijeholt JA, Bulacu M, Marrink SJ, Driessen AJM.** 2010. Immobilization of the plug domain inside the SecY channel allows unrestricted protein translocation. *J Biol Chem* **285**:23747–23754 .

40. **Duong F, Wickner W.** 1999. The PrlA and PrlG phenotypes are caused by a loosened association among the translocase SecYEG subunits. *EMBO J* **18**:3263–3270.

41. **Smith MA, Clemons WM Jr, DeMars CJ, Flower AM.** 2005. Modeling the effects of prl mutations on the *Escherichia coli* SecY complex. *J Bacteriol* **187**:6454–6465.

42. **Silhavy TJ, Mitchell AM.** 2019. Genetic analysis of protein translocation. *Protein J* **38**:217–228.

43. **Osborne RS, Silhavy TJ.** 1993. PrlA suppressor mutations cluster in regions corresponding to three distinct topological domains. *EMBO J* **12**:3391–3398.

44. **Emr SD, Hanley-Way S, Silhavy TJ.** 1981. Suppressor mutations that restore export of a protein with a defective signal sequence. *Cell* **23**:79–88.

45. **de Keyzer J, van der Does C, Swaving J, Driessen AJM.** 2002. The F286Y mutation of

PrlA4 tempers the signal sequence suppressor phenotype by reducing the SecA binding affinity. *FEBS Lett* **510**:17–21.

46. **van der Wolk JP, Fekkes P, Boorsma A, Huie JL, Silhavy TJ, Driessen AJ.** 1998. PrlA4 prevents the rejection of signal sequence defective preproteins by stabilizing the SecA-SecY interaction during the initiation of translocation. *EMBO J* **17**:3631–3639.

47. **Taufik I, Kedrov A, Exterkate M, Driessen AJM.** 2013. Monitoring the activity of single translocons. *J Mol Biol* **425**:4145–4153.

48. **Nouwen N, de Kruijff B, Tommassen J.** 1996. *prlA* suppressors in *Escherichia coli* relieve the proton electrochemical gradient dependency of translocation of wild-type precursors. *Proc Natl Acad Sci U S A* **93**:5953–5957.

49. **Fikes JD, Bassford PJ Jr.** 1989. Novel *secA* alleles improve export of maltose-binding protein synthesized with a defective signal peptide. *J Bacteriol* **171**:402–409.

50. **Stader J, Gansheroff LJ, Silhavy TJ.** 1989. New suppressors of signal-sequence mutations, *prlG*, are linked tightly to the *secE* gene of *Escherichia coli*. *Genes Dev* **3**:1045–1052.

51. **Flower AM, Doebele RC, Silhavy TJ.** 1994. PrlA and PrlG suppressors reduce the requirement for signal sequence recognition. *J Bacteriol* **176**:5607–5614.

52. **Prinz WA, Spiess C, Ehrmann M, Schierle C, Beckwith J.** 1996. Targeting of signal sequenceless proteins for export in *Escherichia coli* with altered protein translocase. *EMBO J* **15**:5209–5217.

53. **Saparov SM, Erlandson K, Cannon K, Schaletzky J, Schulman S, Rapoport TA, Pohl P.** 2007. Determining the conductance of the SecY protein translocation channel for small molecules. *Mol Cell* **26**:501–509.

54. **Tani K, Tokuda H, Mizushima S.** 1990. Translocation of ProOmpA possessing an intramolecular disulfide bridge into membrane vesicles of *Escherichia coli*. Effect of membrane energization. *J Biol Chem* **265**:17341–17347.

55. **Tani K, Mizushima S.** 1991. A chemically cross-linked nonlinear proOmpA molecule can be translocated into everted membrane vesicles of *Escherichia coli* in the presence of the proton motive force. *FEBS Lett* **285**:127–131.

56. **De Keyzer J, Van Der Does C, Driessen AJM.** 2002. Kinetic analysis of the translocation of fluorescent precursor proteins into *Escherichia coli* membrane vesicles. *J Biol Chem* **277**:46059–46065.

57. **Bonardi F, Halza E, Walko M, Du Plessis F, Nouwen N, Feringa BL, Driessen AJM.** 2011. Probing the SecYEG translocation pore size

with preproteins conjugated with sizable rigid spherical molecules. *Proc Natl Acad Sci U S A* **108**:7775–7780.

58. **Hizlan D, Robson A, Whitehouse S, Gold VA, Vonck J, Mills D, Kühlbrandt W, Collinson I.** 2012. Structure of the SecY complex unlocked by a preprotein mimic. *Cell Rep* **1**:21–28.

59. **Breyton C, Haase W, Rapoport TA, Kühlbrandt W, Collinson I.** 2002. Three-dimensional structure of the bacterial protein-translocation complex SecYEG. *Nature* **418**:662–665.

60. **Mitra K, Schaffitzel C, Shaikh T, Tama F, Jenni S, Brooks CL III, Ban N, Frank J, Frank J.** 2005. Structure of the *E. coli* protein-conducting channel bound to a translating ribosome. *Nature* **438**:318–324.

61. **Deville K, Gold VAM, Robson A, Whitehouse S, Sessions RB, Baldwin SA, Radford SE, Collinson I.** 2011. The oligomeric state and arrangement of the active bacterial translocon. *J Biol Chem* **286**:4659–4669.

62. **Das S, Oliver DB.** 2011. Mapping of the SecA·SecY and SecA·SecG interfaces by site-directed in vivo photocross-linking. *J Biol Chem* **286**:12371–12380.

63. **Zheng Z, Blum A, Banerjee T, Wang Q, Dantis V, Oliver D.** 2016. Determination of the oligomeric state of SecYEG protein secretion channel complex using in vivo photo- and disulfide cross-linking. *J Biol Chem* **291**:5997–6010.

64. **Osborne AR, Rapoport TA.** 2007. Protein translocation is mediated by oligomers of the SecY complex with one SecY copy forming the channel. *Cell* **129**:97–110.

65. **Kedrov A, Kusters I, Krasnikov VV, Driessen AJM.** 2011. A single copy of SecYEG is sufficient for preprotein translocation. *EMBO J* **30**:4387–4397.

66. **Tomkiewicz D, Nouwen N, Driessen AJM.** 2007. Pushing, pulling and trapping—modes of motor protein supported protein translocation. *FEBS Lett* **581**:2820–2828.

67. **Findik BT, Smith VF, Randall LL.** 2018. Penetration into membrane of amino-terminal region of SecA when associated with SecYEG in active complexes. *Protein Sci* **27**:681–691.

68. **Huber D, Rajagopalan N, Preissler S, Rocco MA, Merz F, Kramer G, Bukau B.** 2011. SecA interacts with ribosomes in order to facilitate posttranslational translocation in bacteria. *Mol Cell* **41**:343–353.

69. **Sato K, Mori H, Yoshida M, Mizushima S.** 1996. Characterization of a potential catalytic residue, Asp-133, in the high affinity ATP-binding site of *Escherichia coli* SecA, translocation ATPase. *J Biol Chem* **271**:17439–17444.

70. **Hunt JF, Weinkauf S, Henry L, Fak JJ, McNicholas P, Oliver DB, Deisenhofer J.** 2002. Nucleotide control of interdomain interactions in the conformational reaction cycle of SecA. *Science* **297**:2018–2026.

71. **Ye J, Osborne AR, Groll M, Rapoport TA.** 2004. RecA-like motor ATPases—lessons from structures. *Biochim Biophys Acta* **1659**:1–18.

72. **Bauer BW, Rapoport TA.** 2009. Mapping polypeptide interactions of the SecA ATPase during translocation. *Proc Natl Acad Sci U S A* **106**:20800–20805.

73. **Ding H, Mukerji I, Oliver D.** 2003. Nucleotide and phospholipid-dependent control of PPXD and C-domain association for SecA ATPase. *Biochemistry* **42**:13468–13475.

74. **Chada N, Chattrakun K, Marsh BP, Mao C, Bariya P, King GM.** 2018. Single-molecule observation of nucleotide induced conformational changes in basal SecA-ATP hydrolysis. *Sci Adv* **4**:eaat8797.

75. **Papanikolau Y, Papadovasilaki M, Ravelli RBG, McCarthy AA, Cusack S, Economou A, Petratos K.** 2007. Structure of dimeric SecA, the *Escherichia coli* preprotein translocase motor. *J Mol Biol* **366**:1545–1557.

76. **Erlandson KJ, Miller SBM, Nam Y, Osborne AR, Zimmer J, Rapoport TA.** 2008. A role for the two-helix finger of the SecA ATPase in protein translocation. *Nature* **455**:984–987.

77. **Breukink E, Nouwen N, van Raalte A, Mizushima S, Tommassen J, de Kruijff B.** 1995. The C terminus of SecA is involved in both lipid binding and SecB binding. *J Biol Chem* **270**:7902–7907.

78. **Gold VAM, Robson A, Clarke AR, Collinson I.** 2007. Allosteric regulation of SecA: magnesium-mediated control of conformation and activity. *J Biol Chem* **282**:17424–17432.

79. **Lill R, Dowhan W, Wickner W.** 1990. The ATPase activity of SecA is regulated by acidic phospholipids, SecY, and the leader and mature domains of precursor proteins. *Cell* **60**:271–280.

80. **Miller A, Wang L, Kendall DA.** 2002. SecB modulates the nucleotide-bound state of SecA and stimulates ATPase activity. *Biochemistry* **41**:5325–5332.

81. **Gelis I, Bonvin AMJJ, Keramisanou D, Koukaki M, Gouridis G, Karamanou S, Economou A, Kalodimos CG.** 2007. Structural basis for signal-sequence recognition by the translocase motor SecA as determined by NMR. *Cell* **131**:756–769.

82. **van der Wolk JPW, de Wit JG, Driessen AJ.** 1997. The catalytic cycle of the *Escherichia coli* SecA ATPase comprises two distinct preprotein translocation events. *EMBO J* **16**:7297–7304.

83. **Corey RA, Pyle E, Allen WJ, Watkins DW, Casiraghi M, Miroux B, Arechaga I, Politis A, Collinson I.** 2018. Specific cardiolipin-SecY interactions are required for proton-motive force stimulation of protein secretion. *Proc Natl Acad Sci U S A* **115**:7967–7972.

84. **Gold VAM, Robson A, Bao H, Romantsov T, Duong F, Collinson I.** 2010. The action of cardiolipin on the bacterial translocon. *Proc Natl Acad Sci U S A* **107**:10044–10049.

85. **Karamanou S, Vrontou E, Sianidis G, Baud C, Roos T, Kuhn A, Politou AS, Economou A.** 1999. A molecular switch in SecA protein couples ATP hydrolysis to protein transloca-tion. *Mol Microbiol* **34**:1133–1145.

86. **Woodbury RL, Hardy SJ, Randall LL.** 2002. Complex behavior in solution of homodimeric SecA. *Protein Sci* **11**:875–882.

87. **Wang H, Na B, Yang H, Tai PC.** 2008. Additional in vitro and in vivo evidence for SecA functioning as dimers in the membrane: dissociation into monomers is not essential for protein translocation in *Escherichia coli*. *J Bacteriol* **190**:1413–1418.

88. **Kusters I, van den Bogaart G, Kedrov A, Krasnikov V, Fulyani F, Poolman B, Driessen AJM.** 2011. Quaternary structure of SecA in solution and bound to SecYEG probed at the single molecule level. *Structure* **19**:430–439.

89. **Driessen AJ.** 1993. SecA, the peripheral sub-unit of the *Escherichia coli* precursor protein translocase, is functional as a dimer. *Biochem-istry* **32**:13190–13197.

90. **de Keyzer J, van der Sluis EO, Spelbrink REJ, Nijstad N, de Kruijff B, Nouwen N, van der Does C, Driessen AJM.** 2005. Covalently dimerized SecA is functional in protein trans-location. *J Biol Chem* **280**:35255–35260.

91. **Benach J, Chou YT, Fak JJ, Itkin A, Nicolae DD, Smith PC, Wittrock G, Floyd DL, Golsaz CM, Gierasch LM, Hunt JF.** 2003. Phospholipid-induced monomerization and signal-peptide-induced oligomerization of SecA. *J Biol Chem* **278**:3628–3638.

92. **Bu Z, Wang L, Kendall DA.** 2003. Nucleotide binding induces changes in the oligomeric state and conformation of Sec A [*sic*] in a lipid environment: a small-angle neutron-scattering study. *J Mol Biol* **332**:23–30.

93. **Koch S, de Wit JG, Vos I, Birkner JP, Gordiichuk P, Herrmann A, van Oijen AM, Driessen AJM.** 2016. Lipids activate SecA for high affinity binding to the SecYEG complex. *J Biol Chem* **291**:22534–22543.

94. **Jilaveanu LB, Zito CR, Oliver D.** 2005. Dimeric SecA is essential for protein translo-cation. *Proc Natl Acad Sci U S A* **102**:7511–7516.

95. **Karamanou S, Sianidis G, Gouridis G, Pozidis C, Papanikolau Y, Papanikou E, Economou A.** 2005. *Escherichia coli* SecA truncated at its termini is functional and dimeric. *FEBS Lett* **579**:1267–1271.

96. **Banerjee T, Lindenthal C, Oliver D.** 2017. SecA functions in vivo as a discrete anti-parallel dimer to promote protein transport. *Mol Microbiol* **103**:439–451.

97. **Or E, Boyd D, Gon S, Beckwith J, Rapoport T.** 2005. The bacterial ATPase SecA functions as a monomer in protein translocation. *J Biol Chem* **280**:9097–9105.

98. **Gouridis G, Karamanou S, Sardis MF, Schärer MA, Capitani G, Economou A.** 2013. Quaternary dynamics of the SecA motor drive translocase catalysis. *Mol Cell* **52**:655–666.

99. **Fekkes P, de Wit JG, Boorsma A, Friesen RHE, Driessen AJM.** 1999. Zinc stabilizes the SecB binding site of SecA. *Biochemistry* **38**:5111–5116.

100. **Hegde RS, Bernstein HD.** 2006. The surpris-ing complexity of signal sequences. *Trends Biochem Sci* **31**:563–571.

101. **Owji H, Nezafat N, Negahdaripour M, Hajiebrahimi A, Ghasemi Y.** 2018. A compre-hensive review of signal peptides: structure, roles, and applications. *Eur J Cell Biol* **97**:422–441.

102. **Chatzi KE, Sardis MF, Tsirigotaki A, Koukaki M, Šoštarić N, Konijnenberg A, Sobott F, Kalodimos CG, Karamanou S, Economou A.** 2017. Preprotein mature domains contain translocase targeting signals that are essential for secretion. *J Cell Biol* **216**:1357–1369.

103. **Fessl T, Watkins D, Oatley P, Allen WJ, Corey RA, Horne J, Baldwin SA, Radford SE, Collinson I, Tuma R.** 2018. Dynamic action of the Sec machinery during initiation, protein translocation and termination. *eLife* **7**:e35112.

104. **Sardis MF, Tsirigotaki A, Chatzi KE, Portaliou AG, Gouridis G, Karamanou S, Economou A.** 2017. Preprotein conformational dynamics drive bivalent translocase docking and secretion. *Structure* **25**:1056–1067.e6.

105. **Xu Z, Knafels JD, Yoshino K.** 2000. Crystal structure of the bacterial protein export chap-erone SecB. *Nat Struct Biol* **7**:1172–1177.

106. **Crane JM, Suo Y, Lilly AA, Mao C, Hubbell WL, Randall LL.** 2006. Sites of interaction of a precursor polypeptide on the export chaper-one SecB mapped by site-directed spin label-ing. *J Mol Biol* **363**:63–74.

107. **van der Sluis EO, Driessen AJM.** 2006. Stepwise evolution of the Sec machinery in Proteobacteria. *Trends Microbiol* **14**:105–108.

108. **Bechtluft P, van Leeuwen RGH, Tyreman M, Tomkiewicz D, Nouwen N, Tepper HL, Driessen AJM, Tans SJ.** 2007. Direct observa-

tion of chaperone-induced changes in a protein folding pathway. *Science* **318**:1458–1461.

109. **Bauer BW, Shemesh T, Chen Y, Rapoport TA.** 2014. A "push and slide" mechanism allows sequence-insensitive translocation of secretory proteins by the SecA ATPase. *Cell* **157**:1416–1429.

110. **Hendrick JP, Wickner W.** 1991. SecA protein needs both acidic phospholipids and SecY/E protein for functional high-affinity binding to the *Escherichia coli* plasma membrane. *J Biol Chem* **266**:24596–24600.

111. **Floyd JH, You Z, Hsieh Y-H, Ma Y, Yang H, Tai PC.** 2014. The dispensability and requirement of SecA N-terminal aminoacyl residues for complementation, membrane binding, lipid-specific domains and channel activities. *Biochem Biophys Res Commun* **453**:138–142.

112. **Sianidis G, Karamanou S, Vrontou E, Boulias K, Repanas K, Kyrpides N, Politou AS, Economou A.** 2001. Cross-talk between catalytic and regulatory elements in a DEAD motor domain is essential for SecA function. *EMBO J* **20**:961–970.

113. **Chen Y, Bauer BW, Rapoport TA, Gumbart JC.** 2015. Conformational changes of the clamp of the protein translocation ATPase SecA. *J Mol Biol* **427**:2348–2359.

114. **Gold VAM, Whitehouse S, Robson A, Collinson I.** 2013. The dynamic action of SecA during the initiation of protein translocation. *Biochem J* **449**:695–705.

115. **Fak JJ, Itkin A, Ciobanu DD, Lin EC, Song X-J, Chou Y-T, Gierasch LM, Hunt JF.** 2004. Nucleotide exchange from the high-affinity ATP-binding site in SecA is the rate-limiting step in the ATPase cycle of the soluble enzyme and occurs through a specialized conformational state. *Biochemistry* **43**:7307–7327.

116. **Nouwen N, Berrelkamp G, Driessen AJM.** 2009. Charged amino acids in a preprotein inhibit SecA-dependent protein translocation. *J Mol Biol* **386**:1000–1010.

117. **Whitehouse S, Gold VA, Robson A, Allen WJ, Sessions RB, Collinson I.** 2012. Mobility of the SecA 2-helix-finger is not essential for polypeptide translocation via the SecYEG complex. *J Cell Biol* **199**:919–929.

118. **Zhang Q, Lahiri S, Banerjee T, Sun Z, Oliver D, Mukerji I.** 2017. Alignment of the protein substrate hairpin along the SecA two-helix finger primes protein transport in *Escherichia coli*. *Proc Natl Acad Sci U S A* **114**:9343–9348.

119. **Osborne AR, Clemons WM Jr, Rapoport TA.** 2004. A large conformational change of the translocation ATPase SecA. *Proc Natl Acad Sci U S A* **101**:10937–10942.

120. **Catipovic MA, Bauer BW, Loparo JJ, Rapoport TA.** 2019. Protein translocation by the SecA ATPase occurs by a power-stroke mechanism. *EMBO J* **38**:e101140.

121. **Papanikou E, Karamanou S, Baud C, Frank M, Sianidis G, Keramisanou D, Kalodimos CG, Kuhn A, Economou A.** 2005. Identification of the preprotein binding domain of SecA. *J Biol Chem* **280**:43209–43217.

122. **Schiebel E, Driessen AJM, Hartl FU, Wickner W.** 1991. Δ mu H+ and ATP function at different steps of the catalytic cycle of preprotein translocase. *Cell* **64**:927–939.

123. **Allen WJ, Corey RA, Oatley P, Sessions RB, Baldwin SA, Radford SE, Tuma R, Collinson I.** 2016. Two-way communication between SecY and SecA suggests a Brownian ratchet mechanism for protein translocation. *eLife* **5**:e15598.

124. **Botte M, Zaccai NR, Nijeholt JL, Martin R, Knoops K, Papai G, Zou J, Deniaud A, Karuppasamy M, Jiang Q, Roy AS, Schulten K, Schultz P, Rappsilber J, Zaccai G, Berger I, Collinson I, Schaffitzel C.** 2016. A central cavity within the holo-translocon suggests a mechanism for membrane protein insertion. *Sci Rep* **6**:38399.

125. **Simon SM, Peskin CS, Oster GF.** 1992. What drives the translocation of proteins? *Proc Natl Acad Sci U S A* **89**:3770–3774.

126. **Young J, Duong F.** 2019. Investigating the stability of the SecA-SecYEG complex during protein translocation across the bacterial membrane. *J Biol Chem* **294**:3577–3587.

127. **Duong F.** 2003. Binding, activation and dissociation of the dimeric SecA ATPase at the dimeric SecYEG translocase. *EMBO J* **22**:4375–4384.

128. **Or E, Navon A, Rapoport T.** 2002. Dissociation of the dimeric SecA ATPase during protein translocation across the bacterial membrane. *EMBO J* **21**:4470–4479.

129. **Mori H, Ito K.** 2003. Biochemical characterization of a mutationally altered protein translocase: proton motive force stimulation of the initiation phase of translocation. *J Bacteriol* **185**:405–412.

130. **Nishiyama K, Fukuda A, Morita K, Tokuda H.** 1999. Membrane deinsertion of SecA underlying proton motive force-dependent stimulation of protein translocation. *EMBO J* **18**:1049–1058.

131. **Tsukazaki T.** 2018. Structure-based working model of SecDF, a proton-driven bacterial protein translocation factor. *FEMS Microbiol Lett* **365**:fny112.

132. **Furukawa A, Yoshikaie K, Mori T, Mori H, Morimoto YV, Sugano Y, Iwaki S, Minamino T, Sugita Y, Tanaka Y, Tsukazaki T.** 2017.

Tunnel formation inferred from the I-form structures of the proton-driven protein secretion motor SecDF. *Cell Rep* **19:**895–901.

133. **Tsukazaki T, Mori H, Echizen Y, Ishitani R, Fukai S, Tanaka T, Perederina A, Vassylyev DG, Kohno T, Maturana AD, Ito K, Nureki O.** 2011. Structure and function of a membrane component SecDF that enhances protein export. *Nature* **474:**235–238.

134. **Park E, Rapoport TA.** 2012. Mechanisms of Sec61/SecY-mediated protein translocation across membranes. *Annu Rev Biophys* **41:**21–40.

The Two Distinct Types of SecA2-Dependent Export Systems

3

MIRIAM BRAUNSTEIN,[1] BARBARA A. BENSING,[2] and PAUL M. SULLAM[2]

INTRODUCTION

The protein export systems of bacteria deliver proteins from the cytoplasm to the cell envelope or extracellular environment, and in doing so, they play critical roles in bacterial physiology and pathogenesis. In bacteria, the majority of protein export is carried out by the general Sec system (1, 2). The core components of the Sec system are the integral membrane proteins SecY, SecE, and SecG, which form the SecYEG channel through which unfolded proteins traverse the membrane, and the SecA ATPase, which provides energy for export (Fig. 1A). SecA shuttles between the cytoplasm and SecYEG in its role in export. SecDFYajC are auxiliary components that enhance export efficiency. Proteins exported by the Sec pathway are synthesized as preproteins with N-terminal signal peptides that are recognized by the Sec machinery and removed during export to produce the mature protein. Some Gram-positive bacteria, including high-GC Gram-positive actinobacteria such as mycobacteria, possess two SecA proteins. In these cases, SecA (sometimes called SecA1) is the canonical SecA of the Sec pathway, while SecA2 functions in a specialized pathway that exports one or a few proteins. There are at least two

[1]Department of Microbiology and Immunology, University of North Carolina-Chapel Hill, Chapel Hill, NC 27599
[2]Department of Medicine, San Francisco Veterans Affairs Medical Center and the University of California, San Francisco, CA 94121

Protein Secretion in Bacteria
Edited by Maria Sandkvist, Eric Cascales, and Peter J. Christie
© 2019 American Society for Microbiology, Washington, DC
doi:10.1128/microbiolspec.PSIB-0025-2018

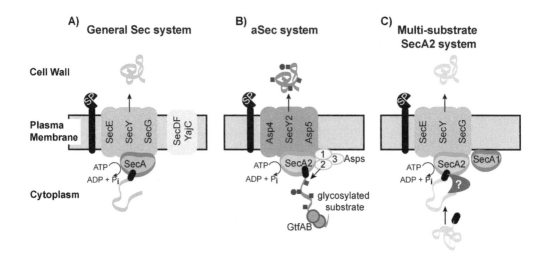

FIGURE 1 Models for the general Sec system, the aSec system, and the multisubstrate SecA2 system. (A) **General Sec system.** SecA uses ATP hydrolysis to export cytoplasmic preproteins through the SecYEG channel in an unfolded state. SecDFYajC are auxiliary components that enhance export efficiency. Sec signal peptides (black rectangle) target preproteins (blue ribbon) for export through SecYEG. Following export across the membrane, the signal peptide is cleaved by a signal peptidase (SP) and the resulting mature protein folds into its proper conformation. **(B) aSec system.** The model depicted is largely based on studies of the *S. gordonii* SecA2 system. Glycosylation of the preprotein (pink ribbon) with GlcNAc (blue squares) and Glc (blue circles) likely occurs cotranslationally. The positively charged N region of the signal peptide (black rectangle) targets the preprotein to anionic phospholipids, which aids the localization with SecA2. Transport through the SecY2/Asp4/5 channel requires a specific sequence in the mature region of the preprotein, as well as Asp1 to Asp3. Asp2 is a bifunctional protein that also mediates O-acetylation of GlcNAc moieties (red square). Cleavage of the signal peptide is thought to be carried out by the general SP. **(C) Multisubstrate SecA2** **system.** The model depicted is largely based on studies of the mycobacterial SecA2 system. SecA2 works with the canonical SecYEG channel and possibly SecA1 to export its specific subset of preproteins (green ribbon). The majority of SecA2 substrates are synthesized as preproteins with a signal peptide (black rectangle) that is cleaved in association with export. The mature domain, not the signal peptide, of a preprotein determines if a protein is exported by this SecA2 system. It is proposed that the mature domain of a SecA2 substrate has the propensity to fold in the cytoplasm and that the role of SecA2 is to facilitate the export of such proteins, in an unfolded state, through the SecYEG channel. Additional factors are likely to work with SecA2 in the pathway (purple symbol). The role of SecA2 in exporting moonlighting proteins that lack signal peptides is unclear and not depicted in the model.

evolutionarily and mechanistically distinct types of SecA2 systems: the accessory Sec (aSec) system, which has also been referred to as the SecA2/SecY2 system, and the multisubstrate SecA2 system, which was initially called the SecA2-only system.

ACCESSORY SEC SYSTEM

Many species of Gram-positive bacteria express an aSec system. Along with SecA2, the aSec system invariably includes SecY2 (a

paralogue of SecY) and three to five accessory Sec proteins (Asps) (Fig. 1B) (3). The latter proteins are essential for substrate transport and are exclusively associated with aSec systems (4, 5). aSec systems transport large, heavily glycosylated cell wall-anchored proteins, known as serine-rich repeat (SRR) glycoproteins (6–8). These substrates undergo extensive O-linked glycosylation intracellularly prior to their transport to the bacterial cell surface, where they function as adhesins important for commensal and pathogenic behavior (9–19).

The gene organization of aSec loci is highly conserved across species and genera (Fig. 2A). Along with the transport components, each aSec locus typically encodes one transported substrate (although up to three have been described) and two or more glycosyltransferase (Gtf) proteins that modify the preprotein in the cytoplasm prior to export (3, 20–22). It is not entirely clear why a dedicated system is necessary for the export of the SRR glycoproteins. One long-standing explanation is that the aSec system transports these unusual substrates because the canonical SecA or SecYEG cannot accommodate glycosylated proteins. Indeed, many aSec substrates cannot undergo canonical Sec transport if glycosylated (23, 24). As discussed below, however, recent studies indicate a more complex role for the aSec system in coordinating transport and posttranslational

modification of the SRR glycoprotein, thereby ensuring proper adhesin function.

Substrates of the aSec System

The SRR glycoproteins comprise a unique family of adhesins that bind a wide range of ligands and impact biofilm formation and virulence (10, 12, 18, 19, 25–29). The adhesins have a conserved domain organization, with a 90-amino-acid signal peptide at the N terminus followed by a short SRR domain, a ligand binding region (BR), a long SRR domain, and a C-terminal LPXTG cell wall anchoring motif (Fig. 3). The BRs can vary significantly, reflecting their considerable repertoire of ligands. For example, several species of oral streptococci express SRR adhesins with "Siglec (sialic acid-binding immunoglobulin-type lectins)-like" binding regions that medi-

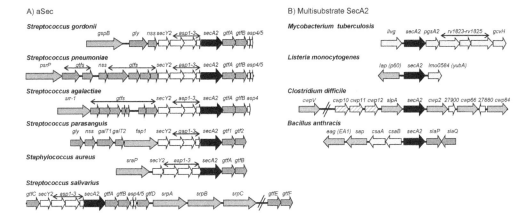

FIGURE 2 Genomic regions encoding aSec and multisubstrate SecA2 proteins. (A) aSec loci. Shown are representative aSec loci in Gram-positive bacteria. The *secA2* gene is shown in black and the other genes encoding core components of the aSec translocase (SecY2 and Asps) are colored yellow. Genes encoding glycosyltransferases (Gtf) and proteins involved in carbohydrate modifications are shown in orange. Genes encoding exported SecA2 substrates are shown in blue. In *Streptococcus parasanguis*, the Asp orthologues are called Gap1 to Gap3. In *Streptococcus salivarius*, the *gtfEF* genes are located distal to the *secA2* locus but are required for the first step of O-GlcNAcylation of the substrate (89) and thus may be functionally analogous to the *gtfAB* pairs found in other aSec loci. **(B) Multisubstrate SecA2 loci.** Shown are representative multisubstrate *secA2* genes and neighboring genomic regions in Gram-positive bacteria. The *secA2* gene is shown in black, and genes encoding SecA2 substrates are shown in blue. Candidate genes for additional SecA2 substrates are shown with blue stripes. Substrates encoded elsewhere in the genome are not shown. Additional proteins with roles in SecA2-dependent export are encoded by genes shown in pink. Genes encoding proteins with no known connection to export are shown in gray.

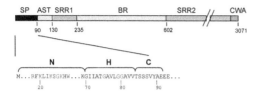

FIGURE 3 GspB domains and features of the N-terminal signal peptide (SP). (Top) Domains of the SRR glycoprotein GspB. AST, aSec transport domain; SRR1 and SRR2, serine-rich repeat regions 1 and 2, respectively; BR, ligand binding region; CWA, cell wall-anchoring domain. The CWA includes a trans-membrane segment, an LPxTG motif, and a charged C-terminal tail (90). **(Bottom) The GspB signal peptide has the tripartite structure of canonical signal peptides: the N-terminal (N), hydrophobic core (H), and cleavage (C) regions.** However, the N region is substantially longer than typical signal peptides and includes a KxYKxGKxW motif (red). Glycine residues in the H region are also indicated in red.

ate binding to sialoglycans (30, 31), while *Streptococcus agalactiae* expresses SRR gly-coproteins that interact with proteins (e.g., human keratin 4 and fibrinogen) (16, 32, 33). This diversity of ligands most likely reflects specific targets for microbial adhesion in different biological niches.

Preprotein Recognition and Trafficking to the aSec System

The preprotein signal peptide of aSec sub-strates has a tripartite structure similar to that of general Sec system substrates, but the N region is approximately three times longer and includes a KxYKxGKxW motif (Fig. 3). This polybasic motif, along with additional basic residues in the extended N region, aids in targeting of the preprotein to anionic phospholipid patches in the membrane and is important for the Asp-independent colo-calization of the preprotein with SecA2 (34). The hydrophobic core (the H region) of the signal peptide is also important for trafficking to the aSec system and contains three glycine residues essential for substrate delivery to the aSec pathway and away from the canonical Sec system.

In addition to the signal peptide, the SRR adhesin GspB of *Streptococcus gordonii* has a specific segment (the accessory Sec transport [AST] domain) at the amino terminus of the mature region that is required for transport. Deletion of the AST domain abolishes aSec export (24, 35), and even single amino acid substitutions within the domain can impair this process. The AST domain interacts directly with SecA2 during transport (36), which affects substrate targeting to the translocon and perhaps opening of the Y2 channel. The requirement of a specific seg-ment in the mature region of the preprotein, along with the involvement of the Asps (see below), is a unique feature of aSec transport that may ensure the selectivity of this path-way for SRR glycoproteins.

The aSec Translocase

SecA2 proteins belonging to the aSec system have a 45-amino-acid truncation of the C-terminal domain (CTD), compared with ca-nonical SecAs, and typically have a proline residue at the C terminus (3) (Fig. 4). These SecA2 proteins have 70% similarity (35 to 40% identity) to SecA of *Escherichia coli*, which includes a high similarity in the preprotein cross-linking domain (PPXD) and the nucle-otide binding motifs NBD1 and NBD2 (3). In *S. gordonii*, SecA2 has a lower basal rate of ATP hydrolysis than its SecA paralogue, and SecA2 requires higher magnesium concentrations for activity (37). These and other findings indicate that streptococcal SecA2 may be more tightly regulated than SecA, which supports the possibility that one or more of the Asps may be required to stimulate ATP binding or hydrolysis, as discussed below.

SecY2 likely forms the transmembrane channel for aSec transport and likely func-tions similarly to SecY (38, 39). The predicted topology of SecY2 is nearly identical to that of SecY (3), even though SecY2 homologues have low primary sequence similarity to the SecY paralogues (20% identity and 60% similarity). Like its paralogue, SecY2 is likely

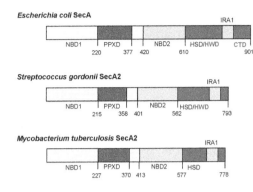

Escherichia coli SecA

Streptococcus gordonii SecA2

Mycobacterium tuberculosis SecA2

FIGURE 4 Domain organization in the canonical SecA of *Escherichia coli* and SecA2 proteins of *S. gordonii* and *M. tuberculosis*. Domains were identified in SecA2 proteins by alignment with *E. coli* SecA using published domain boundaries (91). NBD, nucleotide binding domain; PPXD, preprotein cross-linking domain; HSD, helical scaffold domain; HWD, helical wing domain; IRA, intramolecular regulator of ATPase activity; CTD, C-terminal domain. Compared to the canonical SecA, SecA2 proteins have deletions in the HWD and CTD regions. Amino acid number in the protein sequence is shown below each schematic.

to interact directly with SecA2 to mediate transport (35). It remains unclear how SecY2 can transport an extensively glycosylated protein, in contrast to SecY.

In most species of streptococci with aSec systems, one or two additional small proteins (Asp4 and Asp5) are likely to form a complex with SecY2 *in vivo* (5, 39). Although the roles of Asp4 and Asp5 in transport are uncertain, these proteins are predicted to have structural features resembling those of SecE and SecG of other organisms, respectively, suggesting analogous functions. In complex with SecY2, these proteins enhanced the ATPase activity of streptococcal SecA2 in proteoliposomes, paralleling the effects of SecYEG on SecA (39). Asp4 is partially dispensable for the export of truncated or nonglycosylated GspB variants via the aSec route (40), consistent with a role for Asp4 in stabilizing the open state of the transmembrane channel, rather than a role in the initiation of translocation. Some species lack Asp5 (e.g., *S. agalactiae*) or both Asp4 and Asp5 (e.g.,

Staphylococcus aureus). It is possible that members of the canonical system substitute for these Asps, and indeed, in *S. aureus*, there is evidence for interaction between SecY2 and SecG (41).

Asp1 to Asp3

Asp1 to Asp3 are invariable components of the aSec system and are essential for substrate transport by this pathway. These Asps are located in the cytosol but have an affinity for anionic lipids and can localize as a complex with SecA2 at the inner membrane (38, 42, 43). Although Asp1 to Asp3 lack homology to other transport-associated proteins and their roles in aSec transport are not well defined, their interactions provide some insights as to function. Asp2 and Asp3 directly bind the SRR regions of the GspB preprotein (44). This interaction does not require glycosylation of the SRR domain or specific amino acid motifs. Instead, Asp2 and Asp3 appear to recognize the unstructured or nonfolded sections of the preprotein. Although these Asps bind GspB directly, they do not seem to function as conventional chaperones, since they are not required for GspB stability or targeting to the membrane or translocon (42, 43). However, these Asps augment the physical engagement of the AST domain of substrates with SecA2, as indicated by more extensive AST domain-SecA2 cross-linking *in vivo*, when Asp1 to Asp3 are present (35). Since these interactions are essential for aSec transport, one key role of Asp1 to Asp3 appears to be the enhancement of substrate interactions with the motor protein.

aSec Transport and Posttranslational Modification Are Coordinated Processes

Glycosylation and transport of aSec substrates were initially viewed as independent and sequential processes. Recent studies indicate, however, that these events are coordinated to ensure the proper posttranslational modification and function of the SRR glycoproteins. In addition to its role in transport, Asp2 has been shown to be an acetyltransferase that modifies N-acetylglucosamine moieties on the SRR domains of GspB (45). Targeted mutations of the predicted Asp2 catalytic domain had no effect on transport but abolished acetylation of GspB. Moreover, acetylated GspB was detected only when the glycoprotein had undergone aSec transport, not among cytosolic forms (when aSec transport was blocked) or when GspB was engineered to undergo canonical Sec transport. Thus, Asp2 is a bifunctional protein involved in both the posttranslational modification and transport of SRR glycoproteins. Moreover, these processes appear to be coordinated during the biogenesis of SRR glycoproteins, such that the adhesin is optimally modified for binding. This requirement to couple substrate modification and export may explain the coevolution of the SRR glycoproteins with their specialized glycan modification and export systems.

MULTISUBSTRATE SecA2 SYSTEMS

Multisubstrate SecA2 systems export more than one substrate, although the number of exported substrates is still small compared to that of the general Sec system. The multisubstrate SecA2 systems of *Mycobacterium tuberculosis* (46, 47), *Listeria monocytogenes* (48, 49), and likely *Bacillus anthracis* (50, 51) are required for pathogenesis. In *Corynebacterium glutamicum* (52) and *Clostridium difficile* (53) the multisubstrate systems are essential for bacterial viability. There is no SecY2 in multisubstrate SecA2 systems. Instead, the canonical SecYEG channel is used (Fig. 1C) (54, 55). A common finding across multisubstrate systems is that *secA2* mutations diminish but do not completely abolish export of SecA2 substrates (56–59). Given that SecA2 works with SecYEG, the residual export observed in the absence of SecA2 may be attributable to the general Sec pathway, although this is unproven.

Unlike for aSec systems, phylogenetic analysis of multisubstrate SecA2 proteins and the genomic regions flanking the *secA2* gene do not indicate evolutionary relatedness (Fig. 2B) (3). Thus, there is risk in assuming that there is a single type of multisubstrate SecA2 system with a common mechanism. Nonetheless, there are some intriguing similarities between systems, such that multisubstrate SecA2 systems might be examples of convergent evolution.

Substrates of Multisubstrate SecA2 Systems

Proteomics has been the primary method for identifying substrates of multisubstrate SecA2 systems (48, 56, 58, 60, 61). Proteins exported by multisubstrate systems exhibit a relatively wide variety of functions, with some common themes. Recently, the multisubstrate SecA2 systems of *M. tuberculosis* and *L. monocytogenes* were identified as functioning in RNA secretion as well as protein export (62, 63). While the role for SecA2 in secreting RNA is a complete mystery, this discovery emphasizes the substrate diversity of multisubstrate systems.

Actinobacteria (mycobacteria and corynebacteria)

At least 15 mycobacterial proteins clearly depend on SecA2 for their export to the cell wall or extracellular environment (56, 58, 60). While no corynebacterial SecA2 substrates have been identified, the essentiality of *secA2* in *C. glutamicum* predicts SecA2 substrates with vital functions in this species.

In mycobacteria, one category of SecA2 substrates is cell wall proteins involved in importing solutes, such as solute binding proteins (SBPs) and Mce proteins (56, 60). SBPs deliver solutes to ABC transporters in the membrane (64), and Mce proteins are thought to deliver lipids to Mce transporters (65). A second category of SecA2 substrates is proteins with roles in growth and survival of mycobacteria in macrophages, such as SapM,

PknG, and LipO, which prevent delivery of mycobacteria to phagolysosomes (56, 58, 66), and SodA and KatG, which protect against oxygen radicals (47). As discussed below, peptidoglycan hydrolases are SecA2 substrates in other multisubstrate systems. While this is not clearly the case in mycobacteria, in *M. marinum* a peptidoglycan hydrolase (IipA) was identified as SecA2 dependent (58).

Listeria

In listeriae, the *secA2* gene is adjacent to the gene encoding the p60 protein (Fig. 2B) (57, 67). p60 is one of a group of listeriae SecA2 substrates that are peptidoglycan hydrolases, including NamA (MurA), SspB, and MltD (57, 61, 67, 68). Another functional category of listeriae SecA2 substrates is adhesins, such as Cbp and LAP (61, 69). Finally, similar to the mycobacterial SecA2 system, export of SBPs and superoxide dismutase (SodA) is associated with SecA2 of *L. monocytogenes* (48, 61, 70).

Clostridium

In *C. difficile*, the *secA2* gene is in a locus with genes encoding the major S-layer protein (SlpA) and S-layer-related proteins, called cell wall proteins (CwpV, Cwp2, Cwp66, and Cwp84) (Fig. 2B) (53, 71). SlpA and Cwps are exported as SecA2 substrates (53). The finding that *slpA* is required for *C. difficile* viability explains, at least in part, why the SecA2 system is essential in this bacterium (71). Peptidoglycan hydrolase and adhesin activities have been assigned or predicted for SlpA and/or Cwp proteins (71–74), which is reminiscent of functional categories of listeriae SecA2 substrates.

Bacillus

Similar to the case with *C. difficile,* in *B. anthracis* the SecA2 system exports S-layer proteins (EA1 and Sap) (59), and the *secA2* gene is adjacent to genes encoding these proteins (Fig. 2B). While *B. anthracis* has a *secY2* gene, it is not clustered with *secA2* in the genome and a *secY2* mutant does not exhibit a Sap or EA1 export defect. Thus, there appears to be no SecY2 involvement in SecA2 transport (59). Both Sap and EA1 possess peptidoglycan hydrolase activity (75). Only members of the *Bacillus cereus sensu lato* group have *secA2* or an S-layer (51).

Substrate Recognition by Multisubstrate SecA2 Systems

Most substrates of multisubstrate SecA2 systems have signal peptides that are indistinguishable from canonical Sec signal peptides. Experiments with mycobacteria demonstrate that the signal peptide of a SecA2 substrate is required for export (60, 76). However, the signal peptide does not determine whether a protein is exported by the SecA2 pathway versus another pathway. When signal peptides of SecA2-dependent and SecA1-dependent substrates are swapped, the proteins are still exported by their respective pathways (76). Thus, it is the mature domain of a SecA2-exported protein that determines its transport pathway. These details have been studied only with mycobacterial SecA2 substrates; similar studies are needed for other multisubstrate systems.

In mycobacteria and listeriae there are also examples of proteins lacking signal peptides that depend on SecA2 for export (i.e., PknG, SodA, and KatG in mycobacteria and SodA, LAP, and phosphomannose isomerase in listeriae) (47, 48, 56, 69, 70). For these cases, the proteins are exported as well as localized to the cytoplasm, and they are likely to be "moonlighting" proteins that function in both locations. Nothing is known about the recognition of these proteins by the SecA2 system. Further, it remains possible that the effect of SecA2 on these proteins is indirect. One possibility is that moonlighting proteins might be released from the cytoplasm as a secondary consequence of SecA2-dependent export of peptidoglycan hydrolases that affect cell wall integrity (77).

The Multisubstrate SecA2 Translocase

The mycobacterial SecA2 pathway is the most-studied multisubstrate system, in terms of mechanism. The mycobacterial SecA2 has a role that is distinct from that of SecA1. Even when overexpressed, SecA1 and SecA2 are unable to fulfill the function of one another (78). In addition, *M. tuberculosis* SecA1 and SecA2 share only 38% identity (54% sequence similarity). Thus, it was a surprise to discover broad similarity between the crystal structures of *M. tuberculosis* SecA1 and SecA2 (79). However, compared to SecA1, the CTD of SecA2 is truncated, similar to what was found for SecA2 of the aSec system (Fig. 4) (3). In addition, the helical wing domain (HWD) is missing in the mycobacterial SecA2. The lack of an HWD is a conserved feature of actinobacterial SecA2 proteins, but small HWD truncations may also exist in other SecA2 proteins (79). The significance of CTD and HWD truncations to SecA2 function remains to be investigated.

Like canonical SecAs, SecA2 is an ATPase, and amino acid substitutions in the nucleotide binding domain (NBD1) of SecA2 abolish export (53, 80). However, mycobacterial SecA1 and SecA2 differ in ATPase activity. SecA2 has a lower ATPase rate than SecA1 and also binds ADP and ATP with a higher affinity and releases ADP more slowly (80, 81). Moreover, ADP binding to SecA2 induces a structural rearrangement involving the precursor-binding domain (PPXD) that is not observed in ADP-bound SecA1 or conventional SecA proteins. These differences in nucleotide interactions might reflect the existence of additional proteins that stimulate ATP hydrolysis or ADP release or distinct mechanisms of substrate recognition by SecA2.

Data indicate that SecA2 works with the canonical SecYEG. In mycobacteria and listeriae, suppressors of *secA2* mutants map to the sole *secY* gene in these bacteria, which argues for the canonical SecY being used by SecA2 for export (54, 55). In *C. difficile*, there are also data for SecA2 working with the same SecYEG channel as used by SecA1 (53). Because proteins must be in an unfolded state to transit SecYEG (82, 83), the discovery that SecYEG is used by multisubstrate SecA2 systems implies that the substrates of these systems need to be unfolded for translocation. In mycobacteria, it is demonstrated that the mature domain of a protein dictates the need for SecA2 for export (i.e., not the signal peptide) (76). Further, the mature domain of a mycobacterial SecA2 substrate can be engineered to be exported by the Tat system, a pathway requiring proteins be folded in order to be exported (76). Thus, one possibility is that the mature domain of SecA2 substrates has a propensity to fold or aggregate in the cytoplasm and that the SecA2 system, through currently unknown mechanisms, enables export of such proteins. For example, SecA2 or other players in the multisubstrate system might keep substrates from folding prior to or during export.

There may also be a role for SecA1 in SecA2-dependent export. *M. tuberculosis* SecA1 and SecA2 form heterodimers *in vitro* (84). Additionally, in mycobacteria and listeriae, if SecA1 is depleted or inhibited, SecA2-dependent export is compromised (85, 86). However, further studies are required because the effect of SecA1 on SecA2 export could instead be due to a function of SecA1 in transporting SecYEG proteins to the membrane. In *C. difficile*, SecA1 depletion does not impact SecA2 export, indicating that in this species there is no role for SecA1 in SecA2 transport (53).

Additional Factors Involved in Multisubstrate SecA2 Systems

In mycobacteria, SatS is a cytoplasmic chaperone that works with SecA2 to export a subset of SecA2 substrates (87). SatS stabilizes and prevents aggregation of substrates in the cytoplasm and potentially delivers them to the export machinery. In *B. anthracis* SlaP and SlaQ, which are encoded by genes adjacent to *secA2* (Fig. 2B), are cytoplasmic proteins required for export of SecA2 substrates. The functions of SlaP and SlaQ are unknown (59, 88). In *L. monocytogenes*, the DivIVA protein that recruits proteins to the poles and septum of Gram-positive bacteria is necessary for septal localization and secretion of the p60 and MurA SecA2 substrates (68). The connection between DivIVA and SecA2 export requires further studies to understand.

CONCLUSION

Many Gram-positive bacteria have SecA2 systems that export a small set of proteins and contribute to pathogenesis. However, it is important to recognize that at least two types of SecA2 systems exist (aSec and multi-substrate systems), each with a distinctive mechanism. In the future, it will be important to clarify the defining features of the respective SecA2 substrates as well as the recognition and translocation events of each type of pathway. For multisubstrate systems, in particular, more studies are needed to determine the degree of mechanistic similarity in the absence of evolutionary relatedness.

ACKNOWLEDGMENTS

We gratefully acknowledge Brittany Miller for review of the manuscript.

We acknowledge support from NIH (grant R01AI41513 to B.A.B. and P.M.S. and grant R01AI054540 to M.B.).

CITATION

Braunstein M, Bensing BA, Sullam PM. 2019 The two distinct types of SecA2-dependent export systems. Microbiol Spectrum 7(3): PSIB-0025-2018.

REFERENCES

1. **Tsirigotaki A, De Geyter J, Šoštaric N, Economou A, Karamanou S.** 2017. Protein export through the bacterial Sec pathway. *Nat Rev Microbiol* **15**:21–36.

2. **Crane JM, Randall LL.** 2017. The Sec system: protein export in *Escherichia coli. EcoSal Plus* **7**:ESP-0002-2017.

3. **Bensing BA, Seepersaud R, Yen YT, Sullam PM.** 2014. Selective transport by SecA2: an expanding family of customized motor proteins. *Biochim Biophys Acta* **1843**:1674–1686.

4. **Takamatsu D, Bensing BA, Sullam PM.** 2004. Genes in the accessory *sec* locus of *Streptococcus gordonii* have three functionally distinct effects on the expression of the platelet-binding protein GspB. *Mol Microbiol* **52**:189–203.

5. **Takamatsu D, Bensing BA, Sullam PM.** 2005. Two additional components of the accessory *sec* system mediating export of the *Streptococcus gordonii* platelet-binding protein GspB. *J Bacteriol* **187**:3878–3883.

6. **Bensing BA, Sullam PM.** 2002. An accessory *sec* locus of *Streptococcus gordonii* is required for export of the surface protein GspB and for normal levels of binding to human platelets. *Mol Microbiol* **44**:1081–1094.

7. **Chen Q, Wu H, Fives-Taylor PM.** 2004. Investigating the role of *secA2* in secretion and glycosylation of a fimbrial adhesin in *Streptococcus parasanguis* FW213. *Mol Microbiol* **53**:843–856.

8. **Siboo IR, Chaffin DO, Rubens CE, Sullam PM.** 2008. Characterization of the accessory Sec system of *Staphylococcus aureus. J Bacteriol* **190**:6188–6196.

9. **Mistou MY, Dramsi S, Brega S, Poyart C, Trieu-Cuot P.** 2009. Molecular dissection of the *secA2* locus of group B *Streptococcus* reveals that glycosylation of the Srr1 LPXTG protein is required for full virulence. *J Bacteriol* **191**:4195–4206.

10. **Siboo IR, Chambers HF, Sullam PM.** 2005. Role of SraP, a serine-rich surface protein of *Staphylococcus aureus*, in binding to human platelets. *Infect Immun* **73**:2273–2280.

11. **Seifert KN, Adderson EE, Whiting AA, Bohnsack JF, Crowley PJ, Brady LJ.** 2006. A unique serine-rich repeat protein (Srr-2) and novel surface antigen (epsilon) associated with a virulent lineage of serotype III *Streptococcus agalactiae. Microbiology* **152**:1029–1040.

12. **Xiong YQ, Bensing BA, Bayer AS, Chambers HF, Sullam PM.** 2008. Role of the serine-rich surface glycoprotein GspB of *Streptococcus gordonii* in the pathogenesis of infective endocarditis. *Microb Pathog* **45**:297–301.

13. **Wu H, Zeng M, Fives-Taylor P.** 2007. The glycan moieties and the N-terminal polypep-

tide backbone of a fimbria-associated adhesin, Fap1, play distinct roles in the biofilm development of *Streptococcus parasanguinis*. *Infect Immun* 75:2181–2188.

14. Froeliger EH, Fives-Taylor P. 2001. *Streptococcus parasanguis* fimbria-associated adhesin fap1 is required for biofilm formation. *Infect Immun* 69:2512–2519.

15. Obert C, Sublett J, Kaushal D, Hinojosa E, Barton T, Tuomanen EI, Orihuela CJ. 2006. Identification of a candidate *Streptococcus pneumoniae* core genome and regions of diversity correlated with invasive pneumococcal disease. *Infect Immun* 74:4766–4777.

16. Seo HS, Mu R, Kim BJ, Doran KS, Sullam PM. 2012. Binding of glycoprotein Srr1 of *Streptococcus agalactiae* to fibrinogen promotes attachment to brain endothelium and the development of meningitis. *PLoS Pathog* 8: e1002947.

17. Seo HS, Xiong YQ, Sullam PM. 2013. Role of the serine-rich surface glycoprotein Srr1 of *Streptococcus agalactiae* in the pathogenesis of infective endocarditis. *PLoS One* 8:e64204.

18. Takahashi Y, Takashima E, Shimazu K, Yagishita H, Aoba T, Konishi K. 2006. Contribution of sialic acid-binding adhesin to pathogenesis of experimental endocarditis caused by *Streptococcus gordonii* DL1. *Infect Immun* 74:740–743.

19. van Sorge NM, Quach D, Gurney MA, Sullam PM, Nizet V, Doran KS. 2009. The group B streptococcal serine-rich repeat 1 glycoprotein mediates penetration of the blood-brain barrier. *J Infect Dis* 199:1479–1487.

20. Bensing BA, Gibson BW, Sullam PM. 2004. The *Streptococcus gordonii* platelet binding protein GspB undergoes glycosylation independently of export. *J Bacteriol* 186:638–645.

21. Takamatsu D, Bensing BA, Sullam PM. 2004. Four proteins encoded in the gspB-secY2A2 operon of *Streptococcus gordonii* mediate the intracellular glycosylation of the platelet-binding protein GspB. *J Bacteriol* 186:7100–7111.

22. Zhou M, Wu H. 2009. Glycosylation and biogenesis of a family of serine-rich bacterial adhesins. *Microbiology* 155:317–327.

23. Bensing BA, Takamatsu D, Sullam PM. 2005. Determinants of the streptococcal surface glycoprotein GspB that facilitate export by the accessory Sec system. *Mol Microbiol* 58:1468–1481.

24. Chen Q, Sun B, Wu H, Peng Z, Fives-Taylor PM. 2007. Differential roles of individual domains in selection of secretion route of a *Streptococcus parasanguinis* serine-rich adhesin, Fap1. *J Bacteriol* 189:7610–7617.

25. Sanchez CJ, Shivshankar P, Stol K, Trakhtenbroit S, Sullam PM, Sauer K, Hermans PW, Orihuela CJ. 2010. The pneumococcal serine-rich repeat protein is an intraspecies bacterial adhesin that promotes bacterial aggregation in vivo and in biofilms. *PLoS Pathog* 6:e1001044.

26. Pyburn TM, Bensing BA, Xiong YQ, Melancon BJ, Tomasiak TM, Ward NJ, Yankovskaya V, Oliver KM, Cecchini G, Sulikowski GA, Tyska MJ, Sullam PM, Iverson TM. 2011. A structural model for binding of the serine-rich repeat adhesin GspB to host carbohydrate receptors. *PLoS Pathog* 7:e1002112.

27. Lizcano A, Sanchez CJ, Orihuela CJ. 2012. A role for glycosylated serine-rich repeat proteins in gram-positive bacterial pathogenesis. *Mol Oral Microbiol* 27:257–269.

28. Seo HS, Minasov G, Seepersaud R, Doran KS, Dubrovska I, Shuvalova L, Anderson WF, Iverson TM, Sullam PM. 2013. Characterization of fibrinogen binding by glycoproteins Srr1 and Srr2 of *Streptococcus agalactiae*. *J Biol Chem* 288:35982–35996.

29. Six A, Bellais S, Bouaboud A, Fouet A, Gabriel C, Tazi A, Dramsi S, Trieu-Cuot P, Poyart C. 2015. Srr2, a multifaceted adhesin expressed by ST-17 hypervirulent group B Streptococcus involved in binding to both fibrinogen and plasminogen. *Mol Microbiol* 97:1209–1222.

30. Bensing BA, Khedri Z, Deng L, Yu H, Prakobphol A, Fisher SJ, Chen X, Iverson TM, Varki A, Sullam PM. 2016. Novel aspects of sialoglycan recognition by the Siglec-like domains of streptococcal SRR glycoproteins. *Glycobiology* 26:1222–1234.

31. Deng L, Bensing BA, Thamadilok S, Yu H, Lau K, Chen X, Ruhl S, Sullam PM, Varki A. 2014. Oral streptococci utilize a Siglec-like domain of serine-rich repeat adhesins to preferentially target platelet sialoglycans in human blood. *PLoS Pathog* 10:e1004540.

32. Shivshankar P, Sanchez C, Rose LF, Orihuela CJ. 2009. The *Streptococcus pneumoniae* adhesin PsrP binds to keratin 10 on lung cells. *Mol Microbiol* 73:663–679.

33. Samen U, Eikmanns BJ, Reinscheid DJ, Borges F. 2007. The surface protein Srr-1 of *Streptococcus agalactiae* binds human keratin 4 and promotes adherence to epithelial HEp-2 cells. *Infect Immun* 75:5405–5414.

34. Bensing BA, Siboo IR, Sullam PM. 2007. Glycine residues in the hydrophobic core of the GspB signal sequence route export toward the accessory Sec pathway. *J Bacteriol* 189:3846–3854.

35. **Bensing BA, Sullam PM.** 2010. Transport of preproteins by the accessory Sec system requires a specific domain adjacent to the signal peptide. *J Bacteriol* **192**:4223–4232.

36. **Bensing BA, Yen YT, Seepersaud R, Sullam PM.** 2012. A Specific interaction between SecA2 and a region of the preprotein adjacent to the signal peptide occurs during transport via the accessory Sec system. *J Biol Chem* **287**:24438–24447.

37. **Bensing BA, Sullam PM.** 2009. Characterization of *Streptococcus gordonii* SecA2 as a paralogue of SecA. *J Bacteriol* **191**:3482–3491.

38. **Yen YT, Cameron TA, Bensing BA, Seepersaud R, Zambryski PC, Sullam PM.** 2013. Differential localization of the streptococcal accessory sec components and implications for substrate export. *J Bacteriol* **195**:682–695.

39. **Bandara M, Corey RA, Martin R, Skehel JM, Blocker AJ, Jenkinson HF, Collinson I.** 2016. Composition and activity of the non-canonical Gram-positive SecY2 complex. *J Biol Chem* **291**:21474–21484.

40. **Seepersaud R, Bensing BA, Yen YT, Sullam PM.** 2010. Asp3 mediates multiple protein-protein interactions within the accessory Sec system of *Streptococcus gordonii. Mol Microbiol* **78**:490–505.

41. **Sibbald MJ, Winter T, van der Kooi-Pol MM, Buist G, Tsompanidou E, Bosma T, Schäfer T, Ohlsen K, Hecker M, Antelmann H, Engelmann S, van Dijl JM.** 2010. Synthetic effects of *secG* and *secY2* mutations on exoproteome biogenesis in *Staphylococcus aureus. J Bacteriol* **192**:3788–3800.

42. **Spencer C, Bensing BA, Mishra NN, Sullam PM.** 2019. Membrane trafficking of the bacterial adhesin GspB and the accessory Sec transport machinery. *J Biol Chem* **294**:1502–1515.

43. **Chen Y, Bensing BA, Seepersaud R, Mi W, Liao M, Jeffrey PD, Shajahan A, Sonon RN, Azadi P, Sullam PM, Rapoport TA.** 2018. Unraveling the sequence of cytosolic reactions in the export of GspB adhesin from *Streptococcus gordonii. J Biol Chem* **293**:5360–5373.

44. **Yen YT, Seepersaud R, Bensing BA, Sullam PM.** 2011. Asp2 and Asp3 interact directly with GspB, the export substrate of the *Streptococcus gordonii* accessory Sec system. *J Bacteriol* **193**:3165–3174.

45. **Seepersaud R, Sychantha D, Bensing BA, Clarke AJ, Sullam PM.** 2017. O-acetylation of the serine-rich repeat glycoprotein GspB is coordinated with accessory Sec transport. *PLoS Pathog* **13**:e1006558.

46. **Kurtz S, McKinnon KP, Runge MS, Ting JP, Braunstein M.** 2006. The SecA2 secretion factor of *Mycobacterium tuberculosis* promotes growth in macrophages and inhibits the host immune response. *Infect Immun* **74**:6855–6864.

47. **Braunstein M, Espinosa BJ, Chan J, Belisle JT, Jacobs WR Jr.** 2003. SecA2 functions in the secretion of superoxide dismutase A and in the virulence of *Mycobacterium tuberculosis. Mol Microbiol* **48**:453–464.

48. **Lenz LL, Mohammadi S, Geissler A, Portnoy DA.** 2003. SecA2-dependent secretion of autolytic enzymes promotes *Listeria monocytogenes* pathogenesis. *Proc Natl Acad Sci U S A* **100**:12432–12437.

49. **Chandrabos C, M'Homa Soudja S, Weinrick B, Gros M, Frangaj A, Rahmoun M, Jacobs WR Jr, Lauvau G.** 2015. The p60 and NamA autolysins from *Listeria monocytogenes* contribute to host colonization and induction of protective memory. *Cell Microbiol* **17**:147–163.

50. **Wang YT, Oh SY, Hendrickx AP, Lunderberg JM, Schneewind O.** 2013. *Bacillus cereus* G9241 S-layer assembly contributes to the pathogenesis of anthrax-like disease in mice. *J Bacteriol* **195**:596–605.

51. **Missiakas D, Schneewind O.** 2017. Assembly and function of the *Bacillus anthracis* S-layer. *Annu Rev Microbiol* **71**:79–98.

52. **Caspers M, Freudl R.** 2008. *Corynebacterium glutamicum* possesses two *secA* homologous genes that are essential for viability. *Arch Microbiol* **189**:605–610.

53. **Fagan RP, Fairweather NF.** 2011. *Clostridium difficile* has two parallel and essential Sec secretion systems. *J Biol Chem* **286**:27483–27493.

54. **Ligon LS, Rigel NW, Romanchuk A, Jones CD, Braunstein M.** 2013. Suppressor analysis reveals a role for SecY in the SecA2-dependent protein export pathway of mycobacteria. *J Bacteriol* **195**:4456–4465.

55. **Durack J, Burke TP, Portnoy DA.** 2015. A prl mutation in SecY suppresses secretion and virulence defects of *Listeria monocytogenes secA2* mutants. *J Bacteriol* **197**:932–942.

56. **Feltcher ME, Gunawardena HP, Zulauf KE, Malik S, Griffin JE, Sassetti CM, Chen X, Braunstein M.** 2015. Label-free quantitative proteomics reveals a role for the *Mycobacterium tuberculosis* SecA2 pathway in exporting solute binding proteins and Mce transporters to the cell wall. *Mol Cell Proteomics* **14**:1501–1516.

57. **Lenz LL, Portnoy DA.** 2002. Identification of a second *Listeria secA* gene associated with protein secretion and the rough phenotype. *Mol Microbiol* **45**:1043–1056.

58. van der Woude AD, Stoop EJ, Stiess M, Wang S, Ummels R, van Stempvoort G, Piersma SR, Cascioferro A, Jiménez CR, Houben EN, Luirink J, Pieters J, van der Sar AM, Bitter W. 2014. Analysis of SecA2-dependent substrates in *Mycobacterium marinum* identifies protein kinase G (PknG) as a virulence effector. *Cell Microbiol* 16:280–295.

59. Nguyen-Mau SM, Oh SY, Kern VJ, Missiakas DM, Schneewind O. 2012. Secretion genes as determinants of *Bacillus anthracis* chain length. *J Bacteriol* 194:3841–3850.

60. Gibbons HS, Wolschendorf F, Abshire M, Niederweis M, Braunstein M. 2007. Identification of two *Mycobacterium smegmatis* lipoproteins exported by a SecA2-dependent pathway. *J Bacteriol* 189:5090–5100.

61. Renier S, Chambon C, Viala D, Chagnot C, Hébraud M, Desvaux M. 2013. Exoproteomic analysis of the SecA2-dependent secretion in *Listeria monocytogenes* EGD-e. *J Proteomics* 80:183–195.

62. Cheng Y, Schorey JS. 2018. *Mycobacterium tuberculosis*-induced IFN-β production requires cytosolic DNA and RNA sensing pathways. *J Exp Med* 215:2919–2935.

63. Abdullah Z, Schlee M, Roth S, Mraheil MA, Barchet W, Böttcher J, Hain T, Geiger S, Hayakawa Y, Fritz JH, Civril F, Hopfner KP, Kurts C, Ruland J, Hartmann G, Chakraborty T, Knolle PA. 2012. RIG-I detects infection with live *Listeria* by sensing secreted bacterial nucleic acids. *EMBO J* 31:4153–4164.

64. Cui J, Davidson AL. 2011. ABC solute importers in bacteria. *Essays Biochem* 50:85–99.

65. Casali N, Riley LW. 2007. A phylogenomic analysis of the *Actinomycetales mce* operons. *BMC Genomics* 8:60.

66. Zulauf KE, Sullivan JT, Braunstein M. 2018. The SecA2 pathway of *Mycobacterium tuberculosis* exports effectors that work in concert to arrest phagosome and autophagosome maturation. *PLoS Pathog* 14:e1007011.

67. Mishra KK, Mendonca M, Aroonnual A, Burkholder KM, Bhunia AK. 2011. Genetic organization and molecular characterization of *secA2* locus in *Listeria* species. *Gene* 489:76–85.

68. Halbedel S, Hahn B, Daniel RA, Flieger A. 2012. DivIVA affects secretion of virulence-related autolysins in *Listeria monocytogenes*. *Mol Microbiol* 83:821–839.

69. Burkholder KM, Kim KP, Mishra KK, Medina S, Hahm BK, Kim H, Bhunia AK. 2009. Expression of LAP, a SecA2-dependent secretory protein, is induced under anaerobic environment. *Microbes Infect* 11:859–867.

70. Archambaud C, Nahori MA, Pizarro-Cerda J, Cossart P, Dussurget O. 2006. Control of *Listeria* superoxide dismutase by phosphorylation. *J Biol Chem* 281:31812–31822.

71. Fagan RP, Fairweather NF. 2014. Biogenesis and functions of bacterial S-layers. *Nat Rev Microbiol* 12:211–222.

72. Calabi E, Ward S, Wren B, Paxton T, Panico M, Morris H, Dell A, Dougan G, Fairweather N. 2001. Molecular characterization of the surface layer proteins from *Clostridium difficile*. *Mol Microbiol* 40:1187–1199.

73. Waligora AJ, Hennequin C, Mullany P, Bourlioux P, Collignon A, Karjalainen T. 2001. Characterization of a cell surface protein of *Clostridium difficile* with adhesive properties. *Infect Immun* 69:2144–2153.

74. Bradshaw WJ, Kirby JM, Roberts AK, Shone CC, Acharya KR. 2017. The molecular structure of the glycoside hydrolase domain of Cwp19 from *Clostridium difficile*. *FEBS J* 284:4343–4357.

75. Ahn JS, Chandramohan L, Liou LE, Bayles KW. 2006. Characterization of CidR-mediated regulation in *Bacillus anthracis* reveals a previously undetected role of S-layer proteins as murein hydrolases. *Mol Microbiol* 62:1158–1169.

76. Feltcher ME, Gibbons HS, Ligon LS, Braunstein M. 2013. Protein export by the mycobacterial SecA2 system is determined by the preprotein mature domain. *J Bacteriol* 195:672–681.

77. Ebner P, Gotz F. 2018. Bacterial excretion of cytoplasmic proteins (ECP): occurrence, mechanism, and function. *Trends Microbiol* 10.1016/j.tim.2018.10.006.

78. Braunstein M, Brown AM, Kurtz S, Jacobs WR Jr. 2001. Two nonredundant SecA homologues function in mycobacteria. *J Bacteriol* 183:6979–6990.

79. Swanson S, Ioerger TR, Rigel NW, Miller BK, Braunstein M, Sacchettini JC. 2015. Structural similarities and differences between two functionally distinct SecA proteins: the *Mycobacterium tuberculosis* SecA1 and SecA2. *J Bacteriol* 198:720–730.

80. Hou JM, D'Lima NG, Rigel NW, Gibbons HS, McCann JR, Braunstein M, Teschke CM. 2008. ATPase activity of *Mycobacterium tuberculosis* SecA1 and SecA2 proteins and its importance for SecA2 function in macrophages. *J Bacteriol* 190:4880–4887.

81. D'Lima NG, Teschke CM. 2014. ADP-dependent conformational changes distinguish *Mycobacterium tuberculosis* SecA2 from SecA1. *J Biol Chem* 289:2307–2317.

82. **Bonardi F, Halza E, Walko M, Du Plessis F, Nouwen N, Feringa BL, Driessen AJ.** 2011. Probing the SecYEG translocation pore size with preproteins conjugated with sizable rigid spherical molecules. *Proc Natl Acad Sci U S A* **108:**7775–7780.

83. **Randall LL, Hardy SJ.** 1986. Correlation of competence for export with lack of tertiary structure of the mature species: a study in vivo of maltose-binding protein in *E. coli. Cell* **46:**921–928.

84. **Prabudiansyah I, Kusters I, Driessen AJ.** 2015. In vitro interaction of the housekeeping SecA1 with the accessory SecA2 protein of *Mycobacterium tuberculosis. PLoS One* **10:**e0128788.

85. **Rigel NW, Gibbons HS, McCann JR, McDonough JA, Kurtz S, Braunstein M.** 2009. The accessory SecA2 system of mycobacteria requires ATP binding and the canonical SecA1. *J Biol Chem* **284:**9927–9936.

86. **Halbedel S, Reiss S, Hahn B, Albrecht D, Mannala GK, Chakraborty T, Hain T, Engelmann S, Flieger A.** 2014. A systematic proteomic analysis of *Listeria monocytogenes* house-keeping protein secretion systems. *Mol Cell Proteomics* **13:**3063–3081.

87. **Miller BK, Hughes R, Ligon LS, Rigel NW, Malik S, Anjuwon-Foster BR, Sacchettini JC, Braunstein M.** 2019. *Mycobacterium tuberculosis* SatS is a chaperone for the SecA2 protein export pathway. *eLife* **8:**e40063.

88. **Nguyen-Mau SM, Oh SY, Schneewind DI, Missiakas D, Schneewind O.** 2015. *Bacillus anthracis* SlaQ promotes S-layer protein assembly. *J Bacteriol* **197:**3216–3227.

89. **Couvigny B, Lapaque N, Rigottier-Gois L, Guillot A, Chat S, Meylheuc T, Kulakauskas S, Rohde M, Mistou MY, Renault P, Doré J, Briandet R, Serror P, Guédon E.** 2017. Three glycosylated serine-rich repeat proteins play a pivotal role in adhesion and colonization of the pioneer commensal bacterium, *Streptococcus salivarius. Environ Microbiol* **19:**3579–3594.

90. **Navarre WW, Schneewind O.** 1999. Surface proteins of gram-positive bacteria and mechanisms of their targeting to the cell wall envelope. *Microbiol Mol Biol Rev* **63:**174–229.

91. **Papanikou E, Karamanou S, Economou A.** 2007. Bacterial protein secretion through the translocase nanomachine. *Nat Rev Microbiol* **5:**839–851.

The Conserved Role of YidC in Membrane Protein Biogenesis

4

SRI KARTHIKA SHANMUGAM[1] and ROSS E. DALBEY[1]

INTRODUCTION

Membrane proteins constitute between 20–30% of the cellular proteome (1) and perform critical functions like signal transduction, molecular transport, and cell adhesion. The molecular machineries that catalyze their targeting, insertion, and assembly in the different cellular and subcellular membranes are remarkably conserved. The Sec translocon is responsible for moving the majority of the proteins across/into the bacterial, archaeal, thylakoidal, and endoplasmic reticulum (ER) membranes in an unfolded state (2). In bacteria, it is proposed to form a holocomplex composed of the heterotrimeric protein channel SecYEG and the accessory elements SecDFYajC, SecA ATPase, and YidC (3).

As part of the holocomplex, YidC operates in various capacities ranging from assisting in the membrane insertion process and the lateral clearance of the substrate transmembrane (TM) segments from the channel to serving as a foldase for Sec-dependent proteins (4). In addition to this, YidC facilitates the membrane insertion of small membrane protein substrates independently (5). While larger proteins are typically targeted by the SRP-FtsY partnership to the Sec holotranslocon, smaller substrates that cannot engage

[1]Department of Chemistry and Biochemistry, The Ohio State University, Columbus, OH 43210
Protein Secretion in Bacteria
Edited by Maria Sandkvist, Eric Cascales, and Peter J. Christie
© 2019 American Society for Microbiology, Washington, DC
doi:10.1128/microbiolspec.PSIB-0014-2018

SRP are posttranslationally delivered to YidC (6). However, certain YidC-only substrates like MscL (7) and the tail-anchored proteins TssL (8) and DjlC and Flk (9) employ SRP for targeting.

YidC/Alb3/Oxa1 family proteins are highly conserved insertases that operate in the bacterial, thylakoidal, and mitochondrial inner membranes, respectively (10). Structurally, they are helical bundles formed by 5 core TM segments (Fig. 1). YidC is required for the insertion and assembly of several respiratory and energy-transducing proteins (11), like the subunits of the F_1F_0 ATPase (12), cytochrome o oxidase (13), and NADH dehydrogenase (14). In Gram-negative bacteria, YidC has an additional N-terminal TM segment that acts as a

membrane anchor, followed by a large beta-sandwich fold within the first periplasmic domain (15). Although these regions are largely nonessential for function (16), they have contact sites to SecY (17) and SecDF (18), suggesting a kinetic role in the protein insertion and substrate folding process. Most Gram-positive bacteria possess two paralogs: YidC1 and YidC2. While YidC1 is constitutively expressed, YidC2 gene expression is controlled by a MifM sensor protein in *Bacillus subtilis* (19). Though the paralogs are functionally exchangeable, YidC1 is specifically required for the sporulation process (20).

In archaea, DUF106 protein has a three-TM core with a low structural homology to the bacterial YidC, but its protein insertion

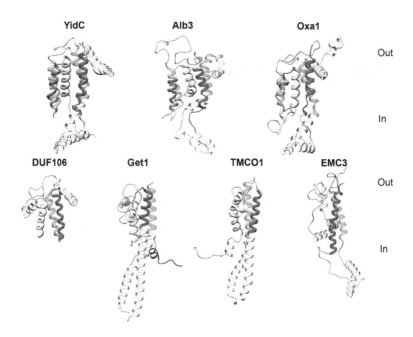

FIGURE 1 YidC family of proteins. (Top) Structural homology in the YidC/Alb3/Oxa1 family, shown by highlighting the conserved TMs in green (TM1), red (TM2), cyan (TM3), purple (TM4), and yellow (TM5). The YidC structure is adapted from the crystal structure solved in *Bacillus halodurans* (PDB code 3WO7); Alb3 and Oxa1 structures are three-dimensional (3D) computational models made using SWISS-MODEL workspace as described in reference 72. (Bottom) Newly identified members of the Oxa1 superfamily, with highlighting of the conserved three TM segments in green (TM1), red (TM2), and yellow (TM3). The archaeal DUF106 structure is adapted from the crystal structure solved in *Methanocaldococcus jannaschii* (PDB code 5C8J). Yeast Get1, human TMCO1, and human EMC3 structures are evolutionary covariance-based 3D models adapted from those described in references 36 and 37. The cytoplasmic regions of these models were modified as described in reference 36.

function remains to be tested (21). Eukaryotes contain multiple YidC paralogs, and some of them can replace *Escherichia coli* YidC at least partially, indicating shared functionality in the cell (22–24). In plants, the paralogs Alb3 and Alb4 exist in the thylakoid membrane of chloroplasts (25, 26). The primary substrates of Alb3 are a subset of the light-harvesting chlorophyll binding protein subunits (27), whereas Alb4 is involved in the biogenesis of chloroplast F_1F_O ATPase assembly (28). A prominent feature of Alb3 is the presence of a long cytoplasmic C-terminal domain which acts as an anchor for SRP43 (29). Both posttranslational and cotranslational targeting occurs, and Alb3 is known to interact with the chloroplast SecYE translocon, like its bacterial counterparts (30). Oxa1 and Oxa2 paralogs are found in the mitochondrial inner membrane of eukaryotic cells (31, 32). Sec is absent in this membrane, so Oxa1 is believed to facilitate the insertion of all mitochondrial DNA-encoded membrane proteins independently (33). Oxa1 has a C-terminal extension which is the ribosome-docking site for translating substrates that are cotranslationally inserted (34). Oxa2 performs similar insertion function for certain respiratory proteins posttranslationally (35).

Until recently, the presence of YidC homologs in the ER was unknown (36). Anghel et al. (37) employed phylogenic homology studies and identified three Oxa1-like highly conserved proteins: TMCO1, ER membrane complex 3 (EMC3), and Get1, which are all involved in the ER membrane protein translocation process in eukaryotes. The study found that Get1 and EMC3 were evolutionarily related to the DUF106 group of proteins. Get1 is a part of the tail-anchored protein insertion complex, and substrates of this pathway have a C-terminally located TM segment that is posttranslationally targeted to the ER membrane (38). EMC3 promotes the cotranslational membrane insertion of multipass ER proteins with charged TM segments (39, 40). TMCO1 is predicted to insert newly synthesized ER membrane proteins cotrans-

lationally, but it also engages with the Sec translocon-like YidC (37).

YidC-ONLY PATHWAY

YidC's function was first annotated in 2000 (5); it was shown to be essential in *E. coli* and required for the insertion of phage proteins Pf3 coat and M13 procoat, which were previously thought to insert by an unassisted mechanism. The minimal functional unit is monomeric (41), even though YidC can dimerize under certain conditions (42). It was shown using reconstituted proteoliposomes that YidC is sufficient for the membrane integration of Pf3 (43). In addition to this, YidC is responsible for the membrane insertion of subunit c of ATP synthase (12), the mechanosensitive channel protein MscL (7), and the C-terminal tail-anchored proteins TssL (44) and DjlC and Flk (9). A common feature of the YidC-only pathway substrates is that they contain short translocated regions followed by one or two TM segments (45).

The crystal structures of YidC from Gram-positive (46) and Gram-negative (47) bacteria uncovered important mechanistic details about its function. The conserved 5-TM core of YidC forms a unique hydrophilic cavity in the inner leaflet facing the cytoplasm but is closed from the periplasmic side. The groove contains a conserved positive charge which was shown to be critical for function in Gram-positive bacteria but not in the Gram-negative homolog (48). Kumazaki et al. showed that MifM substrate could be cross-linked to the groove (46). Hence, it is proposed that the positive charge interacts electrostatically with the charges on the substrate hydrophilic regions to recruit it into the groove and reduce its membrane-crossing distance (Fig. 2). Consistent with this, negative charges on the substrate N-tail or TM segment have been proposed to act as YidC-only pathway determinants (49, 50). The proton motive force is implied to play a role in releasing the hydro-

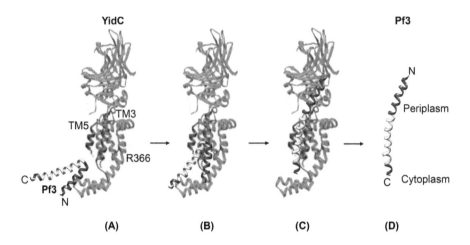

FIGURE 2 Model of YidC-mediated membrane insertion of Pf3 coat protein. This figure is adapted from a review by Kiefer and Kuhn (73). (A) Binding of Pf3 coat protein to YidC. (B) Pf3 TM segment interacts with the cytoplasmic part of the greasy slide, and the N-terminal tail of Pf3 (blue) enters the hydrophilic cavity of YidC possessing the conserved Arg residue (red). (C) Pf3 coat TM segment inserts across the YidC greasy slide formed by TM3 and TM5 (purple) and release of the N-tail into the periplasmic space. (D) Release of Pf3 into the bilayer.

philic domain of the substrate from the groove, but it is unclear whether this occurs and, if so, how it occurs. Further reduction in membrane-crossing distance for the substrate was suggested by molecular dynamics simulation studies (51), which found thinning of the membrane region around YidC.

The major substrate contact sites of YidC are the hydrophobic residues found in TM3 and TM5 that were shown by cross-linking studies to bind the substrate TM segment of Pf3 coat (52) and MscL (53). This suggests that YidC facilitates substrate insertion through hydrophobic interactions via a greasy-sliding mechanism (Fig. 2). In line with this, cryo-electron microscopy studies showed that the TM segment of the F0c substrate is in proximity to the greasy slide (54). Substrate insertion kinetics was studied in real time using time-resolved single-molecule fluorescence resonance energy transfer analysis (55), which showed that the entire process of substrate contact, insertion, and separation from YidC occurred within 20 ms and Pf3 inserted into reconstituted YidC proteoliposomes at the rate of 500 molecules per second.

Another feature of YidC is the cytosolic loops C1 and C2 and the C-terminal tail region, of which the last two constitute the protein docking sites for receiving its translating substrates. The C1 loop forms a helical hairpin that is essential for function (48) and is believed to be highly dynamic based on its relative position in the crystal structures. Cross-linking studies performed by Petriman's group show that the C1 loop interacts with SRP and FtsY, highlighting its role in recruiting substrates (17). Similarly, Geng et al. found that the C2 loop and C-terminal region of YidC provide stable docking sites for ribosome nascent chain complexes (56). These studies define the role played by the different regions of YidC, leading to a better understanding of the mechanism of its insertion function.

YidC-Sec PATHWAY

Substrate specificity studies indicate that YidC has limited potential to function independently and that the translocation of energetically unfavorable regions of substrates requires both YidC and Sec (57). Several essential inner membrane proteins, like ATP synthase subunits a and b (6, 58), subunit II of cytochrome b_0 oxidase (13, 59), TatC (60, 61), and anaerobic respiratory protein NuoK (50), are inserted by the combined efforts of YidC and Sec. This phenomenon may also occur in higher eukaryotes in the ER and thylakoidal membrane, where YidC and Sec homologs are known to interact. The bacterial holotranslocon, made up of SecYEG, SecDFYajC, and YidC, is proposed to be an efficient insertion machine for the membrane protein substrates of the YidC-Sec pathway (62).

SecYEG forms a channel through which substrate polar domains are exported across the membrane, whereas the TM segments exit the channel with the help of YidC via a lateral gate formed by TM2b and TM7 of SecY (63, 64) (Fig. 3). Consistent with this, the lateral gate of SecY can be photo-cross-linked to YidC (65). It is predicted that the greasy slide of YidC might contact the SecY lateral gate and move the substrate TM segment via hydrophobic interactions from within the channel and into the lipid bilayer. Recent insight into how this partnership works has revealed that the first TM of *E. coli* YidC contacts SecY and SecG (17). It is

proposed that this TM may enter the channel and draw the TM segments out through the lateral gate, but this remains to be tested. The study also reported the C1 loop of YidC as a contact site for SecY.

In addition to this, YidC is also known to act as a folding and packaging site for Sec-dependent proteins (66). Nagamori et al. (67) found that LacY protein required YidC to achieve its functional folded form using monoclonal antibodies recognizing specific conformational domains. Strikingly, the translocation of the six periplasmic domains of LacY required only SecYEG, while the folding of the protein was dependent on YidC (68). YidC's role in folding LacY was further explored by Serdiuk et al. by using single-molecule force spectroscopy (69). A mechanical pulling force was applied on a single LacY molecule to unfold it and extract it from a membrane using the stylus of a cantilever. This protein was then slowly allowed to refold into another membrane in the presence of YidC. The study showed that only in the presence of YidC could LacY fold back to its stable form in the membrane.

The accessory element SecDFYajC is believed to promote YidC's interaction with the Sec channel (70). SecDF was shown to contact the periplasmic domain of *E. coli* YidC using

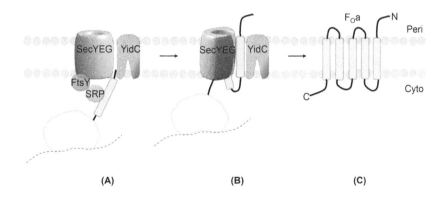

(A)   (B)   (C)

FIGURE 3 Model of the YidC-Sec insertion pathway. (A) The SRP-bound substrate is cotranslationally targeted to the Sec holotranslocon (SecDFYajC [not represented]) via the membrane-associated SRP receptor FtsY. (B) The substrate amino-terminal TM segment inserts at the interface of SecYEG and YidC and the second TM segment initiates C-terminal translocation. (C) The model substrate shown here, F_0a, is inserted into the bilayer.

affinity pull-down experiments (18). This interaction may indicate the shared functional role of SecDF and periplasmic domain of YidC in the substrate folding process. The crystal structure of SecDF and electrophysiological experiments revealed a proton transport mechanism which could provide the energetic driving force for pulling the substrate out of the Sec channel during translocation and prevent its backsliding (71). Substrates of this pathway are targeted to the holotranslocon by SRP and its membrane-associated receptor FtsY for cotranslational insertion (6).

CONCLUSION

Members of the YidC family of proteins catalyze the unfavorable movement of polar domains of membrane proteins across the hydrophobic lipid bilayer and function as chaperones for a subset of proteins to ensure that their functional conformation is reached as well. The complex interplay between the various components of the insertion pathway has been explored in recent research. Though the crystal structure of YidC advanced our understanding of how these insertases function, the exact mechanism by which the substrate moves through YidC or the YidC/Sec holocomplex during the membrane insertion process needs to be elucidated. It is anticipated that the mechanistic details unraveled in *E. coli* could be applied to similar pathways operating in higher organisms due to the conserved nature of the proteins involved.

ACKNOWLEDGMENTS

This work was supported by National Science Foundation grant MCB-1814936 to R.E.D.

CITATION

Shanmugam SK, Dalbey RE. 2019. The conserved role of YidC in membrane protein biogenesis. Microbiol Spectrum 7(1):PSIB-0014-2018.

REFERENCES

1. **Krogh A, Larsson B, von Heijne G, Sonnhammer EL.** 2001. Predicting transmembrane protein topology with a hidden Markov model: application to complete genomes. *J Mol Biol* **305:**567–580.
2. **Pohlschröder M, Prinz WA, Hartmann E, Beckwith J.** 1997. Protein translocation in the three domains of life: variations on a theme. *Cell* **91:**563–566.
3. **Duong F, Wickner W.** 1997. Distinct catalytic roles of the SecYE, SecG and SecDFyajC subunits of preprotein translocase holoenzyme. *EMBO J* **16:**2756–2768.
4. **Dalbey RE, Kuhn A.** 2004. YidC family members are involved in the membrane insertion, lateral integration, folding, and assembly of membrane proteins. *J Cell Biol* **166:**769–774.
5. **Samuelson JC, Chen M, Jiang F, Möller I, Wiedmann M, Kuhn A, Phillips GJ, Dalbey RE.** 2000. YidC mediates membrane protein insertion in bacteria. *Nature* **406:**637–641.
6. **Yi L, Celebi N, Chen M, Dalbey RE.** 2004. Sec/SRP requirements and energetics of membrane insertion of subunits a, b, and c of the *Escherichia coli* F1F0 ATP synthase. *J Biol Chem* **279:**39260–39267.
7. **Facey SJ, Neugebauer SA, Krauss S, Kuhn A.** 2007. The mechanosensitive channel protein MscL is targeted by the SRP to the novel YidC membrane insertion pathway of *Escherichia coli*. *J Mol Biol* **365:**995–1004.
8. **Pross E, Soussoula L, Seitl I, Lupo D, Kuhn A.** 2016. Membrane targeting and insertion of the C-tail protein SciP. *J Mol Biol* **428:**4218–4227.
9. **Peschke M, Le Goff M, Koningstein GM, Karyolaimos A, de Gier JW, van Ulsen P, Luirink J.** 2018. SRP, FtsY, DnaK and YidC are required for the biogenesis of the *E. coli* tail-anchored membrane proteins DjlC and Flk. *J Mol Biol* **430:**389–403.
10. **Hennon SW, Soman R, Zhu L, Dalbey RE.** 2015. YidC/Alb3/Oxa1 family of insertases. *J Biol Chem* **290:**14866–14874.
11. **van der Laan M, Urbanus ML, Ten Hagen-Jongman CM, Nouwen N, Oudega B, Harms N, Driessen AJ, Luirink J.** 2003. A conserved function of YidC in the biogenesis of respiratory chain complexes. *Proc Natl Acad Sci U S A* **100:**5801–5806.
12. **van der Laan M, Bechtluft P, Kol S, Nouwen N, Driessen AJ.** 2004. F1F0 ATP synthase

subunit c is a substrate of the novel YidC pathway for membrane protein biogenesis. *J Cell Biol* **165**:213–222.

13. **du Plessis DJ, Nouwen N, Driessen AJ.** 2006. Subunit a of cytochrome o oxidase requires both YidC and SecYEG for membrane insertion. *J Biol Chem* **281**:12248–12252.

14. **Price CE, Driessen AJ.** 2008. YidC is involved in the biogenesis of anaerobic respiratory complexes in the inner membrane of *Escherichia coli*. *J Biol Chem* **283**:26921–26927.

15. **Ravaud S, Stjepanovic G, Wild K, Sinning I.** 2008. The crystal structure of the periplasmic domain of the *Escherichia coli* membrane protein insertase YidC contains a substrate binding cleft. *J Biol Chem* **283**:9350–9358.

16. **Jiang F, Chen M, Yi L, de Gier JW, Kuhn A, Dalbey RE.** 2003. Defining the regions of *Escherichia coli* YidC that contribute to activity. *J Biol Chem* **278**:48965–48972.

17. **Petriman NA, Jauß B, Hufnagel A, Franz L, Sachelaru I, Drepper F, Warscheid B, Koch HG.** 2018. The interaction network of the YidC insertase with the SecYEG translocon, SRP and the SRP receptor FtsY. *Sci Rep* **8**:578.

18. **Xie K, Kiefer D, Nagler G, Dalbey RE, Kuhn A.** 2006. Different regions of the nonconserved large periplasmic domain of *Escherichia coli* YidC are involved in the SecF interaction and membrane insertase activity. *Biochemistry* **45**:13401–13408.

19. **Chiba S, Ito K.** 2015. MifM monitors total YidC activities of *Bacillus subtilis*, including that of YidC2, the target of regulation. *J Bacteriol* **197**:99–107.

20. **Errington J, Appleby L, Daniel RA, Goodfellow H, Partridge SR, Yudkin MD.** 1992. Structure and function of the spoIIIJ gene of *Bacillus subtilis*: a vegetatively expressed gene that is essential for sigma G activity at an intermediate stage of sporulation. *J Gen Microbiol* **138**:2609–2618.

21. **Borowska MT, Dominik PK, Anghel SA, Kossiakoff AA, Keenan RJ.** 2015. A YidC-like protein in the archaeal plasma membrane. *Structure* **23**:1715–1724.

22. **Jiang F, Yi L, Moore M, Chen M, Rohl T, Van Wijk KJ, De Gier JW, Henry R, Dalbey RE.** 2002. Chloroplast YidC homolog Albino3 can functionally complement the bacterial YidC depletion strain and promote membrane insertion of both bacterial and chloroplast thylakoid proteins. *J Biol Chem* **277**:19281–19288.

23. **Preuss M, Ott M, Funes S, Luirink J, Herrmann JM.** 2005. Evolution of mitochondrial oxa proteins from bacterial YidC. Inherited and acquired functions of a con-served protein insertion machinery. *J Biol Chem* **280**:13004–13011.

24. **van Bloois E, Koningstein G, Bauerschmitt H, Herrmann JM, Luirink J.** 2007. *Saccharomyces cerevisiae* Cox18 complements the essential Sec-independent function of *Escherichia coli* YidC. *FEBS J* **274**:5704–5713.

25. **Sundberg E, Slagter JG, Fridborg I, Cleary SP, Robinson C, Coupland G.** 1997. ALBINO3, an *Arabidopsis* nuclear gene essential for chloroplast differentiation, encodes a chloroplast protein that shows homology to proteins present in bacterial membranes and yeast mitochondria. *Plant Cell* **9**:717–730.

26. **Gerdes L, Bals T, Klostermann E, Karl M, Philippar K, Hünken M, Soll J, Schünemann D.** 2006. A second thylakoid membrane-localized Alb3/OxaI/YidC homologue is involved in proper chloroplast biogenesis in *Arabidopsis thaliana*. *J Biol Chem* **281**:16632–16642.

27. **Woolhead CA, Thompson SJ, Moore M, Tissier C, Mant A, Rodger A, Henry R, Robinson C.** 2001. Distinct Albino3-dependent and -independent pathways for thylakoid membrane protein insertion. *J Biol Chem* **276**:40841–40846.

28. **Benz M, Bals T, Gügel IL, Piotrowski M, Kuhn A, Schünemann D, Soll J, Ankele E.** 2009. Alb4 of *Arabidopsis* promotes assembly and stabilization of a non chlorophyll-binding photosynthetic complex, the CF1CF0-ATP synthase. *Mol Plant* **2**:1410–1424.

29. **Falk S, Ravaud S, Koch J, Sinning I.** 2010. The C terminus of the Alb3 membrane insertase recruits cpSRP43 to the thylakoid membrane. *J Biol Chem* **285**:5954–5962.

30. **Klostermann E, Droste Gen Helling I, Carde JP, Schünemann D.** 2002. The thylakoid membrane protein ALB3 associates with the cpSecY-translocase in *Arabidopsis thaliana*. *Biochem J* **368**:777–781.

31. **Bonnefoy N, Chalvet F, Hamel P, Slonimski PP, Dujardin G.** 1994. OXA1, a *Saccharomyces cerevisiae* nuclear gene whose sequence is conserved from prokaryotes to eukaryotes controls cytochrome oxidase biogenesis. *J Mol Biol* **239**:201–212.

32. **Funes S, Nargang FE, Neupert W, Herrmann JM.** 2004. The Oxa2 protein of *Neurospora crassa* plays a critical role in the biogenesis of cytochrome oxidase and defines a ubiquitous subbranch of the Oxa1/YidC/Alb3 protein family. *Mol Biol Cell* **15**:1853–1861.

33. **Ott M, Herrmann JM.** 2010. Co-translational membrane insertion of mitochondrially encoded proteins. *Biochim Biophys Acta* **1803**:767–775.

34. Jia L, Dienhart M, Schramp M, McCauley M, Hell K, Stuart RA. 2003. Yeast Oxa1 interacts with mitochondrial ribosomes: the importance of the C-terminal region of Oxa1. *EMBO J* **22**:6438–6447.

35. Fiumera HL, Broadley SA, Fox TD. 2007. Translocation of mitochondrially synthesized Cox2 domains from the matrix to the intermembrane space. *Mol Cell Biol* **27**:4664–4673.

36. Chen Y, Dalbey RE. 2018. Oxa1 superfamily: new members found in the ER. *Trends Biochem Sci* **43**:151–153.

37. Anghel SA, McGilvray PT, Hegde RS, Keenan RJ. 2017. Identification of Oxa1 homologs operating in the eukaryotic endoplasmic reticulum. *Cell Rep* **21**:3708–3716.

38. Srivastava R, Zalisko BE, Keenan RJ, Howell SH. 2017. The GET system inserts the tail-anchored protein, SYP72, into endoplasmic reticulum membranes. *Plant Physiol* **173**:1137–1145.

39. Guna A, Volkmar N, Christianson JC, Hegde RS. 2018. The ER membrane protein complex is a transmembrane domain insertase. *Science* **359**:470–473.

40. Shurtleff MJ, Itzhak DN, Hussmann JA, Schirle Oakdale NT, Costa EA, Jonikas M, Weibezahn J, Popova KD, Jan CH, Sinitcyn P, Vembar SS, Hernandez H, Cox J, Burlingame AL, Brodsky JL, Frost A, Borner GH, Weissman JS. 2018. The ER membrane protein complex interacts cotranslationally to enable biogenesis of multipass membrane proteins. *eLife* **7**:7.

41. Spann D, Pross E, Chen Y, Dalbey RE, Kuhn A. 2018. Each protomer of a dimeric YidC functions as a single membrane insertase. *Sci Rep* **8**:589.

42. Boy D, Koch HG. 2009. Visualization of distinct entities of the SecYEG translocon during translocation and integration of bacterial proteins. *Mol Biol Cell* **20**:1804–1815.

43. Serek J, Bauer-Manz G, Struhalla G, van den Berg L, Kiefer D, Dalbey R, Kuhn A. 2004. *Escherichia coli* YidC is a membrane insertase for Sec-independent proteins. *EMBO J* **23**:294–301.

44. Aschtgen MS, Zoued A, Lloubès R, Journet L, Cascales E. 2012. The C-tail anchored TssL subunit, an essential protein of the enteroaggregative *Escherichia coli* Sci-1 type VI secretion system, is inserted by YidC. *Microbiologyopen* **1**:71–82.

45. Dalbey RE, Kuhn A, Zhu L, Kiefer D. 2014. The membrane insertase YidC. *Biochim Biophys Acta* **1843**:1489–1496.

46. Kumazaki K, Chiba S, Takemoto M, Furukawa A, Nishiyama K, Sugano Y, Mori T, Dohmae N, Hirata K, Nakada-Nakura Y, Maturana AD, Tanaka Y, Mori H, Sugita Y, Arisaka F, Ito K, Ishitani R, Tsukazaki T, Nureki O. 2014. Structural basis of Sec-independent membrane protein insertion by YidC. *Nature* **509**:516–520.

47. Kumazaki K, Kishimoto T, Furukawa A, Mori H, Tanaka Y, Dohmae N, Ishitani R, Tsukazaki T, Nureki O. 2014. Crystal structure of *Escherichia coli* YidC, a membrane protein chaperone and insertase. *Sci Rep* **4**:7299.

48. Chen Y, Soman R, Shanmugam SK, Kuhn A, Dalbey RE. 2014. The role of the strictly conserved positively charged residue differs among the Gram-positive, Gram-negative, and chloroplast YidC homologs. *J Biol Chem* **289**:35656–35667.

49. Zhu L, Wasey A, White SH, Dalbey RE. 2013. Charge composition features of model single-span membrane proteins that determine selection of YidC and SecYEG translocase pathways in *Escherichia coli*. *J Biol Chem* **288**:7704–7716.

50. Price CE, Driessen AJ. 2010. Conserved negative charges in the transmembrane segments of subunit K of the NADH:ubiquinone oxidoreductase determine its dependence on YidC for membrane insertion. *J Biol Chem* **285**:3575–3581.

51. Chen Y, Capponi S, Zhu L, Gellenbeck P, Freites JA, White SH, Dalbey RE. 2017. YidC insertase of *Escherichia coli*: water accessibility and membrane shaping. *Structure* **25**:1403–1414.e3.

52. Klenner C, Yuan J, Dalbey RE, Kuhn A. 2008. The Pf3 coat protein contacts TM1 and TM3 of YidC during membrane biogenesis. *FEBS Lett* **582**:3967–3972.

53. Neugebauer SA, Baulig A, Kuhn A, Facey SJ. 2012. Membrane protein insertion of variant MscL proteins occurs at YidC and SecYEG of *Escherichia coli*. *J Mol Biol* **417**:375–386.

54. Kedrov A, Wickles S, Crevenna AH, van der Sluis EO, Buschauer R, Berninghausen O, Lamb DC, Beckmann R. 2016. Structural dynamics of the YidC:ribosome complex during membrane protein biogenesis. *Cell Rep* **17**:2943–2954.

55. Winterfeld S, Ernst S, Börsch M, Gerken U, Kuhn A. 2013. Real time observation of single membrane protein insertion events by the *Escherichia coli* insertase YidC. *PLoS One* **8**: e59023.

56. Geng Y, Kedrov A, Caumanns JJ, Crevenna AH, Lamb DC, Beckmann R, Driessen AJ. 2015. Role of the cytosolic loop C2 and the C terminus of YidC in ribosome binding and insertion activity. *J Biol Chem* **290**:17250–17261.

57. **Soman R, Yuan J, Kuhn A, Dalbey RE.** 2014. Polarity and charge of the periplasmic loop determine the YidC and sec translocase requirement for the M13 procoat lep protein. *J Biol Chem* **289:**1023–1032.

58. **Kol S, Majczak W, Heerlien R, van der Berg JP, Nouwen N, Driessen AJ.** 2009. Subunit a of the F(1)F(0) ATP synthase requires YidC and SecYEG for membrane insertion. *J Mol Biol* **390:**893–901.

59. **Celebi N, Yi L, Facey SJ, Kuhn A, Dalbey RE.** 2006. Membrane biogenesis of subunit II of cytochrome bo oxidase: contrasting requirements for insertion of N-terminal and C-terminal domains. *J Mol Biol* **357:**1428–1436.

60. **Zhu L, Klenner C, Kuhn A, Dalbey RE.** 2012. Both YidC and SecYEG are required for translocation of the periplasmic loops 1 and 2 of the multispanning membrane protein TatC. *J Mol Biol* **424:**354–367.

61. **Welte T, Kudva R, Kuhn P, Sturm L, Braig D, Müller M, Warscheid B, Drepper F, Koch HG.** 2012. Promiscuous targeting of polytopic membrane proteins to SecYEG or YidC by the *Escherichia coli* signal recognition particle. *Mol Biol Cell* **23:**464–479.

62. **Schulze RJ, Komar J, Botte M, Allen WJ, Whitehouse S, Gold VA, Lycklama A, Nijeholt JA, Huard K, Berger I, Schaffitzel C, Collinson I.** 2014. Membrane protein insertion and proton-motive-force-dependent secretion through the bacterial holo-translocon SecYEG-SecDF-YajC-YidC. *Proc Natl Acad Sci U S A* **111:**4844–4849.

63. **Egea PF, Stroud RM.** 2010. Lateral opening of a translocon upon entry of protein suggests the mechanism of insertion into membranes. *Proc Natl Acad Sci U S A* **107:**17182–17187.

64. **Van den Berg B, Clemons WM Jr, Collinson I, Modis Y, Hartmann E, Harrison SC, Rapoport TA.** 2004. X-ray structure of a protein-conducting channel. *Nature* **427:**36–44.

65. **Sachelaru I, Petriman NA, Kudva R, Kuhn P, Welte T, Knapp B, Drepper F, Warscheid B, Koch HG.** 2013. YidC occupies the lateral gate of the SecYEG translocon and is sequentially displaced by a nascent membrane protein. *J Biol Chem* **288:**16295–16307.

66. **Beck K, Eisner G, Trescher D, Dalbey RE, Brunner J, Müller M.** 2001. YidC, an assembly site for polytopic *Escherichia coli* membrane proteins located in immediate proximity to the SecYE translocon and lipids. *EMBO Rep* **2:**709–714.

67. **Nagamori S, Smirnova IN, Kaback HR.** 2004. Role of YidC in folding of polytopic membrane proteins. *J Cell Biol* **165:**53–62.

68. **Zhu L, Kaback HR, Dalbey RE.** 2013. YidC protein, a molecular chaperone for LacY protein folding via the SecYEG protein machinery. *J Biol Chem* **288:**28180–28194.

69. **Serdiuk T, Mari SA, Müller DJ.** 2017. Pull-and-paste of single transmembrane proteins. *Nano Lett* **17:**4478–4488.

70. **Nouwen N, Driessen AJ.** 2002. SecDFyajC forms a heterotetrameric complex with YidC. *Mol Microbiol* **44:**1397–1405.

71. **Tsukazaki T, Mori H, Echizen Y, Ishitani R, Fukai S, Tanaka T, Perederina A, Vassylyev DG, Kohno T, Maturana AD, Ito K, Nureki O.** 2011. Structure and function of a membrane component SecDF that enhances protein export. *Nature* **474:**235–238.

72. **Waterhouse A, Bertoni M, Bienert S, Studer G, Tauriello G, Gumienny R, Heer FT, de Beer TAP, Rempfer C, Bordoli L, Lepore R, Schwede T.** 2018. SWISS-MODEL: homology modelling of protein structures and complexes. *Nucleic Acids Res* **46**(W1):W296–W303.

73. **Kiefer D, Kuhn A.** 2018. YidC-mediated membrane insertion. *FEMS Microbiol Lett* **365:**365.

The Twin-Arginine Pathway for Protein Secretion

KELLY M. FRAIN,[1] JAN MAARTEN VAN DIJL,[2] and COLIN ROBINSON[1]

INTRODUCTION

About 20 to 30% of proteins synthesized in the bacterial cytoplasm are destined for extracytoplasmic locations (1). They pass the cytoplasmic membrane using specialized transport systems, involving gated pores, energy, and signal peptides to direct protein export. Two major protein export systems are known, namely, the general secretory (Sec) pathway and the twin-arginine translocation (Tat) pathway (Fig. 1). Most proteins use the Sec pathway, common to all domains of life. The Tat pathway, the focus of this review, is more exclusive. For example, it has only ~30 native substrates in the Gram-negative bacterium *Escherichia coli*, and it is not universally conserved (2).

TWIN-ARGININE SIGNAL PEPTIDES

Specific N-terminal signal peptides direct proteins to the Sec or Tat pathway. Upon membrane translocation, the signal peptide is removed by signal

[1]The School of Biosciences, University of Kent, Canterbury CT2 7NZ, United Kingdom
[2]University of Groningen, University Medical Center Groningen, Department of Medical Microbiology, Groningen, The Netherlands

Protein Secretion in Bacteria
Edited by Maria Sandkvist, Eric Cascales, and Peter J. Christie
© 2019 American Society for Microbiology, Washington, DC
doi:10.1128/ecosalplus.ESP-0040-2018

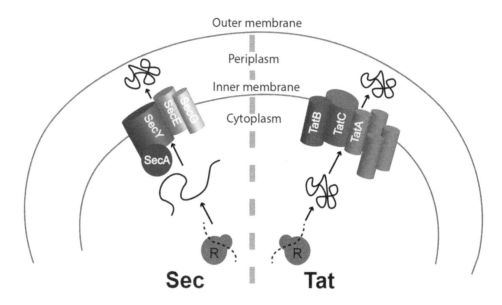

FIGURE 1 Two alternative protein translocation pathways, Sec and Tat. Proteins always originate from translating ribosomes (R). Their N-terminal signal peptide (OmpA or TorA in this overview) directs the nascent polypeptide chain to the correct translocase (Sec or Tat, respectively), which may be aided by chaperones. The unfolded Sec protein is transferred to SecA, where it is threaded through the SecYEG channel in the plasma membrane, powered by repeated cycles of ATP binding and hydrolysis. In the oxidizing periplasm, the unfolded protein assumes its tertiary fully folded state. The Tat-dependently translocated protein is fully folded within the cytoplasm, where it may also acquire its cofactor. Once directed to TatBC, TatA protomers are recruited to translocate the protein across the cytoplasmic membrane. Energy required for this process is created by the PMF. mRNA molecules are schematically represented by an interrupted line, synthesized proteins by uninterrupted lines, and translocase subunits by cylinders.

peptidase to release the mature protein (3–7). Bacterial Sec and Tat signal peptides have a core tripartite structure: a positively charged N-terminal domain, a hydrophobic H domain, and a C-terminal domain (3, 8). The signal peptidase cleavage site, usually Ala-x-Ala, resides in the C domain (Fig. 2). Tat signal peptides are defined by the twin-arginine motif S-R-R-x-F-L-K (x is any polar amino acid), joining the N and H domains (9, 10). The RR residues are essential for protein export (11, 12), although an RR→KR substitution may be tolerated (12–16). Tat signal peptides are around 30 residues long, while Sec signal peptides range between 17 and 24 residues (17). To avoid Sec, Tat signal peptides are less hydrophobic than Sec signal peptides (18) and often contain basic residues as a "Sec avoidance motif" (19, 20).

THE Tat PATHWAY

The Tat pathway was discovered in the 1990s (21, 22). Three thylakoidal membrane proteins were identified as essential for protein translocation via Tat, namely, Tha4, Hcf106, and cpTatC (23–26). Homologues were found in bacteria, archaea, and mitochondria (26, 27). In *E. coli*, these were, respectively, named TatA, TatB, and TatC (9, 28, 29). Their importance for protein export was highlighted by protein mislocalization in mutant strains (27, 28).

Studies on thylakoidal and bacterial Tat pathways showed that they transport complex, fully folded proteins requiring cofactor insertion or oligomerization (30). Accordingly, Tat facilitates many processes, including cell division and cell envelope biogenesis

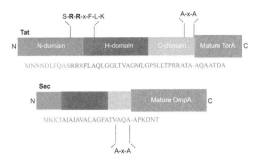

FIGURE 2 Comparison of Tat (TorA) and Sec (OmpA) signal peptides. The structure of Tat and Sec signal peptides includes three regions, namely, a basic N domain, a hydrophobic H domain, and a polar C domain. A signal peptidase cleavage site (AxA) is positioned prior to the mature protein. The amino acid sequences of the TorA and OmpA signal peptides are specified. Tat signal peptides (top) have a consensus motif containing twin arginines, while Sec signal peptides do not contain this motif. Sec signal peptides tend to be shorter, with fewer residues in their N and H domains, than Tat signal peptides.

in *E. coli* (31, 32). Tat mutants of other organisms, e.g., *Agrobacterium* and *Salmonella*, are impaired in quorum sensing, motility, and pathogenesis (33, 34). Unlike Sec, Tat even exports complex proteins from eukaryotes, as first evidenced for tightly folded dihydrofolate reductase or green fluorescent protein provided with Tat signal peptides (35–39).

Tat Genes

The *E. coli* Tat system is constitutively expressed from the *tatABCD* operon, where *tatA* is expressed ~25- and ~50-fold more than *tatB* and *tatC*, respectively (40). Elsewhere in the genome, *tatE* is expressed. Presumably, *tatB* and *tatE* originate from gene duplications of *tatA* (26, 41). Δ*tatABCDE* strains are viable but display pleiotropic phenotypes (42).

Tat systems minimally function with TatA-like and TatC-like proteins, although this varies between organisms (43). The Gram-positive bacterium *Bacillus subtilis* expresses two minimal Tat pathways from distinct *tatAdCd* and *tatAyCy* operons (5). TatAdCd

is coexpressed with its only known substrate, PhoD, during phosphate starvation (44, 45). TatAyCy is expressed constitutively as is the case for its substrates EfeB (YwbN), QcrA, and YkuE (5, 46–49). The *tatAc* gene is constitutively expressed elsewhere in the genome (50).

Escherichia coli Tat Components

TatA channels proteins across the inner membrane. Additionally, *E. coli* has two TatA-like proteins, TatB and TatE, involved in protein translocation (51). In contrast, *B. subtilis* has three TatA proteins, TatAd, TatAy, and TatAc, but no TatB (5, 43, 45). TatA-like proteins can often be interchanged, functioning both within and between bacterial species. In *E. coli*, TatE can replace TatA (41), *B. subtilis* TatAd can replace TatA and TatB (52), and TatAc forms active translocases with TatCd and TatCy (53) that functionally replace *E. coli* TatA and TatE (54). This functional overlap suggests a universal Tat translocation mechanism (52, 55) (Table 1).

TatA-like proteins share structural features, including a short N-terminal domain exposed to the periplasm or cell wall (56), a single transmembrane (TM) helix linked to an amphipathic helix (APH) positioned against the cytoplasmic membrane side (57), and an unstructured cytoplasmic tail (58–61) (Fig. 3). Only a few substitutions in TatA block its export function. In *E. coli* TatA, Gly33 in the "hinge region" connecting the TM helix and the APH is essential for export (62), as is Phe39, which anchors the APH to the membrane (63). Other APH residues and Gln8 in the TM helix are also imperative (64).

TatE shares 57% similarity with TatA (9). Despite its low abundance (40), TatE can replace TatA (41), being a regular constituent of the Tat translocase (65). TatE interacts with Tat signal peptides, partially preventing premature signal peptide cleavage (66). Homologues function in many other bacteria, as exemplified by *B. subtilis* TatAc, showing 45% amino acid sequence similarity to TatE (50).

TABLE 1 Molecular masses of Tat proteins and complexes in *E. coli* and *B. subtilis*[a]

Organism and protein	Molecular mass of gene product (kDa)	Molecular mass of complex (kDa)	Reference
E. coli			
TatA	11	100–500	112
TatB	18	100	109
TatC	30	220	109
TatE	7	50–110	41
TatABC		440–580	162
B. subtilis			
TatAd	7	270*	52
TatCd	27	100	53
TatAy	6	200*	138
TatCy	28	66	53
TatAc	6	100	53
TatAdCd		230	53
TatAyCy		200	53

[a]The predicted molecular masses of *E. coli* and *B. subtilis* Tat proteins are according to gene sizes. The molecular masses of complexes were approximated mostly by blue native gel electrophoresis, except those determined by gel filtration chromatography, indicated with an asterisk.

TatB is functionally different from TatA. It binds Tat signal peptides and subsequent mature proteins. TatB's APH and C-terminal region are longer than those of TatA (63, 67, 68) (Fig. 3), and only mutations in "the hinge" and APH cause translocation defects (69). Single N-terminal amino acid substitutions in TatA allow complementation of Δ*tatB* strains (70, 71). This bifunctionality of TatA mirrors the minimal TatAC systems in Gram-positive bacteria (5, 72).

TatC is the largest Tat protein aiding substrate binding (73, 74). It has 6 TM helices and an N-in, C-in topology (Fig. 3) (75). Crystallization of *Aquifex aeolicus* TatC revealed that this protein resembles a baseball glove or a cupped hand (76). TatC shows restricted structural flexibility and is unlikely to undergo major conformational changes during translocation (77). A notable surface feature of TatC is the conserved Glu165 in *A. aeolicus* or Glu170 in *E. coli* (12, 77). This residue is close to the binding pocket for Tat signal peptides, and its position in the membrane may perturb bilayer structure (78). Other important residues reside in TatC's

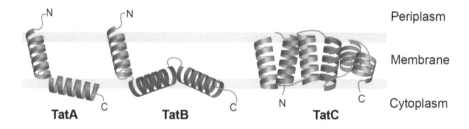

FIGURE 3 Schematic representation of *E. coli* TatABC components in the plasma membrane. Three essential components form the Gram-negative bacterial Tat complex, namely, TatA, TatB, and TatC. TatA/E and TatB have similar topologies in that they have one TM helix domain with a short periplasmic N-terminal region, a tilted APH, and an unstructured C terminus on the cytoplasmic side of the plasma membrane. Notably, TatB is larger than TatA, with a longer C-terminal tail. TatC is significantly bigger, as it contains 6 membrane-embedded helices with both the C- and N-terminal ends residing in the cytoplasm. Helices 5 and 6 do not fully span the membrane, which may contribute to TatC's function.

cytoplasmic N region and first cytoplasmic loop (74, 79).

Bacillus subtilis Tat Components

B. subtilis is the model organism for Gram-positive bacteria. Like all Gram-positive bacteria except *Actinomycetes* (80), *B. subtilis* has only TatA and TatC subunits (5, 45). Accordingly, *B. subtilis* TatA is bifunctional and able to functionally replace both *E. coli* TatA and TatB.

TatAdCd exists as a complex of ~230 kDa, while TatAd alone assembles into complexes of ~270 kDa (52). TatAd has essentially the same structure as *E. coli* TatA (57, 81). Mutations in the N terminus of heterologously expressed TatAd block translocation in Δ*tatB E. coli* cells, implying their importance for TatAd's bifunctionality. Further, electron microscopy (EM) showed that TatAd complexes are too small and homogeneous to serve as pores for passage of differently sized Tat substrates (44). TatCd is very similar to *E. coli* TatC, with 6 TM helices, and it both binds and stabilizes TatAd (82, 83).

TatAyCy and TatAy form complexes of ~200 kDa (43). The conserved hinge region of TatAy is equally as important as in TatAd and *E. coli* TatA (81, 84). The Pro2 residue in the N-terminal extracytoplasmic region of TatAy is key for bifunctionality, and Pro2 mutations interfere with TatAy's protein export function (50, 81, 84). The P2A mutation causes formation of fibril-like TatAy-TatCy assemblies of various lengths, as shown by EM and atomic force microscopy (85). Compared to those of other TatC proteins, TatCy's C terminus is extended by five amino acids that are important for EfeB export (86). Further, depending on the substrate, the N terminus and the first cytoplasmic loop of TatC, TatCd, and TatCy are important for translocation (87).

TatAc is dispensable for Tat function in *B. subtilis* (5). However, when expressed in *E. coli*, TatAc interacts with TatCd and TatCy, facilitating export of Tat substrates, like AmiA, AmiC, and TorA (53). While TatAc can replace TatA and TatB in *E. coli* (54), it cannot replace TatAd or TatAy in *B. subtilis*. Nonetheless, TatAc enhances TatAyCy function in *Bacillus* (50).

Tat Complexes

The largest native Tat substrates are heterodimeric formate dehydrogenases of ~150 kDa (88). However, Tat components themselves are relatively small, so the translocase must coalesce with multiple copies of each individual subunit. Live-cell imaging fluorescence microscopy showed the TatABC system assembling on demand to newly synthesized Tat substrates (89). Two major complexes exist: TatBC and TatA. TatBC forms a receptor that binds Tat substrates at the membrane. This triggers proton motive force (PMF)-dependent TatA recruitment and oligomerization to form an active translocase (90, 91). Starting from substrate-bound TatBC, the translocation event is slow, taking 1 to 3 min (92). Disassembly of TatA from TatBC requires substrate export (93).

TatB and TatC interact in a 1:1 stoichiometry (94), and several TatBC complexes bind substrate(s) (95). Low-resolution EM structures of TatBC revealed a hemispherical morphology, with an internal cavity that could accommodate the signal peptide (73). Seven copies of TatBC fit into the 11- to 17-nm reconstruction, and apparently more than one substrate binds at once (96). Cross-linking showed TatC is the primary interaction site of the signal peptide (91, 97–99), which inserts deeply into TatC by adopting a hairpin-like conformation (76, 97). Subsequently, the signal peptide exposes its C region at the extracytoplasmic membrane side to signal peptidase (100, 101). The signal peptide's H domain interacts with TatB's TM helix (91), and Tat precursors initially bind to TatB at multiple sites (102). Nuclear magnetic resonance (NMR) data indicate that TatB's extended C terminus is flexible to facilitate substrate binding (67). Each TatC monomer has two TatB contact sites (74, 77, 101–103). Intriguingly, the TM helix of TatB seems close to where TatA initiates translocase oligomer assembly, suggesting that TatB is a regulatory surrogate of TatA (104).

Coevolution analysis predicts TatA and TatB binding along TM5/6 of TatC, indicating that a cluster of polar amino acids on TatC forms contact with polar side chains of TatA or TatB (105, 106). Cross-linking studies indicate that upon substrate docking, TatA binding to the TM6 of TatC is reduced, while TatA and TatB binding at the polar site of TatC is increased. As the polar cluster site is adjacent to the docking site of the signal peptide (65), this docking event could cause the conformational rearrangements. Signal peptide binding also alters TatC's resting state arrangement from head-to-tail to tail-to-tail. This opens the complex, allowing TatA access to the vacated polar cluster site and placing TatA adjacent to the concave face of TatC, where TatA nucleation has been suggested (77).

TatA facilitates protein translocation, either by forming a pore (107) or by weakening the membrane (108). While TatA is not obligatory for TatBC formation or substrate docking (109), trace amounts are associated with TatBC, possibly representing nucleation points for TatA oligomers (110, 111). TatA complexes in *E. coli* are in the 100- to 500-kDa size range (112), which can be resolved at 34-kDa intervals, suggesting modular formation of three or four complexes at a time. Self-oligomerization involves TatA's TM helix (58, 67, 113).

MECHANISM OF Tat TRANSLOCATION

The mechanism of Tat translocation is still not fully resolved. The translocation pore model theorizes that TatA forms a channel/pore for protein passage (107). More recent data favor the membrane destabilization model (108, 114).

Low-resolution images of TatA revealed a pore-like complex of various diameters (8.5 to 13 nm) that would accommodate various substrate sizes (107, 115). A lid-like feature was identified which presumably resides on the *cis* side of the inner membrane, as if the transporter were a trap door (116). Thus, TatA's APH could form a hairpin to align with its TM helix, forming a pore for hydrophilic cargo passage (117). Upon substrate docking onto TatBC, TatA protomers would assemble to suit the substrate size in an oligomeric ring format (118, 119). Each TatA protein could be "zipped together" by 7 salt bridges. Indeed, salt dependence for the translocase was reported (84). However, this model is less attractive, as APH insertion into the bilayer was not observed in other studies (56, 120), while solution NMR indicated that the APH fans outwards (58).

The second model proposes TatA complexes weaken/destabilize the membrane due to TatA's short TM domain (58, 108), restricting membrane thickness to its own length. Destabilization only happens when Tat substrate is bound. When no substrate is bound, TatA's APH immerses itself into the membrane, elongating the membrane. Upon substrate association, TatA reorients its APH outside of the membrane (120). The "switch" from one state to another relates also to interactions between substrate and TatA's APH. TatA does not recognize the RR motif, but its APH interacts with the C domain of signal peptides (121, 122), suggesting that the mature protein also interacts with TatA (91, 102, 123).

Energy Requirements

The Tat system is powered by the PMF, consisting of a pH gradient (ΔpH) and the electric potential ($\Delta\psi$) across the membrane, and Tat can use both (124–128). Accordingly, a protein-H^+ antiporter mechanism was suggested (129). It was estimated that a counterflow of 7.9×10^4 protons is needed per protein (130). The PMF is not required for protein targeting or binding to TatBC, but it is necessary for TatA oligomerization and the actual export process (91, 110, 131, 132).

Tat Proofreading

Generally, Tat substrates exit the cytosol only once they are fully folded and contain their cofactor (133–137). The respective proofreading is extremely tight in *B. subtilis* (138, 139) but less stringent in the thylakoidal Tat system, as unfolded proteins are also transported (35). Requirements for conformational stability were studied using nonnative Tat substrates, like PhoA and antibody fragments. These are exported only under oxidizing conditions, under which disulfide bonds form (38, 140–142). Nevertheless, human growth hormone and interferon α2b are also exported via Tat without disulfide bonds formed (37), suggesting that they assume near-native states as shown for *E. coli* CueO, which is exported without its copper cofactor bound (143). Further, unstructured and hydrophilic polypeptides of 100 to 120 residues are tolerated by Tat, but a short hydrophobic stretch stops export (144–146). It thus seems that proofreading senses conformational flexibility. Mutations with less stringent proofreading were identified in *E. coli* TatABC (147), indicating that proofreading is undertaken by the Tat translocon. Proofreading may also involve chaperones known as redox enzyme maturation proteins (148). This is exemplified by TorD, facilitating Tat-dependent export of the trimethylamine-*N*-oxide reductase TorA in *E. coli* (149, 150). Lastly, Tat may recruit quality control proteases to clear the translocase, as evidenced in *B. subtilis* for TatAyCy and the wall protease WprA (151, 152).

FUTURE PERSPECTIVES FOR BIOTECHNOLOGY

Export of fully folded proteins is advantageous for biotechnological applications. Owing to their rapid growth, high yield, ease of scale-up, and cost-effectiveness, ~30% of biotherapeutics are synthesized in bacteria (153). Preferably, these proteins are exported to the oxidizing periplasm with an N-terminal signal peptide. This is the usual approach in *B. subtilis*, where most translocated proteins are secreted into the medium (154, 155). In *E. coli*, export to the periplasm decreases proteolysis and eases downstream purification, as the periplasm only contains 4% of the proteome (156, 157). Here, proteins are released by osmotic shock (158). Recently it was shown that biotherapeutics can be exported via Tat (37, 159). Moreover, Tat overexpression in so-called TatExpress cells resulted in 5-fold-increased export of human growth hormone (160). The "gold standard" of bacterial protein production, protein secretion to the medium, was achieved when *B. subtilis* TatAdCd was expressed in Δ*tatABCDE* strains, as this led to protein release from the periplasm (161). Current efforts to apply Tat in biotechnology focus on different biotherapeutics and increased protein yields. It is anticipated that this area will expand rapidly once the mechanisms underlying Tat-dependent protein export are fully understood.

CITATION

EcoSal Plus 2019; doi:10.1128/ecosalplus.ESP-0040-2018.

REFERENCES

1. **Holland IB.** 2004. Translocation of bacterial proteins—an overview. *Biochim Biophys Acta* **1694:**5–16.
2. **Palmer T, Sargent F, Berks BC.** 2010. The Tat protein export pathway. *EcoSal Plus* **4:**4.3.2.
3. **von Heijne G.** 1990. The signal peptide. *J Membr Biol* **115:**195–201.
4. **Lüke I, Handford JI, Palmer T, Sargent F.** 2009. Proteolytic processing of *Escherichia coli* twin-arginine signal peptides by LepB. *Arch Microbiol* **191:**919–925.
5. **Jongbloed JDH, Grieger U, Antelmann H, Hecker M, Nijland R, Bron S, van Dijl JM.** 2004. Two minimal Tat translocases in *Bacillus*. *Mol Microbiol* **54:**1319–1325.

6. Dalbey RE, Wang P, van Dijl JM. 2012. Membrane proteases in the bacterial protein secretion and quality control pathway. *Microbiol Mol Biol Rev* **76**:311–330.

7. Sakaguchi M, Tomiyoshi R, Kuroiwa T, Mihara K, Omura T. 1992. Functions of signal and signal-anchor sequences are determined by the balance between the hydrophobic segment and the N-terminal charge. *Proc Natl Acad Sci U S A* **89**:16–19.

8. Berks BC. 1996. A common export pathway for proteins binding complex redox cofactors? *Mol Microbiol* **22**:393–404.

9. Sargent F, Bogsch EG, Stanley NR, Wexler M, Robinson C, Berks BC, Palmer T. 1998. Overlapping functions of components of a bacterial Sec-independent protein export pathway. *EMBO J* **17**:3640–3650.

10. Berks BC, Palmer T, Sargent F. 2003. The Tat protein translocation pathway and its role in microbial physiology. *Adv Microb Physiol* **47**:187–254.

11. Stanley NR, Palmer T, Berks BC. 2000. The twin arginine consensus motif of Tat signal peptides is involved in Sec-independent protein targeting in *Escherichia coli*. *J Biol Chem* **275**:11591–11596.

12. Buchanan G, Sargent F, Berks BC, Palmer T. 2001. A genetic screen for suppressors of *Escherichia coli* Tat signal peptide mutations establishes a critical role for the second arginine within the twin-arginine motif. *Arch Microbiol* **177**:107–112.

13. Chaddock AM, Mant A, Karnauchov I, Brink S, Herrmann RG, Klösgen RB, Robinson C. 1995. A new type of signal peptide: central role of a twin-arginine motif in transfer signals for the delta pH-dependent thylakoidal protein translocase. *EMBO J* **14**:2715–2722.

14. Halbig D, Hou B, Freudl R, Sprenger GA, Klösgen RB. 1999. Bacterial proteins carrying twin-R signal peptides are specifically targeted by the delta pH-dependent transport machinery of the thylakoid membrane system. *FEBS Lett* **447**:95–98.

15. Hinsley AP, Stanley NR, Palmer T, Berks BC. 2001. A naturally occurring bacterial Tat signal peptide lacking one of the 'invariant' arginine residues of the consensus targeting motif. *FEBS Lett* **497**:45–49.

16. DeLisa MP, Samuelson P, Palmer T, Georgiou G. 2002. Genetic analysis of the twin arginine translocator secretion pathway in bacteria. *J Biol Chem* **277**:29825–29831.

17. Tullman-Ercek D, DeLisa MP, Kawarasaki Y, Iranpour P, Ribnicky B, Palmer T, Georgiou G. 2007. Export pathway selectivity of *Escherichia coli* twin arginine translocation signal peptides. *J Biol Chem* **282**:8309–8316.

18. Cristóbal S, de Gier JW, Nielsen H, von Heijne G. 1999. Competition between Sec- and TAT-dependent protein translocation in *Escherichia coli*. *EMBO J* **18**:2982–2990.

19. Bogsch E, Brink S, Robinson C. 1997. Pathway specificity for a delta pH-dependent precursor thylakoid lumen protein is governed by a 'Sec-avoidance' motif in the transfer peptide and a 'Sec-incompatible' mature protein. *EMBO J* **16**:3851–3859.

20. Mould RM, Robinson C. 1991. A proton gradient is required for the transport of two lumenal oxygen-evolving proteins across the thylakoid membrane. *J Biol Chem* **266**:12189–12193.

21. Klösgen RB, Brock IW, Herrmann RG, Robinson C. 1992. Proton gradient-driven import of the 16 kDa oxygen-evolving complex protein as the full precursor protein by isolated thylakoids. *Plant Mol Biol* **18**:1031–1034.

22. Clark SA, Theg SM. 1997. A folded protein can be transported across the chloroplast envelope and thylakoid membranes. *Mol Biol Cell* **8**:923–934.

23. Mori H, Summer EJ, Ma X, Cline K. 1999. Component specificity for the thylakoidal Sec and ΔpH-dependent protein transport pathways. *J Cell Biol* **146**:45–56.

24. Settles AM, Yonetani A, Baron A, Bush DR, Cline K, Martienssen R. 1997. Sec-independent protein translocation by the maize Hcf106 protein. *Science* **278**:1467–1470.

25. Cline K, Mori H. 2001. Thylakoid DeltapH-dependent precursor proteins bind to a cpTatC-Hcf106 complex before Tha4-dependent transport. *J Cell Biol* **154**:719–729.

26. Wu LF, Ize B, Chanal A, Quentin Y, Fichant G. 2000. Bacterial twin-arginine signal peptide-dependent protein translocation pathway: evolution and mechanism. *J Mol Microbiol Biotechnol* **2**:179–189.

27. Bogsch EG, Sargent F, Stanley NR, Berks BC, Robinson C, Palmer T. 1998. An essential component of a novel bacterial protein export system with homologues in plastids and mitochondria. *J Biol Chem* **273**:18003–18006.

28. Sargent F, Stanley NR, Berks BC, Palmer T. 1999. Sec-independent protein translocation in *Escherichia coli*. A distinct and pivotal role for the TatB protein. *J Biol Chem* **274**:36073–36082.

29. Palmer T, Sargent F, Berks BC. 2005. Export of complex cofactor-containing proteins by the bacterial Tat pathway. *Trends Microbiol* **13**:175–180.

30. **Rodrigue A, Chanal A, Beck K, Müller M, Wu LF.** 1999. Co-translocation of a periplasmic enzyme complex by a hitchhiker mechanism through the bacterial Tat pathway. *J Biol Chem* **274:**13223–13228.

31. **Bernhardt TG, de Boer PAJ.** 2003. The *Escherichia coli* amidase AmiC is a periplasmic septal ring component exported via the twin-arginine transport pathway. *Mol Microbiol* **48:**1171–1182.

32. **Ize B, Stanley NR, Buchanan G, Palmer T.** 2003. Role of the *Escherichia coli* Tat pathway in outer membrane integrity. *Mol Microbiol* **48:**1183–1193.

33. **Ding Z, Christie PJ.** 2003. *Agrobacterium tumefaciens* twin-arginine-dependent translocation is important for virulence, flagellation, and chemotaxis but not type IV secretion. *J Bacteriol* **185:**760–771.

34. **Craig M, Sadik AY, Golubeva YA, Tidhar A, Slauch JM.** 2013. Twin-arginine translocation system (*tat*) mutants of *Salmonella* are attenuated due to envelope defects, not respiratory defects. *Mol Microbiol* **89:**887–902.

35. **Hynds PJ, Robinson D, Robinson C.** 1998. The Sec-independent twin-arginine translocation system can transport both tightly folded and malfolded proteins across the thylakoid membrane. *J Biol Chem* **273:**34868–34874.

36. **Santini CL, Bernadac A, Zhang M, Chanal A, Ize B, Blanco C, Wu LF.** 2001. Translocation of jellyfish green fluorescent protein via the Tat system of *Escherichia coli* and change of its periplasmic localization in response to osmotic up-shock. *J Biol Chem* **276:**8159–8164.

37. **Alanen HI, Walker KL, Lourdes Velez Suberbie M, Matos CFRO, Bönisch S, Freedman RB, Keshavarz-Moore E, Ruddock LW, Robinson C.** 2015. Efficient export of human growth hormone, interferon α2b and antibody fragments to the periplasm by the *Escherichia coli* Tat pathway in the absence of prior disulfide bond formation. *Biochim Biophys Acta* **1853:**756–763.

38. **DeLisa MP, Tullman D, Georgiou G.** 2003. Folding quality control in the export of proteins by the bacterial twin-arginine translocation pathway. *Proc Natl Acad Sci U S A* **100:**6115–6120.

39. **Richter S, Brüser T.** 2005. Targeting of unfolded PhoA to the TAT translocon of *Escherichia coli*. *J Biol Chem* **280:**42723–42730.

40. **Jack RL, Sargent F, Berks BC, Sawers G, Palmer T.** 2001. Constitutive expression of *Escherichia colitat* genes indicates an important role for the twin-arginine translocase during aerobic and anaerobic growth. *J Bacteriol* **183:**1801–1804.

41. **Baglieri J, Beck D, Vasisht N, Smith CJ, Robinson C.** 2012. Structure of TatA paralog, TatE, suggests a structurally homogeneous form of Tat protein translocase that transports folded proteins of differing diameter. *J Biol Chem* **287:**7335–7344.

42. **Stanley NR, Findlay K, Berks BC, Palmer T.** 2001. *Escherichia coli* strains blocked in Tat-dependent protein export exhibit pleiotropic defects in the cell envelope. *J Bacteriol* **183:**139–144.

43. **Goosens VJ, Monteferrante CG, van Dijl JM.** 2014. The Tat system of Gram-positive bacteria. *Biochim Biophys Acta* **1843:**1698–1706.

44. **Ridder AN, de Jong EJ, Jongbloed JD, Kuipers OP, van Dijl JM, Robinson C, Smith C.** 2009. Subcellular localization of TatAd of *Bacillus subtilis* depends on the presence of TatCd or TatCy. *J Bacteriol* **191:**4410–4418.

45. **Jongbloed JDH, Martin U, Antelmann H, Hecker M, Tjalsma H, Venema G, Bron S, van Dijl JM, Müller J.** 2000. TatC is a specificity determinant for protein secretion via the twin-arginine translocation pathway. *J Biol Chem* **275:**41350–41357.

46. **van der Ploeg R, Mäder U, Homuth G, Schaffer M, Denham EL, Monteferrante CG, Miethke M, Marahiel MA, Harwood CR, Winter T, Hecker M, Antelmann H, van Dijl JM.** 2011. Environmental salinity determines the specificity and need for Tat-dependent secretion of the YwbN protein in *Bacillus subtilis*. *PLoS One* **6:**e18140.

47. **Monteferrante CG, Miethke M, van der Ploeg R, Glasner C, van Dijl JM.** 2012. Specific targeting of the metallophosphoesterase YkuE to the *Bacillus* cell wall requires the twin-arginine translocation system. *J Biol Chem* **287:**29789–29800.

48. **Miethke M, Monteferrante CG, Marahiel MA, van Dijl JM.** 2013. The *Bacillus subtilis* EfeUOB transporter is essential for high-affinity acquisition of ferrous and ferric iron. *Biochim Biophys Acta* **1833:**2267–2278.

49. **Goosens VJ, Otto A, Glasner C, Monteferrante CC, van der Ploeg R, Hecker M, Becher D, van Dijl JM.** 2013. Novel twin-arginine translocation pathway-dependent phenotypes of *Bacillus subtilis* unveiled by quantitative proteomics. *J Proteome Res* **12:**796–807.

50. **Goosens VJ, De-San-Eustaquio-Campillo A, Carballido-López R, van Dijl JM.** 2015. A Tat ménage à trois—the role of *Bacillus subtilis* TatAc in twin-arginine protein translocation. *Biochim Biophys Acta* **1853**(10 Part A):2745–2753.

51. **Sargent F, Gohlke U, De Leeuw E, Stanley NR, Palmer T, Saibil HR, Berks BC.** 2001.

Purified components of the *Escherichia coli* Tat protein transport system form a double-layered ring structure. *Eur J Biochem* **268**:3361–3367.

52. **Barnett JP, Eijlander RT, Kuipers OP, Robinson C.** 2008. A minimal Tat system from a gram-positive organism: a bifunctional TatA subunit participates in discrete TatAC and TatA complexes. *J Biol Chem* **283**:2534–2542.

53. **Monteferrante CG, Baglieri J, Robinson C, van Dijl JM.** 2012. TatAc, the third TatA subunit of *Bacillus subtilis*, can form active twin-arginine translocases with the TatCd and TatCy subunits. *Appl Environ Microbiol* **78**:4999–5001.

54. **Beck D, Vasisht N, Baglieri J, Monteferrante CG, van Dijl JM, Robinson C, Smith CJ.** 2013. Ultrastructural characterisation of *Bacillus subtilis* TatA complexes suggests they are too small to form homooligomeric translocation pores. *Biochim Biophys Acta* **1833**:1811–1819.

55. **Porcelli I, de Leeuw E, Wallis R, van den Brink-van der Laan E, de Kruijff B, Wallace BA, Palmer T, Berks BC.** 2002. Characterization and membrane assembly of the TatA component of the *Escherichia coli* twin-arginine protein transport system. *Biochemistry* **41**:13690–13697.

56. **Koch S, Fritsch MJ, Buchanan G, Palmer T.** 2012. *Escherichia coli* TatA and TatB proteins have N-out, C-in topology in intact cells. *J Biol Chem* **287**:14420–14431.

57. **Hu Y, Zhao E, Li H, Xia B, Jin C.** 2010. Solution NMR structure of the TatA component of the twin-arginine protein transport system from gram-positive bacterium *Bacillus subtilis*. *J Am Chem Soc* **132**:15942–15944.

58. **Rodriguez F, Rouse SL, Tait CE, Harmer J, De Riso A, Timmel CR, Sansom MS, Berks BC, Schnell JR.** 2013. Structural model for the protein-translocating element of the twin-arginine transport system. *Proc Natl Acad Sci U S A* **110**:E1092–E1101.

59. **Müller SD, De Angelis AA, Walther TH, Grage SL, Lange C, Opella SJ, Ulrich AS.** 2007. Structural characterization of the pore forming protein TatAd of the twin-arginine translocase in membranes by solid-state 15N-NMR. *Biochim Biophys Acta* **1768**:3071–3079.

60. **Walther TH, Grage SL, Roth N, Ulrich AS.** 2010. Membrane alignment of the pore-forming component TatA(d) of the twin-arginine translocase from *Bacillus subtilis* resolved by solid-state NMR spectroscopy. *J Am Chem Soc* **132**:15945–15956.

61. **Lange C, Müller SD, Walther TH, Bürck J, Ulrich AS.** 2007. Structure analysis of the

protein translocating channel TatA in membranes using a multi-construct approach. *Biochim Biophys Acta* **1768**:2627–2634.

62. **Barrett CML, Ray N, Thomas JD, Robinson C, Bolhuis A.** 2003. Quantitative export of a reporter protein, GFP, by the twin-arginine translocation pathway in *Escherichia coli*. *Biochem Biophys Res Commun* **304**:279–284.

63. **Hicks MG, de Leeuw E, Porcelli I, Buchanan G, Berks BC, Palmer T.** 2003. The *Escherichia coli* twin-arginine translocase: conserved residues of TatA and TatB family components involved in protein transport. *FEBS Lett* **539**:61–67.

64. **Greene NP, Porcelli I, Buchanan G, Hicks MG, Schermann SM, Palmer T, Berks BC.** 2007. Cysteine scanning mutagenesis and disulfide mapping studies of the TatA component of the bacterial twin arginine translocase. *J Biol Chem* **282**:23937–23945.

65. **Eimer E, Fröbel J, Blümmel AS, Müller M.** 2015. TatE as a regular constituent of bacterial twin-arginine protein translocases. *J Biol Chem* **290**:29281–29289.

66. **Eimer E, Kao WC, Fröbel J, Blümmel AS, Hunte C, Müller M.** 2018. Unanticipated functional diversity among the TatA-type components of the Tat protein translocase. *Sci Rep* **8**:1326.

67. **Zhang Y, Wang L, Hu Y, Jin C.** 2014. Solution structure of the TatB component of the twin-arginine translocation system. *Biochim Biophys Acta* **1838**:1881–1888.

68. **Lee PA, Buchanan G, Stanley NR, Berks BC, Palmer T.** 2002. Truncation analysis of TatA and TatB defines the minimal functional units required for protein translocation. *J Bacteriol* **184**:5871–5879.

69. **Barrett CML, Mathers JE, Robinson C.** 2003. Identification of key regions within the *Escherichia coli* TatAB subunits. *FEBS Lett* **537**:42–46.

70. **Blaudeck N, Kreutzenbeck P, Müller M, Sprenger GA, Freudl R.** 2005. Isolation and characterization of bifunctional *Escherichia coli* TatA mutant proteins that allow efficient Tat-dependent protein translocation in the absence of TatB. *J Biol Chem* **280**:3426–3432.

71. **Barrett CML, Freudl R, Robinson C.** 2007. Twin arginine translocation (Tat)-dependent export in the apparent absence of TatABC or TatA complexes using modified *Escherichia coli* TatA subunits that substitute for TatB. *J Biol Chem* **282**:36206–36213.

72. **Jongbloed JDH, van der Ploeg R, van Dijl JM.** 2006. Bifunctional TatA subunits in minimal Tat protein translocases. *Trends Microbiol* **14**:2–4.

73. **Tarry MJ, Schäfer E, Chen S, Buchanan G, Greene NP, Lea SM, Palmer T, Saibil HR, Berks BC.** 2009. Structural analysis of substrate binding by the TatBC component of the twin-arginine protein transport system. *Proc Natl Acad Sci U S A* **106:**13284–13289.

74. **Kneuper H, Maldonado B, Jäger F, Krehenbrink M, Buchanan G, Keller R, Müller M, Berks BC, Palmer T.** 2012. Molecular dissection of TatC defines critical regions essential for protein transport and a TatB-TatC contact site. *Mol Microbiol* **85:**945–961.

75. **Behrendt J, Standar K, Lindenstrauss U, Brüser T.** 2004. Topological studies on the twin-arginine translocase component TatC. *FEMS Microbiol Lett* **234:**303–308.

76. **Ramasamy S, Abrol R, Suloway CJM, Clemons WM Jr.** 2013. The glove-like structure of the conserved membrane protein TatC provides insight into signal sequence recognition in twin-arginine translocation. *Structure* **21:**777–788.

77. **Rollauer SE, Tarry MJ, Graham JE, Jääskeläinen M, Jäger F, Johnson S, Krehenbrink M, Liu SM, Lukey MJ, Marcoux J, McDowell MA, Rodriguez F, Roversi P, Stansfeld PJ, Robinson CV, Sansom MS, Palmer T, Högbom M, Berks BC, Lea SM.** 2012. Structure of the TatC core of the twin-arginine protein transport system. *Nature* **492:**210–214.

78. **Blümmel AS, Drepper F, Knapp B, Eimer E, Warscheid B, Müller M, Fröbel J.** 2017. Structural features of the TatC membrane protein that determine docking and insertion of a twin-arginine signal peptide. *J Biol Chem* **292:**21320–21329.

79. **Holzapfel E, Eisner G, Alami M, Barrett CML, Buchanan G, Lüke I, Betton JM, Robinson C, Palmer T, Moser M, Müller M.** 2007. The entire N-terminal half of TatC is involved in twin-arginine precursor binding. *Biochemistry* **46:**2892–2898.

80. **Schaerlaekens K, Schierová M, Lammertyn E, Geukens N, Anné J, Van Mellaert L.** 2001. Twin-arginine translocation pathway in *Streptomyces lividans*. *J Bacteriol* **183:**6727–6732.

81. **Barnett JP, Lawrence J, Mendel S, Robinson C.** 2011. Expression of the bifunctional *Bacillus subtilis* TatAd protein in *Escherichia coli* reveals distinct TatA/B-family and TatB-specific domains. *Arch Microbiol* **193:**583–594.

82. **Schreiber S, Stengel R, Westermann M, Volkmer-Engert R, Pop OI, Müller JP.** 2006. Affinity of TatCd for TatAd elucidates its receptor function in the *Bacillus subtilis* twin arginine translocation (Tat) translocase system. *J Biol Chem* **281:**19977–19984.

83. **Nolandt OV, Walther TH, Roth S, Bürck J, Ulrich AS.** 2009. Structure analysis of the membrane protein TatC(d) from the Tat system of *B. subtilis* by circular dichroism. *Biochim Biophys Acta* **1788:**2238–2244.

84. **van der Ploeg R, Barnett JP, Vasisht N, Goosens VJ, Pöther DC, Robinson C, van Dijl JM.** 2011. Salt sensitivity of minimal twin arginine translocases. *J Biol Chem* **286:**43759–43770.

85. **Patel R, Vasilev C, Beck D, Monteferrante CG, van Dijl JM, Hunter CN, Smith C, Robinson C.** 2014. A mutation leading to super-assembly of twin-arginine translocase (Tat) protein complexes. *Biochim Biophys Acta* **1843:**1978–1986.

86. **Eijlander RT, Jongbloed JDH, Kuipers OP.** 2009. Relaxed specificity of the *Bacillus subtilis* TatAdCd translocase in Tat-dependent protein secretion. *J Bacteriol* **191:**196–202.

87. **Eijlander RT, Kolbusz MA, Berendsen EM, Kuipers OP.** 2009. Effects of altered TatC proteins on protein secretion efficiency via the twin-arginine translocation pathway of *Bacillus subtilis*. *Microbiology* **155:**1776–1785.

88. **Berks BC, Sargent F, Palmer T.** 2000. The Tat protein export pathway. *Mol Microbiol* **35:**260–274.

89. **Rose P, Fröbel J, Graumann PL, Müller M.** 2013. Substrate-dependent assembly of the Tat translocase as observed in live *Escherichia coli* cells. *PLoS One* **8:**e69488.

90. **Berks BC, Lea SM, Stansfeld PJ.** 2014. Structural biology of Tat protein transport. *Curr Opin Struct Biol* **27:**32–37.

91. **Alami M, Lüke I, Deitermann S, Eisner G, Koch HG, Brunner J, Müller M.** 2003. Differential interactions between a twin-arginine signal peptide and its translocase in *Escherichia coli*. *Mol Cell* **12:**937–946.

92. **Celedon JM, Cline K.** 2012. Stoichiometry for binding and transport by the twin arginine translocation system. *J Cell Biol* **197:**523–534.

93. **Alcock F, Baker MAB, Greene NP, Palmer T, Wallace MI, Berks BC.** 2013. Live cell imaging shows reversible assembly of the TatA component of the twin-arginine protein transport system. *Proc Natl Acad Sci U S A* **110:**E3650–E3659.

94. **Bolhuis A, Mathers JE, Thomas JD, Barrett CM, Robinson C, Robinson C.** 2001. TatB and TatC form a functional and structural unit of the twin-arginine translocase from *Escherichia coli*. *J Biol Chem* **276:**20213–20219.

95. **Behrendt J, Brüser T.** 2014. The TatBC complex of the Tat protein translocase in *Escherichia coli* and its transition to the substrate-bound TatABC complex. *Biochemistry* **53:**2344–2354.

96. **Ma X, Cline K.** 2010. Multiple precursor proteins bind individual Tat receptor complexes and are collectively transported. *EMBO J* 29:1477–1488.

97. **Gérard F, Cline K.** 2006. Efficient twin arginine translocation (Tat) pathway transport of a precursor protein covalently anchored to its initial cpTatC binding site. *J Biol Chem* 281:6130–6135.

98. **Kreutzenbeck P, Kröger C, Lausberg F, Blaudeck N, Sprenger GA, Freudl R.** 2007. *Escherichia coli* twin arginine (Tat) mutant translocases possessing relaxed signal peptide recognition specificities. *J Biol Chem* 282:7903–7911.

99. **Zoufaly S, Fröbel J, Rose P, Flecken T, Maurer C, Moser M, Müller M.** 2012. Mapping precursor-binding site on TatC subunit of twin arginine-specific protein translocase by site-specific photo cross-linking. *J Biol Chem* 287:13430–13441.

100. **Fröbel J, Rose P, Lausberg F, Blümmel A-S, Freudl R, Müller M.** 2012. Transmembrane insertion of twin-arginine signal peptides is driven by TatC and regulated by TatB. *Nat Commun* 3:1311.

101. **Blümmel AS, Haag LA, Eimer E, Müller M, Fröbel J.** 2015. Initial assembly steps of a translocase for folded proteins. *Nat Commun* 6:7234.

102. **Maurer C, Panahandeh S, Jungkamp A-C, Moser M, Müller M.** 2010. TatB functions as an oligomeric binding site for folded Tat precursor proteins. *Mol Biol Cell* 21:4151–4161.

103. **Lee PA, Orriss GL, Buchanan G, Greene NP, Bond PJ, Punginelli C, Jack RL, Sansom MSP, Berks BC, Palmer T.** 2006. Cysteine-scanning mutagenesis and disulfide mapping studies of the conserved domain of the twin-arginine translocase TatB component. *J Biol Chem* 281:34072–34085.

104. **Cline K.** 2015. Mechanistic aspects of folded protein transport by the twin arginine translocase (Tat). *J Biol Chem* 290:16530–16538.

105. **Alcock F, Stansfeld PJ, Basit H, Habersetzer J, Baker MA, Palmer T, Wallace MI, Berks BC.** 2016. Assembling the Tat protein translocase. *eLife* 5:e20718.

106. **Habersetzer J, Moore K, Cherry J, Buchanan G, Stansfeld PJ, Palmer T.** 2017. Substrate-triggered position switching of TatA and TatB during Tat transport in *Escherichia coli*. *Open Biol* 7:8.

107. **Gohlke U, Pullan L, McDevitt CA, Porcelli I, de Leeuw E, Palmer T, Saibil HR, Berks BC.** 2005. The TatA component of the twin-arginine protein transport system forms channel complexes of variable diameter. *Proc Natl Acad Sci U S A* 102:10482–10486.

108. **Brüser T, Sanders C.** 2003. An alternative model of the twin arginine translocation system. *Microbiol Res* 158:7–17.

109. **Orriss GL, Tarry MJ, Ize B, Sargent F, Lea SM, Palmer T, Berks BC.** 2007. TatBC, TatB, and TatC form structurally autonomous units within the twin arginine protein transport system of *Escherichia coli*. *FEBS Lett* 581:4091–4097.

110. **Aldridge C, Ma X, Gerard F, Cline K.** 2014. Substrate-gated docking of pore subunit Tha4 in the TatC cavity initiates Tat translocase assembly. *J Cell Biol* 205:51–65.

111. **Hauer RS, Schlesier R, Heilmann K, Dittmar J, Jakob M, Klösgen RB.** 2013. Enough is enough: TatA demand during Tat-dependent protein transport. *Biochim Biophys Acta* 1833:957–965.

112. **Oates J, Barrett CML, Barnett JP, Byrne KG, Bolhuis A, Robinson C.** 2005. The *Escherichia coli* twin-arginine translocation apparatus incorporates a distinct form of TatABC complex, spectrum of modular TatA complexes and minor TatAB complex. *J Mol Biol* 346:295–305.

113. **White GF, Schermann SM, Bradley J, Roberts A, Greene NP, Berks BC, Thomson AJ.** 2010. Subunit organization in the TatA complex of the twin arginine protein translocase: a site-directed EPR spin labeling study. *J Biol Chem* 285:2294–2301.

114. **Hou B, Heidrich ES, Mehner-Breitfeld D, Brüser T.** 2018. The TatA component of the twin-arginine translocation system locally weakens the cytoplasmic membrane of *Escherichia coli* upon protein substrate binding. *J Biol Chem* 293:7592–7605.

115. **Sargent F, Berks BC, Palmer T.** 2006. Pathfinders and trailblazers: a prokaryotic targeting system for transport of folded proteins. *FEMS Microbiol Lett* 254:198–207.

116. **Gouffi K, Gérard F, Santini CL, Wu LF.** 2004. Dual topology of the *Escherichia coli* TatA protein. *J Biol Chem* 279:11608–11615.

117. **Walther TH, Gottselig C, Grage SL, Wolf M, Vargiu AV, Klein MJ, Vollmer S, Prock S, Hartmann M, Afonin S, Stockwald E, Heinzmann H, Nolandt OV, Wenzel W, Ruggerone P, Ulrich AS.** 2013. Folding and self-assembly of the TatA translocation pore based on a charge zipper mechanism. *Cell* 152:316–326.

118. **Dabney-Smith C, Mori H, Cline K.** 2006. Oligomers of Tha4 organize at the thylakoid Tat translocase during protein transport. *J Biol Chem* 281:5476–5483.

119. **Chan CS, Zlomislic MR, Tieleman DP, Turner RJ.** 2007. The TatA subunit of *Escherichia coli*

twin-arginine translocase has an N-in topology. *Biochemistry* **46:**7396–7404.

120. **Aldridge C, Storm A, Cline K, Dabney-Smith C.** 2012. The chloroplast twin arginine transport (Tat) component, Tha4, undergoes conformational changes leading to Tat protein transport. *J Biol Chem* **287:**34752–34763.

121. **Taubert J, Hou B, Risselada HJ, Mehner D, Lünsdorf H, Grubmüller H, Brüser T.** 2015. TatBC-independent TatA/Tat substrate interactions contribute to transport efficiency. *PLoS One* **10:**e0119761.

122. **Pal D, Fite K, Dabney-Smith C.** 2013. Direct interaction between a precursor mature domain and transport component Tha4 during twin arginine transport of chloroplasts. *Plant Physiol* **161:**990–1001.

123. **Taubert J, Brüser T.** 2014. Twin-arginine translocation-arresting protein regions contact TatA and TatB. *Biol Chem* **395:**827–836.

124. **Mould RM, Shackleton JB, Robinson C.** 1991. Transport of proteins into chloroplasts. Requirements for the efficient import of two lumenal oxygen-evolving complex proteins into isolated thylakoids. *J Biol Chem* **266:**17286–17289.

125. **Mori H, Cline K.** 2002. A twin arginine signal peptide and the pH gradient trigger reversible assembly of the thylakoid [Δ]pH/Tat translocase. *J Cell Biol* **157:**205–210.

126. **DeLisa MP, Lee P, Palmer T, Georgiou G.** 2004. Phage shock protein PspA of *Escherichia coli* relieves saturation of protein export via the Tat pathway. *J Bacteriol* **186:**366–373.

127. **Finazzi G, Chasen C, Wollman FA, de Vitry C.** 2003. Thylakoid targeting of Tat passenger proteins shows no delta pH dependence in vivo. *EMBO J* **22:**807–815.

128. **Braun NA, Davis AW, Theg SM.** 2007. The chloroplast Tat pathway utilizes the transmembrane electric potential as an energy source. *Biophys J* **93:**1993–1998.

129. **Musser SM, Theg SM.** 2000. Proton transfer limits protein translocation rate by the thylakoid DeltapH/Tat machinery. *Biochemistry* **39:**8228–8233.

130. **Alder NN, Theg SM.** 2003. Energetics of protein transport across biological membranes. A study of the thylakoid DeltapH-dependent/cpTat pathway. *Cell* **112:**231–242.

131. **Gérard F, Cline K.** 2007. The thylakoid proton gradient promotes an advanced stage of signal peptide binding deep within the Tat pathway receptor complex. *J Biol Chem* **282:**5263–5272.

132. **Bageshwar UK, Musser SM.** 2007. Two electrical potential-dependent steps are required for transport by the *Escherichia coli* Tat machinery. *J Cell Biol* **179:**87–99.

133. **Halbig D, Wiegert T, Blaudeck N, Freudl R, Sprenger GA.** 1999. The efficient export of NADP-containing glucose-fructose oxidoreductase to the periplasm of *Zymomonas mobilis* depends both on an intact twin-arginine motif in the signal peptide and on the generation of a structural export signal induced by cofactor binding. *Eur J Biochem* **263:**543–551.

134. **Matos CFRO, Robinson C, Di Cola A.** 2008. The Tat system proofreads FeS protein substrates and directly initiates the disposal of rejected molecules. *EMBO J* **27:**2055–2063.

135. **Goosens VJ, Monteferrante CG, van Dijl JM.** 2014. Co-factor insertion and disulfide bond requirements for twin-arginine translocase-dependent export of the *Bacillus subtilis* Rieske protein QcrA. *J Biol Chem* **289:**13124–13131.

136. **Sanders C, Wethkamp N, Lill H.** 2001. Transport of cytochrome c derivatives by the bacterial Tat protein translocation system. *Mol Microbiol* **41:**241–246.

137. **Sutherland GA, Grayson KJ, Adams NBP, Mermans DMJ, Jones AS, Robertson AJ, Auman DB, Brindley AA, Sterpone F, Tuffery P, Derreumaux P, Dutton PL, Robinson C, Hitchcock A, Hunter CN.** 2018. Probing the quality control mechanism of the *Escherichia coli* twin-arginine translocase with folding variants of a *de novo*-designed heme protein. *J Biol Chem* **293:**6672–6681.

138. **Barnett JP, van der Ploeg R, Eijlander RT, Nenninger A, Mendel S, Rozeboom R, Kuipers OP, van Dijl JM, Robinson C.** 2009. The twin-arginine translocation (Tat) systems from *Bacillus subtilis* display a conserved mode of complex organization and similar substrate recognition requirements. *FEBS J* **276:**232–243.

139. **Kolkman MA, van der Ploeg R, Bertels M, van Dijk M, van der Laan J, van Dijl JM, Ferrari E.** 2008. The twin-arginine signal peptide of *Bacillus subtilis* YwbN can direct either Tat- or Sec-dependent secretion of different cargo proteins: secretion of active subtilisin via the *B. subtilis* Tat pathway. *Appl Environ Microbiol* **74:**7507–7513.

140. **Austerberry JI, Dajani R, Panova S, Roberts D, Golovanov AP, Pluen A, van der Walle CF, Uddin S, Warwicker J, Derrick JP, Curtis R.** 2017. The effect of charge mutations on the stability and aggregation of a human single chain Fv fragment. *Eur J Pharm Biopharm* **115:**18–30.

141. **Jones AS, Austerberry JI, Dajani R, Warwicker J, Curtis R, Derrick JP, Robinson**

C. 2016. Proofreading of substrate structure by the twin-arginine translocase is highly dependent on substrate conformational flexibility but surprisingly tolerant of surface charge and hydrophobicity changes. *Biochim Biophys Acta* **1863**:3116–3124.

142. **Panahandeh S, Maurer C, Moser M, DeLisa MP, Müller M.** 2008. Following the path of a twin-arginine precursor along the TatABC translocase of *Escherichia coli*. *J Biol Chem* **283**:33267–33275.

143. **Stolle P, Hou B, Brüser T.** 2016. The Tat substrate CueO is transported in an incomplete folding state. *J Biol Chem* **291**:13520–13528.

144. **Cline K, McCaffery M.** 2007. Evidence for a dynamic and transient pathway through the TAT protein transport machinery. *EMBO J* **26**:3039–3049.

145. **Richter S, Lindenstrauss U, Lücke C, Bayliss R, Brüser T.** 2007. Functional Tat transport of unstructured, small, hydrophilic proteins. *J Biol Chem* **282**:33257–33264.

146. **Lindenstrauss U, Brüser T.** 2009. Tat transport of linker-containing proteins in *Escherichia coli*. *FEMS Microbiol Lett* **295**:135–140.

147. **Rocco MA, Waraho-Zhmayev D, DeLisa MP.** 2012. Twin-arginine translocase mutations that suppress folding quality control and permit export of misfolded substrate proteins. *Proc Natl Acad Sci U S A* **109**:13392–13397.

148. **Turner RJ, Papish AL, Sargent F.** 2004. Sequence analysis of bacterial redox enzyme maturation proteins (REMPs). *Can J Microbiol* **50**:225–238.

149. **Tranier S, Mortier-Barrière I, Ilbert M, Birck C, Iobbi-Nivol C, Méjean V, Samama JP.** 2002. Characterization and multiple molecular forms of TorD from *Shewanella massilia*, the putative chaperone of the molybdoenzyme TorA. *Protein Sci* **11**:2148–2157.

150. **Chan CS, Chang L, Rommens KL, Turner RJ.** 2009. Differential interactions between Tat-specific redox enzyme peptides and their chaperones. *J Bacteriol* **191**:2091–2101.

151. **Monteferrante CG, MacKichan C, Marchadier E, Prejean MV, Carballido-López R, van Dijl JM.** 2013. Mapping the twin-arginine protein translocation network of *Bacillus subtilis*. *Proteomics* **13**:800–811.

152. **Krishnappa L, Monteferrante CG, van Dijl JM.** 2012. Degradation of the twin-arginine translocation substrate YwbN by extracytoplasmic proteases of *Bacillus subtilis*. *Appl Environ Microbiol* **78**:7801–7804.

153. **Overton TW.** 2014. Recombinant protein production in bacterial hosts. *Drug Discov Today* **19**:590–601.

154. **Pooley HM, Merchante R, Karamata D.** 1996. Overall protein content and induced enzyme components of the periplasm of *Bacillus subtilis*. *Microb Drug Resist* **2**:9–15.

155. **van Dijl JM, Braun PG, Robinson C, Quax WJ, Antelmann H, Hecker M, Müller J, Tjalsma H, Bron S, Jongbloed JD.** 2002. Functional genomic analysis of the *Bacillus subtilis* Tat pathway for protein secretion. *J Biotechnol* **98**:243–254.

156. **Dröge MJ, Boersma YL, Braun PG, Buining RJ, Julsing MK, Selles KGA, van Dijl JM, Quax WJ.** 2006. Phage display of an intracellular carboxylesterase of *Bacillus subtilis*: comparison of Sec and Tat pathway export capabilities. *Appl Environ Microbiol* **72**:4589–4595.

157. **Goosens VJ, van Dijl JM.** 2017. Twin-arginine protein translocation. *Curr Top Microbiol Immunol* **404**:69–94.

158. **French C, Keshavarz-Moore E, Ward JM.** 1996. Development of a simple method for the recovery of recombinant proteins from the *Escherichia coli* periplasm. *Enzyme Microb Technol* **19**:332–338.

159. **Matos CFRO, Robinson C, Alanen HI, Prus P, Uchida Y, Ruddock LW, Freedman RB, Keshavarz-Moore E.** 2014. Efficient export of prefolded, disulfide-bonded recombinant proteins to the periplasm by the Tat pathway in *Escherichia coli* CyDisCo strains. *Biotechnol Prog* **30**:281–290.

160. **Browning DF, Richards KL, Peswani AR, Roobol J, Busby SJW, Robinson C.** 2017. *Escherichia coli* "TatExpress" strains super-secrete human growth hormone into the bacterial periplasm by the Tat pathway. *Biotechnol Bioeng* **114**:2828–2836.

161. **Albiniak AM, Matos CFRO, Branston SD, Freedman RB, Keshavarz-Moore E, Robinson C.** 2013. High-level secretion of a recombinant protein to the culture medium with a *Bacillus subtilis* twin-arginine translocation system in *Escherichia coli*. *FEBS J* **280**:3810–3821.

162. **Behrendt J, Lindenstrauss U, Brüser T.** 2007. The TatBC complex formation suppresses a modular TatB-multimerization in *Escherichia coli*. *FEBS Lett* **581**:4085–4090.

Lipoproteins and Their Trafficking to the Outer Membrane

6

MARCIN GRABOWICZ[1,2,3]

Lipoproteins are a family of secreted proteins that are acylated after their translocation across the plasma membrane (1–3). Acylation spatially confines lipoproteins by anchoring them into membranes. Lipoproteins are bioinformatically identifiable by the highly conserved lipobox motif in their short signal peptides (4). Within the lipobox is a cleavage site for signal peptidase II (SPII; Lsp). Immediately adjacent is an invariant Cys residue which is the target of acylation reactions. Most lipoproteins are secreted from the cytosol via the SecYEG translocon (5–7), though secretion via the twin-arginine transport (Tat) system has also been identified (8–11). Following translocation, the inner membrane (IM) enzyme Lgt attaches a diacyl moiety to the lipobox Cys of prolipoproteins via a thioester linkage (Fig. 1) (12–14). The diacylated product is a substrate for Lsp, which releases the apolipoprotein from its signal peptide (Fig. 1) (15–17). The diacylated Cys residue then becomes the first amino acid of the lipoprotein (Cys^{+1}). In Gram-negative bacteria, a third acyl group is attached by the enzyme Lnt to the Cys^{+1} amino group (which was made available following Lsp cleavage) (Fig. 1) (18–21). The acyl chain donors in Lgt and Lnt reactions are plasma membrane phospholipids (Fig. 1). Gram-

[1]Emory Antibiotic Resistance Center, Emory University School of Medicine, Atlanta, GA 30322
[2]Department of Microbiology & Immunology, Emory University School of Medicine, Atlanta, GA 30322
[3]Division of Infectious Diseases, Department of Medicine, Emory University School of Medicine, Atlanta, GA 30322

Protein Secretion in Bacteria
Edited by Maria Sandkvist, Eric Cascales, and Peter J. Christie
© 2019 American Society for Microbiology, Washington, DC
doi:10.1128/ecosalplus.ESP-0038-2018

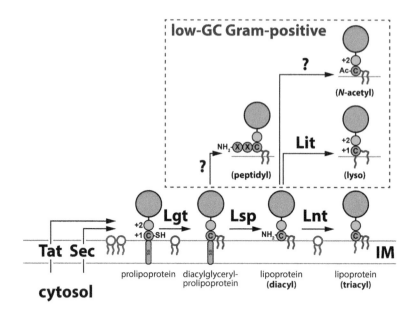

FIGURE 1 Posttranslocation lipoprotein maturation. Secreted lipoproteins are first diacylated at an invariant Cys residue by Lgt using resident phospholipids as acyl donors. The signal sequence is then cleaved by the peptidase Lsp to yield diacyl-form lipoproteins. In almost all Gram-negative bacteria, Lnt attaches another acyl chain to the amino group of Cys^{+1} to yield triacyl-form lipoproteins. Low-GC Gram-negative bacteria can also produce peptidyl forms (likely due to an Lsp-type enzyme that yields Cys^{+3}), as well as N-acetyl and lyso forms that are derived from diacyl lipoproteins. Triacyl- and lyso-lipoproteins can efficiently interact with LolCDE for trafficking to the OM in Gram-negative organisms. Diacyl-form lipoproteins can be trafficked to the OM via LolDF.

negative bacteria produce triacylated lipoproteins; *lnt*, *lsp*, and *lgt* are therefore conserved and essential in the majority of these organisms. Low-GC Gram-positive bacteria lack *lnt* homologs and generate considerable diversity in lipoprotein acylation; in addition to the triacyl form, these bacteria can variously generate diacyl, lyso, peptidyl, and N-acetyl lipoprotein forms (22, 23) (Fig. 1). How such diversity is generated largely awaits discovery, although recent progress has identified the enzyme, Lit, that is responsible for producing lyso-form lipoproteins in *Enterococcus faecalis* and *Bacillus cereus* (24).

In Gram-positive bacteria, lipoprotein maturation completes their biogenesis. In Gram-negative organisms, many lipoproteins await a new journey. These diderm bacteria traffic many of their lipoproteins from the plasma IM to the outer membrane (OM) (3, 25). The model organism *Escherichia coli* targets almost

90% of the lipoprotein species it produces to the OM (26). In fact, the very first lipoprotein identified was Lpp, a highly abundant *E. coli* OM lipoprotein (27). Lpp forms a covalent C-terminal attachment to the cell wall peptidoglycan (PG) and functions as an architectural element in the cell envelope that ensures accurate spacing between the OM and cell wall (28–30). Lpp that is mislocalized to the IM also forms PG cross-links, but these are lethally toxic for *E. coli* (31). Hence, in order to avoid such toxicity, trafficking of lipoproteins to the OM must be highly efficient. Lipoproteins face daunting hurdles to reaching the OM: the highly hydrophobic acyl moieties must leave a favorable IM lipid bilayer, cross an adverse aqueous periplasmic environment, and then be inserted into the OM bilayer. Some lipoproteins are subsequently translocated from the periplasm and across the OM to become surface exposed.

THE Lol PATHWAY

The major trafficking route that brings lipoproteins from the IM to the OM was discovered entirely by the lab of Hajime Tokuda, who named this pathway Lol (localization of lipoproteins) (Fig. 2) (1, 32). The Lol pathway has components in each compartment of the cell envelope: LolCDE, an ATP-binding cassette (ABC) transporter in the IM; LolA, a soluble chaperone protein in the periplasm; and LolB, itself a lipoprotein, at the OM. In model organisms with a full LolABCDE pathway, mature lipoproteins are extracted from the IM by LolCDE, transferred to LolA, and shuttled to the OM, where LolB receives and then anchors them into the bilayer. Each of the Lol proteins was found to be essential for viability of wild-type *E. coli* (33–35). However, recent work has unexpectedly discovered conditions under which both *lolA*

and *lolB* can be deleted in *E. coli*, revealing that at least one other unknown trafficking route can deliver lipoproteins to the OM (discussed below) (36). Additionally, phylogenetic analysis suggests that several Gram-negative genera do not encode a *lolB* homolog. The emergence of LolAB-independent trafficking has prompted a reassessment of our understanding of how lipoproteins reach the OM.

GETTING TO THE OM

Amino Acid Targeting Signals

Given that all lipoproteins mature in the IM, the Gram-negative cell must first decide which of these will be targeted for trafficking to the OM and which will remain in the IM. Gene fusion experiments proved that targeting

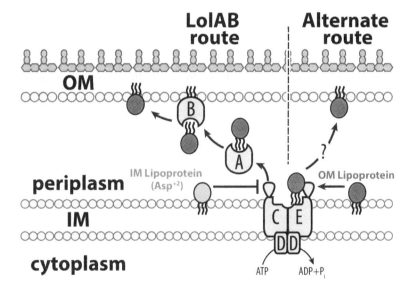

FIGURE 2 Trafficking routes for OM lipoproteins. The OM lipoprotein trafficking routes of *E. coli* are shown. Once mature, OM-targeted lipoproteins engage with the LolCDE transporter in the IM. LolE interacts with lipoproteins and LolC recruits the periplasmic chaperone protein LolA. At the expense of ATP hydrolysis by LolD, the LolCDE complex extracts lipoproteins from the IM bilayer and transfers them to LolA. Lipoproteins are shuttled through the periplasm in a LolA-bound complex. At the OM, the lipoprotein LolB receives LolA-bound client lipoproteins and anchors them into the OM bilayer. Since ΔlolAB mutants are viable, an alternate trafficking route must exist that can traffic essential OM lipoproteins to support cell viability. LolCDE remains essential in such ΔlolAB mutants, suggesting that lipoproteins originate from this complex and are then trafficked to the OM via an unknown mechanism.

signals were encoded within lipoproteins themselves; heterologous proteins could be targeted to the OM by fusing them to N-terminal sequences of OM lipoproteins (37). Studies of *E. coli* soon revealed an elegantly simple OM targeting signal. In these bacteria, the identity of the second amino acid (adjacent to the lipidated Cys^{+1}) determines localization. The presence of Asp^{+2} retains lipoproteins in the IM (37–39). Most other residues result in OM localization (39). This targeting mechanism became known as the "+2 rule." There is compelling biochemical evidence that IM retention is caused by strong electrostatic interactions between Asp^{+2} and the head-groups of anionic IM phospholipids (40). These interactions likely prevent Asp^{+2} lipoproteins from interacting with the LolCDE transporter (41). Hence, the Asp^{+2} targeting mechanism has been coined "Lol avoidance." Alternate +2 residues can also cause IM retention of *E. coli* lipoproteins, including Trp, Phe, Pro, Gly, and Tyr (42, 43), though none of these alternate retention signals are found in native *E. coli* lipoproteins. Given their chemistry, it is clear that the alternate retention signals do not cause IM retention via the same electrostatic mechanism as Asp^{+2}. Most likely, alternate signals cause aberrant or inefficient interactions between lipoproteins and the LolCDE complex (44). *Pseudomonas aeruginosa* employs more complex retention signals that also involve the +3 and +4 residues (45–47). How these signals function is unclear; Asp^{+2} does function in this organism as a potent retention signal, though it is found infrequently among pseudomonal lipoproteins (45–47). More distantly related organisms employ altogether different targeting signals. For example, *Borrelia burgdorferi* relies on an acidic N-terminal linker region to retain lipoproteins in the IM (48, 49). It is remarkable that highly disparate lipoprotein targeting strategies appear to have evolved. Presumably, each strategy reflects the different cell envelope compositions and structures of diverse organisms. It is possible that the common goal of all these strategies is "Lol avoidance"—preventing lipoproteins that should be retained in the IM from interacting with the Lol ABC transporter—although this hypothesis awaits clear confirmation in many organisms.

Lipoprotein Acylation

The acylation state of lipoproteins is important for trafficking to the OM. For example, in *Yersinia pestis*, lack of modification by Lnt is proposed to act as a retention signal that prevents lipoprotein release from the IM (50). This proposal is based on the finding that the LolCDE complex in *E. coli* has very low affinity for diacyl-form lipoproteins. In wild-type *E. coli*, Lnt is an essential protein, but *lnt* can be deleted if LolCDE is highly overexpressed (51). Recent work suggests that Lnt N-acylation of Cys^{+1} is perhaps the key determinant for lipoprotein interaction with LolCDE. Armbruster and Meredith were able to complement the lethal loss of Lnt in *E. coli* by expressing a transacylase, Lit, from the low-GC Gram-positive organisms *Enterococcus faecalis* and *Bacillus cereus* (24). Lit removes one of the two thioester-linked acyl chains generated by the Lgt modification and attaches it to the Cys^{+1} amino group (Fig. 1) (24). The complementation of Lnt with Lit was successful without requiring LolCDE overproduction, and OM lipoprotein trafficking remained efficient enough to support viability (24). The requirement of LolCDE for lipoprotein N-acylation—the final maturation step—seems to serve as a secretion checkpoint mechanism that avoids premature trafficking of earlier maturation intermediates. Yet some Gram-negative species do not produce a LolCDE complex; rather than having a heterodimer of the IM LolC and LolE proteins, these organisms produce a homodimer of LolF proteins (52). LolF appears to be a hybrid of LolC and LolE proteins, containing key motifs from both (52). Bacteria that produce an IM complex of LolDF are suggested to not require Lnt for viability (52). Indeed, this was directly demonstrated for

Francisella tularensis, Neisseria meningitidis, Acinetobacter baylyi, and *Acinetobacter baumannii,* from which *lnt* has successfully been deleted (52, 53). However, it should be noted that each of these organisms natively produces triacyl-form lipoproteins, at least when grown under laboratory conditions (52, 53). An intriguing possibility is that in some bacteria *N*-acylation may be a regulatable process that is linked to pathogenesis (52). In any case, it appears that LolDF complexes recognize triacylated lipoproteins but do not require *N*-acylation for lipoprotein trafficking.

DEPARTING THE IM VIA LolCDE

In the *E. coli* LolCDE complex, lipoprotein clients seem to interact primarily with LolE (54). This conclusion is based on site-specific photocross-linking at LolE, which can capture abundant clients *in vivo* (54). Meanwhile, LolC does not seem to interact with clients, despite having a hydrophobic cavity similar to that of LolE (54). The key role for LolC appears to be in recruiting LolA to the IM complex (54–57). LolC and LolE each contain one large periplasmic domain that is homologous between the proteins. LolA can be captured only at the LolC loop (54). Indeed, the specific LolA interaction with the LolC periplasmic domain was recently confirmed in a cocrystal structure and by biochemical methods (57). So LolC and LolE contribute to different functions in the early trafficking step: LolE recruits incoming lipoprotein clients, and LolC recruits the chaperone. In LolDF complexes, each of the LolF monomers in the homodimeric complexes must perform both of these functions. Arguably, segregating these functions between LolC and LolE may increase trafficking efficiency by generating a unidirectional flow of clients through the complex.

LolD is the ATPase that powers the LolCDE transporter (34, 58, 59). Recruitment of LolA does not require ATP binding or hydrolysis. Likely, the energy released by

ATP hydrolysis is needed for the unfavorable step of extracting the acyl chains from the IM bilayer. However, current *in vitro* evidence suggests that the initial step of ATP binding alters the LolCDE-client complex in a way that makes the lipoprotein removable with detergent (58). This finding implies that lipoproteins are extracted upon ATP binding by LolD. In this case, the hydrolysis step should be important for the subsequent release reaction that transfers lipoproteins to LolA, for resetting the LolCDE complex, or for both activities.

TRAFFICKING TO THE OM VIA LolAB

LolA adopts an incomplete β-barrel structure with an enclosed hydrophobic cavity (60). LolA is recruited to LolC via a recently identified hook-and-pad interaction (57). The hook is a solvent exposed β-hairpin loop extending from the LolC periplasmic domain (57). The pad consists of three residues in the LolC periplasmic domain to which LolA binds (57). Interactions with both regions of LolC are involved in recruiting LolA (57). The periplasmic domains of LolC and LolA share sequence homology, and lipoprotein transfer is suggested to occur by the hydrophobic cavities lining up in a mouth-to-mouth orientation (56). The function of LolA must be to shield the acyl chains of its client from the aqueous periplasm. Yet *E. coli* LolA structures suggest that its hydrophobic cavity might not accommodate all three acyl chains of client lipoproteins (60). An alternate proposal for lipoprotein binding suggests that some acyl chains might bind hydrophobic patches on the surface of LolA (61).

At the OM, LolB receives lipoproteins from LolA and completes the trafficking route in *E. coli* by anchoring the lipoprotein into the OM bilayer (62). LolB is structurally similar to LolA (60). Surprisingly, LolB acylation is not required for its anchoring activity. A freely soluble, periplasmic LolB (termed mLolB; generated by replacing the native lipobox-containing signal sequence) remains able to receive lipoproteins from LolA and anchor them to membranes (63). However, mLolB perceives the periplasmic phospholipid headgroups of the IM and OM as equivalent, and it inserts lipoproteins into both the IM and the OM (63). Misinserted lipoproteins reenter the LolCDE transporter and try once more to reach the OM. Such a trafficking pathway is clearly inefficient; however, mLolB can complement inactivation of the native *lolB* gene if Lpp is either deleted or prevented from forming lethal PG cross-links from the IM (63). By anchoring LolB into the OM as a lipoprotein, the Lol pathway ensures accurate and unidirectional trafficking. A loop of LolB is important for the anchoring reaction, though the mechanism of anchoring remains unknown (64). Curiously, many Gram-negative organisms natively lack any *lolB* homolog (3). How such bacteria complete the trafficking pathway is an outstanding question. However, evidence from the artificial mLolB system suggests that an OM-localized lipoprotein membrane transferase might not be a strict requirement for trafficking.

TESTING THE ESSENTIALITY OF LolA AND LolB

Soon after they were discovered, LolA and LolB were identified as essential proteins for *E. coli*. Neither *lolA* nor *lolB* could be deleted, and depleting levels of either protein caused a decrease in cell viability (35, 62). LolCDE were discovered later and likewise determined to be essential (34). The finding that the Lol pathway is essential may have been initially puzzling—at the time, the only known essential OM lipoprotein was LolB itself. LolB depletion studies with *E. coli* revealed that lipoproteins mislocalize to the IM and also accumulate in the periplasm, complexed with LolA (33). The lethality of mislocalized Lpp was already known, but this did not explain Lol pathway essentiality since

lolA, *lolB*, and *lolCDE* remained essential even when *lpp* was deleted (33, 35, 65). Hence, the reasonable conclusion was made that Lol proteins were essential because mislocalization of some lipoproteins may be toxic or may severely perturb the cell envelope (33). Essential OM lipoprotein clients were discovered in subsequent years and seemed to rationalize Lol protein essentiality. BamD and LptE are essential components of the OM assembly machinery that fold β-barrel OM proteins (the Bam machine) and transport lipopolysaccharide (the Lpt system), respectively (66–69). Accessory BamBCE OM lipoproteins and cell wall synthesis-regulating LpoAB OM lipoproteins are also collectively essential in *E. coli* to build a robust cell envelope, and combination mutants are lethal (70–74). It seemed that the Lol proteins were essential because they needed to deliver critical Bam, Lpt, and Lpo lipoproteins to the OM.

ALTERNATE TRAFFICKING ROUTE(S) FOR OM LIPOPROTEINS

Recently, the underlying reasons for LolA and LolB essentiality were directly tested (36). While conditions permitting deletion of both *lolA* and *lolB* were identified, *lolCDE* could not be deleted (36). Thus, the LolCDE complex is fundamentally required for all routes of lipoprotein trafficking. These findings revealed that LolAB are not truly essential for trafficking (36). Rather, their essential function in wild-type cells is to provide an efficient trafficking route that mitigates toxicities caused by OM lipoproteins mislocalizing to the IM. When the LolAB route was depleted, two OM-targeted lipoproteins were found kill the cell, most likely by accumulating in the IM: Lpp, which (as discussed above) forms toxic PG cross-links from the IM, and OsmB, which may form pores across the IM that dissipate the proton motive force, killing the cell (36). There must be at least one other alternate trafficking route that can perform the essential task of bringing lipoproteins from LolCDE, through the periplasm, and into the OM in Δ*lolAB* mutants. Indeed, the Bam lipoproteins were directly shown to reach the OM even when LolAB were absent (36). Remarkably, *Helicobacter pylori*, which lacks a *lolB* homolog, also appears to tolerate inactivation of *lolA* (75). Therefore, it is tempting to speculate that the same alternate trafficking route that functions in *E. coli* Δ*lolAB* cells is required to support essential OM lipoprotein trafficking in *H. pylori* Δ*lolA* mutants. If this is true, the alternate trafficking route may even be ancestral to the LolAB route. LolAB may have emerged to provide increased trafficking efficiency and capacity, a requirement for evolving OM lipoproteins (such as Lpp and OsmB) whose activity at the OM is beneficial but whose accumulation in the IM is potently toxic. Indeed, Lpp and OsmB are narrowly conserved to a subset of Gram-negative bacteria that possess both *lolA* and *lolB*.

SUMMARY

Efforts in recent years have yielded considerable insights into the maturation of lipoproteins and, in Gram-negative bacteria, their trafficking towards the OM. New questions have emerged: how is acylation diversity achieved in Gram-positive bacteria, and why is the triacyl form the apparent default among Gram-negative organisms? How are lipoproteins trafficked when the known LolAB route is inactivated? How does lipoprotein trafficking occur in organisms lacking *lolB*? Moreover, what are the molecular mechanisms that underlie the highly efficient trafficking via LolAB? As answers to these questions are found, we will be rewarded with an increasingly sophisticated and comprehensive understanding of lipoprotein biogenesis and trafficking. Given that the OM is a major barrier against antibiotics and that OM lipoproteins are essential for OM assembly, insights into trafficking may prove invaluable to the goal of developing new drugs to treat increasingly antibiotic-resistant Gram-negative infections.

ACKNOWLEDGMENTS

I thank Kerrie May and the anonymous reviewers for comments that have improved this manuscript.

This work was supported by institutional startup funding from Emory University.

CITATION

EcoSal Plus 2019; doi:10.1128/ecosalplus. ESP-0038-2018.

REFERENCES

1. **Narita S-I, Tokuda H.** 2017. Bacterial lipoproteins; biogenesis, sorting and quality control. *Biochim Biophys Acta Mol Cell Biol Lipids* **1862:**1414–1423.
2. **Buddelmeijer N.** 2015. The molecular mechanism of bacterial lipoprotein modification—how, when and why? *FEMS Microbiol Rev* **39:**246–261.
3. **Grabowicz M.** 2018. Lipoprotein transport: greasing the machines of outer membrane biogenesis. *Bioessays* **40:**e1700187.
4. **Babu MM, Priya ML, Selvan AT, Madera M, Gough J, Aravind L, Sankaran K.** 2006. A database of bacterial lipoproteins (DOLOP) with functional assignments to predicted lipoproteins. *J Bacteriol* **188:**2761–2773.
5. **Hayashi S, Wu HC.** 1985. Accumulation of prolipoprotein in *Escherichia coli* mutants defective in protein secretion. *J Bacteriol* **161:**949–954.
6. **Sugai M, Wu HC.** 1992. Export of the outer membrane lipoprotein is defective in *secD, secE,* and *secF* mutants of *Escherichia coli. J Bacteriol* **174:**2511–2516.
7. **Fröderberg L, Houben ENG, Baars L, Luirink J, de Gier J-W.** 2004. Targeting and translocation of two lipoproteins in *Escherichia coli* via the SRP/Sec/YidC pathway. *J Biol Chem* **279:**31026–31032.
8. **Thompson BJ, Widdick DA, Hicks MG, Chandra G, Sutcliffe IC, Palmer T, Hutchings MI.** 2010. Investigating lipoprotein biogenesis and function in the model Gram-positive bacte-

rium *Streptomyces coelicolor. Mol Microbiol* **77**:943–957.

9. **Widdick DA, Hicks MG, Thompson BJ, Tschumi A, Chandra G, Sutcliffe IC, Brülle JK, Sander P, Palmer T, Hutchings MI.** 2011. Dissecting the complete lipoprotein biogenesis pathway in *Streptomyces scabies. Mol Microbiol* **80**:1395–1412.

10. **Shruthi H, Anand P, Murugan V, Sankaran K.** 2010. Twin arginine translocase pathway and fast-folding lipoprotein biosynthesis in *E. coli*: interesting implications and applications. *Mol Biosyst* **6**:999–1007.

11. **Randall LB, Dobos K, Papp-Wallace KM, Bonomo RA, Schweizer HP.** 2015. Membrane-bound PenA β-lactamase of *Burkholderia pseudomallei. Antimicrob Agents Chemother* **60**:1509–1514.

12. **Tokunaga M, Tokunaga H, Wu HC.** 1982. Post-translational modification and processing of *Escherichia coli* prolipoprotein *in vitro. Proc Natl Acad Sci U S A* **79**:2255–2259.

13. **Sankaran K, Wu HC.** 1994. Lipid modification of bacterial prolipoprotein. Transfer of diacylglyceryl moiety from phosphatidylglycerol. *J Biol Chem* **269**:19701–19706.

14. **Mao G, Zhao Y, Kang X, Li Z, Zhang Y, Wang X, Sun F, Sankaran K, Zhang XC.** 2016. Crystal structure of *E. coli* lipoprotein diacylglyceryl transferase. *Nat Commun* **7**:10198.

15. **Vogeley L, El Arnaout T, Bailey J, Stansfeld PJ, Boland C, Caffrey M.** 2016. Structural basis of lipoprotein signal peptidase II action and inhibition by the antibiotic globomycin. *Science* **351**:876–880.

16. **Tokunaga M, Loranger JM, Wu HC.** 1984. Prolipoprotein modification and processing enzymes in *Escherichia coli. J Biol Chem* **259**:3825–3830.

17. **Inouye S, Franceschini T, Sato M, Itakura K, Inouye M.** 1983. Prolipoprotein signal peptidase of *Escherichia coli* requires a cysteine residue at the cleavage site. *EMBO J* **2**:87–91.

18. **Gupta SD, Gan K, Schmid MB, Wu HC.** 1993. Characterization of a temperature-sensitive mutant of *Salmonella typhimurium* defective in apolipoprotein N-acyltransferase. *J Biol Chem* **268**:16551–16556.

19. **Noland CL, Kattke MD, Diao J, Gloor SL, Pantua H, Reichelt M, Katakam AK, Yan D, Kang J, Zilberleyb I, Xu M, Kapadia SB, Murray JM.** 2017. Structural insights into lipoprotein N-acylation by *Escherichia coli* apolipoprotein N-acyltransferase. *Proc Natl Acad Sci U S A* **114**:E6044–E6053.

20. **Wiktor M, Weichert D, Howe N, Huang C-Y, Olieric V, Boland C, Bailey J, Vogeley L, Stansfeld PJ, Buddelmeijer N, Wang M, Caffrey M.** 2017. Structural insights into the mechanism of the membrane integral N-acyltransferase step in bacterial lipoprotein synthesis. *Nat Commun* **8**:15952.

21. **Vidal-Ingigliardi D, Lewenza S, Buddelmeijer N.** 2007. Identification of essential residues in apolipoprotein N-acyl transferase, a member of the CN hydrolase family. *J Bacteriol* **189**:4456–4464.

22. **Nakayama H, Kurokawa K, Lee BL.** 2012. Lipoproteins in bacteria: structures and bio-synthetic pathways. *FEBS J* **279**:4247–4268.

23. **Kurokawa K, Ryu K-H, Ichikawa R, Masuda A, Kim M-S, Lee H, Chae J-H, Shimizu T, Saitoh T, Kuwano K, Akira S, Dohmae N, Nakayama H, Lee BL.** 2012. Novel bacterial lipoprotein structures conserved in low-GC content gram-positive bacteria are recognized by Toll-like receptor 2. *J Biol Chem* **287**:13170–13181.

24. **Armbruster KM, Meredith TC.** 2017. Identification of the lyso-form N-acyl intramolecular transferase in low-GC Firmicutes. *J Bacteriol* **199**:e00099-17.

25. **Konovalova A, Silhavy TJ.** 2015. Outer membrane lipoprotein biogenesis: Lol is not the end. *Philos Trans R Soc Lond B Biol Sci* **370**:20150030.

26. **Horler RSP, Butcher A, Papangelopoulos N, Ashton PD, Thomas GH.** 2009. EchoLOCATION: an *in silico* analysis of the subcellular locations of *Escherichia coli* proteins and comparison with experimentally derived locations. *Bioinformatics* **25**:163–166.

27. **Hantke K, Braun V.** 1973. Covalent binding of lipid to protein. Diglyceride and amide-linked fatty acid at the N-terminal end of the murein-lipoprotein of the *Escherichia coli* outer membrane. *Eur J Biochem* **34**:284–296.

28. **Braun V, Rehn K.** 1969. Chemical characterization, spatial distribution and function of a lipoprotein (murein-lipoprotein) of the *E. coli* cell wall. The specific effect of trypsin on the membrane structure. *Eur J Biochem* **10**:426–438.

29. **Asmar AT, Ferreira JL, Cohen EJ, Cho S-H, Beeby M, Hughes KT, Collet J-F.** 2017. Communication across the bacterial cell envelope depends on the size of the periplasm. *PLoS Biol* **15**:e2004303.

30. **Cohen EJ, Ferreira JL, Ladinsky MS, Beeby M, Hughes KT.** 2017. Nanoscale-length control of the flagellar driveshaft requires hitting the tethered outer membrane. *Science* **356**:197–200.

31. **Yakushi T, Tajima T, Matsuyama S, Tokuda H.** 1997. Lethality of the covalent linkage between mislocalized major outer membrane

lipoprotein and the peptidoglycan of *Escherichia coli*. *J Bacteriol* **179:**2857–2862.

32. **Okuda S, Tokuda H.** 2011. Lipoprotein sorting in bacteria. *Annu Rev Microbiol* **65:**239–259.

33. **Tanaka K, Matsuyama S-I, Tokuda H.** 2001. Deletion of *lolB*, encoding an outer membrane lipoprotein, is lethal for *Escherichia coli* and causes accumulation of lipoprotein localization intermediates in the periplasm. *J Bacteriol* **183:**6538–6542.

34. **Yakushi T, Masuda K, Narita S, Matsuyama S, Tokuda H.** 2000. A new ABC transporter mediating the detachment of lipid-modified proteins from membranes. *Nat Cell Biol* **2:**212–218.

35. **Tajima T, Yokota N, Matsuyama S, Tokuda H.** 1998. Genetic analyses of the *in vivo* function of LolA, a periplasmic chaperone involved in the outer membrane localization of *Escherichia coli* lipoproteins. *FEBS Lett* **439:**51–54.

36. **Grabowicz M, Silhavy TJ.** 2017. Redefining the essential trafficking pathway for outer membrane lipoproteins. *Proc Natl Acad Sci U S A* **114:**4769–4774.

37. **Yamaguchi K, Yu F, Inouye M.** 1988. A single amino acid determinant of the membrane localization of lipoproteins in *E. coli*. *Cell* **53:**423–432.

38. **Gennity JM, Inouye M.** 1991. The protein sequence responsible for lipoprotein membrane localization in *Escherichia coli* exhibits remarkable specificity. *J Biol Chem* **266:**16458–16464.

39. **Terada M, Kuroda T, Matsuyama S-I, Tokuda H.** 2001. Lipoprotein sorting signals evaluated as the LolA-dependent release of lipoproteins from the cytoplasmic membrane of *Escherichia coli*. *J Biol Chem* **276:**47690–47694.

40. **Hara T, Matsuyama S, Tokuda H.** 2003. Mechanism underlying the inner membrane retention of *Escherichia coli* lipoproteins caused by Lol avoidance signals. *J Biol Chem* **278:**40408–40414.

41. **Masuda K, Matsuyama S, Tokuda H.** 2002. Elucidation of the function of lipoprotein-sorting signals that determine membrane localization. *Proc Natl Acad Sci U S A* **99:**7390–7395.

42. **Seydel A, Gounon P, Pugsley AP.** 1999. Testing the '+2 rule' for lipoprotein sorting in the *Escherichia coli* cell envelope with a new genetic selection. *Mol Microbiol* **34:**810–821.

43. **Lewenza S, Vidal-Ingigliardi D, Pugsley AP.** 2006. Direct visualization of red fluorescent lipoproteins indicates conservation of the membrane sorting rules in the family *Enterobacteriaceae*. *J Bacteriol* **188:**3516–3524.

44. **Sakamoto C, Satou R, Tokuda H, Narita S.** 2010. Novel mutations of the LolCDE complex causing outer membrane localization of lipo-proteins despite their inner membrane-retention signals. *Biochem Biophys Res Commun* **401:**586–591.

45. **Lewenza S, Mhlanga MM, Pugsley AP.** 2008. Novel inner membrane retention signals in *Pseudomonas aeruginosa* lipoproteins. *J Bacteriol* **190:**6119–6125.

46. **Narita S, Tokuda H.** 2007. Amino acids at positions 3 and 4 determine the membrane specificity of *Pseudomonas aeruginosa* lipoproteins. *J Biol Chem* **282:**13372–13378.

47. **Tanaka S-Y, Narita S, Tokuda H.** 2007. Characterization of the *Pseudomonas aeruginosa* Lol system as a lipoprotein sorting mechanism. *J Biol Chem* **282:**13379–13384.

48. **Schulze RJ, Zückert WR.** 2006. *Borrelia burgdorferi* lipoproteins are secreted to the outer surface by default. *Mol Microbiol* **59:**1473–1484.

49. **Kumru OS, Schulze RJ, Rodnin MV, Ladokhin AS, Zückert WR.** 2011. Surface localization determinants of *Borrelia* OspC/Vsp family lipoproteins. *J Bacteriol* **193:**2814–2825.

50. **Silva-Herzog E, Ferracci F, Jackson MW, Joseph SS, Plano GV.** 2008. Membrane localization and topology of the *Yersinia pestis* YscJ lipoprotein. *Microbiology* **154:**593–607.

51. **Narita S, Tokuda H.** 2011. Overexpression of LolCDE allows deletion of the *Escherichia coli* gene encoding apolipoprotein N-acyltransferase. *J Bacteriol* **193:**4832–4840.

52. **LoVullo ED, Wright LF, Isabella V, Huntley JF, Pavelka MS Jr.** 2015. Revisiting the Gram-negative lipoprotein paradigm. *J Bacteriol* **197:**1705–1715.

53. **Gwin CM, Prakash N, Christian Belisario J, Haider L, Rosen ML, Martinez LR, Rigel NW.** 2018. The apolipoprotein N-acyl transferase Lnt is dispensable for growth in *Acinetobacter* species. *Microbiology* **164:**1547–1556.

54. **Mizutani M, Mukaiyama K, Xiao J, Mori M, Satou R, Narita S, Okuda S, Tokuda H.** 2013. Functional differentiation of structurally similar membrane subunits of the ABC transporter LolCDE complex. *FEBS Lett* **587:**23–29.

55. **Okuda S, Watanabe S, Tokuda H.** 2008. A short helix in the C-terminal region of LolA is important for the specific membrane localization of lipoproteins. *FEBS Lett* **582:**2247–2251.

56. **Okuda S, Tokuda H.** 2009. Model of mouth-to-mouth transfer of bacterial lipoproteins through inner membrane LolC, periplasmic LolA, and outer membrane LolB. *Proc Natl Acad Sci U S A* **106:**5877–5882.

57. **Kaplan E, Greene NP, Crow A, Koronakis V.** 2018. Insights into bacterial lipoprotein trafficking from a structure of LolA bound to the

LolC periplasmic domain. *Proc Natl Acad Sci U S A* **115:**E7389–E7397.

58. **Ito Y, Kanamaru K, Taniguchi N, Miyamoto S, Tokuda H.** 2006. A novel ligand bound ABC transporter, LolCDE, provides insights into the molecular mechanisms underlying membrane detachment of bacterial lipoproteins. *Mol Microbiol* **62:**1064–1075.

59. **Taniguchi N, Tokuda H.** 2008. Molecular events involved in a single cycle of ligand transfer from an ATP binding cassette transporter, LolCDE, to a molecular chaperone, LolA. *J Biol Chem* **283:**8538–8544.

60. **Takeda K, Miyatake H, Yokota N, Matsuyama S, Tokuda H, Miki K.** 2003. Crystal structures of bacterial lipoprotein localization factors, LolA and LolB. *EMBO J* **22:**3199–3209.

61. **Remans K, Pauwels K, van Ulsen P, Buts L, Cornelis P, Tommassen J, Savvides SN, Decanniere K, Van Gelder P.** 2010. Hydrophobic surface patches on LolA of *Pseudomonas aeruginosa* are essential for lipoprotein binding. *J Mol Biol* **401:**921–930.

62. **Matsuyama S, Yokota N, Tokuda H.** 1997. A novel outer membrane lipoprotein, LolB (HemM), involved in the LolA (p20)-dependent localization of lipoproteins to the outer membrane of *Escherichia coli. EMBO J* **16:**6947–6955.

63. **Tsukahara J, Mukaiyama K, Okuda S, Narita S, Tokuda H.** 2009. Dissection of LolB function—lipoprotein binding, membrane targeting and incorporation of lipoproteins into lipid bilayers. *FEBS J* **276:**4496–4504.

64. **Hayashi Y, Tsurumizu R, Tsukahara J, Takeda K, Narita S, Mori M, Miki K, Tokuda H.** 2014. Roles of the protruding loop of factor B essential for the localization of lipoproteins (LolB) in the anchoring of bacterial triacylated proteins to the outer membrane. *J Biol Chem* **289:**10530–10539.

65. **Narita S, Tanaka K, Matsuyama S, Tokuda H.** 2002. Disruption of *lolCDE*, encoding an ATP-binding cassette transporter, is lethal for *Escherichia coli* and prevents release of lipoproteins from the inner membrane. *J Bacteriol* **184:**1417–1422.

66. **Malinverni JC, Werner J, Kim S, Sklar JG, Kahne D, Misra R, Silhavy TJ.** 2006. YfiO stabilizes the YaeT complex and is essential for outer membrane protein assembly in *Escherichia coli. Mol Microbiol* **61:**151–164.

67. **Wu T, McCandlish AC, Gronenberg LS, Chng S-S, Silhavy TJ, Kahne D.** 2006. Identification of a protein complex that assembles lipopolysaccharide in the outer membrane of *Escherichia coli. Proc Natl Acad Sci U S A* **103:**11754–11759.

68. **Konovalova A, Kahne DE, Silhavy TJ.** 2017. Outer membrane biogenesis. *Annu Rev Microbiol* **71:**539–556.

69. **Okuda S, Sherman DJ, Silhavy TJ, Ruiz N, Kahne D.** 2016. Lipopolysaccharide transport and assembly at the outer membrane: the PEZ model. *Nat Rev Microbiol* **14:**337–345.

70. **Lupoli TJ, Lebar MD, Markovski M, Bernhardt T, Kahne D, Walker S.** 2014. Lipoprotein activators stimulate *Escherichia coli* penicillin-binding proteins by different mechanisms. *J Am Chem Soc* **136:**52–55.

71. **Paradis-Bleau C, Markovski M, Uehara T, Lupoli TJ, Walker S, Kahne DE, Bernhardt TG.** 2010. Lipoprotein cofactors located in the outer membrane activate bacterial cell wall polymerases. *Cell* **143:**1110–1120.

72. **Typas A, Banzhaf M, van den Berg van Saparoea B, Verheul J, Biboy J, Nichols RJ, Zietek M, Beilharz K, Kannenberg K, von Rechenberg M, Breukink E, den Blaauwen T, Gross CA, Vollmer W.** 2010. Regulation of peptidoglycan synthesis by outer-membrane proteins. *Cell* **143:**1097–1109.

73. **Misra R, Stikeleather R, Gabriele R.** 2015. In vivo roles of BamA, BamB and BamD in the biogenesis of BamA, a core protein of the β-barrel assembly machine of *Escherichia coli. J Mol Biol* **427:**1061–1074.

74. **Rigel NW, Schwalm J, Ricci DP, Silhavy TJ.** 2012. BamE modulates the *Escherichia coli* beta-barrel assembly machine component BamA. *J Bacteriol* **194:**1002–1008.

75. **Chalker AF, Minehart HW, Hughes NJ, Koretke KK, Lonetto MA, Brinkman KK, Warren PV, Lupas A, Stanhope MJ, Brown JR, Hoffman PS.** 2001. Systematic identification of selective essential genes in *Helicobacter pylori* by genome prioritization and allelic replacement mutagenesis. *J Bacteriol* **183:**1259–1268.

7

Protein Secretion in Spirochetes

WOLFRAM R. ZÜCKERT[1]

INTRODUCTION

Spirochetes form a distinct bacterial phylum of slender, diderm (dual-membrane) bacteria that exhibit either a coiled "corkscrew" or flat-wave "serpentine" morphology. These distinct phenotypes are at least partly due to various numbers of periplasmic flagella that are inserted subterminally at both poles of the bacteria, wrapping around the protoplasmic cylinder and often overlapping in the middle of the cell. Coordinated rotation of the flagellar bands or bundles, sometimes referred to as axial filaments, leads to rotation of the cell cylinder and cellular motility that is particularly prominent in viscous environments.

Spirochetes have evolved to occupy a variety of niches, from living freely in soil and marine sediments to becoming fully host adapted as commensals in the guts of termites and other arthropods, or as often pathogenic obligate parasites of vertebrates. Three spirochetal genera, *Borrelia*, *Leptospira*, and *Treponema*, have garnered significant attention due to their medical importance and have emerged as unique spirochetal model systems in studies of microbial physiology and pathogenesis. *Borrelia burgdorferi* (recently reclassified as *Borreliella*

[1]Department of Microbiology, Molecular Genetics and Immunology, University of Kansas School of Medicine, Kansas City, KS 66160
Protein Secretion in Bacteria
Edited by Maria Sandkvist, Eric Cascales, and Peter J. Christie
© 2019 American Society for Microbiology, Washington, DC
doi:10.1128/microbiolspec.PSIB-0026-2019

burgdorferi [1]), together with other closely related species, causes tick-borne Lyme disease in temperate climates of the Northern Hemisphere; a distinct group of *Borrelia* species causes endemic tick-borne relapsing fever in the western United States and Africa and outbreaks of louse-borne relapsing fever among confined human populations. *Leptospira interrogans* and other pathogenic species are the causative agents of leptospirosis, a global zoonosis that is mainly transmitted through direct or indirect contact with soil or water contaminated with animal urine, fluids, or tissues; other leptospires are either organisms with intermediate pathogenicity able to colonize animals or free-living saprophytes (e.g., *Leptospira biflexa*) growing in moist environmental settings. *Treponema pallidum* subsp. *pallidum* (hereafter referred to as *T. pallidum*) is the causative agent of venereal syphilis, an ancient sexually transmitted human infection that has persisted in the human population to the present day; other nonvenereal treponemes are also transmitted by direct human-to-human contact, and *Treponema denticola* is part of the complex human oral flora and a common periodontal pathogen. Other treponemal species colonize vertebrates and insects and can be found in aquatic and soil environments.

This review covers our current understanding of protein secretion mechanisms in *B. burgdorferi*, *L. interrogans*, and *T. pallidum*. Despite their common phylogenetic classification as spirochetes, these three species differ significantly in their envelope structure, particularly when looking at the composition of the bacterial surface as the interface with the host. The generation and maintenance of these different interfaces throughout transmission and subsequent infection require protein secretion pathways that have been evolutionarily adapted and fine-tuned for efficiency. Our current understanding of these mechanisms is most detailed for *B. burgdorferi*, in part due to the earlier availability of robust genetic tools. But recent technological advances are likely to further stimulate the analysis of

leptospiral and treponemal protein secretion systems as well.

SPIROCHETAL CELL ENVELOPE ULTRASTRUCTURES

Cryo-electron tomography studies on *Borrelia*, *Leptospira*, and *Treponema* have provided a comprehensive and remarkably detailed overview of the different envelope ultrastructures (2–6). The cross-referencing of these micrographic data with detailed protein localization studies now allows us to catalog the requirement for protein secretion pathways in the different systems.

Common to all three model spirochetes is a diderm envelope architecture similar to that of Gram-negative bacteria, with an inner cytoplasmic membrane (IM), an outer membrane (OM), and a periplasmic space in between (Fig. 1). A thin periplasmic peptidoglycan cell wall layer appears to be more closely associated with the IM than the OM, so as not to interfere with the movement of the periplasmic flagella that can be visualized between the cell wall and the OM. The OM of both pathogenic and saprophytic *Leptospira* is considerably thicker than that of *Borrelia* or *Treponema*, mainly due to the presence of a protease-resistant layer of lipopolysaccharide (LPS) in the surface leaflet (6). In contrast, the *Borrelia* OM lacks LPS but contains abundant OM proteins (OMPs), including a particularly large array of surface lipoproteins that can be removed by proteolytic surface shaving (3, 7). LPS is also missing from the envelopes of treponemes, with the *T. pallidum* OM containing only low-density integral membrane proteins and a few surface lipoproteins (8) and *T. denticola* likewise displaying only distinct outer sheath surface proteins (9). Thus, the *Leptospira* OM most closely conforms to a typical Gram-negative envelope structure, with LPS molecules serving as major and serotype-defining surface antigens. The envelope structures of *Borrelia* and *Treponema*, however, diverge significantly

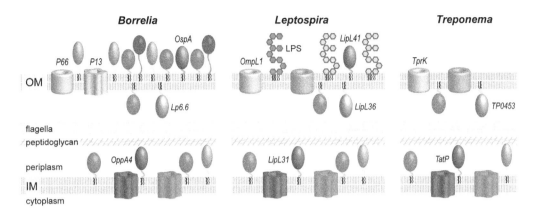

FIGURE 1 Envelope structures of model spirochetes *Borrelia*, *Leptospira*, and *Treponema*. Spirochetes have a common diderm envelope structure with an inner membrane (IM) and outer membrane (OM) and a periplasmic space in between that contains a thin peptidoglycan cell wall and periplasmic flagella. A major difference between the three genera is seen in the OM: *Borrelia* displays a limited set of integral OM proteins (OMPs), among them the unusual porin P13 with a predicted α-helical TM topology, but a large variable set of surface lipoproteins. The *Leptospira* OM most closely resembles a Gram-negative OM, with lipopolysaccharide (LPS) being the major component of the surface leaflet, complemented by a large number of OMPs and a limited set of surface lipoproteins. *Treponema* expresses only rare OMPs with limited surface exposure. Some of the model proteins under study are labeled in italics. See text for details.

from those of more classical diderm bacteria, with consequences for bacterial virulence mechanisms and protein secretion pathway requirements. *T. pallidum*'s paucity of surface-exposed proteins contributes to it being a "stealth pathogen" that can disseminate and cause symptomatic tissue damage while evading host immune responses (8). The peculiar abundance of *Borrelia* surface lipoproteins (7), which serve as immunodominant antigens and drive transmission, colonization, dissemination, and persistence (10, 11), puts the focus squarely on how this set of proteins reaches the outer leaflet of the OM.

Sec-MEDIATED EXPORT AND IM INSERTION

Bioinformatic evidence in the current inventory of clusters of orthologous groups (12, 13) suggests that spirochetal protein export from the cytoplasm through the IM follows the general secretory pathway that has been described and amply investigated in other diderm bacterial models (Fig. 2). The deter-mined *Borrelia*, *Treponema*, and *Leptospira* genomes (14–18) encode all the Sec proteins that are needed to form a functional translocon. The presence of homologs for the signal recognition particle Ffh and its receptor, FtsY, but absence of the cytoplasmic SecB chaperone suggest that proteins can be exported cotranslationally, i.e., upon docking of the translating ribosome to the Sec translocon via Ffh-Fts. SecB's function in posttranslational export may be taken over by other cytoplasmic folding chaperones, such as GroEL. There is no evidence for a functional twin-arginine transport (TAT) pathway in spirochetes that would allow for export of folded proteins through the IM; note that the described *T. pallidum* Tat proteins are unrelated to TAT but rather part of a tripartite ATP-dependent periplasmic-type transporter system that seems to facilitate import of hydrophobic molecules through the periplasm and the IM (8, 19).

All three model spirochete genomes encode homologs of the IM protein insertase YidC. Thus, the mechanism of release of integral IM proteins from the Sec translocon

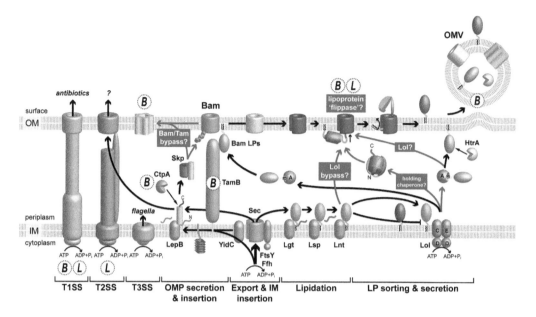

FIGURE 2 Generalized model of spirochetal protein secretion. The left half shows type I to III secretion systems (T1SS, T2SS, and T3SS) as well as the pathways involved in the secretion of nonlipidated membrane and periplasmic proteins. The right half shows the pathways for lipoprotein (LP) modification, sorting, and secretions. Pathway components or mechanisms that appear unique to a particular spirochetal genus are labeled by stippled red circles with the genus initial: *B*, *Borrelia*; *L*, *Leptospira*; and *T*, *Treponema*. For example, a potential T2SS has been identified only for *Leptospira*. The pathways in blue delineate the current alternative periplasmic and OM mechanisms that may be involved in the secretion of α-helical integral OMPs and surface lipoproteins in *Borrelia*; of note, *Leptospira* could also take advantage of its T2SS to secrete surface lipoproteins (23, 95). Release of outer membrane vesicles (OMVs) has been observed and studied in *Borrelia*. See text for details.

into the IM is likely similar to that of well-studied Gram-negative model bacteria. At the same time, IM proteins with simple transmembrane (TM) topologies, such as the abundant and highly immunogenic *B. burgdorferi* Bdr proteins with their single C-terminal TM domains (20–22), may be inserted into the IM by YidC in a Sec-independent manner (Fig. 2).

SIGNAL PEPTIDE PROCESSING

Like in other bacterial systems, two distinct signal peptidases appear responsible for cleaving the N-terminal Sec-dependent signal peptides once exported spirochetal proteins have reached the periplasm (Fig. 2). Signal I peptidase LepB removes the signal peptides

from soluble periplasmic or secreted proteins and integral OMPs. Signal II peptidase Lsp removes signal peptides from lipoproteins that will ultimately remain associated with the IM or transported further to the periplasmic and surface leaflets of the OM (to be discussed below). Interestingly, *B. burgdorferi* encodes three LepB homologs that—like sets of multiple LepBs in some Gram-positive bacteria—may have different substrate specificities. *B. burgdorferi* LepB2, encoded by chromosomal open reading frame (ORF) BB0031, is closest to a canonical LepB, with an N-terminal two-TM anchor domain and an appropriately localized periplasmic active site. LepB3 (BB0263) similarly is predicted to have all enzymatically relevant residues located in the periplasm, anchored by a single N-terminal TM domain. LepB1 (BB0030), immediately upstream of LepB2, however, appears to be a signal peptidase I-like protein with a four-TM domain topology that lacks the canonical active-site residues. A conditional knockout of *lepB1* led to a marked growth defect with signs of significant envelope instability (M. K. You, C. R. Whetstine, and W. R. Zückert, unpublished results), but the protein's precise function remains unknown.

Sec-DEPENDENT AND -INDEPENDENT SECRETION

Each of the three spirochetal model genera shows *in silico* evidence of distinct subsets of Sec-dependent type II (T2SS) or Sec-independent type I (T1SS) and type III (T3SS) secretion system homologs (12, 13) (Fig. 2). T2SSs span the periplasm and the OM as a major terminal branch of the general secretory pathway. There is no *in silico* evidence of T2SS components in either *Borrelia* or *Treponema* (12, 13). An identified T2SS in *Leptospira* appears to be incomplete, and its function in leptospiral protein secretion remains to be determined (23). Homologs of T1SS components, which together secrete ions, carbohydrates, drugs, or proteins in other bacterial systems, have been identified in *Borrelia* and *Leptospira* but not in *Treponema* (23). The *B. burgdorferi* T1SS is homologous to the Tol system in other diderm bacteria and consists of a trimeric OM channel-tunnel protein (BesC), a periplasmic membrane fusion protein (BesA), and an IM ATP-dependent export pump (BesB). A BesC-deficient mutant was noninfectious in a mouse model of infection and was more sensitive to several antibiotics, detergent, and DNA-intercalating agents. Recombinant BesC formed pores in lipid bilayers, and modeling of the *B. burgdorferi* Bes complex on the Tol structure suggested a less stable interaction of BesA with BesC due to a truncated α-helical interaction domain in BesA (24). Together, these findings suggest that the *B. burgdorferi* Bes system functions at least in part as an efflux pump for noxious compounds. No protein substrates for the *B. burgdorferi* Bes system have been identified. T3SS homologs have been identified in all three spirochetal model genera (12, 13) but appear limited to the IM protein complexes that export components of the periplasmic flagella, as shown for *B. burgdorferi* (25).

LIPOPROTEIN MODIFICATION

Lipoproteins can be found in all three possible compartments of the spirochetal envelope: in the periplasmic leaflet of the IM, the periplasmic leaflet of the OM, and the surface leaflet of the OM. Based on *in silico* predictions, *T. pallidum* encodes about 20 lipoproteins (16), while *L. interrogans* and *B. burgdorferi* encode about 180 and 130 lipoproteins, respectively (14, 15, 18, 26, 27); in *B. burgdorferi*, this amounts to about 8% of the genome's coding capacity. *T. pallidum* lipoproteins have been localized to the periplasm, where they have been shown to function as part of nutrient import machineries (19). TP0453 is an OM low-abundancy amphiphilic lipoprotein that is thought to additionally interact with the lipid bilayer through amphipathic helices (28). Another *T. pallidum* lipoprotein, encoded by ORF TP0435, has shown some very limited surface exposure when expressed heterologously in *B. burgdorferi* (29), but the biological significance of this finding remains to be determined. Similarly, only about half a dozen lipoproteins have been localized to the surface of *L. interrogans* (23), with an initial vaccine target, LipL32, recently being relegated to the periplasm (30). The most comprehensive localization study was carried out with *B. burgdorferi* by combining the analysis of an epitope-tagged lipoprotein expression library with quantitative proteomics. Of the 125 *B. burgdorferi* lipoproteins that were examined, 86 localized to the surface; only 8 of the 39 periplasmic lipoproteins were anchored in the OM (7). This unique abundance of surface-displayed *Borrelia* lipoproteins highlights their pivotal role in interactions with vector and host animals (10, 11) and also points to the requirement for an efficient lipoprotein secretion pathway that extends through the OM to the bacterial surface (discussed below).

All three model spirochetes encode the known components of a diderm lipoprotein modification machinery (Fig. 2). In a first committing step, lipoprotein diacylglyceryl transferase Lgt catalyzes the attachment of a diacylglycerol moiety to the side chain of a conserved cysteine within an internal N-terminal "lipobox" motif. Next, lipoprotein signal peptidase Lsp (also called signal peptidase II) recognizes the lipobox residues to remove the signal peptide. Finally, N-acyltransferase Lnt adds another acyl chain to the now available amino-terminal cysteine, leading to N-terminally triacylated lipoproteins that are at least temporarily positioned in the periplasmic leaflet of the IM (31–34). Spirochetal Lsp appears to recognize lipobox motifs in lipoprotein signal peptides that are more relaxed than those found in other diderm systems (26, 34). Yet as for other bacteria, spirochetal Lsp activity can be blocked by the antibiotic globomycin, a noncompetitive Lsp inhibitor (35–37).

LIPOPROTEIN PERIPLASMIC TRANSPORT AND SECRETION

One well-defined route for fully modified lipoproteins to the OM involves interactions with the Lol pathway. *B. burgdorferi*, *L. interrogans*, and *T. pallidum* all encode a partial Lol pathway that lacks the OM lipoprotein receptor/insertase LolB found only in beta- and gammaproteobacteria (34) (Fig. 2). In *B. burgdorferi*, conditional mutants in the LolD ATPase and the periplasmic lipoprotein carrier LolA are lethal, indicating that the pathway is essential (K. M. Bridges, A. S. Dowdell, S. Chen, J. L. Kueker, J. Liu, J. S. Blevins, and W. R. Zückert, unpublished results). Any direct or indirect involvement of the Lol pathway in surface lipoprotein secretion is pending experimental investigation. Several independent lines of evidence point toward a periplasmic mechanism that maintains *B. burgdorferi* surface lipoproteins in a translocation-competent conformation by preventing them from folding prematurely in the periplasm (38–41). The commonly disordered N-terminal "tether" peptides of surface lipoproteins are intimately involved in this process, as tether deletion and substitution mutants of several surface lipoproteins were specifically deficient in OM translocation (41–43). This suggests a periplasmic "holding" mechanism that engages surface-targeted prolipoprotein peptides as they emerge tether first from the IM Sec complex (Fig. 2). Surface lipoproteins are then thought to be delivered in an at least partially unfolded conformation to an OM lipoprotein flippase, which facilitates translocation through the OM and leads to the ultimate anchoring of surface lipoproteins in the surface leaflet of the OM (38, 39, 41) (Fig. 2). Intriguingly, *B. burgdorferi* encodes distant homologs of five of the seven known LPS secretion pathway components—despite a lack of LPS. BB0838, a homolog of the OM LPS transporter LptD (44, 45), has been localized to the OM (46) and is likely essential due to a lack of Tn insertions (47, 48). BB0838 could be thus considered an orphan OM translocase in search of substrates, with the most likely substrates being either surface glycolipids found in the *B. burgdorferi* OM (49–51) or surface lipoproteins. If the latter is the case, this would represent a paradigm-shifting substrate expansion for the LptD family, and the main questions would turn to whether the Lol, Lpt, or a yet-to-be-discovered pathway feeds into this unique OM mechanism (Fig. 2).

INSERTION AND TOPOLOGY OF INTEGRAL OMPS

Spirochetal OMs contain various sets of integral, membrane-spanning OMPs. Underscoring the Gram-negative-like features of the leptospiral envelope, the *L. interrogans* OM is predicted to contain over 180 OMPs (27), including a leptospiral TolC homolog, the iron transporter FecA, and the OmpL1 porin (23). In contrast, only 9 Tpr (*Treponema pallidum* repeat) proteins and BamA, a central component of the β-barrel assembly machinery (BAM), have been described for the OM of *T. pallidum*; 9 additional hypothetical OMPs have been postulated (52). In *B. burgdorferi*, computational analysis predicts 41 OMPs (46), with OM localization confirmed experimentally for a subset of them, including the porins P13 and P66, BesC, BamA, and the LptD homolog mentioned above. P13 is an unusual OM porin in that it is predicted to assume an α-helical conformation (53–58) (Fig. 1 and 2). It is C-terminally processed by a periplasmic protease, CtpA (59) (Fig. 2), and found in multiprotein complexes with several surface lipoproteins (60). In purified form, P13 forms heterogeneous homo-oligomeric pores in lipid bilayers with no apparent gating mechanism or substrate specificity (55, 56, 61). Eight P13 paralogs are found on various *B. burgdorferi* plasmids, and one of them is upregulated in a P13 deletion mutant (55), suggesting that the expression of at least one P13 family protein is essential for cellular survival. A second well-studied porin, P66, assumes a more traditional β-barrel conformation and, in addition to forming a pore (54, 62–64), also functions as an adhesin that facilitates dissemination (65, 66).

The roles of some of the periplasmic and OM secretion pathway components for OMPs have been studied in *B. burgdorferi*. A Skp homolog is expressed downstream from BamA, but its function remains to be determined (67). As in some other bacteria, the *B. burgdorferi* HtrA chaperone/protease homolog (BB0104) has specific activity in the *Borrelia* periplasm, at the OM and as a secreted protein (68). In concert with the above-mentioned CtpA, HtrA is involved in the proteolytic maturation of BB0323, a periplasmic OM lipoprotein that is required for envelope stability and contains a C-terminal peptidoglycan-binding LysM domain (69, 70); other identified HtrA substrates are involved in *Borrelia* chemotaxis and motility (71). At the same time, HtrA has been shown to degrade components of host extracellular matrix components (72).

Insertion of OMPs into the OMs of diderm bacteria is generally mediated by the BAM complex, and Bam protein homologs are present in all three model spirochetes. Studies with *B. burgdorferi* have shown that its BamA homolog (BB0795) is associated with two periplasmic OM lipoproteins, BamB (BB0028) and BamD (BB0324) (73). Depletion of BamA led to a noticeable decrease of selected OMPs, including P66 (74); the insertion mechanism of P13 remains unresolved and may bypass BAM like α-helical OMPs do in other systems (75) (Fig. 2). Together, these findings support the fundamental conservation of OMP assembly mechanisms in diderm bacteria. In an apparent divergence, however, the usually essential BamD was dispensable for borrelial growth, while depletion of the generally auxiliary BamB led to a detectable growth defect (76–78). *B. burgdorferi* BamA also was shown to interact with a TamB (BB0794) homolog that in other bacteria is part of a parallel OMP translocation and assembly module pathway (67, 79). This indicates that *B. burgdorferi* has adapted modules of separate secretion machineries to function in a single hybrid pathway (80). Intriguingly, BamA depletion also led to a reduction of lipoproteins in the OM (74). Together, these data are compatible with a likely indirect BAM dependency of the surface lipoprotein flipping process (Fig. 2).

RELEASE OF OUTER MEMBRANE VESICLES AND INTERACTIONS WITH HOST CELLS

In stark contrast to other bacterial systems, secretion and release of soluble spirochetal proteins into the milieu has been observed rather sporadically. This may be a function of the limited secretory capabilities of spirochetes, as detailed above, but requires further investigation. For example, the exoproteome of pathogenic *Leptospira* species includes proteases such as thermolysin, but it remains unclear how these proteins are secreted (81).

Borrelia OM vesicles (OMVs) have been shown to be released near the sites of cell division (2), and shedding can be observed both in culture and *in vivo*, e.g., within ticks taking a blood meal (82). OMVs, which primarily contain OMPs but also periplasmic, IM, and cytoplasmic components, are released under natural conditions from other diderm bacteria and have been shown to play roles in virulence and immunomodulation (83). The precise roles of OMVs in the pathogenesis of *Borrelia* infections remain to be defined. One surprising *B. burgdorferi* OMV-associated protein is enolase, a "moonlighting" signal peptide-less protein that is nevertheless exported and ultimately becomes accessible to the bacterial milieu. There, enolase is thought to work as one of the spirochete's many plasminogen-binding proteins that facilitate pericellular proteolysis and thereby aid in nutrient acquisition and dissemination (84–86).

Low-level exchange of cholesterol lipids between *B. burgdorferi* and eukaryotic cells by direct contact or via release and capture of OMVs has been observed in tissue culture experiments (87). In one direction, this exchange allows for the acquisition of membrane lipids by a bacterium that is unable to synthesize its own. In the other direction, the transfer of bacterial lipids and glycolipids to host cells could result in antigen presentation and immune recognition that may contribute to the pathogenesis of Lyme disease through cell and tissue damage. Since at least a subset of *B. burgdorferi* surface lipoproteins are found associated with OM lipid rafts (88, 89), it is entirely conceivable that bacterial proteins are transferred to host cells via the same mechanism as well, thus extending spirochetal secretion pathways to eukaryotic cells.

CONCLUSIONS

Since the first description of "corkscrew-like animalcules in foul water" in 1773, the first definition of *spirochaeta* in 1838, and the ultimate discovery of *Spirochaeta pallida* (now *T. pallidum*) in syphilitic lesions in 1905 (90), spirochetes have become increasingly fascinating constituents of the global pan-microbiome. *B. burgdorferi* and *T. pallidum* were among the first fully sequenced bacterial pathogens (14, 16), and their genomes were as noteworthy for what was missing compared to other bacteria, e.g., an LPS biosynthesis pathway, as for the presence of numerous hypothetical and sometimes paralogous genes of unknown function. Since then, the development of robust genetic tools (91) has driven significant dissection of molecular mechanisms—including protein secretion—in *B. burgdorferi* and, by extension, in some other *Borrelia* species. This has led to findings that have stimulated a reevaluation of previously overgeneralized microbial processes. Studies of *Leptospira* are now catching up, due in no small part to the use of the genetically more amenable saprophytic *L. biflexa* system and development of improved genetic tools, including conjugation of *Escherichia coli* to *L. interrogans* (92). At the same time, molecular investigations of *T. pallidum* that often had to resort to heterologous systems because of the inability to grow the spirochete *ex vivo* are now bound to be transformed by a novel tissue culture-based culture system (93). Enteric spirochetes belonging to the genus *Brachyspira*, long recognized as veterinary pathogens, are now emerging as human pathogens in at-risk populations (94), beckoning further study of virulence mechanisms. As spirochetal research enters the postgenomic era and diversity in our understanding of both physiological and pathogenic microbial mechanisms is growing, we would be wise to expect one or another paradigm-shifting discovery.

ACKNOWLEDGMENTS

I thank all current and former laboratory members, mentors, collaborators, and colleagues for their contributions to understanding spirochetal protein secretion. I am indebted to the editors for their infinite patience.

This work was supported in part by unrestricted University of Kansas School of Medicine funds and by NIH grants AI129522, AI133056, and AI139956.

CITATION

Zückert WR. 2019. Protein secretion in spirochetes. Microbiol Spectrum 7(3):PSIB-0026-2019.

REFERENCES

1. **Barbour AG.** 2018. Borreliaceae. *In* Whitman WB, Rainey F, Kämpfer P, Trujillo M, Chun J, DeVos P, Hedlund B, Dedysh S (ed), *Bergey's Manual of Systematics of Archaea and Bacteria*. Wiley, New York, NY.

2. **Kudryashev M, Cyrklaff M, Baumeister W, Simon MM, Wallich R, Frischknecht F.** 2009. Comparative cryo-electron tomography of pathogenic Lyme disease spirochetes. *Mol Microbiol* **71:**1415–1434.

3. **Liu J, Lin T, Botkin DJ, McCrum E, Winkler H, Norris SJ.** 2009. Intact flagellar motor of *Borrelia burgdorferi* revealed by cryo-electron tomography: evidence for stator ring curvature and rotor/C-ring assembly flexion. *J Bacteriol* **191:**5026–5036.

4. **Izard J, Renken C, Hsieh CE, Desrosiers DC, Dunham-Ems S, La Vake C, Gebhardt LL, Limberger RJ, Cox DL, Marko M, Radolf JD.** 2009. Cryo-electron tomography elucidates the molecular architecture of *Treponema pallidum*, the syphilis spirochete. *J Bacteriol* **191:**7566–7580.

5. **Liu J, Howell JK, Bradley SD, Zheng Y, Zhou ZH, Norris SJ.** 2010. Cellular architecture of *Treponema pallidum*: novel flagellum, periplasmic cone, and cell envelope as revealed by cryo electron tomography. *J Mol Biol* **403:**546–561.

6. **Raddi G, Morado DR, Yan J, Haake DA, Yang XF, Liu J.** 2012. Three-dimensional structures of pathogenic and saprophytic *Leptospira* species revealed by cryo-electron tomography. *J Bacteriol* **194:**1299–1306.

7. **Dowdell AS, Murphy MD, Azodi C, Swanson SK, Florens L, Chen S, Zückert WR.** 2017. Comprehensive spatial analysis of the *Borrelia burgdorferi* lipoproteome reveals a compartmentalization bias toward the bacterial surface. *J Bacteriol* **199:**00658-16.

8. **Radolf JD, Deka RK, Anand A, Šmajs D, Norgard MV, Yang XF.** 2016. *Treponema pallidum*, the syphilis spirochete: making a living as a stealth pathogen. *Nat Rev Microbiol* **14:**744–759.

9. **Dashper SG, Seers CA, Tan KH, Reynolds EC.** 2011. Virulence factors of the oral spirochete *Treponema denticola. J Dent Res* **90:**691–703.

10. **Radolf JD, Caimano MJ, Stevenson B, Hu LT.** 2012. Of ticks, mice and men: understanding the dual-host lifestyle of Lyme disease spirochaetes. *Nat Rev Microbiol* **10:**87–99.

11. **Steere AC, Strle F, Wormser GP, Hu LT, Branda JA, Hovius JW, Li X, Mead PS.** 2016. Lyme borreliosis. *Nat Rev Dis Primers* **2:**16090.

12. **Natale DA, Galperin MY, Tatusov RL, Koonin EV.** 2000. Using the COG database to improve gene recognition in complete genomes. *Genetica* **108:**9–17.

13. **Tatusov RL, Galperin MY, Natale DA, Koonin EV.** 2000. The COG database: a tool for genome-scale analysis of protein functions and evolution. *Nucleic Acids Res* **28:**33–36.

14. **Fraser CM, Casjens S, Huang WM, Sutton GG, Clayton R, Lathigra R, White O, Ketchum KA, Dodson R, Hickey EK, Gwinn M, Dougherty B, Tomb JF, Fleischmann RD, Richardson D, Peterson J, Kerlavage AR, Quackenbush J, Salzberg S, Hanson M, van Vugt R, Palmer N, Adams MD, Gocayne J, Weidman J, Utterback T, Watthey L, McDonald L, Artiach P, Bowman C, Garland S, Fuji C, Cotton MD, Horst K, Roberts K, Hatch B, Smith HO, Venter JC.** 1997. Genomic sequence of a Lyme disease spirochaete, *Borrelia burgdorferi. Nature* **390:**580–586.

15. **Casjens S, Palmer N, van Vugt R, Huang WM, Stevenson B, Rosa P, Lathigra R, Sutton G, Peterson J, Dodson RJ, Haft D, Hickey E, Gwinn M, White O, Fraser CM.** 2000. A bacterial genome in flux: the twelve linear and nine circular extrachromosomal DNAs in an infectious isolate of the Lyme disease spirochete *Borrelia burgdorferi. Mol Microbiol* **35:**490–516.

16. **Fraser CM, Norris SJ, Weinstock GM, White O, Sutton GG, Dodson R, Gwinn M, Hickey EK, Clayton R, Ketchum KA, Sodergren E,**

Hardham JM, McLeod MP, Salzberg S, Peterson J, Khalak H, Richardson D, Howell JK, Chidambaram M, Utterback T, McDonald L, Artiach P, Bowman C, Cotton MD, Fujii C, Garland S, Hatch B, Horst K, Roberts K, Sandusky M, Weidman J, Smith HO, Venter JC. 1998. Complete genome sequence of *Treponema pallidum*, the syphilis spirochete. *Science* **281**:375–388.

17. Seshadri R, Myers GS, Tettelin H, Eisen JA, Heidelberg JF, Dodson RJ, Davidsen TM, DeBoy RT, Fouts DE, Haft DH, Selengut J, Ren Q, Brinkac LM, Madupu R, Kolonay J, Durkin SA, Daugherty SC, Shetty J, Shvartsbeyn A, Gebregeorgis E, Geer K, Tsegaye G, Malek J, Ayodeji B, Shatsman S, McLeod MP, Smajs D, Howell JK, Pal S, Amin A, Vashisth P, McNeill TZ, Xiang Q, Sodergren E, Baca E, Weinstock GM, Norris SJ, Fraser CM, Paulsen IT. 2004. Comparison of the genome of the oral pathogen *Treponema denticola* with other spirochete genomes. *Proc Natl Acad Sci U S A* **101**:5646–5651.

18. Nascimento AL, Ko AI, Martins EA, Monteiro-Vitorello CB, Ho PL, Haake DA, Verjovski-Almeida S, Hartskeerl RA, Marques MV, Oliveira MC, Menck CF, Leite LC, Carrer H, Coutinho LL, Degrave WM, Dellagostin OA, El-Dorry H, Ferro ES, Ferro MI, Furlan LR, Gamberini M, Giglioti EA, Góes-Neto A, Goldman GH, Goldman MH, Harakava R, Jerônimo SM, Junqueira-de-Azevedo IL, Kimura ET, Kuramae EE, Lemos EG, Lemos MV, Marino CL, Nunes LR, de Oliveira RC, Pereira GG, Reis MS, Schriefer A, Siqueira WJ, Sommer P, Tsai SM, Simpson AJ, Ferro JA, Camargo LE, Kitajima JP, Setubal JC, Van Sluys MA. 2004. Comparative genomics of two *Leptospira interrogans* serovars reveals novel insights into physiology and pathogenesis. *J Bacteriol* **186**:2164–2172.

19. Deka RK, Brautigam CA, Goldberg M, Schuck P, Tomchick DR, Norgard MV. 2012. Structural, bioinformatic, and in vivo analyses of two *Treponema pallidum* lipoproteins reveal a unique TRAP transporter. *J Mol Biol* **416**:678–696.

20. Zückert WR, Meyer J, Barbour AG. 1999. Comparative analysis and immunological characterization of the *Borrelia* Bdr protein family. *Infect Immun* **67**:3257–3266.

21. Roberts DM, Theisen M, Marconi RT. 2000. Analysis of the cellular localization of Bdr paralogs in *Borrelia burgdorferi*, a causative agent of Lyme disease: evidence for functional diversity. *J Bacteriol* **182**:4222–4226.

22. Barbour AG, Jasinskas A, Kayala MA, Davies DH, Steere AC, Baldi P, Felgner PL. 2008. A genome-wide proteome array reveals a limited set of immunogens in natural infections of humans and white-footed mice with *Borrelia burgdorferi*. *Infect Immun* **76**:3374–3389.

23. Haake DA, Zückert WR. 2015. The leptospiral outer membrane. *Curr Top Microbiol Immunol* **387**:187–221.

24. Bunikis I, Denker K, Ostberg Y, Andersen C, Benz R, Bergström S. 2008. An RND-type efflux system in *Borrelia burgdorferi* is involved in virulence and resistance to antimicrobial compounds. *PLoS Pathog* **4**:e1000009.

25. Lin T, Gao L, Zhao X, Liu J, Norris SJ. 2015. Mutations in the *Borrelia burgdorferi* flagellar type III secretion system genes fliH and fliI profoundly affect spirochete flagellar assembly, morphology, motility, structure, and cell division. *mBio* **6**:e00579-15.

26. Setubal JC, Reis M, Matsunaga J, Haake DA. 2006. Lipoprotein computational prediction in spirochaetal genomes. *Microbiology* **152**:113–121.

27. Pinne M, Matsunaga J, Haake DA. 2012. Leptospiral outer membrane protein microarray, a novel approach to identification of host ligand-binding proteins. *J Bacteriol* **194**:6074–6087.

28. Hazlett KR, Cox DL, Decaffmeyer M, Bennett MP, Desrosiers DC, La Vake CJ, La Vake ME, Bourell KW, Robinson EJ, Brasseur R, Radolf JD. 2005. TP0453, a concealed outer membrane protein of *Treponema pallidum*, enhances membrane permeability. *J Bacteriol* **187**:6499–6508.

29. Chan K, Nasereddin T, Alter L, Centurion-Lara A, Giacani L, Parveen N. 2016. *Treponema pallidum* lipoprotein TP0435 expressed in *Borrelia burgdorferi* produces multiple surface/periplasmic isoforms and mediates adherence. *Sci Rep* **6**:25593.

30. Pinne M, Haake DA. 2013. LipL32 is a subsurface lipoprotein of *Leptospira interrogans*: presentation of new data and reevaluation of previous studies. *PLoS One* **8**:e51025.

31. Belisle JT, Brandt ME, Radolf JD, Norgard MV. 1994. Fatty acids of *Treponema pallidum* and *Borrelia burgdorferi* lipoproteins. *J Bacteriol* **176**:2151–2157.

32. Brandt ME, Riley BS, Radolf JD, Norgard MV. 1990. Immunogenic integral membrane proteins of *Borrelia burgdorferi* are lipoproteins. *Infect Immun* **58**:983–991.

33. Beermann C, Lochnit G, Geyer R, Groscurth P, Filgueira L. 2000. The lipid component of lipoproteins from *Borrelia burgdorferi*: structural analysis, antigenicity, and presentation via human dendritic cells. *Biochem Biophys Res Commun* **267**:897–905.

34. Zückert WR. 2014. Secretion of bacterial lipoproteins: through the cytoplasmic mem-

brane, the periplasm and beyond. *Biochim Biophys Acta* **1843:**1509–1516.

35. **Erdile LF, Brandt MA, Warakomski DJ, Westrack GJ, Sadziene A, Barbour AG, Mays JP.** 1993. Role of attached lipid in immunogenicity of *Borrelia burgdorferi* OspA. *Infect Immun* **61:**81–90.

36. **Shang ES, Summers TA, Haake DA.** 1996. Molecular cloning and sequence analysis of the gene encoding LipL41, a surface-exposed lipoprotein of pathogenic *Leptospira* species. *Infect Immun* **64:**2322–2330.

37. **Swancutt MA, Radolf JD, Norgard MV.** 1990. The 34-kilodalton membrane immunogen of *Treponema pallidum* is a lipoprotein. *Infect Immun* **58:**384–392.

38. **Chen S, Kumru OS, Zückert WR.** 2011. Determination of *Borrelia* surface lipoprotein anchor topology by surface proteolysis. *J Bacteriol* **193:**6379–6383.

39. **Chen S, Zückert WR.** 2011. Probing the *Borrelia burgdorferi* surface lipoprotein secretion pathway using a conditionally folding protein domain. *J Bacteriol* **193:**6724–6732.

40. **Kumru OS, Schulze RJ, Rodnin MV, Ladokhin AS, Zückert WR.** 2011. Surface localization determinants of *Borrelia* OspC/Vsp family lipoproteins. *J Bacteriol* **193:**2814–2825.

41. **Schulze RJ, Chen S, Kumru OS, Zückert WR.** 2010. Translocation of *Borrelia burgdorferi* surface lipoprotein OspA through the outer membrane requires an unfolded conformation and can initiate at the C-terminus. *Mol Microbiol* **76:**1266–1278.

42. **Kumru OS, Schulze RJ, Slusser JG, Zückert WR.** 2010. Development and validation of a FACS-based lipoprotein localization screen in the Lyme disease spirochete *Borrelia burgdorferi*. *BMC Microbiol* **10:**277.

43. **Schulze RJ, Zückert WR.** 2006. *Borrelia burgdorferi* lipoproteins are secreted to the outer surface by default. *Mol Microbiol* **59:**1473–1484.

44. **Li X, Gu Y, Dong H, Wang W, Dong C.** 2015. Trapped lipopolysaccharide and LptD intermediates reveal lipopolysaccharide translocation steps across the *Escherichia coli* outer membrane. *Sci Rep* **5:**11883.

45. **Konovalova A, Kahne DE, Silhavy TJ.** 2017. Outer membrane biogenesis. *Annu Rev Microbiol* **71:**539–556.

46. **Kenedy MR, Scott EJ II, Shrestha B, Anand A, Iqbal H, Radolf JD, Dyer DW, Akins DR.** 2016. Consensus computational network analysis for identifying candidate outer membrane proteins from *Borrelia* spirochetes. *BMC Microbiol* **16:**141.

47. **Lin T, Gao L, Zhang C, Odeh E, Jacobs MB, Coutte L, Chaconas G, Philipp MT, Norris SJ.** 2012. Analysis of an ordered, comprehensive STM mutant library in infectious *Borrelia burgdorferi*: insights into the genes required for mouse infectivity. *PLoS One* **7:**e47532.

48. **Lin T, Troy EB, Hu LT, Gao L, Norris SJ.** 2014. Transposon mutagenesis as an approach to improved understanding of *Borrelia* pathogenesis and biology. *Front Cell Infect Microbiol* **4:**63.

49. **Hossain H, Wellensiek HJ, Geyer R, Lochnit G.** 2001. Structural analysis of glycolipids from *Borrelia burgdorferi*. *Biochimie* **83:**683–692.

50. **Ben-Menachem G, Kubler-Kielb J, Coxon B, Yergey A, Schneerson R.** 2003. A newly discovered cholesteryl galactoside from *Borrelia burgdorferi*. *Proc Natl Acad Sci U S A* **100:**7913–7918.

51. **Schröder NW, Eckert J, Stübs G, Schumann RR.** 2008. Immune responses induced by spirochetal outer membrane lipoproteins and glycolipids. *Immunobiology* **213:**329–340.

52. **Cox DL, Luthra A, Dunham-Ems S, Desrosiers DC, Salazar JC, Caimano MJ, Radolf JD.** 2010. Surface immunolabeling and consensus computational framework to identify candidate rare outer membrane proteins of *Treponema pallidum*. *Infect Immun* **78:**5178–5194.

53. **Sadziene A, Thomas DD, Barbour AG.** 1995. *Borrelia burgdorferi* mutant lacking Osp: biological and immunological characterization. *Infect Immun* **63:**1573–1580.

54. **Pinne M, Thein M, Denker K, Benz R, Coburn J, Bergström S.** 2007. Elimination of channel-forming activity by insertional inactivation of the p66 gene in *Borrelia burgdorferi*. *FEMS Microbiol Lett* **266:**241–249.

55. **Pinne M, Östberg Y, Comstedt P, Bergström S.** 2004. Molecular analysis of the channel-forming protein P13 and its paralogue family 48 from different Lyme disease *Borrelia* species. *Microbiology* **150:**549–559.

56. **Östberg Y, Pinne M, Benz R, Rosa P, Bergström S.** 2002. Elimination of channel-forming activity by insertional inactivation of the p13 gene in *Borrelia burgdorferi*. *J Bacteriol* **184:**6811–6819.

57. **Nilsson CL, Cooper HJ, Håkansson K, Marshall AG, Ostberg Y, Lavrinovicha M, Bergström S.** 2002. Characterization of the P13 membrane protein of *Borrelia burgdorferi* by mass spectrometry. *J Am Soc Mass Spectrom* **13:**295–299.

58. **Noppa L, Östberg Y, Lavrinovicha M, Bergström S.** 2001. P13, an integral membrane protein of *Borrelia burgdorferi*, is C-terminally

processed and contains surface-exposed domains. *Infect Immun* **69**:3323–3334.

59. **Östberg Y, Carroll JA, Pinne M, Krum JG, Rosa P, Bergström S.** 2004. Pleiotropic effects of inactivating a carboxyl-terminal protease, CtpA, in *Borrelia burgdorferi*. *J Bacteriol* **186**:2074–2084.

60. **Yang X, Promnares K, Qin J, He M, Shroder DY, Kariu T, Wang Y, Pal U.** 2011. Characterization of multiprotein complexes of the *Borrelia burgdorferi* outer membrane vesicles. *J Proteome Res* **10**:4556–4566.

61. **Bárcena-Uribarri I, Thein M, Barbot M, Sans-Serramitjana E, Bonde M, Mentele R, Lottspeich F, Bergström S, Benz R.** 2014. Study of the protein complex, pore diameter, and pore-forming activity of the *Borrelia burgdorferi* P13 porin. *J Biol Chem* **289**:18614–18624.

62. **Bunikis J, Barbour AG.** 1999. Access of antibody or trypsin to an integral outer membrane protein (P66) of *Borrelia burgdorferi* is hindered by Osp lipoproteins. *Infect Immun* **67**:2874–2883.

63. **Bunikis J, Noppa L, Ostberg Y, Barbour AG, Bergström S.** 1996. Surface exposure and species specificity of an immunoreactive domain of a 66-kilodalton outer membrane protein (P66) of the *Borrelia* spp. that cause Lyme disease. *Infect Immun* **64**:5111–5116.

64. **Skare JT, Mirzabekov TA, Shang ES, Blanco DR, Erdjument-Bromage H, Bunikis J, Bergström S, Tempst P, Kagan BL, Miller JN, Lovett MA.** 1997. The Oms66 (p66) protein is a *Borrelia burgdorferi* porin. *Infect Immun* **65**:3654–3661.

65. **Ristow LC, Bonde M, Lin YP, Sato H, Curtis M, Wesley E, Hahn BL, Fang J, Wilcox DA, Leong JM, Bergström S, Coburn J.** 2015. Integrin binding by *Borrelia burgdorferi* P66 facilitates dissemination but is not required for infectivity. *Cell Microbiol* **17**:1021–1036.

66. **Coburn J, Cugini C.** 2003. Targeted mutation of the outer membrane protein P66 disrupts attachment of the Lyme disease agent, *Borrelia burgdorferi*, to integrin alphavbeta3. *Proc Natl Acad Sci U S A* **100**:7301–7306.

67. **Iqbal H, Kenedy MR, Lybecker M, Akins DR.** 2016. The TamB ortholog of *Borrelia burgdorferi* interacts with the β-barrel assembly machine (BAM) complex protein BamA. *Mol Microbiol* **102**:757–774.

68. **Gherardini FC.** 2013. *Borrelia burgdorferi* HtrA may promote dissemination and irritation. *Mol Microbiol* **90**:209–213.

69. **Stewart PE, Hoff J, Fischer E, Krum JG, Rosa PA.** 2004. Genome-wide transposon mutagenesis of *Borrelia burgdorferi* for identification of phenotypic mutants. *Appl Environ Microbiol* **70**:5973–5979.

70. **Kariu T, Yang X, Marks CB, Zhang X, Pal U.** 2013. Proteolysis of BB0323 results in two polypeptides that impact physiologic and infectious phenotypes in *Borrelia burgdorferi*. *Mol Microbiol* **88**:510–522.

71. **Coleman JL, Crowley JT, Toledo AM, Benach JL.** 2013. The HtrA protease of *Borrelia burgdorferi* degrades outer membrane protein BmpD and chemotaxis phosphatase CheX. *Mol Microbiol* **88**:619–633.

72. **Russell TM, Delorey MJ, Johnson BJ.** 2013. *Borrelia burgdorferi* BbHtrA degrades host ECM proteins and stimulates release of inflammatory cytokines in vitro. *Mol Microbiol* **90**:241–251.

73. **Lenhart TR, Kenedy MR, Yang X, Pal U, Akins DR.** 2012. BB0324 and BB0028 are constituents of the *Borrelia burgdorferi* β-barrel assembly machine (BAM) complex. *BMC Microbiol* **12**:60.

74. **Lenhart TR, Akins DR.** 2010. *Borrelia burgdorferi* locus BB0795 encodes a BamA orthologue required for growth and efficient localization of outer membrane proteins. *Mol Microbiol* **75**:692–709.

75. **Dunstan RA, Hay ID, Wilksch JJ, Schittenhelm RB, Purcell AW, Clark J, Costin A, Ramm G, Strugnell RA, Lithgow T.** 2015. Assembly of the secretion pores GspD, Wza and CsgG into bacterial outer membranes does not require the Omp85 proteins BamA or TamA. *Mol Microbiol* **97**:616–629.

76. **Misra R, Stikeleather R, Gabriele R.** 2015. In vivo roles of BamA, BamB and BamD in the biogenesis of BamA, a core protein of the β-barrel assembly machine of *Escherichia coli*. *J Mol Biol* **427**:1061–1074.

77. **Gunasinghe SD, Shiota T, Stubenrauch CJ, Schulze KE, Webb CT, Fulcher AJ, Dunstan RA, Hay ID, Naderer T, Whelan DR, Bell TDM, Elgass KD, Strugnell RA, Lithgow T.** 2018. The WD40 protein BamB mediates coupling of BAM complexes into assembly precincts in the bacterial outer membrane. *Cell Rep* **23**:2782–2794.

78. **Dunn JP, Kenedy MR, Iqbal H, Akins DR.** 2015. Characterization of the β-barrel assembly machine accessory lipoproteins from *Borrelia burgdorferi*. *BMC Microbiol* **15**:70.

79. **Stubenrauch CJ, Lithgow T.** 2019. The TAM: a translocation and assembly module of the β-barrel assembly machinery in bacterial outer membranes. *EcoSal Plus* **8**:ESP-0036-2018.

80. **Stubenrauch C, Grinter R, Lithgow T.** 2016. The modular nature of the β-barrel assembly machinery, illustrated in *Borrelia burgdorferi*. *Mol Microbiol* **102**:753–756.

81. **Picardeau M.** 2017. Virulence of the zoonotic agent of leptospirosis: still terra incognita? *Nat Rev Microbiol* **15:**297–307.

82. **Dunham-Ems SM, Caimano MJ, Pal U, Wolgemuth CW, Eggers CH, Balic A, Radolf JD.** 2009. Live imaging reveals a biphasic mode of dissemination of *Borrelia burgdorferi* within ticks. *J Clin Invest* **119:**3652–3665.

83. **Ellis TN, Kuehn MJ.** 2010. Virulence and immunomodulatory roles of bacterial outer membrane vesicles. *Microbiol Mol Biol Rev* **74:**81–94.

84. **Floden AM, Watt JA, Brissette CA.** 2011. *Borrelia burgdorferi* enolase is a surface-exposed plasminogen binding protein. *PLoS One* **6:** e27502.

85. **Nogueira SV, Smith AA, Qin JH, Pal U.** 2012. A surface enolase participates in *Borrelia burgdorferi*-plasminogen interaction and contributes to pathogen survival within feeding ticks. *Infect Immun* **80:**82–90.

86. **Toledo A, Coleman JL, Kuhlow CJ, Crowley JT, Benach JL.** 2012. The enolase of *Borrelia burgdorferi* is a plasminogen receptor released in outer membrane vesicles. *Infect Immun* **80:**359–368.

87. **Crowley JT, Toledo AM, LaRocca TJ, Coleman JL, London E, Benach JL.** 2013. Lipid exchange between *Borrelia burgdorferi* and host cells. *PLoS Pathog* **9:**e1003109.

88. **Toledo A, Crowley JT, Coleman JL, LaRocca TJ, Chiantia S, London E, Benach JL.** 2014. Selective association of outer surface lipoproteins with the lipid rafts of *Borrelia burgdorferi*. *mBio* **5:** e00899-14.

89. **Toledo A, Pérez A, Coleman JL, Benach JL.** 2015. The lipid raft proteome of *Borrelia burgdorferi*. *Proteomics* **15:**3662–3675.

90. **Ward HB.** 1908. The spirochetes and their relationship to other organisms. *Am Nat* **42:**374–387.

91. **Drecktrah D, Samuels DS.** 2018. Genetic manipulation of *Borrelia* spp. *Curr Top Microbiol Immunol* **415:**113–140.

92. **Picardeau M.** 2015. Genomics, proteomics, and genetics of leptospira. *Curr Top Microbiol Immunol* **387:**43–63.

93. **Edmondson DG, Hu B, Norris SJ.** 2018. Long-term *in vitro* culture of the syphilis spirochete *Treponema pallidum* subsp. *pallidum*. *mBio* **9:** e01153-18.

94. **Hampson DJ.** 2017. The spirochete *Brachyspira pilosicoli*, enteric pathogen of animals and humans. *Clin Microbiol Rev* **31:**e00087-17.

95. **Haake DA, Zückert WR.** 2018. Spirochetal lipoproteins in pathogenesis and immunity. *Curr Top Microbiol Immunol* **415:**239–271.

Outer Membrane Protein Insertion by the β-barrel Assembly Machine

8

DANTE P. RICCI[1] and THOMAS J. SILHAVY[2]

INTRODUCTION

The presence in Gram-negative bacteria of an extracytoplasmic outer membrane (OM), which is distinct from the inner membrane (IM) both in constitution and in function, presents a complex topological problem, as all proteinaceous and lipidic OM components are synthesized cytoplasmically (1). In order to reach their destination in the growing OM, these components must translocate across the IM and traverse the aqueous, crowded periplasmic space. This problem is solved through a series of semi-independent and highly conserved transport pathways that coordinate the efficient delivery and integration of all OM constituents.

OM-specific lipopolysaccharide (LPS) is trafficked via a multicomponent trans-envelope protein bridge (2, 3), the LPS transport (Lpt) pathway, which terminates at an OM-integral translocase (LptDE) that incorporates free LPS into the outer leaflet of the OM (4). A transport system that enables retrograde (OM to IM) phospholipid transport has been described (5–10), but the mechanism of anterograde (IM to OM) transport is mysterious and represents an area of active investigation. Periplasmic lipoproteins, which can be anchored

[1]Department of Early Research, Achaogen, Inc., South San Francisco, CA 94080
[2]Department of Molecular Biology, Princeton University, Princeton, NJ 08544
Protein Secretion in Bacteria
Edited by Maria Sandkvist, Eric Cascales, and Peter J. Christie
© 2019 American Society for Microbiology, Washington, DC
doi:10.1128/ecosalplus.ESP-0035-2018

to either membrane via N-terminal lipid moieties, are sorted in a sequence-dependent manner to the OM via the Lol system, which extracts IM-associated lipoproteins and shuttles them to the OM via a soluble periplasmic carrier (11). Finally, integral OM β-barrel proteins (OMPs) are translocated in an unfolded form across the IM, ferried to the OM in a nonnative but folding-competent state via a diverse network of periplasmic chaperones, and integrated into the OM in a manner dependent on a ubiquitous, essential multiprotein complex known as the β-barrel assembly machine (Bam).

Despite the functional and structural heterogeneity observed across OMP families, the *in vivo* folding and membrane integration of all OM β-barrel proteins require Bam, an OM-associated heteromeric complex composed of BamA (itself a β-barrel protein) and a variable number of OM lipoproteins (BamB to -F) that bind to and act in concert with BamA to drive the OMP assembly process (1, 4, 12, 13). At present, definitive roles cannot yet be unambiguously assigned to the individual components, and the general mechanism of BamA-dependent OMP assembly remains elusive and controversial. Here we offer a compendious review of recent inquiry into the mechanism of OM β-barrel folding and its catalysis by the Bam complex.

Bam COMPLEX CONSTITUENTS

BamA, the central component of the Bam complex, is composed of a C-terminal β-barrel domain and an N-terminal periplasmic domain that serves as the physical hub of a functional network that includes substrates (14–18), accessory lipoproteins (19–26), periplasmic chaperones (27–29), and proteases (30). This fishhook-shaped domain is typically subdivided into five structurally homologous POTRA domains (31, 32) that may nucleate the early formation of OMP secondary structure through β-strand augmentation (26, 32, 33) and which, together with the Bam

lipoproteins, form a cavernous periplasmic ring that circumscribes the vestibule of the BamA β-barrel lumen (24, 34). The OM-integral BamA β-barrel domain is atypical and can be distinguished from canonical membrane β-barrels in three fundamental ways: (i) the interstrand hydrogen bond network that seals β1 and β16 (to complete the barrel) is metastable, allowing transient destabilization of this seam and reversible lateral opening of the barrel (35–38); (ii) the β-barrel forms an unusually narrow protein-lipid interface adjacent to the seam that is thought to physically alter the local properties of the OM (37, 39–41); and (iii) a conserved, essential latching loop (L6) internally braces the BamA barrel interior and globally stabilizes the otherwise thermolabile β-barrel domain (42–46), likely compensating for the metastability at the β1-β16 seam (Fig. 1). The mechanistic implications of these features are discussed in an ensuing section.

BamA is the central catalyst of OMP insertion, but the partner lipoprotein BamD also plays an apparently essential role in the process (13, 14, 16, 17, 20, 47–51). BamD is a solenoid protein thought to serve as a generic receptor for substrate OMPs through recognition of the β-signal, a semidegenerate C-terminal peptide motif common to all prokaryotic and eukaryotic OMPs (14, 52–60). BamD plays a critical role in the binding and OM localization of OMPs (including BamA), although BamB, a β-propeller protein, has been proposed to perform an overlapping function for some Bam substrates (17, 26, 47). Additionally, commensurate with its role as an OMP receptor, BamD has been implicated in regulation of the conformational dynamics and activity of BamA during the OMP assembly cycle, tentatively linking the recognition and binding of nascent OMP C-termini to conformational changes in BamA that enable OM insertion (14, 48, 61, 62).

The remaining Bam lipoproteins (BamB, BamC, BamE, and the BamC-like α-proteobacterial protein BamF) are variably conserved, are not central to the mechanism of

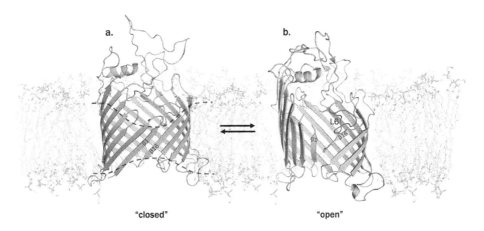

FIGURE 1 Unique features of the BamA β-barrel domain. (a) The BamA β-barrel (pink) is asymmetric, with one face forming a narrow protein-lipid interface (approximated by the dashed line) that is thought to physically alter the local properties of the bilayer (green). **(b)** The activated BamA β-barrel undergoes a dramatic conformational rearrangement that disrupts the continuous β-barrel structure, separates the β-strands comprising the lateral gate (β1 and β16), and exposes an aqueous channel that spans the membrane. Additionally, a highly conserved extracellular loop (L6, blue) internally braces and globally stabilizes the β-barrel domain and compensates for the instability introduced by the conformational dynamics. This image was generated using PDB structures 4K3B (left) and 5EKQ (right).

Bam-catalyzed OMP assembly, and instead play accessory roles that enhance the efficiency of the process for at least a subset of OMPs (13, 63). BamCE associate indirectly with BamA via BamD and play an adjunctive role in the regulation of BamA dynamics and function, potentially by stabilizing the interaction between BamA and BamD following each round of OMP assembly (20, 21, 47, 49, 61, 64–67). BamB interacts with BamA in a BamD-independent manner through direct contacts with multiple POTRA domains (20–22, 25, 26) and has been shown to contribute significantly to the efficiency of substrate assembly for certain OMPs (17, 47, 68–75). We have proposed that BamB (together with the OMP chaperone SurA) influences substrate flux to ensure streamlined assembly of both high-abundance targets (e.g., OmpA and porins) and low-abundance, high-priority targets (e.g., LptDE) (28). Intriguing recent work has also uncovered a role for BamB in the sequestration of Bam complexes into tightly clustered "assembly precincts" that are proposed to accelerate

the assembly and multimerization of abundant OMP species (75).

Bam AS A FOLDING CHAPERONE AND CATALYST OF OM PROTEIN INSERTION

A wide variety of OM β-barrel proteins can autonomously fold in hydrophobic environments in accordance with Anfinsen's dogma. Decades of *in vitro* studies using model OMPs have revealed the following key observations (Fig. 2):

1. In the appropriate hydrophobic context, OMPs spontaneously and rapidly fold into extremely stable membrane integral species (76–89).
2. Native OM phospholipids impose a kinetic barrier to OMP assembly (a critical phenomenon that prevents the lethal assembly of OM β-barrels into the IM) (40, 41, 90–92).
3. The kinetics of OMP assembly can be dramatically accelerated by altering

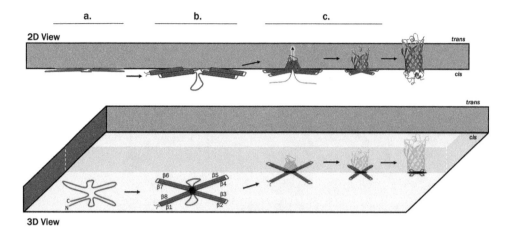

FIGURE 2 Proposed unassisted folding model for a membrane β-barrel protein. (a) A nascent eight-stranded OMP rapidly adsorbs to the *cis* surface (green) of the lipid bilayer, where hydrophobic lipid-facing side chains begin to penetrate into the membrane and β-strands assume a cloverleaf-like circular arrangement according to their relative position in the folded protein. **(b)** β-hairpins begin to form as the *trans* ends of the TM β-strands, oriented toward the center of the cloverleaf, plunge into the lipid phase. **(c)** Hydrogen bonds form between neighboring β-hairpins as they enter the membrane, stabilizing the native fold in concert with membrane insertion. The highlighted dashed line indicates the proposed path of the leading (*trans*) ends of the β-strands.

physical properties of the membrane so as to induce local defects (e.g., bilayer thinning or increased curvature) (39, 54, 93, 94).

4. OMP folding is a concerted process in which barrel formation and insertion happen concurrently and in which all β-strands integrate simultaneously rather than sequentially (79, 95–99).

The fact that β-barrel folding and insertion occur spontaneously implies a role for Bam in accelerating the intrinsic folding kinetics of OMPs, akin to classical folding chaperones. However, alternative pathways for *in vivo* folding have been surmised that involve the formation of transient chimeric BamA:OMP barrels (37, 100) or elongated substrate β-sheets nucleated at BamA β1 (101) as necessary intermediates. These models have explanatory power but are problematic from a thermodynamic perspective and require unnecessary invention of a distinct and strictly BamA-dependent folding pathway for β-barrel proteins (102). A recent biochemical analysis of mitochondrial β-barrel insertion by the BamA homolog Sam50 yielded observations consistent with a β-strand exchange model (100); however, the use of nonnative, truncated substrates complicates interpretation, and the available evidence does not rule out more parsimonious alternatives. We argue that existing evidence better comports with a view of Bam as an OM-adjacent Anfinsen cage that positions client proteins for OM insertion, prevents aggregation, degradation, and off-pathway misfolding, and accelerates the native OMP folding reaction. Numerous studies have established the role of lipid bilayer defects in the acceleration of β-barrel folding kinetics, and BamA has been shown to give rise to such defects (36, 37, 39–41, 103). These observations inform a simple model in which Bam complexes effectively localize client proteins to OM "entry points" generated by the local defects imposed by BamA itself, removing the primary barrier to rapid OMP folding and enabling OM biogenesis on physiologically relevant timescales (15, 36, 90, 104).

The importance of the BamA lateral gate is a matter of ongoing debate. Artificial locking of this seam is lethal *in vivo* (34, 105) and moderately impairs OmpT folding kinetics into proteoliposomes (38), but it has no effect on the *in vitro* assembly of OmpA or OmpX, both small OMPs with minimal loop structure (36, 41). This incongruity might be reconciled if certain Bam substrates do not require opening of the gate whereas others do, such as those with large barrels or extensive hydrophilic extracellular loops (45, 106), which could avoid the entropic penalty associated with membrane translocation by traversing the OM through the hydrophilic lumen of the open BamA barrel, with the gate serving as a transient "slit" allowing passage of loops attached to transmembrane β-strands that are integrating into the lipid phase. This bears some resemblance to the mechanism of polytopic IM protein assembly by the SecYEG insertase/translocase, in which transmembrane segments diffuse into the membrane adjacent to a lateral gate and periplasmic loops are translocated through the activated, ungated SecYEG pore (107–109). However, it is also clear that specific loops from certain substrates are buried within the lumen of the barrel during folding, potentially scaffolding barrel formation and driving the maturation process, likely rendering an assisted loop translocation process dispensable.

The requirement for both rapid OM growth and a reliably impermeable OM represents an intriguing paradox and raises the possibility that local defects induced by BamA form only when OMP insertion is imminent, as a constitutively "open" complex would likely generate membrane instability and allow indiscriminate diffusion across the OM. This paradox implies tight regulation of Bam activity and a concerted mechanism for Bam activation that is linked to substrate recognition (14, 16, 48, 62). Consistent with this notion, there is mounting evidence that the BamA barrel exists in equilibrium between two conformations: a "closed" state in which extracellular loops form a dome that stabilizes

a thermostable and complete β-barrel, and an "open" state in which extracellular loops reorient and the N-terminal β-strands of the barrel are dramatically wrenched outward, begetting a thermolabile, incomplete β-barrel with an exposed aqueous pore (21, 34, 37). This equilibrium can apparently be altered through mutations in L6 that prevent loop latching and destabilize the barrel (42, 43, 110), or through mutations in the BamCDE subcomplex that are presumed to slow the restoration of the closed conformation following a round of OMP assembly (61). Together with observations linking barrel-proximal POTRA 5, lipoproteins, and L6 to conformational dynamics within the BamA barrel (14, 38, 44, 61, 62, 64, 65, 111, 112), these findings extend the mechanistic model to include a role for nascent OMPs as homotropic allosteric activators of Bam that initiate an activation cascade resulting in transient opening of the BamA barrel and concomitant OM integration of folding OMPs.

A MODEL FOR BamA-ASSISTED β-BARREL FOLDING

In light of the emerging mechanistic details reviewed above, the *in vivo* β-barrel assembly process can be summarized as follows (Fig. 3). Unfolded OMPs are transferred from periplasmic chaperones to Bam, where they are scaffolded through direct interaction with the POTRAs and one or more Bam lipoproteins (18, 23, 24, 26). β-strands of substrate OMPs may be organized circumferentially according to their relative positions in the final folded β-barrel structure (Fig. 2, item b) (18, 113–115). Recognition of the β-signal by BamD triggers a conformational rearrangement in the complex that includes rotation of the POTRAs away from the BamA barrel lumen and/or adoption of an open barrel conformation (14, 21, 34). Local perturbation of the OM, induced by BamA and exacerbated by conformational dynamics within the activated β-barrel domain, allows rapid, spontaneous folding and

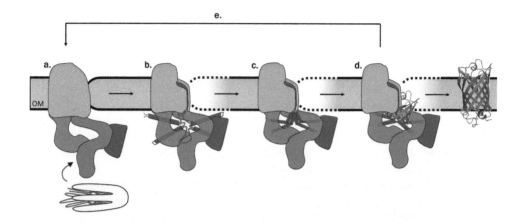

FIGURE 3 Proposed Bam-assisted folding model. (a) Nascent OMPs (red), maintained in a folding-competent state by periplasmic chaperones (green), are transferred to the Bam complex (blue). **(b)** Client proteins associate with multiple epitopes on Bam, potentially stimulating formation of early β-structure and orienting circularly-arranged β-strands/hairpins toward the presumptive substrate exit pore. Recognition of conserved OMP motifs triggers a conformational change in BamA that exposes the barrel lumen and destabilizes the lateral gate, further perturbing the local membrane environment and generating an OM integration path for OMP substrates. **(c)** The Bam complex prevents aggregation, protects substrates from proteolysis, and lowers the kinetic barrier to OM integration to enable rapid OMP folding along the native pathway. **(d)** OMPs spontaneously fold into the locally destabilized membrane, with the exposed BamA lumen potentially accommodating the folding barrel and/or secreted extracellular domains of client proteins. **(e)** Release of substrate from the complex prompts restoration of the closed, inert state of the complex to enable an ensuing round of assembly.

simultaneous OM insertion of OMP substrates along the native folding pathway, with long, hydrophilic loops potentially threaded through the BamA barrel lumen during OMP folding via the lateral slit. Although the precise translocation path of BamA substrates remains undefined, the orientation of the periplasmic ring-like Bam apparatus relative to the OM raises the possibility that substrates are funneled through a proteinaceous aperture adjacent to the BamA barrel that could serve as both a substrate exit channel and a site of OM insertion (24). Following substrate release, the closed conformation of the BamA barrel is restored through conformational dynamics among the Bam lipoproteins and BamA, priming the complex for an ensuing round of assembly (61, 116). We anticipate that the availability of *in vitro* reconstitution systems, comprehensive panels of informative mutants, high-resolution imaging techniques, sensitive biophysical assays and models, and an abun-

dance of structural data will enable a detailed analysis of the Bam catalytic mechanism and the potential variations on the generic theme presented here for diverse OMP substrates.

ACKNOWLEDGMENTS

The authors thank fellow lab members for suggestions and critical reading of the manuscript. T.J.S. was supported by a grant from NIGMS (R35GM118024).

CITATION

EcoSal Plus 2019; doi:10.1128/ecosalplus.ESP-0035-2018.

REFERENCES

1. **Konovalova A, Kahne DE, Silhavy TJ.** 2017. Outer membrane biogenesis. *Annu Rev Microbiol* **71:**539–556.

2. **Okuda S, Freinkman E, Kahne D.** 2012. Cytoplasmic ATP hydrolysis powers transport of lipopolysaccharide across the periplasm in *E. coli. Science* **338:**1214–1217.

3. **Sherman DJ, Xie R, Taylor RJ, George AH, Okuda S, Foster PJ, Needleman DJ, Kahne D.** 2018. Lipopolysaccharide is transported to the cell surface by a membrane-to-membrane protein bridge. *Science* **359:**798–801.

4. **Botos I, Noinaj N, Buchanan SK.** 2017. Insertion of proteins and lipopolysaccharide into the bacterial outer membrane. *Philos Trans R Soc Lond B Biol Sci* **372:**20160224.

5. **Sutterlin HA, Shi H, May KL, Miguel A, Khare S, Huang KC, Silhavy TJ.** 2016. Disruption of lipid homeostasis in the Gram-negative cell envelope activates a novel cell death pathway. *Proc Natl Acad Sci U S A* **113:** E1565–E1574.

6. **Malinverni JC, Silhavy TJ.** 2009. An ABC transport system that maintains lipid asymmetry in the gram-negative outer membrane. *Proc Natl Acad Sci U S A* **106:**8009–8014.

7. **Abellón-Ruiz J, Kaptan SS, Baslé A, Claudi B, Bumann D, Kleinekathöfer U, van den Berg B.** 2017. Structural basis for maintenance of bacterial outer membrane lipid asymmetry. *Nat Microbiol* **2:**1616–1623.

8. **Chong Z-S, Woo W-F, Chng S-S.** 2015. Osmoporin OmpC forms a complex with MlaA to maintain outer membrane lipid asymmetry in *Escherichia coli. Mol Microbiol* **98:**1133–1146.

9. **Powers MJ, Trent MS.** 2018. Phospholipid retention in the absence of asymmetry strengthens the outer membrane permeability barrier to last-resort antibiotics. *Proc Natl Acad Sci U S A* **115:** E8518–E8527.

10. **Ekiert DC, Bhabha G, Isom GL, Greenan G, Ovchinnikov S, Henderson IR, Cox JS, Vale RD.** 2017. Architectures of lipid transport systems for the bacterial outer membrane. *Cell* **169:**273–285.e17.

11. **Okuda S, Tokuda H.** 2011. Lipoprotein sorting in bacteria. *Annu Rev Microbiol* **65:**239–259.

12. **Ranava D, Caumont-Sarcos A, Albenne C, Ieva R.** 2018. Bacterial machineries for the assembly of membrane-embedded β-barrel proteins. *FEMS Microbiol Lett* **365:**fny087.

13. **Anwari K, Webb CT, Poggio S, Perry AJ, Belousoff M, Celik N, Ramm G, Lovering A, Sockett RE, Smit J, Jacobs-Wagner C, Lithgow T.** 2012. The evolution of new lipoprotein subunits of the bacterial outer membrane BAM complex. *Mol Microbiol* **84:**832–844.

14. **Lee J, Sutterlin HA, Wzorek JS, Mandler MD, Hagan CL, Grabowicz M, Tomasek D, May MD, Hart EM, Silhavy TJ, Kahne D.** 2018. Substrate binding to BamD triggers a conformational change in BamA to control membrane insertion. *Proc Natl Acad Sci U S A* **115:**2359–2364.

15. **Lee J, Xue M, Wzorek JS, Wu T, Grabowicz M, Gronenberg LS, Sutterlin HA, Davis RM, Ruiz N, Silhavy TJ, Kahne DE.** 2016. Characterization of a stalled complex on the β-barrel assembly machine. *Proc Natl Acad Sci U S A* **113:**8717–8722.

16. **Hagan CL, Wzorek JS, Kahne D.** 2015. Inhibition of the β-barrel assembly machine by a peptide that binds BamD. *Proc Natl Acad Sci U S A* **112:**2011–2016.

17. **Hagan CL, Westwood DB, Kahne D.** 2013. Bam lipoproteins assemble BamA *in vitro. Biochemistry* **52:**6108–6113.

18. **Ieva R, Tian P, Peterson JH, Bernstein HD.** 2011. Sequential and spatially restricted interactions of assembly factors with an autotransporter beta domain. *Proc Natl Acad Sci U S A* **108:**E383–E391.

19. **Vuong P, Bennion D, Mantei J, Frost D, Misra R.** 2008. Analysis of YfgL and YaeT interactions through bioinformatics, mutagenesis, and biochemistry. *J Bacteriol* **190:**1507–1517.

20. **Malinverni JC, Werner J, Kim S, Sklar JG, Kahne D, Misra R, Silhavy TJ.** 2006. YfiO stabilizes the YaeT complex and is essential for outer membrane protein assembly in *Escherichia coli. Mol Microbiol* **61:**151–164.

21. **Bakelar J, Buchanan SK, Noinaj N.** 2016. The structure of the β-barrel assembly machinery complex. *Science* **351:**180–186.

22. **Chen Z, Zhan LH, Hou HF, Gao ZQ, Xu JH, Dong C, Dong YH.** 2016. Structural basis for the interaction of BamB with the POTRA3-4 domains of BamA. *Acta Crystallogr D Struct Biol* **72:**236–244.

23. **Bergal HT, Hopkins AH, Metzner SI, Sousa MC.** 2016. The structure of a BamA-BamD fusion illuminates the architecture of the β-barrel assembly machine core. *Structure* **24:**243–251.

24. **Han L, Zheng J, Wang Y, Yang X, Liu Y, Sun C, Cao B, Zhou H, Ni D, Lou J, Zhao Y, Huang Y.** 2016. Structure of the BAM complex and its implications for biogenesis of outer-membrane proteins. *Nat Struct Mol Biol* **23:**192–196.

25. **Jansen KB, Baker SL, Sousa MC.** 2015. Crystal structure of BamB bound to a periplasmic domain fragment of BamA, the central component of the β-barrel assembly machine. *J Biol Chem* **290:**2126–2136.

26. **Heuck A, Schleiffer A, Clausen T.** 2011. Augmenting β-augmentation: structural basis of how BamB binds BamA and may support

folding of outer membrane proteins. *J Mol Biol* **406:**659–666.

27. **Bennion D, Charlson ES, Coon E, Misra R.** 2010. Dissection of β-barrel outer membrane protein assembly pathways through characterizing BamA POTRA 1 mutants of *Escherichia coli. Mol Microbiol* **77:**1153–1171.

28. **Ricci DP, Schwalm J, Gonzales-Cope M, Silhavy TJ.** 2013. The activity and specificity of the outer membrane protein chaperone SurA are modulated by a proline isomerase domain. *mBio* **4:**e00540-13.

29. **Patel GJ, Kleinschmidt JH.** 2013. The lipid bilayer-inserted membrane protein BamA of *Escherichia coli* facilitates insertion and folding of outer membrane protein A from its complex with Skp. *Biochemistry* **52:**3974–3986.

30. **Narita S, Masui C, Suzuki T, Dohmae N, Akiyama Y.** 2013. Protease homolog BepA (YfgC) promotes assembly and degradation of β-barrel membrane proteins in *Escherichia coli. Proc Natl Acad Sci U S A* **110:**E3612–E3621.

31. **Gatzeva-Topalova PZ, Warner LR, Pardi A, Sousa MC.** 2010. Structure and flexibility of the complete periplasmic domain of BamA: the protein insertion machine of the outer membrane. *Structure* **18:**1492–1501.

32. **Kim S, Malinverni JC, Sliz P, Silhavy TJ, Harrison SC, Kahne D.** 2007. Structure and function of an essential component of the outer membrane protein assembly machine. *Science* **317:**961–964.

33. **Remaut H, Waksman G.** 2006. Protein-protein interaction through beta-strand addition. *Trends Biochem Sci* **31:**436–444.

34. **Gu Y, Li H, Dong H, Zeng Y, Zhang Z, Paterson NG, Stansfeld PJ, Wang Z, Zhang Y, Wang W, Dong C.** 2016. Structural basis of outer membrane protein insertion by the BAM complex. *Nature* **531:**64–69.

35. **Lundquist K, Bakelar J, Noinaj N, Gumbart JC.** 2018. C-terminal kink formation is required for lateral gating in BamA. *Proc Natl Acad Sci U S A* **115:**E7942–E7949.

36. **Doerner PA, Sousa MC.** 2017. Extreme dynamics in the BamA β-barrel seam. *Biochemistry* **56:**3142–3149.

37. **Noinaj N, Kuszak AJ, Gumbart JC, Lukacik P, Chang H, Easley NC, Lithgow T, Buchanan SK.** 2013. Structural insight into the biogenesis of β-barrel membrane proteins. *Nature* **501:**385–390.

38. **Iadanza MG, Higgins AJ, Schiffrin B, Calabrese AN, Brockwell DJ, Ashcroft AE, Radford SE, Ranson NA.** 2016. Lateral opening in the intact β-barrel assembly machinery captured by cryo-EM. *Nat Commun* **7:**12865.

39. **Danoff EJ, Fleming KG.** 2015. Membrane defects accelerate outer membrane β-barrel protein folding. *Biochemistry* **54:**97–99.

40. **Gessmann D, Chung YH, Danoff EJ, Plummer AM, Sandlin CW, Zaccai NR, Fleming KG.** 2014. Outer membrane β-barrel protein folding is physically controlled by periplasmic lipid head groups and BamA. *Proc Natl Acad Sci U S A* **111:**5878–5883.

41. **Schiffrin B, Calabrese AN, Higgins AJ, Humes JR, Ashcroft AE, Kalli AC, Brockwell DJ, Radford SE.** 2017. Effects of periplasmic chaperones and membrane thickness on bama-catalyzed outer-membrane protein folding. *J Mol Biol* **429:**3776–3792.

42. **Leonard-Rivera M, Misra R.** 2012. Conserved residues of the putative L6 loop of *Escherichia coli* BamA play a critical role in the assembly of β-barrel outer membrane proteins, including that of BamA itself. *J Bacteriol* **194:**4662–4668.

43. **Dwyer RS, Ricci DP, Colwell LJ, Silhavy TJ, Wingreen NS.** 2013. Predicting functionally informative mutations in *Escherichia coli* BamA using evolutionary covariance analysis. *Genetics* **195:**443–455.

44. **Thoma J, Sun Y, Ritzmann N, Müller DJ.** 2018. POTRA domains, extracellular lid, and membrane composition modulate the conformational stability of the β barrel assembly factor BamA. *Structure* **26:**987–996.e3.

45. **Wzorek JS, Lee J, Tomasek D, Hagan CL, Kahne DE.** 2017. Membrane integration of an essential β-barrel protein prerequires burial of an extracellular loop. *Proc Natl Acad Sci U S A* **114:**2598–2603.

46. **Hartmann J-B, Zahn M, Burmann IM, Bibow S, Hiller S.** 2018. Sequence-specific solution NMR assignments of the β-barrel insertase BamA to monitor its conformational ensemble at the atomic level. *J Am Chem Soc* **140:**11252–11260.

47. **Misra R, Stikeleather R, Gabriele R.** 2015. In vivo roles of BamA, BamB and BamD in the biogenesis of BamA, a core protein of the β-barrel assembly machine of *Escherichia coli. J Mol Biol* **427:**1061–1074.

48. **Ricci DP, Hagan CL, Kahne D, Silhavy TJ.** 2012. Activation of the *Escherichia coli* β-barrel assembly machine (Bam) is required for essential components to interact properly with substrate. *Proc Natl Acad Sci U S A* **109:**3487–3491.

49. **Sikora AE, Wierzbicki IH, Zielke RA, Ryner RF, Korotkov KV, Buchanan SK, Noinaj N.** 2018. Structural and functional insights into the role of BamD and BamE within the β-barrel assembly machinery in *Neisseria gonorrhoeae. J Biol Chem* **293:**1106–1119.

50. **Mahoney TF, Ricci DP, Silhavy TJ.** 2016. Classifying β-barrel assembly substrates by manipulating essential Bam complex members. *J Bacteriol* **198:**1984–1992.

51. **Rossiter AE, Leyton DL, Tveen-Jensen K, Browning DF, Sevastsyanovich Y, Knowles TJ, Nichols KB, Cunningham AF, Overduin M, Schembri MA, Henderson IR.** 2011. The essential β-barrel assembly machinery complex components BamD and BamA are required for autotransporter biogenesis. *J Bacteriol* **193:**4250–4253.

52. **Kutik S, Stojanovski D, Becker L, Becker T, Meinecke M, Krüger V, Prinz C, Meisinger C, Guiard B, Wagner R, Pfanner N, Wiedemann N.** 2008. Dissecting membrane insertion of mitochondrial beta-barrel proteins. *Cell* **132:**1011–1024.

53. **Walther DM, Papic D, Bos MP, Tommassen J, Rapaport D.** 2009. Signals in bacterial beta-barrel proteins are functional in eukaryotic cells for targeting to and assembly in mitochondria. *Proc Natl Acad Sci U S A* **106:**2531–2536.

54. **Iyer BR, Zadafiya P, Vetal PV, Mahalakshmi R.** 2017. Energetics of side-chain partitioning of β-signal residues in unassisted folding of a transmembrane β-barrel protein. *J Biol Chem* **292:**12351–12365.

55. **Paramasivam N, Habeck M, Linke D.** 2012. Is the C-terminal insertional signal in Gram-negative bacterial outer membrane proteins species-specific or not? *BMC Genomics* **13:**510.

56. **Kozjak-Pavlovic V, Ott C, Götz M, Rudel T.** 2011. Neisserial Omp85 protein is selectively recognized and assembled into functional complexes in the outer membrane of human mitochondria. *J Biol Chem* **286:**27019–27026.

57. **Müller JEN, Papic D, Ulrich T, Grin I, Schütz M, Oberhettinger P, Tommassen J, Linke D, Dimmer KS, Autenrieth IB, Rapaport D.** 2011. Mitochondria can recognize and assemble fragments of a beta-barrel structure. *Mol Biol Cell* **22:**1638–1647.

58. **Robert V, Volokhina EB, Senf F, Bos MP, Van Gelder P, Tommassen J.** 2006. Assembly factor Omp85 recognizes its outer membrane protein substrates by a species-specific C-terminal motif. *PLoS Biol* **4:**e377.

59. **Albrecht R, Zeth K.** 2011. Structural basis of outer membrane protein biogenesis in bacteria. *J Biol Chem* **286:**27792–27803.

60. **Sandoval CM, Baker SL, Jansen K, Metzner SI, Sousa MC.** 2011. Crystal structure of BamD: an essential component of the β-barrel assembly machinery of gram-negative bacteria. *J Mol Biol* **409:**348–357.

61. **Rigel NW, Ricci DP, Silhavy TJ.** 2013. Conformation-specific labeling of BamA and suppressor analysis suggest a cyclic mechanism for β-barrel assembly in *Escherichia coli. Proc Natl Acad Sci U S A* **110:**5151–5156.

62. **McCabe AL, Ricci D, Adetunji M, Silhavy TJ.** 2017. Conformational changes that coordinate the activity of BamA and BamD allowing β-barrel assembly. *J Bacteriol* **199:**e00373-17.

63. **Webb CT, Heinz E, Lithgow T.** 2012. Evolution of the β-barrel assembly machinery. *Trends Microbiol* **20:**612–620.

64. **Tellez R Jr, Misra R.** 2012. Substitutions in the BamA β-barrel domain overcome the conditional lethal phenotype of a Δ*bamB* Δ*bamE* strain of *Escherichia coli. J Bacteriol* **194:**317–324.

65. **Rigel NW, Schwalm J, Ricci DP, Silhavy TJ.** 2012. BamE modulates the *Escherichia coli* beta-barrel assembly machine component BamA. *J Bacteriol* **194:**1002–1008.

66. **Sklar JG, Wu T, Gronenberg LS, Malinverni JC, Kahne D, Silhavy TJ.** 2007. Lipoprotein SmpA is a component of the YaeT complex that assembles outer membrane proteins in *Escherichia coli. Proc Natl Acad Sci U S A* **104:**6400–6405.

67. **Kim KH, Kang H-S, Okon M, Escobar-Cabrera E, McIntosh LP, Paetzel M.** 2011. Structural characterization of *Escherichia coli* BamE, a lipoprotein component of the β-barrel assembly machinery complex. *Biochemistry* **50:**1081–1090.

68. **Ureta AR, Endres RG, Wingreen NS, Silhavy TJ.** 2007. Kinetic analysis of the assembly of the outer membrane protein LamB in *Escherichia coli* mutants each lacking a secretion or targeting factor in a different cellular compartment. *J Bacteriol* **189:**446–454.

69. **Charlson ES, Werner JN, Misra R.** 2006. Differential effects of *yfgL* mutation on *Escherichia coli* outer membrane proteins and lipopolysaccharide. *J Bacteriol* **188:**7186–7194.

70. **Ruiz N, Falcone B, Kahne D, Silhavy TJ.** 2005. Chemical conditionality: a genetic strategy to probe organelle assembly. *Cell* **121:**307–317.

71. **Hsieh P-F, Hsu C-R, Chen C-T, Lin T-L, Wang J-T.** 2016. The *Klebsiella pneumoniae* YfgL (BamB) lipoprotein contributes to outer membrane protein biogenesis, type-1 fimbriae expression, anti-phagocytosis, and in vivo virulence. *Virulence* **7:**587–601.

72. **Hagan CL, Kim S, Kahne D.** 2010. Reconstitution of outer membrane protein assembly from purified components. *Science* **328:**890–892.

73. **Palomino C, Marín E, Fernández LÁ.** 2011. The fimbrial usher FimD follows the SurA-BamB pathway for its assembly in the outer membrane of *Escherichia coli. J Bacteriol* **193:**5222–5230.

74. **Schwalm J, Mahoney TF, Soltes GR, Silhavy TJ.** 2013. Role for Skp in LptD assembly in *Escherichia coli. J Bacteriol* **195:**3734–3742.

75. **Gunasinghe SD, Shiota T, Stubenrauch CJ, Schulze KE, Webb CT, Fulcher AJ, Dunstan RA, Hay ID, Naderer T, Whelan DR, Bell TDM, Elgass KD, Strugnell RA, Lithgow T.** 2018. The WD40 protein BamB mediates coupling of BAM complexes into assembly precincts in the bacterial outer membrane. *Cell Rep* **23:**2782–2794.

76. **Huysmans GHM, Radford SE, Brockwell DJ, Baldwin SA.** 2007. The N-terminal helix is a post-assembly clamp in the bacterial outer membrane protein PagP. *J Mol Biol* **373:**529–540.

77. **Hou VC, Moe GR, Raad Z, Wuorimaa T, Granoff DM.** 2003. Conformational epitopes recognized by protective anti-neisserial surface protein A antibodies. *Infect Immun* **71:**6844–6849.

78. **Chaturvedi D, Mahalakshmi R.** 2013. Methionine mutations of outer membrane protein X influence structural stability and beta-barrel unfolding. *PLoS One* **8:**e79351.

79. **Kleinschmidt JH, den Blaauwen T, Driessen AJ, Tamm LK.** 1999. Outer membrane protein A of *Escherichia coli* inserts and folds into lipid bilayers by a concerted mechanism. *Biochemistry* **38:**5006–5016.

80. **Johansson MU, Alioth S, Hu K, Walser R, Koebnik R, Pervushin K.** 2007. A minimal transmembrane beta-barrel platform protein studied by nuclear magnetic resonance. *Biochemistry* **46:**1128–1140.

81. **Surrey T, Schmid A, Jähnig F.** 1996. Folding and membrane insertion of the trimeric β-barrel protein OmpF. *Biochemistry* **35:**2283–2288.

82. **Jansen C, Wiese A, Reubsaet L, Dekker N, de Cock H, Seydel U, Tommassen J.** 2000. Biochemical and biophysical characterization of in vitro folded outer membrane porin PorA of *Neisseria meningitidis. Biochim Biophys Acta* **1464:**284–298.

83. **de Cock H, Hendriks R, de Vrije T, Tommassen J.** 1990. Assembly of an in vitro synthesized *Escherichia coli* outer membrane porin into its stable trimeric configuration. *J Biol Chem* **265:**4646–4651.

84. **Otzen DE, Andersen KK.** 2013. Folding of outer membrane proteins. *Arch Biochem Biophys* **531:**34–43.

85. **Conlan S, Bayley H.** 2003. Folding of a monomeric porin, OmpG, in detergent solution. *Biochemistry* **42:**9453–9465.

86. **Mogensen JE, Kleinschmidt JH, Schmidt MA, Otzen DE.** 2005. Misfolding of a bacterial autotransporter. *Protein Sci* **14:**2814–2827.

87. **Kramer RA, Zandwijken D, Egmond MR, Dekker N.** 2000. In vitro folding, purification and characterization of *Escherichia coli* outer membrane protease ompT. *Eur J Biochem* **267:**885–893.

88. **Pocanschi CL, Apell H-J, Puntervoll P, Høgh B, Jensen HB, Welte W, Kleinschmidt JH.** 2006. The major outer membrane protein of *Fusobacterium nucleatum* (FomA) folds and inserts into lipid bilayers via parallel folding pathways. *J Mol Biol* **355:**548–561.

89. **Dekker N, Merck K, Tommassen J, Verheij HM.** 1995. In vitro folding of *Escherichia coli* outer-membrane phospholipase A. *Eur J Biochem* **232:**214–219.

90. **Fleming KG.** 2015. A combined kinetic push and thermodynamic pull as driving forces for outer membrane protein sorting and folding in bacteria. *Philos Trans R Soc Lond B Biol Sci* **370:**20150026.

91. **Peterson JH, Plummer AM, Fleming KG, Bernstein HD.** 2017. Selective pressure for rapid membrane integration constrains the sequence of bacterial outer membrane proteins. *Mol Microbiol* **106:**777–792.

92. **Grabowicz M, Koren D, Silhavy TJ.** 2016. The CpxQ sRNA negatively regulates Skp to prevent mistargeting of β-barrel outer membrane proteins into the cytoplasmic membrane. *mBio* **7:**e00312-16.

93. **Maurya SR, Chaturvedi D, Mahalakshmi R.** 2013. Modulating lipid dynamics and membrane fluidity to drive rapid folding of a transmembrane barrel. *Sci Rep* **3:**1989.

94. **Horne JE, Radford SE.** 2016. A growing toolbox of techniques for studying β-barrel outer membrane protein folding and biogenesis. *Biochem Soc Trans* **44:**802–809.

95. **Kleinschmidt JH.** 2015. Folding of β-barrel membrane proteins in lipid bilayers—unassisted and assisted folding and insertion. *Biochim Biophys Acta* **1848:**1927–1943.

96. **Danoff EJ, Fleming KG.** 2017. Novel kinetic intermediates populated along the folding pathway of the transmembrane β-barrel OmpA. *Biochemistry* **56:**47–60.

97. **Huysmans GHM, Baldwin SA, Brockwell DJ, Radford SE.** 2010. The transition state for folding of an outer membrane protein. *Proc Natl Acad Sci U S A* **107:**4099–4104.

98. **Tamm LK, Hong H, Liang B.** 2004. Folding and assembly of beta-barrel membrane proteins. *Biochim Biophys Acta* **1666:**250–263.

99. **Thoma J, Bosshart P, Pfreundschuh M, Müller DJ.** 2012. Out but not in: the large transmembrane β-barrel protein FhuA unfolds but cannot refold via β-hairpins. *Structure* 20:2185–2190.

100. **Höhr AIC, Lindau C, Wirth C, Qiu J, Stroud DA, Kutik S, Guiard B, Hunte C, Becker T, Pfanner N, Wiedemann N.** 2018. Membrane protein insertion through a mitochondrial β-barrel gate. *Science* 359:eaah6834.

101. **Schiffrin B, Brockwell DJ, Radford SE.** 2017. Outer membrane protein folding from an energy landscape perspective. *BMC Biol* 15:123.

102. **Plummer AM, Fleming KG.** 2016. From chaperones to the membrane with a BAM! *Trends Biochem Sci* 41:872–882.

103. **Fleming PJ, Patel DS, Wu EL, Qi Y, Yeom MS, Sousa MC, Fleming KG, Im W.** 2016. BamA POTRA domain interacts with a native lipid membrane surface. *Biophys J* 110:2698–2709.

104. **Michalik M, Orwick-Rydmark M, Habeck M, Alva V, Arnold T, Linke D.** 2017. An evolutionarily conserved glycine-tyrosine motif forms a folding core in outer membrane proteins. *PLoS One* 12:e0182016.

105. **Noinaj N, Kuszak AJ, Balusek C, Gumbart JC, Buchanan SK.** 2014. Lateral opening and exit pore formation are required for BamA function. *Structure* 22:1055–1062.

106. **Hagan CL, Silhavy TJ, Kahne D.** 2011. β-Barrel membrane protein assembly by the Bam complex. *Annu Rev Biochem* 80:189–210.

107. **Botte M, Zaccai NR, Nijeholt JLÀ, Martin R, Knoops K, Papai G, Zou J, Deniaud A, Karuppasamy M, Jiang Q, Roy AS, Schulten K, Schultz P, Rappsilber J, Zaccai G, Berger I, Collinson I, Schaffitzel C.** 2016. A central cavity within the holo-translocon suggests a mechanism for membrane protein insertion. *Sci Rep* 6:38399.

108. **Li L, Park E, Ling J, Ingram J, Ploegh H, Rapoport TA.** 2016. Crystal structure of a substrate-engaged SecY protein-translocation channel. *Nature* 531:395–399.

109. **Dalbey RE, Wang P, Kuhn A.** 2011. Assembly of bacterial inner membrane proteins. *Annu Rev Biochem* 80:161–187.

110. **Ni D, Wang Y, Yang X, Zhou H, Hou X, Cao B, Lu Z, Zhao X, Yang K, Huang Y.** 2014. Structural and functional analysis of the β-barrel domain of BamA from *Escherichia coli*. *FASEB J* 28:2677–2685.

111. **Noinaj N, Gumbart JC, Buchanan SK.** 2017. The β-barrel assembly machinery in motion. *Nat Rev Microbiol* 15:197–204.

112. **Sinnige T, Weingarth M, Daniëls M, Boelens R, Bonvin AMJJ, Houben K, Baldus M.** 2015. Conformational plasticity of the POTRA 5 domain in the outer membrane protein assembly factor BamA. *Structure* 23:1317–1324.

113. **Albenne C, Ieva R.** 2017. Job contenders: roles of the β-barrel assembly machinery and the translocation and assembly module in autotransporter secretion. *Mol Microbiol* 106:505–517.

114. **Pavlova O, Peterson JH, Ieva R, Bernstein HD.** 2013. Mechanistic link between β barrel assembly and the initiation of autotransporter secretion. *Proc Natl Acad Sci U S A* 110:E938–E947.

115. **Ieva R, Skillman KM, Bernstein HD.** 2008. Incorporation of a polypeptide segment into the beta-domain pore during the assembly of a bacterial autotransporter. *Mol Microbiol* 67:188–201.

116. **Plummer AM, Fleming KG.** 2015. BamA alone accelerates outer membrane protein folding in vitro through a catalytic mechanism. *Biochemistry* 54:6009–6011.

The TAM: A Translocation and Assembly Module of the β-barrel Assembly Machinery in Bacterial Outer Membranes

9

CHRISTOPHER J. STUBENRAUCH[1] and TREVOR LITHGOW[1]

INTRODUCTION

The vast majority of integral membrane proteins residing within the outer membrane of Gram-negative bacteria adopt a β-barrel architecture. Mechanistically, how these proteins fold remains uncertain, but the process requires assistance from at least two nanomachines: the translocation and assembly module (TAM) and the β-barrel assembly machinery (BAM) complex (1–3). But whether the TAM and the BAM complex collaborate or act independently on each nascent membrane protein substrate arriving at the outer membrane has yet to be determined. The TAM is comprised of two subunits: TamA, an integral outer membrane protein (2, 4, 5), and TamB, an inner membrane-anchored protein (2). The BAM complex is variable in composition between genera, and it is comprised of 2 to 5 accessory lipoproteins attached to an integral outer membrane protein, BamA (1, 6, 7). Our current understanding of the BAM complex is the focus of chapter 8.

TamA is structurally, functionally, and evolutionarily related to BamA, but while TamA is found almost exclusively in Proteobacteria, BamA is found in most, if not all, diderms (2, 7–9). TamB is also distributed among most didermic bacterial lineages (9) and is potentially under positive selection (10). This led Heinz et al. (9) to speculate that the original TAM and BAM complex may not

[1]Infection & Immunity Program, Biomedicine Discovery Institute, and Department of Microbiology, Monash University, Clayton 3800, Australia

Protein Secretion in Bacteria
Edited by Maria Sandkvist, Eric Cascales, and Peter J. Christie
© 2019 American Society for Microbiology, Washington, DC
doi:10.1128/ecosalplus.ESP-0036-2018

have been separate but rather comprised of an inner membrane TamB-like component and an outer membrane BamA-like component, much like the extant BamABD-TamB complex observed in the spirochete *Borrelia burgdorferi* (11, 12). Philosophically, and with support of this evidence, the β-barrel assembly machinery is a broad term for the core BAM complex and additional modules such as the TAM which, depending on the bacterial lineage, may or may not segregate when the outer membrane is solubilized with detergent for analysis by native gel electrophoresis or immunoprecipitation.

In the model bacterium *Escherichia coli*, under laboratory growth conditions, the BAM complex is essential for viability (1, 13, 14) and is about 20 times (15) to 40 times (16) more abundant than the TAM. Under these laboratory conditions, the BAM complex is reported to assemble the majority of outer membrane proteins in *E. coli* and other bacterial species (17). For growth within a host, the BAM complex is more abundant than the TAM (18, 19), but mutants compromised in the function of the TAM show severe virulence and/or colonization defects (2, 20–23). While it has been tempting to speculate that β-barrel assembly falls predominantly under the instruction of the TAM during infection, there is a growing appreciation that the BAM complex and the TAM may not act independently. Instead, multiple BAM complexes can assemble β-barrel proteins cooperatively (24) and/or in conjunction with the TAM (3). This review focuses primarily on the specific contribution of the TAM to β-barrel assembly, irrespective of potential higher-order complexity in the membrane assembly process.

STRUCTURE, FUNCTION, AND MECHANISM OF THE TAM

Structure and Function of the TAM

TamA is the archetype of a distinct bacterial subfamily from the Omp85/TpsB superfamily

of proteins (8). The hallmark of this superfamily is its C-terminal β-barrel transmembrane domain that forms the basis of its ability to translocate substrate across, or assemble and insert substrates into, the outer membrane. Distinct subfamilies may be characterized according to the composition of their N-terminal domain, usually in the form of various combinations of polypeptide transport-associated (POTRA) domains (8, 25). TamA has been well characterized structurally, with five structures obtained through X-ray crystallography or nuclear magnetic resonance spectroscopy (26, 27). It contains three POTRA domains, which reside within the periplasm, connected to a C-terminal 16-stranded β-barrel transmembrane domain (26, 27) (Fig. 1a).

The structure of TamB has proven significantly more difficult to characterize. Algorithms that predict secondary structure indicate that TamB is largely comprised of β-strands, to the point that it was incorrectly predicted to be either a hemolysin-like type 5 secretion system effector with β-helical characteristics (5, 28) or a transmembrane β-barrel itself (29). These two possibilities have been discounted through topological analysis that showed that TamB is signal anchored in the inner membrane, with the bulk of it residing within the periplasm (2), and through a crystal structure comprising about 13% of the periplasmic domain that demonstrated that it adopts a novel β-taco fold: an elongated taco shell-shaped molecule consisting entirely of β-sheet and random coil, with a striking interior surface contributed entirely by aliphatic and aromatic residues (30) (Fig. 1b). Additionally, using the Phyre2 homology server (31), it was predicted that a further 69% of the periplasmic portion of TamB will adopt this β-taco fold with 97 to 99% certainty (Fig. 1b), leaving the C-terminal 124 amino acids, which are likely involved in directly binding TamA (2). This feature would be sufficient in size and hydrophobicity to accommodate a polypeptide strand (30) and invites speculation that the interior cavity would enable TamB to chaperone nascent

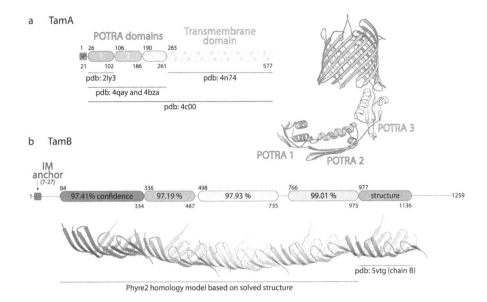

FIGURE 1 Domain architecture and structure of the TAM. (a) Nascent TamA contains a signal peptide (SP) that is cleaved during translocation across the inner membrane (IM). The mature protein comprises residues 22 to 577 (2), with three POTRA domains (numbered from the N terminus) connected to a 16-stranded β-barrel. (Left) Domain distribution of TamA showing the protein data bank (pdb) entries for the five solved structures and their respective amino acid coverage. (Right) A ribbon diagram of the full-length TamA (pdb code 4c00). (b) TamB is tethered to the IM through an IM anchor, but the bulk of the protein resides within the periplasm. (Bottom) A ribbon diagram of the solved structure of TamB with its pdb code indicated. Using the Phyre2 homology server (31), additional structures were determined as shown by modeling the remaining region of TamB on the solved structure. (Top) The domain distribution of TamB with the confidence in the predicted structures indicated, as determined by the Phyre2 homology server (31). Residue numbers are indicated above (first residue) and below (last residue) the domain in question.

outer membrane proteins across the periplasm to TamA in the outer membrane.

TamA (about 60 kDa) and TamB (about 150 kDa) tether together within the periplasm to form a stable intermembrane-spanning complex, which may be purified from bacterial membranes using blue native polyacrylamide gel electrophoresis (BN-PAGE) or immunoprecipitation (2, 27, 32, 33), and under these conditions the TAM from *E. coli* does not copurify with the BAM complex. Deletion of either one (or both) of the components abolishes the activity of the TAM. While the stoichiometry of TamA:TamB is presumed to be 1:1, the oligomerization state of an active TAM remains unclear (2). The TAM migrates as a 440-kDa species on BN-PAGE, whereas TamA migrates as a mixture between 66- and 140-kDa species (2, 33). Presumably, TamA exists in an equilibrium between monomeric and dimeric forms, and while the size of the TAM is suggestive of a dimer of dimers (about 420 kDa), whether the active TAM might also have additional protein components remains to be determined.

Mechanistic Insight into the Role of the TamA Transmembrane Domain

The role of the transmembrane domain in substrate insertion has been a contentious topic in recent years, particularly for the related protein BamA, ranging from actively engaging substrate to indirectly facilitating substrate insertion (17). Two structural features within the β-barrel of TamA underpin the proposed mechanisms for TamA-dependent substrate insertion: a lopsided aromatic girdle and the presence of a lateral gate (27, 33) (Fig. 2). The strong hydrogen bonding network typically observed between adjacent β-strands within a β-barrel is significantly reduced between the first and last β-strands of TamA (33) (Fig. 2a). Molecular dynamics simulations indicate a relatively low energetic barrier to separate the first and last β-strands of TamA (33), suggesting a lateral gating mechanism remi-

niscent of the lateral gates utilized by the SecYEG translocon at the inner membrane (34) or LptDE complex in the outer membrane (35) to secrete substrate from the lumen into the membrane.

Aromatic girdles are a general feature of β-barrel proteins and may be observed as a dual ring of aromatic residues encircling the top and bottom of a given transmembrane domain (36). For the majority of β-barrel proteins, the distance between these aromatic rings is evenly distributed, but for TamA the

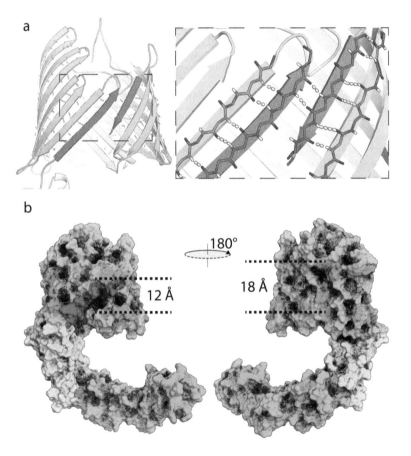

FIGURE 2 Structural features of the TamA β-barrel. Ribbon diagram (a) and surface structures (b) of TamA (pdb code 4c00). (a) Interstrand hydrogen bonding (yellow dashed lines) within the TamA β-barrel is prominent between all strands, except the first and last strands that form the lateral gate (shown in purple), where only 2 hydrogen bonds are observed. Inset, close-up view of the hydrogen bonding network involving the main chain atoms of the first and last strands, where the atoms are colored as follows: C, gray; H, white; N, blue; and O, red. (b) Aromatic residues within the β-barrel are shown in pink, and the likely positions of the lipid head groups at, or opposite, the lateral gate (shown in purple) are indicated by black dashed lines. The distances between the dashed lines are as per Selkrig et al. (27).

aromatic girdle is thinner near the lateral gate (12 Å) and thicker on the opposite side (18 Å) (27) (Fig. 2b). This lopsidedness within the aromatic girdle has been shown through molecular dynamics simulations with BamA to cause a local thinning of the membrane near the lateral gate and a three-fold increase in lipid disorder (37). In an analogous scenario, similar β-barrel mechanics would likely also distort the TamA-lipid interface near the lateral gate, thereby lowering the activation energy required for protein insertion.

Whether the lateral gate is required for substrate secretion or as a mechanism to further increase lipid disorder has yet to be distinguished experimentally. However, recent affinity profiling against a peptide array comprising the 24 β-strands of FimD using three natively folded TamA constructs (full length, transmembrane domain only, and POTRA domains only) (33) has demonstrated that the transmembrane domain of TamA directly engages its substrate. The peptide binding profile was shown to be specific to TamA, where the transmembrane domain alone was significantly better than either the POTRA domains alone or the full-length protein at engaging the substrate. Furthermore, Bamert et al. (33) found that the sixth β-strand of FimD could reproducibly crosslink to the barrel lumen of TamA. This would point toward a mechanism favoring substrate threading through the barrel lumen and direct insertion into the lipid bilayer through the lateral gate, which was subsequently confirmed by molecular dynamics simulations of the sixth β-strand of FimD readily entering the barrel lumen of TamA (33).

Mechanistic Insight into the Role of the TamA POTRA Domains

The function of POTRA domains varies depending on the particular protein in question. In the case of TamA, its POTRA domains have been shown to bind the substrate, but they are primarily involved in binding its partner protein, TamB (2, 27, 32, 33). To determine whether the POTRA domains are mechanistically important for TAM function, Shen et al. (32) designed an artificial system to reconstitute TamA within a planar lipid environment comprised of model lipids. To do this, they purified a histidine-tagged variant of TamA (this mutation in extracellular loop 8 does not affect function [32]) with detergent and attached it onto a nickel-nitrilotriacetic acid-functionalized gold surface. After replacing the detergent with model lipids, they were able to assess the binding of the partner protein TamB and/or the substrate protein Ag43 using either (i) quartz crystal microbalance with dissipation (QCM-D), to assess whether binding occurred, or (ii) magnetic contrast neutron reflectometry (MCNR), to structurally characterize each domain (27, 32).

Using MCNR, the TamA POTRA domains were shown to extend 44 Å from the planar lipid bilayer, in what would topologically be considered the periplasm (32). When TamB was added, QCM-D and MCNR analyses revealed that (i) TamB bound to TamA specifically and (ii) TamB extended 160 Å further from the POTRA domains (32). The size of TamB measured using MCNR was independently reproducible through atomic force microscopy and dynamic light scattering techniques (32), so when considering the size of the periplasm (210 Å) measured using transmission electron cryomicroscopy (38), the 204-Å length of the POTRA-TamB complex measured by MCNR (32) approximates the distance between the inner and outer membranes.

On addition of a urea-denatured substrate protein, MCNR analyses demonstrated that the TamA POTRA domains would extend from 44 Å to 77 Å upon substrate binding (32). Considering that the flexibility, and at times rigidity, between adjacent POTRA domains is important for the function of proteins within the Omp85/TpsB superfamily (39–45), Selkrig et al. (27) sought to delineate the cause of the 33-Å POTRA domain movement through small-angle X-ray scattering.

Ultimately, they found that POTRA domains 1 to 3 form a rigid body and inferred that the observed flexibility demonstrated upon substrate binding is due to an inherent flexibility between the barrel and POTRA domain 3. This flexibility between the most C-terminal POTRA domain and the barrel seems to be a consistent theme within the Omp85/TpsB superfamily (46, 47), potentially explaining why this POTRA domain is the most conserved POTRA domain among the Omp85/TpsB superfamily (8, 25).

From these data, a mechanism starts to emerge where once TamA binds substrate, the POTRA domains of TamA would act as a lever arm and press onto the inner membrane-bound TamB. The pressure exerted onto TamB is unlikely to overcome the turgor pressure of the inner membrane, so the force generated will either (i) shift TamB by tilting or bending it or (ii) cause TamA to "push"

away from the inner membrane, locally raising and distorting the outer membrane (27). Given the rigidity of the peptidoglycan layer that likely reduces the propensity for TamB to shift or bend laterally, we favor the second scenario, in which the lipids within the thinner membrane at the site of the lateral gate would potentially undergo an increase in disorder beyond that caused by the lopsided aromatic girdle of TamA (Fig. 3). Once the substrate disengages from the TAM, the pressure exerted by the POTRA domains would be released, and the position of the TAM can be reset to a substrate-ready conformation.

CONCLUSION

Mechanistic details of TAM-dependent substrate assembly are still being determined experimentally, and several questions remain

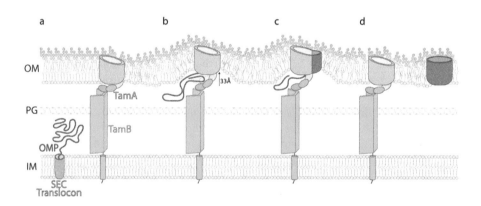

FIGURE 3 Proposed mechanism of TAM-dependent substrate insertion. For the purpose of clarity, the BAM complex is not shown. As noted in the text, it remains unclear whether the TAM and the BAM complex collaborate or act independently on each molecule of their outer membrane protein substrates. (a) A substrate is translocated across the inner membrane (IM) via the Sec translocon. The TAM is in a substrate-ready conformation, where the structural features of the TamA β-barrel (highlighted in Fig. 2) cause membrane thinning and increased lipid disorder near the TamA lateral gate. (b) The substrate makes its way across the periplasm and engages the TAM. This causes a structural change in TamA, whereby its POTRA domains extend 33 Å from the β-barrel domain (27, 32). TamB is rigidified by the peptidoglycan (PG) layer and the turgor pressure of the IM, so the movement of the POTRA domains causes a local raising of the outer membrane (OM) that increases lipid disorder and lowers the activation energy required for substrate insertion. (c) The substrate passes through the lumen of TamA and uses the β-strands comprising the TamA lateral gate as a template to fold into a β-barrel itself. (d) The substrate is released into the OM, allowing the POTRA domains to retract 33 Å so that the TAM is reset into its substrate-ready conformation.

as to the contribution of other factors toward the function of the TAM. How and where does substrate engage the TAM to initiate its lever arm mechanism step of β-barrel biogenesis? How does TamB thread through the peptidoglycan layer during its assembly? Recently, it was demonstrated that *Klebsiella* spp. encode a BamA paralogue, BamK, that can functionally replace BamA in the BAM complex when certain environmental conditions are met (48). Perhaps there are additional assembly factors that may assist (or replace) the TAM components, or novel nanomachines like the TAM that assemble β-barrel proteins. One thing is for certain: this is an exciting and rapidly developing area of research.

ACKNOWLEDGMENTS

We thank Rebecca Bamert, Iain Hay, Von Torres, and Grishma Vadlamani for their comments on the manuscript.

Work in this laboratory is supported by the Australian Research Council (FL130100038) and National Health & Medical Research Council (APP1092262).

CITATION

EcoSal Plus 2019; doi:10.1128/ecosalplus. ESP-0036-2018.

REFERENCES

1. **Wu T, Malinverni J, Ruiz N, Kim S, Silhavy TJ, Kahne D.** 2005. Identification of a multicomponent complex required for outer membrane biogenesis in *Escherichia coli. Cell* **121**:235–245.
2. **Selkrig J, Mosbahi K, Webb CT, Belousoff MJ, Perry AJ, Wells TJ, Morris F, Leyton DL, Totsika M, Phan MD, Celik N, Kelly M, Oates C, Hartland EL, Robins-Browne RM, Ramarathinam SH, Purcell AW, Schembri MA, Strugnell RA, Henderson IR, Walker D, Lithgow T.** 2012. Discovery of an archetypal protein transport system in bacterial outer membranes. *Nat Struct Mol Biol* **19**:506–510, S1.
3. **Stubenrauch C, Belousoff MJ, Hay ID, Shen H-H, Lillington J, Tuck KL, Peters KM, Phan MD, Lo AW, Schembri MA, Strugnell RA, Waksman G, Lithgow T.** 2016. Effective assembly of fimbriae in *Escherichia coli* depends on the translocation assembly module nanomachine. *Nat Microbiol* **1**:16064.
4. **Marani P, Wagner S, Baars L, Genevaux P, de Gier JW, Nilsson I, Casadio R, von Heijne G.** 2006. New *Escherichia coli* outer membrane proteins identified through prediction and experimental verification. *Protein Sci* **15**:884–889.
5. **Stegmeier JF, Glück A, Sukumaran S, Mäntele W, Andersen C.** 2007. Characterisation of YtfM, a second member of the Omp85 family in *Escherichia coli. Biol Chem* **388**:37–46.
6. **Sklar JG, Wu T, Gronenberg LS, Malinverni JC, Kahne D, Silhavy TJ.** 2007. Lipoprotein SmpA is a component of the YaeT complex that assembles outer membrane proteins in *Escherichia coli. Proc Natl Acad Sci U S A* **104**:6400–6405.
7. **Webb CT, Heinz E, Lithgow T.** 2012. Evolution of the β-barrel assembly machinery. *Trends Microbiol* **20**:612–620.
8. **Heinz E, Lithgow T.** 2014. A comprehensive analysis of the Omp85/TpsB protein superfamily structural diversity, taxonomic occurrence, and evolution. *Front Microbiol* **5**:370.
9. **Heinz E, Selkrig J, Belousoff MJ, Lithgow T.** 2015. Evolution of the translocation and assembly module (TAM). *Genome Biol Evol* **7**:1628–1643.
10. **Xu Z, Chen H, Zhou R.** 2011. Genome-wide evidence for positive selection and recombination in *Actinobacillus pleuropneumoniae. BMC Evol Biol* **11**:203.
11. **Iqbal H, Kenedy MR, Lybecker M, Akins DR.** 2016. The TamB ortholog of *Borrelia burgdorferi* interacts with the β-barrel assembly machine (BAM) complex protein BamA. *Mol Microbiol* **102**:757–774.
12. **Stubenrauch C, Grinter R, Lithgow T.** 2016. The modular nature of the β-barrel assembly machinery, illustrated in *Borrelia burgdorferi. Mol Microbiol* **102**:753–756.
13. **Onufryk C, Crouch ML, Fang FC, Gross CA.** 2005. Characterization of six lipoproteins in the sigmaE regulon. *J Bacteriol* **187**:4552–4561.
14. **Werner J, Misra R.** 2005. YaeT (Omp85) affects the assembly of lipid-dependent and lipid-independent outer membrane proteins of *Escherichia coli. Mol Microbiol* **57**:1450–1459.
15. **Li GW, Burkhardt D, Gross C, Weissman JS.** 2014. Quantifying absolute protein synthesis rates reveals principles underlying allocation of cellular resources. *Cell* **157**:624–635.
16. **Wiśniewski JR, Rakus D.** 2014. Multi-enzyme digestion FASP and the 'Total Protein Approach'-

based absolute quantification of the *Escherichia coli* proteome. *J Proteomics* **109**:322–331.

17. **Noinaj N, Gumbart JC, Buchanan SK.** 2017. The β-barrel assembly machinery in motion. *Nat Rev Microbiol* **15**:197–204.

18. **Damron FH, Oglesby-Sherrouse AG, Wilks A, Barbier M.** 2016. Dual-seq transcriptomics reveals the battle for iron during *Pseudomonas aeruginosa* acute murine pneumonia. *Sci Rep* **6**:39172.

19. **Nuss AM, Beckstette M, Pimenova M, Schmühl C, Opitz W, Pisano F, Heroven AK, Dersch P.** 2017. Tissue dual RNA-seq allows fast discovery of infection-specific functions and riboregulators shaping host-pathogen transcriptomes. *Proc Natl Acad Sci U S A* **114**:E791–E800.

20. **Struve C, Forestier C, Krogfelt KA.** 2003. Application of a novel multi-screening signature-tagged mutagenesis assay for identification of *Klebsiella pneumoniae* genes essential in colonization and infection. *Microbiology* **149**:167–176.

21. **Burall LS, Harro JM, Li X, Lockatell CV, Himpsl SD, Hebel JR, Johnson DE, Mobley HL.** 2004. *Proteus mirabilis* genes that contribute to pathogenesis of urinary tract infection: identification of 25 signature-tagged mutants attenuated at least 100-fold. *Infect Immun* **72**:2922–2938.

22. **Kelly M, Hart E, Mundy R, Marchès O, Wiles S, Badea L, Luck S, Tauschek M, Frankel G, Robins-Browne RM, Hartland EL.** 2006. Essential role of the type III secretion system effector NleB in colonization of mice by *Citrobacter rodentium*. *Infect Immun* **74**:2328–2337.

23. **Brooks JF II, Gyllborg MC, Cronin DC, Quillin SJ, Mallama CA, Foxall R, Whistler C, Goodman AL, Mandel MJ.** 2014. Global discovery of colonization determinants in the squid symbiont *Vibrio fischeri*. *Proc Natl Acad Sci U S A* **111**:17284–17289.

24. **Gunasinghe SD, Shiota T, Stubenrauch CJ, Schulze KE, Webb CT, Fulcher AJ, Dunstan RA, Hay ID, Naderer T, Whelan DR, Bell TDM, Elgass KD, Strugnell RA, Lithgow T.** 2018. The WD40 protein BamB mediates coupling of BAM complexes into assembly precincts in the bacterial outer membrane. *Cell Rep* **23**:2782–2794.

25. **Arnold T, Zeth K, Linke D.** 2010. Omp85 from the thermophilic cyanobacterium *Thermosynechococcus elongatus* differs from proteobacterial Omp85 in structure and domain composition. *J Biol Chem* **285**:18003–18015.

26. **Gruss F, Zähringer F, Jakob RP, Burmann BM, Hiller S, Maier T.** 2013. The structural basis of autotransporter translocation by TamA. *Nat Struct Mol Biol* **20**:1318–1320.

27. **Selkrig J, Belousoff MJ, Headey SJ, Heinz E, Shiota T, Shen H-H, Beckham SA, Bamert RS, Phan MD, Schembri MA, Wilce MC, Scanlon MJ, Strugnell RA, Lithgow T.** 2015. Conserved features in TamA enable interaction with TamB to drive the activity of the translocation and assembly module. *Sci Rep* **5**:12905.

28. **Chaudhuri RR, Sebaihia M, Hobman JL, Webber MA, Leyton DL, Goldberg MD, Cunningham AF, Scott-Tucker A, Ferguson PR, Thomas CM, Frankel G, Tang CM, Dudley EG, Roberts IS, Rasko DA, Pallen MJ, Parkhill J, Nataro JP, Thomson NR, Henderson IR.** 2010. Complete genome sequence and comparative metabolic profiling of the prototypical enteroaggregative *Escherichia coli* strain 042. *PLoS One* **5**:e8801.

29. **Babu M, Díaz-Mejía JJ, Vlasblom J, Gagarinova A, Phanse S, Graham C, Yousif F, Ding H, Xiong X, Nazarians-Armavil A, Alamgir M, Ali M, Pogoutse O, Pe'er A, Arnold R, Michaut M, Parkinson J, Golshani A, Whitfield C, Wodak SJ, Moreno-Hagelsieb G, Greenblatt JF, Emili A.** 2011. Genetic interaction maps in *Escherichia coli* reveal functional crosstalk among cell envelope biogenesis pathways. *PLoS Genet* **7**:e1002377.

30. **Josts I, Stubenrauch CJ, Vadlamani G, Mosbahi K, Walker D, Lithgow T, Grinter R.** 2017. The structure of a conserved domain of TamB reveals a hydrophobic beta taco fold. *Structure* **25**:1898–1906.e5.

31. **Kelley LA, Mezulis S, Yates CM, Wass MN, Sternberg MJ.** 2015. The Phyre2 web portal for protein modeling, prediction and analysis. *Nat Protoc* **10**:845–858.

32. **Shen H-H, Leyton DL, Shiota T, Belousoff MJ, Noinaj N, Lu J, Holt SA, Tan K, Selkrig J, Webb CT, Buchanan SK, Martin LL, Lithgow T.** 2014. Reconstitution of a nanomachine driving the assembly of proteins into bacterial outer membranes. *Nat Commun* **5**:5078.

33. **Bamert RS, Lundquist K, Hwang H, Webb CT, Shiota T, Stubenrauch CJ, Belousoff MJ, Goode RJA, Schittenhelm RB, Zimmerman R, Jung M, Gumbart JC, Lithgow T.** 2017. Structural basis for substrate selection by the translocation and assembly module of the β-barrel assembly machinery. *Mol Microbiol* **106**:142–156.

34. **du Plessis DJ, Berrelkamp G, Nouwen N, Driessen AJ.** 2009. The lateral gate of SecYEG opens during protein translocation. *J Biol Chem* **284**:15805–15814.

35. **Dong H, Xiang Q, Gu Y, Wang Z, Paterson NG, Stansfeld PJ, He C, Zhang Y, Wang W, Dong C.** 2014. Structural basis for outer membrane lipopolysaccharide insertion. *Nature* **511:**52–56.

36. **Schulz GE.** 2002. The structure of bacterial outer membrane proteins. *Biochim Biophys Acta* **1565:**308–317.

37. **Noinaj N, Kuszak AJ, Gumbart JC, Lukacik P, Chang H, Easley NC, Lithgow T, Buchanan SK.** 2013. Structural insight into the biogenesis of β-barrel membrane proteins. *Nature* **501:** 385–390.

38. **Matias VR, Al-Amoudi A, Dubochet J, Beveridge TJ.** 2003. Cryo-transmission electron microscopy of frozen-hydrated sections of *Escherichia coli* and *Pseudomonas aeruginosa*. *J Bacteriol* **185:**6112–6118.

39. **Gatzeva-Topalova PZ, Walton TA, Sousa MC.** 2008. Crystal structure of YaeT: conformational flexibility and substrate recognition. *Structure* **16:**1873–1881.

40. **Ward R, Zoltner M, Beer L, El Mkami H, Henderson IR, Palmer T, Norman DG.** 2009. The orientation of a tandem POTRA domain pair, of the beta-barrel assembly protein BamA, determined by PELDOR spectroscopy. *Structure* **17:**1187–1194.

41. **Gatzeva-Topalova PZ, Warner LR, Pardi A, Sousa MC.** 2010. Structure and flexibility of the complete periplasmic domain of BamA: the protein insertion machine of the outer membrane. *Structure* **18:**1492–1501.

42. **Koenig P, Mirus O, Haarmann R, Sommer MS, Sinning I, Schleiff E, Tews I.** 2010. Conserved properties of polypeptide transport-associated (POTRA) domains derived from cyanobacterial Omp85. *J Biol Chem* **285:**18016–18024.

43. **Dastvan R, Brouwer EM, Schuetz D, Mirus O, Schleiff E, Prisner TF.** 2016. Relative orientation of POTRA domains from cyanobacterial Omp85 studied by pulsed EPR spectroscopy. *Biophys J* **110:**2195–2206.

44. **Fleming PJ, Patel DS, Wu EL, Qi Y, Yeom MS, Sousa MC, Fleming KG, Im W.** 2016. BamA POTRA domain interacts with a native lipid membrane surface. *Biophys J* **110:**2698–2709.

45. **Warner LR, Gatzeva-Topalova PZ, Doerner PA, Pardi A, Sousa MC.** 2017. Flexibility in the periplasmic domain of BamA is important for function. *Structure* **25:**94–106.

46. **Albrecht R, Schütz M, Oberhettinger P, Faulstich M, Bermejo I, Rudel T, Diederichs K, Zeth K.** 2014. Structure of BamA, an essential factor in outer membrane protein biogenesis. *Acta Crystallogr D Biol Crystallogr* **70:**1779–1789.

47. **Guérin J, Saint N, Baud C, Meli AC, Etienne E, Locht C, Vezin H, Jacob-Dubuisson F.** 2015. Dynamic interplay of membrane-proximal POTRA domain and conserved loop L6 in Omp85 transporter FhaC. *Mol Microbiol* **98:**490–501.

48. **Vergel L, Torres V, Heinz E, Stubenrauch CJ, Wilksch JJ, Cao H, Yang J, Clements A, Dunstan RA, Alcock F, Webb CT, Dougan G, Strugnell RA, Hay ID, Lithgow T.** 2018. An investigation into the Omp85 protein BamK in hypervirulent *Klebsiella pneumoniae*, and its role in outer membrane biogenesis. *Mol Microbiol* **109:**584–599.

10

The Dynamic Structures of the Type IV Pilus

MATTHEW MCCALLUM,[1,2] LORI L. BURROWS,[3] and P. LYNNE HOWELL[1,2]

INTRODUCTION

The fundamental type IVa pilus (T4aP)-like architecture includes a retractable pilus fiber, a motor, an alignment subcomplex, and—in Gram-negative bacteria—an outer membrane secretin pore (Fig. 1). The pilus is an extracellular polymer of pilins. Pilins subunits are stored in the inner membrane and the motor powers their polymerization (extension) and depolymerization (retraction) at the pilus base. The alignment subcomplex connects the secretin with the motor and controls pilus dynamics. Finally, the secretin pore allows the pilus to extend through the outer membrane. Since publication of previous T4aP reviews (1–4), discoveries made using cryo-electron microscopy (cryo-EM), cryo-electron tomography (cryo-ET), X-ray crystallography, and nuclear magnetic resonance (NMR) have dramatically reshaped our understanding of T4P-like systems. Here we put these discoveries in context with the structure and function of the T4aP, using the *Pseudomonas aeruginosa* T4aP system nomenclature.

[1]Department of Biochemistry, University of Toronto, Toronto, ON M5S 1A8, Canada
[2]Program in Molecular Medicine, Peter Gilgan Centre for Research and Learning, The Hospital for Sick Children, Toronto, ON M5G 0A4, Canada
[3]Department of Biochemistry and Biomedical Sciences and the Michael G. DeGroote Institute for Infectious Disease Research, McMaster University, Hamilton, ON L8N 3Z5, Canada

Protein Secretion in Bacteria
Edited by Maria Sandkvist, Eric Cascales, and Peter J. Christie
© 2019 American Society for Microbiology, Washington, DC
doi:10.1128/microbiolspec.PSIB-0006-2018

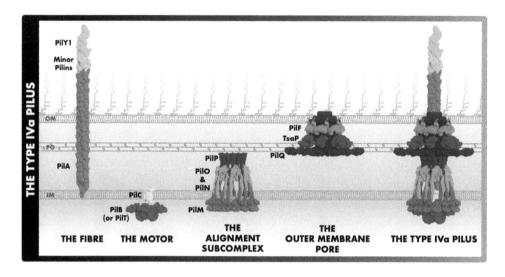

FIGURE 1 Subcomplexes of the T4aP. The protein structures portrayed reflect the full-length structure predictions and their predicted location in the T4aP. This figure is largely consistent with the previously published working model of the *M. xanthus* T4aP (43). Due to limited information, there is uncertainty regarding the locations of PilF, TsaP, PilY1, and the minor pilins.

THE FIBER

The Major Subunit, PilA

The major pilin PilA is the most abundant subunit in the pilus fiber. X-ray crystallography (5–18) and NMR (19–21) structures revealed four typical features: an elongated S-shaped N-terminal α-helix (α1N), a ~4-stranded antiparallel β-sheet, the αβ-loop, and the D-loop (Fig. 2) (Table 1). The αβ-loop and D-loop vary in sequence, structure, and posttranslational modifications between strains and species (7, 13, 18, 22–27). Together, the antiparallel β-sheet, αβ-loop, and D-loop create a globular domain from which the α1N helix protrudes. The hydrophobic α1N helix retains pilins in the inner membrane prior to pilin polymerization and after pilin depolymerization. The peptidase PilD hydrolyzes the cytoplasmic leader sequence of nascent pilins and then methylates the new N terminus (28).

Recent 5- to 8-Å resolution cryo-EM maps of pilus fibers from *Neisseria* and *P. aeru-*

ginosa (29, 30) largely validated older, lower-resolution models (6, 7, 13, 21, 31). With a helical rise of ~10 Å and 80 to 100° twist, the α1N helices bundle to form a hydrophobic core, while Glu-5 forms a salt bridge with the positively charged N-terminal amine of distal PilA molecules (6, 7, 13, 21, 29, 30). The αβ-loop and D-loop are surface exposed in these models, consistent with their sequence variability and posttranslational modifications that facilitate bacteriophage and immune evasion (6, 7, 24, 25). The 5- to 8-Å cryo-EM maps revealed that the segment between conserved Gly-14 and Pro-22 in the α1N helix was unexpectedly disordered in polymerized filaments (29, 30). This disordered region might allow the pilus to reversibly stretch to three times its original length (29, 32). Based on circular-dichroism analyses (33), the α1 helix is structured in pilin monomers (29, 30). Thus, relative to PilA monomers, the α1N in pilus fibers is stretched and may be under tension, or prestress—an architectural concept reviewed here (34), potentially explaining why the pilus appears to act more like a rod than a rope in micrograph animations (35). Rod-like behavior could facilitate preferential adhesion of the pilus tip to substrates and twitching motility (36).

Minor Pilins

The low-abundance minor pilins are similar in architecture to PilA (37, 38). PilV, PilW, and PilX are functionally equivalent to GspI, GspJ, and GspK of the type II secretion system (T2SS), respectively (16, 39, 40). Crystallographic analysis of the globular domains of GspIJK heterotrimers revealed a helical arrangement of the three components with GspK at the tip (Fig. 2) (41). The bulky globular domain of GspK may hinder its insertion anywhere but at the pilus tip (41). This suggests that the GspIJK trimer, and by extension the PilVWX trimer, is located at the pilus tip, though PilX is less bulky than GspK (41). By self-assembling a short stem, minor pilins are thought to prime pilus assembly by reducing the energy barrier to extraction of pilins from the membrane (39, 41–43). Minor pilins FimU and PilE connect PilVWX to the PilA fiber (40, 44), and because they have the α1N helix-destabilizing Pro-22 residue missing in PilVXW (39, 40), they may initiate α1N helix melting during pilus assembly. Some T4aP include additional minor pilins with specialized binding capabilities, like ComP in *Neisseria*, which binds DNA to promote uptake (38, 45–48).

PilY1

PilY1 is an adhesin that likely localizes to the pilus tip with PilVWX (39, 49, 50). The N-terminal region of PilY1 is variable and important for PilY1 adhesive capacities (49, 51). Crystallographic analyses revealed that the conserved C-terminal domain of PilY1 is a 7-bladed β-propeller domain with a calcium binding motif (51). Manipulation of this motif reduces PilY1-based adhesion in *Neisseria gonorrhoeae* (52) and causes retraction defects in *P. aeruginosa* (51).

THE MOTOR

PilC

PilC is a 3-pass inner membrane protein with two homologous globular cytoplasmic domains (53, 54) and, in many proteobacteria, a small cytoplasmic N-terminal domain with predicted ββαβ topology (Fig. 2). Cryo-electron tomography (cryo-ET) of the *Myxococcus xanthus* T4aP machinery suggests that PilC is a dimer and localized in the pore of cytoplasmic hexameric ATPases PilB and PilT, while its transmembrane segments may interact with PilA (43). Given this organization, PilC might be rotated by PilB and PilT to catalyze PilA polymerization and depolymerization, respectively (43, 53). *In vitro* analyses support direct interactions between PilB and PilC, PilT and PilC, and PilA and PilC (53, 55–57). Purified PilC is dimeric and tetrameric, consistent with a 22-Å-resolution cryo-EM analysis of PilC from *Neisseria meningitidis* (54, 57, 58). The N-terminal domain of PilC from *Thermus thermophilus* was crystallized as an asymmetric homodimer, and mutating this dimer interface ablated *in vitro* tetramers but not dimers (54).

PilB and PilT

C_2-symmetric structures of PilB from *T. thermophilus* (PilB^{Tt}) bound to the slowly hydrolyzable ATP analog ATPγS (59), PilB from *Geobacter metallireducens* (PilB^{Gm}) bound to ADP plus the nonhydrolyzable ATP analog AMP-PNP (60), and apo-PilB from *Geobacter sulfurreducens* (PilB^{Gs}) (61) were recently determined. The symmetry of PilB^{Tt} initially suggested a C_2 rotary mechanism with a predicted counterclockwise pore rotation (59). In contrast, the two structures of PilB^{Gm} representing pre- and posthydrolysis states indicated a clockwise pore rotation (60) (Fig. 2) and that four of the six ATPγS molecules in the lower-resolution PilB^{Tt} hexamer should have been modeled as ADP and magnesium. We proposed that

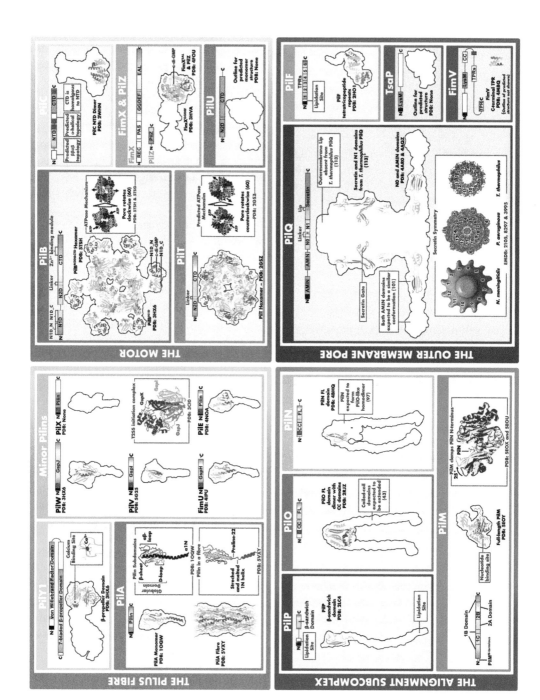

clockwise rotation of the PilB pore may move PilC clockwise in 60° increments to facilitate the polymerization of PilA into the right-handed helix observed by cryo-EM (29, 30, 60). Note that hydrolysis of ATP by PilB gives the impression of pore rotation, though PilB does not rotate during this process. Recent 8-Å resolution cryo-EM analysis of PilB^{Tt} showed a domain N terminal to the motor domains, the N1D, in a position that may block PilC from directly contacting the pore (62). Thus, the N1D might participate directly in the motor function of PilB (62). Alternatively, the N1Ds could regulate whether the PilB^{Tt} pore is available for binding PilC. Consistent with a rotary model, the PilB homolog from the archaellum FlaI is proposed to facilitate archaellum spinning; consistent with a C_2-symmetric mechanism, purified FlaI, PilB, and PilT have two free ATP binding sites (61, 63–68).

The N1D of PilB is composed of two subdomains, N1D_N and N1D_C (69). The N1D_N subdomain of PilB binds the biofilm-related second messenger, c-di-GMP (69–71); the N1D_C subdomain also contacts c-di-GMP (69) (Fig. 2). A complex consisting of N1D_C of the T2SS PilB homolog, GspE, bound to the cytoplasmic domain of GspL (PilM) was crystallized (72, 73), suggesting a similar interaction in the T4aP (55). Therefore, c-di-GMP binding to PilB might regulate the PilB interaction with PilM and thus PilB engagement with the T4aP. PilB homologs from *P. aeruginosa*, *Xanthomonas campestris*, and *Xanthomonas axonopodis* lack an obvious c-di-GMP-binding motif in their N1D_N subdomains, though they interact with c-di-GMP-binding FimX, which also interacts with PilB-binding PilZ (74–76).

PilT is homologous to the motor domains of PilB and powers pilus depolymerization/retraction with extraordinary forces (35, 77–81). In addition to PilT, its paralogs in some T4aP systems, such as PilU, are also functionally significant (82–84). Pilus retraction is essential for several T4aP functions, including twitching motility, competence, and phage infection (77, 85, 86). In contrast to PilB, structures of PilT exhibit dissimilar rotational symmetries (87, 88). We applied the direction of ATP binding and hydrolysis of PilB to the structure of C_2-symmetric PilT from *Aquifex aeolicus* (89) (Fig. 2). This analysis suggested

---

FIGURE 2 The structures of the type IV pilus. The four subcomplexes are split into four quadrants, which are further subdivided into individual proteins in boxes colored to correspond to Fig. 1. In the linear domain architecture, domains are displayed to scale as blocks colored to indicate known structures (rainbow colors), segments with high-confidence structure predictions (gray), unknown structures (white), transmembrane segments (diagonal bars), hydrolyzed signal peptides (black), or predicted/known disorder (black line). The known or predicted domain name is written; if a domain has no name, it could not be predicted. In the black outlined cartoon structures, a black outline of the predicted (127–129) full-length homology model is shown to scale for reference. Known structures are displayed as cartoons in rainbow colors corresponding to the colors shown in the linear domain architecture. A short description of the rainbow-colored cartoon structure and the PDB accession code are written in black font. Since the black outline is a *P. aeruginosa* structure prediction while the cartoons correspond to structures sometimes determined in other species, the black outline and cartoons may not fully match. Note that the PDB coordinate file for PilQ from *T. thermophilus* (marked with an asterisk) was obtained from the authors of reference 113 and used here with their permission; only the secretin and adjacent N1 domain (N5 in *T. thermophilus*) are shown here, as the other *T. thermophilus* PilQ domains are divergent or atypical compared to those in *P. aeruginosa*. (The N1 domain is also named N2, N3, N4, or N5 in systems or species where the N1 domain is duplicated.) No black outline is shown for FimV, as high-confidence structure prediction was not possible for most of this component. Unexpectedly, most of TsaP was predicted (127) with high confidence to be structurally similar to the protein with PDB code 3SLU. Gray boxes note interesting features of the protein or other relevant structures; structures in gray boxes are not to scale.

TABLE 1 List of available T4aP structures[a]

Pseudomonas equivalent	Name	Species	PDB (EMDB) code(s)	Reference(s)
The pilus fiber				
PilY1	PilY1	*Pseudomonas aeruginosa* PAO1	3HX6	51
PilV	Tt1218	*Thermus thermophilus*	5G25	16
PilW	Tt1219	*Thermus thermophilus*	5G23, 5G24	16
FimU	FimU	*Pseudomonas aeruginosa* PAO1	4IPU, 4IPV	39
PilE	PilE	*Pseudomonas aeruginosa* PAO1	4NOA	40
	PilX	*Neisseria meningitidis*	2OPD, 2OPE	45
	PilV	*Neisseria meningitidis*	5V23,[b] 5V0M[b]	None
PilA	PilA	*Acinetobacter baumannii*	4XA2, 5VAW,[b] 5IHJ, 5CFV	8
	PilA	*Geobacter sulfurreducens*	2M7G	19
	PilA	*Pseudomonas aeruginosa* PAK	1DZO, 1OQW, 1X6P/Q/R/ X/Y/Z	6, 12, 17
	PilA	*Pseudomonas aeruginosa* K122-4	1HPW, 1QVE, 1RG0	14, 21
	PilA	*Pseudomonas aeruginosa* Pa110594	JYZ, 3JZZ	5
	PilA	*Pseudomonas aeruginosa* consensus	2PY0	11
	PilA1	*Clostridium difficile*	4TSM, 4OGM, 4PE2	9
	FimA	*Dichelobacter nodosus*	3SOK	10
	PilBac1	*Shewanella oneidensis*	4D40, 4US7	15
	PilE	*Francisella tularensis*	3SOJ	10
	PilE	*Neisseria gonorrhoeae*	1AY2, 2HI2, 2PIL	7, 13, 18
	PilE	*Neisseria meningitidis*	5JW8, 4V1J[b]	29
	Tt1221	*Thermus thermophilus* HB8	4BHR	94
PilA fiber	PilA	*Pseudomonas aeruginosa* PAK	5VXY (8740)	30
	PilA4	*Thermus thermophilus*	None (3024)	119
	PilE	*Neisseria gonorrhoeae*	5VXX (8739)	30
	PilE	*Neisseria gonorrhoeae*	2HIL (1236)	7
	PilE	*Neisseria meningitidis*	5KUA (8287)	29
None known	ComP	*Neisseria meningitidis*	2M3K, 5HZ7	38, 48
	ComP	*Neisseri subflava*	2NBA	38
	PilJ	*Clostridium difficile*	4IXJ	37
	Tt1222	*Thermus thermophilus* HB8	5G2F	16
The motor				
PilB	PilB	*Geobacter metallireducens*	5TSG, 5TSH	60
	PilF	*Thermus thermophilus*	5IT5, 6EJF (3882), 6F8L (4194), (2222), (2223)	59, 62, 130
	PilB	*Geobacter sulfurreducens*	5ZFR	61
PilB (N1D)	MshE	*Vibrio cholerae* serotype O1	5HTL	69
PilT	PilT	*Pseudomonas aeruginosa* PAO1	3JVU, 3JVV	88
	PilT	*Aquifex aeolicus* strain VF5	2EWV, 2EWW, 2EYU, 2GSZ	89
	PilT4	*Geobacter sulfurreducens*	5ZFQ	61
	PilT2	*Thermus thermophilus*	5FL3[b]	None
PilC	PilC	*Thermus thermophilus*	2WHN	54
FimX	FimX	*Pseudomonas aeruginosa* PAO1	4AG0, 3HV9, 3HVA, 3HV8, 4AFY, 4J40	135, 136
	FimX	*Xanthomonas* sp.	4FOK, 4F3H	76, 137

(Continued on next page)

TABLE 1 **List of available T4aP structures**[a] *(Continued)*

Pseudomonas equivalent	Name	Species	PDB (EMDB) code(s)	Reference(s)
PilZ	PilZ	*Xanthomonas* sp.	3CNR, 3DSG	75, 138
FimX and PilZ	FimX and PilZ	*Xanthomonas* sp.	4FOU, 4F48	76, 137
The alignment complex				
PilM	PilM	*Pseudomonas aeruginosa* PAO1	5EOX, 5EOY, 5EQ6	55
PilM and PilN^{cyto}	PilM and PilN^{cyto}	*Pseudomonas aeruginosa* PAO1	5EOU	55
	PilM and PilN^{cyto}	*Thermus thermophilus*	2YCH	92
PilN	PilN	*Thermus thermophilus*	4BHQ	94
PilO	PilO	*Pseudomonas aeruginosa* PAO1	2RJZ, 5UVR	95, 96
PilMN, PilMNO, and PilMNOA	PilMN, PilMNO, and PilMNOA	*Thermus thermophilus*	None[c] (4157, 4157, 4159)	94
PilP	PilP	*Pseudomonas aeruginosa* PAO1	2LC4, 2Y4X,[b] 2Y4Y[b]	100
	PilP	*Neisseria meningitidis*	2IVW	99
The outer membrane pore				
PilQ	PilQ	*Pseudomonas aeruginosa* PAO1	None (8297)	114
	PilQ	*Thermus thermophilus*	None[c] (3985, 3995, 3996, 3997, 3998)	113
	PilQ	*Candidatus pelagibacter*	None (8330)	117
PilQ and PilP	PilQ and PilP	*Neisseria meningitidis*	4AV2 (2105)	101
PilQ (N0 and linker from N0 to N1)	PilQ (N0 and linker from N0 to N1)	*Neisseria meningitidis*	4AR0	101
PilQ (AMIN)	PilQ (β2)	*Neisseria meningitidis*	4AQZ	101
PilF	PilF	*Pseudomonas aeruginosa* PAO1	2HO1, 2FI7	134, 139
PilF	PilW	*Neisseria meningitidis*	2VQ2	126
FimV	FimV	*Pseudomonas aeruginosa* PAO1	4MBQ, 4MAL	140
In situ structures of the T4aP		*Myxococcus xanthus*	3JC8, 3JC9 (3247–3264)	43
		Thermus thermophilus	None (8224, 3021–3024)	118, 119

[a]These are the published T4aP structures, not those from related T2SS, T4bP, Com pilus, Tad/Flp pilus, or archaellum.
[b]Peer-reviewed article describing the structure has not been published yet.
[c]A model could be built into the density, but the model was not published to the Protein Data Bank.

that the pore and thus PilC would rotate counterclockwise to facilitate PilA depolymermization and pilus retraction (60). Intriguingly, the T4aP of *Vibrio cholerae* was recently reported to retract with low speed and force in a PilT knockout mutant (35), and the type IVb pilus (T4bP) systems of *V. cholerae* and the Tad pilus of *Caulobacter crescentus*, which lack retraction ATPases, were also reported to retract with low speed and force (90, 91). Thus, PilT might simply enhance the speed and force of retraction (35). The energy for retraction in the absence of

PilT was proposed to be potential energy stored in the pilus (90), possibly elastic tension in the melted α1N helices of PilA (see above).

THE ALIGNMENT SUBCOMPLEX

PilM

Cryo-ET of *M. xanthus* indicated that PilM forms a ring on the inner leaflet of the inner membrane surrounding PilB or PilT (43), consistent with evidence that PilM binds to both PilB and PilT (53, 55, 57). The X-ray crystallographic structures of PilM from *T. thermophilus* and *P. aeruginosa* bound to the first eight residues of PilN have been solved, revealing that PilM is structurally similar to FtsA (55, 92). Interestingly, the T2SS homolog of PilM, GspL, is equivalent to a fusion of PilM and PilN (73, 92, 93). In the T4aP there is functional relevance for discrete PilM and PilN proteins, since PilN binding favors PilM-PilB interactions while reducing PilM-PilT interactions in bacterial two-hybrid experiments (55). Likewise, PilM subdomain 1C turns on a hinge to bind PilN (55) (Fig. 2), and this hinge is the predicted site for PilM-PilB interactions based on the homologous interactions in the T2SS (55, 72, 73). Thus, PilN binding to PilM might influence which ATPase is associated with the T4aP (55). Consistent with proposed conformational changes, the diameter of the PilM ring of *M. xanthus* is wider in the piliated than in the nonpiliated state (43).

PilN and PilO

PilN and PilO are structurally similar inner membrane proteins with a cytoplasmic N-terminal peptide of ~20 residues, a transmembrane domain followed by a coiled-coil domain, and a C-terminal globular domain with a ferredoxin-like fold (94–96). They form an oligomeric cage-like ring in the inner membrane and periplasm (43). The ferredoxin-like domains from *T. thermophilus* PilN and *P. aeruginosa* PilO crystallized as homodimers (94–96). One PilO structure from *P. aeruginosa* revealed a distinct PilO ferredoxin-like domain homodimer interface (95), and *in vivo* cysteine disulfide cross-linking studies of PilN and PilO homodimers and heterodimers are consistent with this interface (95–97). Cross-linked heterodimers, but not homodimers, interfere with T4aP function, suggesting that the homodimeric interface is stable while the heterodimeric interface may be dynamic (97). Mutagenesis studies suggest that heterodimerization

occurs mainly through the coiled-coil domains, and coiled-coil mutations lead to pilus extension and retraction defects (98). Thus, PilN and PilO heterodimers are proposed to interact with PilM to influence which ATPase is bound (43, 55, 98).

PilP

PilP is an inner membrane lipoprotein with a partially disordered N-terminal region followed by a globular β-sandwich domain (99, 100). The partially disordered region binds to PilN and PilO heterodimers but not homodimers (100). Since heterodimeric PilN and PilO interactions are dynamic, the PilP interaction with PilN and PilO may also be dynamic (98). NMR analysis and pulldown experiments demonstrated that the β-domain of PilP also interacts with the N0 domain of PilQ, as do the homologous domains in the T2SS and T4bP (101–104). Thus, the PilMNOP subcomplex links the cytoplasmic ATPases to PilQ (43, 102). The dynamic interactions predicted for PilMNOP may also transduce signals. In *P. aeruginosa*, surface sensing is thought to initiate with PilY1, proceed through PilMNOP, require PilT, and ultimately activate the diguanylate cyclase SadC (105–108).

THE OUTER MEMBRANE PORE

PilQ

The secretin domain of PilQ forms the outer membrane pore, and until recently this domain resisted structure determination. The structures of the other PilQ subdomains were determined by NMR: one of two peptidoglycan-binding AMIN (β) domains located near the N terminus, the N0 domain, and the N1 domain (101). In *P. aeruginosa*, the AMIN domains localize PilQ to sites of cell division for preinstallation of the T4aP complex into the nascent septa of the daughter cells (109). Since the N0 domain of PilQ is bound by PilP (see above), PilQ recruits and localizes PilMNOP (110, 111). In the absence of an outer membrane, Gram-positive bacteria with T4aP mostly lack PilP and PilQ homologs (112).

Using single-particle cryo-EM, ~19-Å-, 7.4-Å-, and 7.0-Å-resolution maps of full-length PilQ from *N. meningitidis*, *P. aeruginosa*, and *T. thermophilus*, respectively, have been determined (101, 113, 114). Two-dimensional (2D) cryo-EM top views are also available for outer membrane-embedded *N. gonorrhoeae* and *N. meningitidis* PilQ (115, 116). These maps are 12-, 13-, or 14-fold symmetric, consistent with dodecamers, tridecamers, or tetradecamers (101, 113, 114, 117). Given that PilQ ultimately connects to hexameric ATPases in the cytoplasm via the alignment subcomplex, there may be stoichiometry mismatches between subcomplexes in some T4aP systems.

Two internal gates have been identified in PilQ: the secretin gate and the periplasmic gate (43, 101, 113, 114, 118, 119), although some cryo-EM maps are missing one or the other. Model building in the 7.0-Å-resolution *T. thermophilus* PilQ cryo-EM map was made possible by using homology models of new 3- to 4-Å-resolution GspD cryo-EM models (113, 120–123). Based on these models, it is clear how the secretin gate prevents leakage of molecules in the absence of the pilus (113, 124). The function of the periplasmic gate is less clear. The linker between the N0 domain and N1 domain forms the periplasmic gate in *T. thermophilus* PilQ, though this portion of PilQ shows limited homology to more typical T4aP systems (113). Since the gates must reorient during pilus extension and after retraction, it is conceivable that PilP binding to the N0 domain could sense these movements and transmit a signal to the cytoplasm via PilMNO.

PilF, TsaP, and FimV

The PG-binding protein FimV is widely distributed in T4aP systems (125) and was originally thought to be a PilQ-stabilizing protein, as mutants had reduced levels of PilQ (126). Recent data suggest that FimV plays a role in localizing PilQ and contributes to the expression of multiple T4aP proteins via a cAMP-dependent surface sensing mechanism (109, 127–130).

The PilF pilotin protein is required for stability, outer membrane localization, and multimerization of PilQ (131–134). PilF is a six-tetratricopeptide repeat lipoprotein localized to the outer membrane (126, 134). In the cryo-EM 2D class averages of *Neisseria* PilQ and the 3D reconstruction of *P. aeruginosa* PilQ, 7-fold symmetric spokes were detected around 14-fold symmetric PilQ (114–116). In the *P. aeruginosa* 3D reconstruction, the spokes are localized to the inner leaflet of the outer membrane (114). Density consistent with these spokes was also present in cryo-ET images of *M. xanthus* T4aP and the 3D reconstruction of *N. meningitidis* PilQ (43, 101). Similar spokes in the T2SS correspond to the unrelated T2SS pilotin (120), suggesting the T4aP spokes could be PilF. Indeed, the spoke symmetry suggests that the spokes may simultaneously bind at least two PilQ protomers in the assembled secretin, potentially stabilizing nascent PilQ oligomers to assist PilQ multimerization. Alternatively, it has been proposed that the peptidoglycan-binding protein TsaP may form the spokes, since deleting TsaP from *M. xanthus* or *N. gonorrhoeae* causes the spokes to disappear (43, 116). It is possible that these spokes comprise multiple proteins.

CONCLUSION

Recent advances in cryo-EM and cryo-ET allow us to put crystallographic and NMR structures of individual T4aP components into biological context (Fig. 1). With these advances, new mysteries have emerged. For example, how does PilC interface with PilA and the pilus fiber in a way that facilitates extension and retraction yet opposes pilus shedding? Do the proposed dynamics in PilMNOPQ form a mechanical signal cascade? Structural insights from the T4aP will help rationalize new findings and expedite our understanding of other T4P-like systems.

ACKNOWLEDGMENTS

M.M. has been funded by graduate scholarships from the Canadian Institutes for Health Research (CIHR) and the Province of Ontario. P.L.H. is the recipient of a Canada Research Chair. This work was supported by grant MOP 93585 from CIHR to L.L.B. and P.L.H.

CITATION

McCallum M, Burrows LL, Howell PL. 2019. The dynamic structures of the type IV pilus. Microbiol Spectrum 7(2):PSIB-0006-2018.

REFERENCES

1. **Craig L, Li J.** 2008. Type IV pili: paradoxes in form and function. *Curr Opin Struct Biol* **18:**267–277.
2. **Leighton TL, Buensuceso RN, Howell PL, Burrows LL.** 2015. Biogenesis of *Pseudomonas aeruginosa* type IV pili and regulation of their function. *Environ Microbiol* **17:**4148–4163.
3. **Ayers M, Howell PL, Burrows LL.** 2010. Architecture of the type II secretion and type IV pilus machineries. *Future Microbiol* **5:**1203–1218.
4. **Hospenthal MK, Costa TRD, Waksman G.** 2017. A comprehensive guide to pilus biogenesis in Gram-negative bacteria. *Nat Rev Microbiol* **15:**365–379.
5. **Nguyen Y, Jackson SG, Aidoo F, Junop M, Burrows LL.** 2010. Structural characterization of novel *Pseudomonas aeruginosa* type IV pilins. *J Mol Biol* **395:**491–503.
6. **Craig L, Taylor RK, Pique ME, Adair BD, Arvai AS, Singh M, Lloyd SJ, Shin DS, Getzoff ED, Yeager M, Forest KT, Tainer JA.** 2003. Type IV pilin structure and assembly: X-ray and EM analyses of *Vibrio cholerae* toxin-coregulated pilus and *Pseudomonas aeruginosa* PAK pilin. *Mol Cell* **11:**1139–1150.
7. **Craig L, Volkmann N, Arvai AS, Pique ME, Yeager M, Egelman EH, Tainer JA.** 2006. Type IV pilus structure by cryo-electron microscopy and crystallography: implications for pilus assembly and functions. *Mol Cell* **23:**651–662.
8. **Piepenbrink KH, Lillehoj E, Harding CM, Labonte JW, Zuo X, Rapp CA, Munson RS Jr, Goldblum SE, Feldman MF, Gray JJ, Sundberg EJ.** 2016. Structural diversity in the type IV pili of multidrug-resistant *Acinetobacter*. *J Biol Chem* **291:**22924–22935.
9. **Piepenbrink KH, Maldarelli GA, Martinez de la Peña CF, Dingle TC, Mulvey GL, Lee A, von Rosenvinge E, Armstrong GD, Donnenberg MS, Sundberg EJ.** 2015. Structural and evolutionary analyses show unique stabilization strategies in the type IV pili of *Clostridium difficile*. *Structure* **23:**385–396.
10. **Hartung S, Arvai AS, Wood T, Kolappan S, Shin DS, Craig L, Tainer JA.** 2011. Ultrahigh resolution and full-length pilin structures with insights for filament assembly, pathogenic functions, and vaccine potential. *J Biol Chem* **286:**44254–44265.
11. **Kao DJ, Churchill ME, Irvin RT, Hodges RS.** 2007. Animal protection and structural studies of a consensus sequence vaccine targeting the receptor binding domain of the type IV pilus of *Pseudomonas aeruginosa*. *J Mol Biol* **374:**426–442.
12. **Hazes B, Sastry PA, Hayakawa K, Read RJ, Irvin RT.** 2000. Crystal structure of *Pseudomonas aeruginosa* PAK pilin suggests a main-chain-dominated mode of receptor binding. *J Mol Biol* **299:**1005–1017.
13. **Parge HE, Forest KT, Hickey MJ, Christensen DA, Getzoff ED, Tainer JA.** 1995. Structure of the fibre-forming protein pilin at 2.6 A resolution. *Nature* **378:**32–38.
14. **Audette GF, Irvin RT, Hazes B.** 2004. Crystallographic analysis of the *Pseudomonas aeruginosa* strain K122-4 monomeric pilin reveals a conserved receptor-binding architecture. *Biochemistry* **43:**11427–11435.
15. **Gorgel M, Ulstrup JJ, Bøggild A, Jones NC, Hoffmann SV, Nissen P, Boesen T.** 2015. High-resolution structure of a type IV pilin from the metal-reducing bacterium *Shewanella oneidensis*. *BMC Struct Biol* **15:**4.
16. **Karuppiah V, Thistlethwaite A, Derrick JP.** 2016. Structures of type IV pilins from *Thermus thermophilus* demonstrate similarities with type II secretion system pseudopilins. *J Struct Biol* **196:**375–384.
17. **Dunlop KV, Irvin RT, Hazes B.** 2005. Pros and cons of cryocrystallography: should we also collect a room-temperature data set? *Acta Crystallogr D Biol Crystallogr* **61:**80–87.

18. **Forest KT, Dunham SA, Koomey M, Tainer JA.** 1999. Crystallographic structure reveals phosphorylated pilin from *Neisseria*: phosphoserine sites modify type IV pilus surface chemistry and fibre morphology. *Mol Microbiol* **31**:743–752.

19. **Reardon PN, Mueller KT.** 2013. Structure of the type IVa major pilin from the electrically conductive bacterial nanowires of *Geobacter sulfurreducens*. *J Biol Chem* **288**:29260–29266.

20. **Nguyen Y, Boulton S, McNicholl ET, Akimoto M, Harvey H, Aidoo F, Melacini G, Burrows LL.** 2018. A highly dynamic loop of the *Pseudomonas aeruginosa* PA14 type IV pilin is essential for pilus assembly. *ACS Infect Dis* **4**:936–943.

21. **Keizer DW, Slupsky CM, Kalisiak M, Campbell AP, Crump MP, Sastry PA, Hazes B, Irvin RT, Sykes BD.** 2001. Structure of a pilin monomer from *Pseudomonas aeruginosa*: implications for the assembly of pili. *J Biol Chem* **276**:24186–24193.

22. **Hegge FT, Hitchen PG, Aas FE, Kristiansen H, Løvold C, Egge-Jacobsen W, Panico M, Leong WY, Bull V, Virji M, Morris HR, Dell A, Koomey M.** 2004. Unique modifications with phosphocholine and phosphoethanolamine define alternate antigenic forms of *Neisseria gonorrhoeae* type IV pili. *Proc Natl Acad Sci USA* **101**:10798–10803.

23. **Jennings MP, Jen FE, Roddam LF, Apicella MA, Edwards JL.** 2011. *Neisseria gonorrhoeae* pilin glycan contributes to CR3 activation during challenge of primary cervical epithelial cells. *Cell Microbiol* **13**:885–896.

24. **Harvey H, Bondy-Denomy J, Marquis H, Sztanko KM, Davidson AR, Burrows LL.** 2018. *Pseudomonas aeruginosa* defends against phages through type IV pilus glycosylation. *Nat Microbiol* **3**:47–52.

25. **Gault J, Ferber M, Machata S, Imhaus AF, Malosse C, Charles-Orszag A, Millien C, Bouvier G, Bardiaux B, Péhau-Arnaudet G, Klinge K, Podglajen I, Ploy MC, Seifert HS, Nilges M, Chamot-Rooke J, Duménil G.** 2015. *Neisseria meningitidis* type IV pili composed of sequence invariable pilins are masked by multisite glycosylation. *PLoS Pathog* **11**:e1005162.

26. **Tan RM, Kuang Z, Hao Y, Lee F, Lee T, Lee RJ, Lau GW.** 2015. Type IV pilus glycosylation mediates resistance of *Pseudomonas aeruginosa* to opsonic activities of the pulmonary surfactant protein A. *Infect Immun* **83**:1339–1346.

27. **Kus JV, Kelly J, Tessier L, Harvey H, Cvitkovitch DG, Burrows LL.** 2008. Modification of *Pseudomonas aeruginosa* Pa5196 type IV pilins at multiple sites with d-Ara*f* by a novel GT-C family arabinosyltransferase, TfpW. *J Bacteriol* **190**:7464–7478.

28. **LaPointe CF, Taylor RK.** 2000. The type 4 prepilin peptidases comprise a novel family of aspartic acid proteases. *J Biol Chem* **275**:1502–1510.

29. **Kolappan S, Coureuil M, Yu X, Nassif X, Egelman EH, Craig L.** 2016. Structure of the *Neisseria meningitidis* type IV pilus. *Nat Commun* **7**:13015.

30. **Wang F, Coureuil M, Osinski T, Orlova A, Altindal T, Gesbert G, Nassif X, Egelman EH, Craig L.** 2017. Cryoelectron microscopy reconstructions of the *Pseudomonas aeruginosa* and *Neisseria gonorrhoeae* type IV pili at subnanometer resolution. *Structure* **25**:1423–1435.e4.

31. **Folkhard W, Marvin DA, Watts TH, Paranchych W.** 1981. Structure of polar pili from *Pseudomonas aeruginosa* strains K and O. *J Mol Biol* **149**:79–93.

32. **Egelman EH.** 2017. Cryo-EM of bacterial pili and archaeal flagellar filaments. *Curr Opin Struct Biol* **46**:31–37.

33. **Watts TH, Scraba DG, Paranchych W.** 1982. Formation of 9-nm filaments from pilin monomers obtained by octyl-glucoside dissociation of *Pseudomonas aeruginosa* pili. *J Bacteriol* **151**:1508–1513.

34. **Ingber DE, Wang N, Stamenovic D.** 2014. Tensegrity, cellular biophysics, and the mechanics of living systems. *Rep Prog Phys* **77**:046603.

35. **Ellison CK, Dalia TN, Vidal Ceballos A, Wang JC, Biais N, Brun YV, Dalia AB.** 2018. Retraction of DNA-bound type IV competence pili initiates DNA uptake during natural transformation in *Vibrio cholerae*. *Nat Microbiol* **3**:773–780.

36. **de Haan HW.** 2016. Modeling and simulating the dynamics of type IV pili extension of *Pseudomonas aeruginosa*. *Biophys J* **111**:2263–2273.

37. **Piepenbrink KH, Maldarelli GA, de la Peña CF, Mulvey GL, Snyder GA, De Masi L, von Rosenvinge EC, Günther S, Armstrong GD, Donnenberg MS, Sundberg EJ.** 2014. Structure of *Clostridium difficile* PilJ exhibits unprecedented divergence from known type IV pilins. *J Biol Chem* **289**:4334–4345.

38. **Berry JL, Xu Y, Ward PN, Lea SM, Matthews SJ, Pelicic V.** 2016. A comparative structure/function analysis of two type IV pilin DNA receptors defines a novel mode of DNA binding. *Structure* **24**:926–934.

39. **Nguyen Y, Sugiman-Marangos S, Harvey H, Bell SD, Charlton CL, Junop MS, Burrows LL.** 2015. *Pseudomonas aeruginosa* minor pilins prime type IVa pilus assembly and promote surface display of the PilY1 adhesin. *J Biol Chem* **290**:601–611.

40. **Nguyen Y, Harvey H, Sugiman-Marangos S, Bell SD, Buensuceso RN, Junop MS, Burrows**

LL. 2015. Structural and functional studies of the *Pseudomonas aeruginosa* minor pilin, PilE. *J Biol Chem* 290:26856–26865.

41. **Korotkov KV, Hol WG.** 2008. Structure of the GspK-GspI-GspJ complex from the enterotoxigenic *Escherichia coli* type 2 secretion system. *Nat Struct Mol Biol* 15:462–468.

42. **Cisneros DA, Bond PJ, Pugsley AP, Campos M, Francetic O.** 2012. Minor pseudopilin self-assembly primes type II secretion pseudopilus elongation. *EMBO J* 31:1041–1053.

43. **Chang YW, Rettberg LA, Treuner-Lange A, Iwasa J, Søgaard-Andersen L, Jensen GJ.** 2016. Architecture of the type IVa pilus machine. *Science* 351:aad2001.

44. **Giltner CL, Habash M, Burrows LL.** 2010. *Pseudomonas aeruginosa* minor pilins are incorporated into type IV pili. *J Mol Biol* 398:444–461.

45. **Helaine S, Dyer DH, Nassif X, Pelicic V, Forest KT.** 2007. 3D structure/function analysis of PilX reveals how minor pilins can modulate the virulence properties of type IV pili. *Proc Natl Acad Sci U S A* 104:15888–15893.

46. **Wolfgang M, van Putten JP, Hayes SF, Koomey M.** 1999. The comP locus of *Neisseria gonorrhoeae* encodes a type IV prepilin that is dispensable for pilus biogenesis but essential for natural transformation. *Mol Microbiol* 31:1345–1357.

47. **Brown DR, Helaine S, Carbonnelle E, Pelicic V.** 2010. Systematic functional analysis reveals that a set of seven genes is involved in fine-tuning of the multiple functions mediated by type IV pili in *Neisseria meningitidis*. *Infect Immun* 78:3053–3063.

48. **Cehovin A, Simpson PJ, McDowell MA, Brown DR, Noschese R, Pallett M, Brady J, Baldwin GS, Lea SM, Matthews SJ, Pelicic V.** 2013. Specific DNA recognition mediated by a type IV pilin. *Proc Natl Acad Sci U S A* 110:3065–3070.

49. **Johnson MD, Garrett CK, Bond JE, Coggan KA, Wolfgang MC, Redinbo MR.** 2011. *Pseudomonas aeruginosa* PilY1 binds integrin in an RGD- and calcium-dependent manner. *PLoS One* 6:e29629.

50. **Rudel T, Scheurerpflug I, Meyer TF.** 1995. *Neisseria* PilC protein identified as type-4 pilus tip-located adhesin. *Nature* 373:357–359.

51. **Orans J, Johnson MD, Coggan KA, Sperlazza JR, Heiniger RW, Wolfgang MC, Redinbo MR.** 2010. Crystal structure analysis reveals *Pseudomonas* PilY1 as an essential calcium-dependent regulator of bacterial surface motility. *Proc Natl Acad Sci U S A* 107:1065–1070.

52. **Cheng Y, Johnson MD, Burillo-Kirch C, Mocny JC, Anderson JE, Garrett CK, Redinbo MR, Thomas CE.** 2013. Mutation of

the conserved calcium-binding motif in *Neisseria gonorrhoeae* PilC1 impacts adhesion but not piliation. *Infect Immun* 81:4280–4289.

53. **Takhar HK, Kemp K, Kim M, Howell PL, Burrows LL.** 2013. The platform protein is essential for type IV pilus biogenesis. *J Biol Chem* 288:9721–9728.

54. **Karuppiah V, Hassan D, Saleem M, Derrick JP.** 2010. Structure and oligomerization of the PilC type IV pilus biogenesis protein from *Thermus thermophilus*. *Proteins* 78:2049–2057.

55. **McCallum M, Tammam S, Little DJ, Robinson H, Koo J, Shah M, Calmettes C, Moraes TF, Burrows LL, Howell PL.** 2016. PilN binding modulates the structure and binding partners of the *Pseudomonas aeruginosa* type IVa pilus protein PilM. *J Biol Chem* 291:11003–11015.

56. **Georgiadou M, Castagnini M, Karimova G, Ladant D, Pelicic V.** 2012. Large-scale study of the interactions between proteins involved in type IV pilus biology in *Neisseria meningitidis*: characterization of a subcomplex involved in pilus assembly. *Mol Microbiol* 84:857–873.

57. **Bischof LF, Friedrich C, Harms A, Søgaard-Andersen L, van der Does C.** 2016. The type IV pilus assembly ATPase PilB of *Myxococcus xanthus* interacts with the inner membrane platform protein PilC and the nucleotide-binding protein PilM. *J Biol Chem* 291:6946–6957.

58. **Collins RF, Saleem M, Derrick JP.** 2007. Purification and three-dimensional electron microscopy structure of the *Neisseria meningitidis* type IV pilus biogenesis protein PilG. *J Bacteriol* 189:6389–6396.

59. **Mancl JM, Black WP, Robinson H, Yang Z, Schubot FD.** 2016. Crystal structure of a type IV pilus assembly ATPase: insights into the molecular mechanism of PilB from *Thermus thermophilus*. *Structure* 24:1886–1897.

60. **McCallum M, Tammam S, Khan A, Burrows LL, Howell PL.** 2017. The molecular mechanism of the type IVa pilus motors. *Nat Commun* 8:15091.

61. **Solanki V, Kapoor S, Thakur KG.** 2018. Structural insights into the mechanism of type IVa pilus extension and retraction ATPase motors. *FEBS J* 285:3402–3421.

62. **Collins R, Karuppiah V, Siebert CA, Dajani R, Thistlethwaite A, Derrick JP.** 2018. Structural cycle of the *Thermus thermophilus* PilF ATPase: the powering of type IVa pilus assembly. *Sci Rep* 8:14022.

63. **Kinosita Y, Uchida N, Nakane D, Nishizaka T.** 2016. Direct observation of rotation and steps of the archaellum in the swimming halophilic archaeon *Halobacterium salinarum*. *Nat Microbiol* 1:16148.

64. **Reindl S, Ghosh A, Williams GJ, Lassak K, Neiner T, Henche AL, Albers SV, Tainer JA.** 2013. Insights into FlaI functions in archaeal motor assembly and motility from structures, conformations, and genetics. *Mol Cell* **49**:1069–1082.

65. **Alam M, Oesterhelt D.** 1984. Morphology, function and isolation of halobacterial flagella. *J Mol Biol* **176**:459–475.

66. **Marwan W, Alam M, Oesterhelt D.** 1991. Rotation and switching of the flagellar motor assembly in *Halobacterium halobium*. *J Bacteriol* **173**:1971–1977.

67. **Shahapure R, Driessen RP, Haurat MF, Albers SV, Dame RT.** 2014. The archaellum: a rotating type IV pilus. *Mol Microbiol* **91**:716–723.

68. **Chaudhury P, van der Does C, Albers SV.** 2018. Characterization of the ATPase FlaI of the motor complex of the *Pyrococcus furiosus* archaellum and its interactions between the ATP-binding protein FlaH. *Peer J* **6**:e4984.

69. **Wang YC, Chin KH, Tu ZL, He J, Jones CJ, Sanchez DZ, Yildiz FH, Galperin MY, Chou SH.** 2016. Nucleotide binding by the wide-spread high-affinity cyclic di-GMP receptor MshEN domain. *Nat Commun* **7**:12481.

70. **Hendrick WA, Orr MW, Murray SR, Lee VT, Melville SB.** 2017. Cyclic di-GMP binding by an assembly ATPase (PilB2) and control of type IV pilin polymerization in the Gram-positive pathogen *Clostridium perfringens*. *J Bacteriol* **199**:e00034-17.

71. **Jones CJ, Utada A, Davis KR, Thongsomboon W, Zamorano Sanchez D, Banakar V, Cegelski L, Wong GC, Yildiz FH.** 2015. C-di-GMP regulates motile to sessile transition by modulating MshA pili biogenesis and near-surface motility behavior in *Vibrio cholerae*. *PLoS Pathog* **11**:e1005068.

72. **Lu C, Korotkov KV, Hol WGJ.** 2014. Crystal structure of the full-length ATPase GspE from the *Vibrio vulnificus* type II secretion system in complex with the cytoplasmic domain of GspL. *J Struct Biol* **187**:223–235.

73. **Abendroth J, Murphy P, Sandkvist M, Bagdasarian M, Hol WG.** 2005. The X-ray structure of the type II secretion system complex formed by the N-terminal domain of EpsE and the cytoplasmic domain of EpsL of *Vibrio cholerae*. *J Mol Biol* **348**:845–855.

74. **Jain R, Sliusarenko O, Kazmierczak BI.** 2017. Interaction of the cyclic-di-GMP binding protein FimX and the type 4 pilus assembly ATPase promotes pilus assembly. *PLoS Pathog* **13**:e1006594.

75. **Guzzo CR, Salinas RK, Andrade MO, Farah CS.** 2009. PILZ protein structure and interac-tions with PILB and the FIMX EAL domain: implications for control of type IV pilus biogenesis. *J Mol Biol* **393**:848–866.

76. **Chin KH, Kuo WT, Yu YJ, Liao YT, Yang MT, Chou SH.** 2012. Structural polymorphism of c-di-GMP bound to an EAL domain and in complex with a type II PilZ-domain protein. *Acta Crystallogr D Biol Crystallogr* **68**:1380–1392.

77. **Merz AJ, So M, Sheetz MP.** 2000. Pilus retraction powers bacterial twitching motility. *Nature* **407**:98–102.

78. **Maier B, Potter L, So M, Long CD, Seifert HS, Sheetz MP.** 2002. Single pilus motor forces exceed 100 pN. *Proc Natl Acad Sci U S A* **99**:16012–16017.

79. **Clausen M, Koomey M, Maier B.** 2009. Dynamics of type IV pili is controlled by switching between multiple states. *Biophys J* **96**:1169–1177.

80. **Beaussart A, Baker AE, Kuchma SL, El-Kirat-Chatel S, O'Toole GA, Dufrêne YF.** 2014. Nanoscale adhesion forces of *Pseudomonas aeruginosa* type IV pili. *ACS Nano* **8**:10723–10733.

81. **Clausen M, Jakovljevic V, Søgaard-Andersen L, Maier B.** 2009. High-force generation is a conserved property of type IV pilus systems. *J Bacteriol* **191**:4633–4638.

82. **Chiang P, Habash M, Burrows LL.** 2005. Disparate subcellular localization patterns of *Pseudomonas aeruginosa* type IV pilus ATPases involved in twitching motility. *J Bacteriol* **187**:829–839.

83. **Chiang P, Sampaleanu LM, Ayers M, Pahuta M, Howell PL, Burrows LL.** 2008. Functional role of conserved residues in the characteristic secretion NTPase motifs of the *Pseudomonas aeruginosa* type IV pilus motor proteins PilB, PilT and PilU. *Microbiology* **154**:114–126.

84. **Kurre R, Höne A, Clausen M, Meel C, Maier B.** 2012. PilT2 enhances the speed of gonococcal type IV pilus retraction and of twitching motility. *Mol Microbiol* **86**:857–865.

85. **Bradley DE.** 1972. Shortening of *Pseudomonas aeruginosa* pili after RNA-phage adsorption. *J Gen Microbiol* **72**:303–319.

86. **Wolfgang M, Lauer P, Park HS, Brossay L, Hébert J, Koomey M.** 1998. PilT mutations lead to simultaneous defects in competence for natural transformation and twitching motility in piliated *Neisseria gonorrhoeae*. *Mol Microbiol* **29**:321–330.

87. **Forest KT, Satyshur KA, Worzalla GA, Hansen JK, Herdendorf TJ.** 2004. The pilus-retraction protein PilT: ultrastructure of the biological assembly. *Acta Crystallogr D Biol Crystallogr* **60**:978–982.

88. **Misic AM, Satyshur KA, Forest KT.** 2010. *P. aeruginosa* PilT structures with and without nucleotide reveal a dynamic type IV pilus retraction motor. *J Mol Biol* **400:**1011–1021.

89. **Satyshur KA, Worzalla GA, Meyer LS, Heiniger EK, Aukema KG, Misic AM, Forest KT.** 2007. Crystal structures of the pilus retraction motor PilT suggest large domain movements and subunit cooperation drive motility. *Structure* **15:**363–376.

90. **Ng D, Harn T, Altindal T, Kolappan S, Marles JM, Lala R, Spielman I, Gao Y, Hauke CA, Kovacikova G, Verjee Z, Taylor RK, Biais N, Craig L.** 2016. The *Vibrio cholerae* minor pilin TcpB initiates assembly and retraction of the toxin-coregulated pilus. *PLoS Pathog* **12:**e1006109.

91. **Ellison CK, Kan J, Dillard RS, Kysela DT, Ducret A, Berne C, Hampton CM, Ke Z, Wright ER, Biais N, Dalia AB, Brun YV.** 2017. Obstruction of pilus retraction stimulates bacterial surface sensing. *Science* **358:**535–538.

92. **Karuppiah V, Derrick JP.** 2011. Structure of the PilM-PilN inner membrane type IV pilus biogenesis complex from *Thermus thermophilus*. *J Biol Chem* **286:**24434–24442.

93. **Abendroth J, Bagdasarian M, Sandkvist M, Hol WG.** 2004. The structure of the cytoplasmic domain of EpsL, an inner membrane component of the type II secretion system of *Vibrio cholerae*: an unusual member of the actin-like ATPase superfamily. *J Mol Biol* **344:**619–633.

94. **Karuppiah V, Collins RF, Thistlethwaite A, Gao Y, Derrick JP.** 2013. Structure and assembly of an inner membrane platform for initiation of type IV pilus biogenesis. *Proc Natl Acad Sci U S A* **110:**E4638–E4647.

95. **Sampaleanu LM, Bonanno JB, Ayers M, Koo J, Tammam S, Burley SK, Almo SC, Burrows LL, Howell PL.** 2009. Periplasmic domains of *Pseudomonas aeruginosa* PilN and PilO form a stable heterodimeric complex. *J Mol Biol* **394:**143–159.

96. **Leighton TL, Mok MC, Junop MS, Howell PL, Burrows LL.** 2018. Conserved, unstructured regions in *Pseudomonas aeruginosa* PilO are important for type IVa pilus function. *Sci Rep* **8:**2600.

97. **Leighton TL, Yong DH, Howell PL, Burrows LL.** 2016. Type IV pilus alignment subcomplex proteins PilN and PilO form homo- and heterodimers in vivo. *J Biol Chem* **291:**19923–19938.

98. **Leighton TL, Dayalani N, Sampaleanu LM, Howell PL, Burrows LL.** 2015. A novel role for PilNO in type IV pilus retraction revealed by

alignment subcomplex mutations. *J Bacteriol* **197:**2229–2238.

99. **Golovanov AP, Balasingham S, Tzitzilonis C, Goult BT, Lian LY, Homberset H, Tønjum T, Derrick JP.** 2006. The solution structure of a domain from the *Neisseria meningitidis* lipoprotein PilP reveals a new beta-sandwich fold. *J Mol Biol* **364:**186–195.

100. **Tammam S, Sampaleanu LM, Koo J, Sundaram P, Ayers M, Chong PA, Forman-Kay JD, Burrows LL, Howell PL.** 2011. Characterization of the PilN, PilO and PilP type IVa pilus subcomplex. *Mol Microbiol* **82:**1496–1514.

101. **Berry JL, Phelan MM, Collins RF, Adomavicius T, Tønjum T, Frye SA, Bird L, Owens R, Ford RC, Lian LY, Derrick JP.** 2012. Structure and assembly of a trans-periplasmic channel for type IV pili in *Neisseria meningitidis*. *PLoS Pathog* **8:**e1002923.

102. **Tammam S, Sampaleanu LM, Koo J, Manoharan K, Daubaras M, Burrows LL, Howell PL.** 2013. PilMNOPQ from the *Pseudomonas aeruginosa* type IV pilus system form a transenvelope protein interaction network that interacts with PilA. *J Bacteriol* **195:**2126–2135.

103. **Korotkov KV, Johnson TL, Jobling MG, Pruneda J, Pardon E, Héroux A, Turley S, Steyaert J, Holmes RK, Sandkvist M, Hol WG.** 2011. Structural and functional studies on the interaction of GspC and GspD in the type II secretion system. *PLoS Pathog* **7:**e1002228.

104. **Chang YW, Kjær A, Ortega DR, Kovacikova G, Sutherland JA, Rettberg LA, Taylor RK, Jensen GJ.** 2017. Architecture of the *Vibrio cholerae* toxin-coregulated pilus machine revealed by electron cryotomography. *Nat Microbiol* **2:**16269.

105. **Kuchma SL, Ballok AE, Merritt JH, Hammond JH, Lu W, Rabinowitz JD, O'Toole GA.** 2010. Cyclic-di-GMP-mediated repression of swarming motility by *Pseudomonas aeruginosa*: the *pilY1* gene and its impact on surface-associated behaviors. *J Bacteriol* **192:**2950–2964.

106. **Luo Y, Zhao K, Baker AE, Kuchma SL, Coggan KA, Wolfgang MC, Wong GC, O'Toole GA.** 2015. A hierarchical cascade of second messengers regulates *Pseudomonas aeruginosa* surface behaviors. *mBio* **6:**e02456-14.

107. **Siryaporn A, Kuchma SL, O'Toole GA, Gitai Z.** 2014. Surface attachment induces *Pseudomonas aeruginosa* virulence. *Proc Natl Acad Sci U S A* **111:**16860–16865.

108. **Rodesney CA, Roman B, Dhamani N, Cooley BJ, Katira P, Touhami A, Gordon VD.** 2017. Mechanosensing of shear by *Pseudomonas aeruginosa* leads to increased levels of the

cyclic-di-GMP signal initiating biofilm development. *Proc Natl Acad Sci U S A* **114:**5906–5911.

109. **Carter T, Buensuceso RN, Tammam S, Lamers RP, Harvey H, Howell PL, Burrows LL.** 2017. The type IVa pilus machinery is recruited to sites of future cell division. *mBio* **8:**e02103-16.

110. **Seitz P, Blokesch M.** 2013. DNA-uptake machinery of naturally competent *Vibrio cholerae*. *Proc Natl Acad Sci U S A* **110:**17987–17992.

111. **Friedrich C, Bulyha I, Søgaard-Andersen L.** 2014. Outside-in assembly pathway of the type IV pilus system in *Myxococcus xanthus*. *J Bacteriol* **196:**378–390.

112. **Imam S, Chen Z, Roos DS, Pohlschröder M.** 2011. Identification of surprisingly diverse type IV pili, across a broad range of gram-positive bacteria. *PLoS One* **6:**e28919.

113. **D'Imprima E, Salzer R, Bhaskara RM, Sánchez R, Rose I, Kirchner L, Hummer G, Kühlbrandt W, Vonck J, Averhoff B.** 2017. Cryo-EM structure of the bifunctional secretin complex of *Thermus thermophilus*. *eLife* **6:**e30483.

114. **Koo J, Lamers RP, Rubinstein JL, Burrows LL, Howell PL.** 2016. Structure of the *Pseudomonas aeruginosa* type IVa pilus secretin at 7.4 Å. *Structure* **24:**1778–1787.

115. **Jain S, Mościcka KB, Bos MP, Pachulec E, Stuart MC, Keegstra W, Boekema EJ, van der Does C.** 2011. Structural characterization of outer membrane components of the type IV pili system in pathogenic *Neisseria*. *PLoS One* **6:**e16624.

116. **Siewering K, Jain S, Friedrich C, Webber-Birungi MT, Semchonok DA, Binzen I, Wagner A, Huntley S, Kahnt J, Klingl A, Boekema EJ, Søgaard-Andersen L, van der Does C.** 2014. Peptidoglycan-binding protein TsaP functions in surface assembly of type IV pili. *Proc Natl Acad Sci USA* **111:**E953–E961.

117. **Zhao X, Schwartz CL, Pierson J, Giovannoni SJ, McIntosh JR, Nicastro D.** 2017. Three-dimensional structure of the ultraoligotrophic marine bacterium "*Candidatus* Pelagibacter ubique." *Appl Environ Microbiol* **83:**e02807-16.

118. **Salzer R, D'Imprima E, Gold VA, Rose I, Drechsler M, Vonck J, Averhoff B.** 2016. Topology and structure/function correlation of ring- and gate-forming domains in the dynamic secretin complex of *Thermus thermophilus*. *J Biol Chem* **291:**14448–14456.

119. **Gold VA, Salzer R, Averhoff B, Kühlbrandt W.** 2015. Structure of a type IV pilus machinery in the open and closed state. *eLife* **4:**e07380.

120. **Yin M, Yan Z, Li X.** 2018. Structural insight into the assembly of the type II secretion system pilotin-secretin complex from enterotoxigenic *Escherichia coli*. *Nat Microbiol* **3:**581–587.

121. **Hay ID, Belousoff MJ, Dunstan RA, Bamert RS, Lithgow T.** 2018. Structure and membrane topography of the vibrio-type secretin complex from the type 2 secretion system of enteropathogenic *Escherichia coli*. *J Bacteriol* **200:**e00521-17.

122. **Hay ID, Belousoff MJ, Lithgow T.** 2017. Structural basis of type 2 secretion system engagement between the inner and outer bacterial membranes. *mBio* **8:**e01344-17.

123. **Yan Z, Yin M, Xu D, Zhu Y, Li X.** 2017. Structural insights into the secretin translocation channel in the type II secretion system. *Nat Struct Mol Biol* **24:**177–183.

124. **Seike K, Yasuda M, Hatazaki K, Mizutani K, Yuhara K, Ito Y, Fujimoto Y, Ito S, Tsuchiya T, Yokoi S, Nakano M, Deguchi T.** 2016. Novel *penA* mutations identified in *Neisseria gonorrhoeae* with decreased susceptibility to ceftriaxone isolated between 2000 and 2014 in Japan. *J Antimicrob Chemother* **71:**2466–2470.

125. **Semmler AB, Whitchurch CB, Leech AJ, Mattick JS.** 2000. Identification of a novel gene, *fimV*, involved in twitching motility in *Pseudomonas aeruginosa*. *Microbiology* **146:**1321–1332.

126. **Trindade MB, Job V, Contreras-Martel C, Pelicic V, Dessen A.** 2008. Structure of a widely conserved type IV pilus biogenesis factor that affects the stability of secretin multimers. *J Mol Biol* **378:**1031–1039.

127. **Kelley LA, Mezulis S, Yates CM, Wass MN, Sternberg MJ.** 2015. The Phyre2 web portal for protein modeling, prediction and analysis. *Nat Protoc* **10:**845–858.

128. **Yang J, Yan R, Roy A, Xu D, Poisson J, Zhang Y.** 2015. The I-TASSER Suite: protein structure and function prediction. *Nat Methods* **12:**7–8.

129. **Kim DE, Chivian D, Baker D.** 2004. Protein structure prediction and analysis using the Robetta server. *Nucleic Acids Res* **32:**W526–W531.

130. **Collins RF, Hassan D, Karuppiah V, Thistlethwaite A, Derrick JP.** 2013. Structure and mechanism of the PilF DNA transformation ATPase from *Thermus thermophilus*. *Biochem J* **450:**417–425.

131. **Koo J, Burrows LL, Howell PL.** 2012. Decoding the roles of pilotins and accessory proteins in secretin escort services. *FEMS Microbiol Lett* **328:**1–12.

132. **Nudleman E, Wall D, Kaiser D.** 2006. Polar assembly of the type IV pilus secretin in *Myxococcus xanthus*. *Mol Microbiol* **60:**16–29.

133. **Rumszauer J, Schwarzenlander C, Averhoff B.** 2006. Identification, subcellular localization and functional interactions of PilMNOWQ

and PilA4 involved in transformation competency and pilus biogenesis in the thermophilic bacterium *Thermus thermophilus* HB27. *FEBS J* **273:**3261–3272.

134. **Koo J, Tammam S, Ku SY, Sampaleanu LM, Burrows LL, Howell PL.** 2008. PilF is an outer membrane lipoprotein required for multimerization and localization of the *Pseudomonas aeruginosa* type IV pilus secretin. *J Bacteriol* **190:**6961–6969.

135. **Robert-Paganin J, Nonin-Lecomte S, Réty S.** 2012. Crystal structure of an EAL domain in complex with reaction product 5′-pGpG. *PLoS One* **7:**e52424.

136. **Navarro MV, De N, Bae N, Wang Q, Sondermann H.** 2009. Structural analysis of the GGDEF-EAL domain-containing c-di-GMP receptor FimX. *Structure* **17:**1104–1116.

137. **Guzzo CR, Dunger G, Salinas RK, Farah CS.** 2013. Structure of the PilZ-FimXEAL-c-di-GMP complex responsible for the regulation of bacterial type IV pilus biogenesis. *J Mol Biol* **425:**2174–2197.

138. **Li TN, Chin KH, Liu JH, Wang AH, Chou SH.** 2009. XC1028 from *Xanthomonas campestris* adopts a PilZ domain-like structure without a c-di-GMP switch. *Proteins* **75:**282–288.

139. **Kim K, Oh J, Han D, Kim EE, Lee B, Kim Y.** 2006. Crystal structure of PilF: functional implication in the type 4 pilus biogenesis in *Pseudomonas aeruginosa*. *Biochem Biophys Res Commun* **340:**1028–1038.

140. **Buensuceso RN, Nguyen Y, Zhang K, Daniel-Ivad M, Sugiman-Marangos SN, Fleetwood AD, Zhulin IB, Junop MS, Howell PL, Burrows LL.** 2016. The conserved tetratricopeptide repeat-containing C-terminal domain of *Pseudomonas aeruginosa* FimV is required for its cyclic AMP-dependent and -independent functions. *J Bacteriol* **198:**2263–2274.

11

Gram-Positive Type IV Pili and Competence

SANDRA MUSCHIOL,[1,2] MARIE-STEPHANIE ASCHTGEN,[1,2]
PRIYANKA NANNAPANENI,[1,2] and BIRGITTA HENRIQUES-NORMARK[1,2]

T4P ARE PRESENT IN A WIDE RANGE OF GRAM-POSITIVE BACTERIA

In recent years, numerous complete bacterial genome sequences became available and led to the identification of surprisingly diverse type IV pili (T4P) across a broad range of Gram-positive bacteria. The genes encoding T4P components cluster together in distinct loci, and three subsets of T4P loci present in Gram-positive bacteria have been described: *(i) pil* (pilin) loci, *(ii) com* (competence) loci, and *(iii) tad* (tight adherence) loci (1). Interestingly, they are not mutually exclusive. In fact, many Gram-positive bacteria harbor a combination of *pil*, *com*, and *tad* loci, suggesting diverse functional roles for T4P in Gram-positive bacteria (1). *pil* loci are commonly found in *Clostridium* spp., and clostridial T4P are best studied for *Clostridium perfringens* and *Clostridium difficile* (2, 3). *com* loci are widespread in Firmicutes, among Bacillales and Lactobacillales. *tad* loci are present in archaea and Gram-negative and Gram-positive bacteria. Proteins of the *tad* system assemble adhesive fimbrial low-molecular-weight protein (Flp) pili that are largely unexplored in Gram-positive bacteria (4). *pil* and *tad* loci are not extensively discussed here. Instead we focus on the *com* operon present in the human respiratory pathogen *Streptococcus pneumoniae* and

[1]Department of Microbiology, Tumor and Cell Biology
[2]Department of Clinical Microbiology, Karolinska University Hospital, 171 77 Stockholm, Sweden
Protein Secretion in Bacteria
Edited by Maria Sandkvist, Eric Cascales, and Peter J. Christie
© 2019 American Society for Microbiology, Washington, DC
doi:10.1128/microbiolspec.PSIB-0011-2018

containing the genes involved in the formation of the pneumococcal type IV pilus, also referred to as competence pilus or transformation pilus.

MORPHOLOGY AND FUNCTION OF T4P OF GRAM-POSITIVE SPECIES

T4P are long surface-exposed, thin, flexible filaments that have been studied thoroughly in a few Gram-negative species, as they constitute important virulence factors and play essential roles in diverse cellular processes (5, 6). Architecturally, T4P can be several micrometers in length and are composed of multiple subunits of the major pilin protein tightly packed in a helical arrangement to form the pilus filament.

The first evidence that Gram-positive bacteria possess T4P-like structures came from studies with *Bacillus subtilis*, which possesses a classical *com* operon and the major pilin ComGC (7–9). It was later reported that competent *B. subtilis* could assemble a polymeric complex composed of ComGC subunits with an estimated overall length of 40 to 100 nm (10). This structure was referred to as competence pseudopilus but has never been visualized. So far, T4P have been observed by electron microscopy only in a few Gram-positive bacteria, such as *C. perfringens* (11), *C. difficile* (12, 13), *Ruminococcus albus* (14), and naturally competent streptococci, including the respiratory pathogen *S. pneumoniae* (15, 16) and *Streptococcus sanguinis* (17, 18). The competence pilus of *S. pneumoniae* is morphologically very similar to T4P found in Gram-negative bacteria. It can be several micrometers in length, is 6 to 7 nm in width (16), and is highly flexible (Fig. 1). While competent *S. pneumoniae* usually expresses only one type IV pilus per cell, clostridial species, *R. albus*, and *S. sanguinis* can express multiple T4P on their surfaces. *S. sanguinis* is unique among streptococci since it has both a *pil* locus and a *com* system; thus, it could potentially assemble two types of T4P. Flp pili were

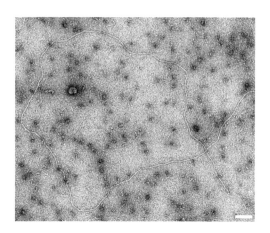

FIGURE 1 **Type IV competence pilus produced by *S. pneumoniae* R6. The pilus was visualized by transmission electron microscopy after negative staining with uranyl acetate. Scale bar, 100 nm.**

observed by immunofluorescence microscopy on the cell surface of *Micrococcus luteus* (19).

Although much remains to be learned about these organelles, their widespread distribution and genetic diversity suggest a variety of functions for T4P also in Gram-positive bacteria. In the current literature, three main T4P-associated functions have been reported, including adhesion, motility, and natural transformation. Briefly, T4P of competent *S. pneumoniae* can directly bind DNA, and pilus-deficient mutants are nontransformable (15). Similarly, Flp pili were required for natural transformation of *M. luteus* (19). In clostridia, T4P have been shown to be involved in twitching motility and biofilm formation (2, 3), and in the case of *R. albus*, T4P bind to cellulose in the gastrointestinal tract of ruminant animals (14, 20). Lastly, *S. sanguinis* produces retractable T4P that can power intense surface-associated motility (18).

TYPE IV COMPETENCE PILI AND NATURAL TRANSFORMATION IN *STREPTOCOCCUS PNEUMONIAE*

S. pneumoniae (the pneumococcus) is a commensal of the human nasopharynx. However, when given the opportunity, it can turn into a life-threatening respiratory pathogen causing severe infectious diseases, such as otitis media, pneumonia, sepsis, and meningitis.

Pneumococci are naturally transformable and can take up and integrate extracellular DNA into their genomes. The ability to take up DNA is transient and referred to as competence state, a highly regulated process. Pneumococcal competence is initiated by competence-stimulating peptide (CSP), a peptide pheromone (21, 22). Secreted CSP is sensed by the bacteria and triggers an activation cascade that results in expression of the competence control operon *comCDE* and *comX* (23, 24). The alternative sigma factor ComX controls expression of all the genes (except *endA*) that are required for transformation, including those involved in the formation of the pneumococcal competence pilus and the DNA uptake machinery called the transformasome (25).

The type IV pilus in *S. pneumoniae* is encoded by the *comG* operon, which is composed of seven genes, namely, *comGA*, *comGB*, *comGC*, *comGD*, *comGE*, *comGF*, and *comGG*. All of them are essential for transformation (26). The first gene of the operon, *comGA*, encodes an ATPase that is required for pilus assembly, as pneumococci deficient in ComGA are nonpiliated and therefore nontransformable (15). Similarly, ComGA of *B. subtilis* is required for binding and transport of transforming DNA (27). *comGB* encodes an integral membrane protein that likely forms the base structure for pilus assembly. The remaining genes in the operon, *comGC* to *-G*, encode five pilin proteins, with ComGC being the major pilin and ComGD to *-G* the minor pilins. The gene for the prepilin peptidase PilD is, unlike in bacteria possessing *pil* or *tad* operons, not part of the *comG* operon and is found elsewhere in the genome (1). The pneumococcal competence system is closely related to other T4P and type II secretion (T2S) systems in Gram-negative bacteria and archaealla, suggesting a common evolutionary ancestor (28–31).

Type IV Pilin Processing and Pneumococcal Pilins

T4P are polymerized from a pool of pilin proteins, which are synthesized as prepilins and need to be processed by the prepilin peptidase prior to pilus assembly. Type IV pilins are characterized by their small size (less than 200 amino acids) and a well-defined class III signal peptide composed of a hydrophilic leader peptide ending on a glycine or alanine followed by a stretch of hydrophobic residues, including a negatively charged glutamate residue at position 5 (Glu5) in mature pilins (6). The conserved, hydrophobic N-terminal domain of mature pilins forms an extended α-helix and protrudes from the more variable C-terminal head domain that can exhibit high sequence variety (28).

The major pilin, ComGC, in *S. pneumoniae* is a small protein (108 amino acids) that fulfills the typical criteria of a classical type IV pilin as mentioned above. It is processed *in vivo* in competent pneumococci (15) and can be cleaved *in vitro* in *Escherichia coli* coexpressing ComGC and the prepilin peptidase, PilD (16). Moreover, the Glu5 residue of mature pneumococcal ComGC is essential for competence pilus assembly and transformation (15). The pneumococcal minor pilins differ substantially in size (86 to 137 amino acids) but share the conserved N-terminal domain. All minor pilins except ComGG have the typical Glu5 residue. Interestingly, ComGG is the largest among the pneumococcal pilins and has a hydrophobic residue at position 5. These two features are characteristic for other minor pilins and pseudopilins that belong to the GspK family and are present in other T4P systems and T2S systems in Gram-negative bacteria. The minor pilin GspK of the *E. coli* T2S system was shown to form a tertiary complex with GspI and GspJ, presumably forming a cap at the tip of the pseudopilus (32). Based on these data, it was speculated that ComGG and its homolog PilK in *C. difficile*, which also lacks the Glu5 residue, could function as initiator pilins to start pilus polymerization and hence would be localized at the pilus tip (3).

Competence Pilus Assembly and DNA Uptake

Type IV pilus biogenesis has been widely studied in Gram-negative bacteria and can be briefly summarized as follows. T4P subunits are synthesized as precursors that after processing are assembled into helical filaments by a dedicated filament assembly machinery in the cytoplasmic membrane. The energy for this process is provided by a conserved traffic ATPase. Pilus formation is initiated on top of a membrane protein, which functions as a building platform, and pilin subunits are added from the base into the growing filament. Many Gram-negative bacteria also have specific ATPases that catalyze pilus disassembly or retraction, which is crucial for T4P functions, such as DNA uptake in the naturally transformable species of the genera *Neisseria* (33, 34) and *Vibrio* (35, 36).

Our understanding of the pneumococcal competence pilus and its biogenesis is still limited. The major pilin ComGC was shown to form the backbone of the pilus (15, 16), and bacteria deficient in ComGC are non-piliated and nontransformable, suggesting that competence pili are essential for transformation. Based on DNA binding experiments with native pili, the competence pilus likely functions as the primary DNA receptor on the surface of competent pneumococci (15). Once captured, DNA must be passed on to the pneumococcal transformasome that facilitates DNA uptake into the cell and integration into the genome (Fig. 2). How exactly DNA is trapped by the pilus remains elusive; however, the process is very efficient, as up to 5.8% of the pneumococcal genome can be transferred to a single recipient cell through 20 recombination events (37). Another puzzling aspect is how DNA enters through the capsule and the thick layer of peptidoglycan. While other competent Gram-negative bacteria rely on retrac-

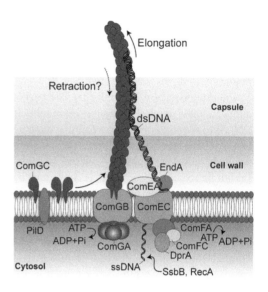

FIGURE 2 Schematic representation of the DNA uptake machinery in competent *S. pneumoniae.* Upon competence induction, the pneumococcal competence pilus composed of ComGC is made and captures extracellular, transforming DNA (15). Captured DNA is passed to the DNA receptor ComEA and the transmembrane channel protein ComEC. This step is possibly mediated by an unknown retraction ATPase and subsequent pilus retraction or yet another undescribed mechanism. Double-stranded DNA (dsDNA) is cleaved by the endonuclease EndA and single-stranded DNA (ssDNA) enters the cytoplasm through the ComEC pore driven by the DNA-dependent ATPase ComFA (48). ComFA forms a complex with ComFC and DprA (49) and together with SsbB and RecA help in stabilization and processing of internalized ssDNA to facilitate genetic exchange (24).

tion ATPases, no additional ATPase that could somehow be involved in the retraction of the competence pilus has been identified in the pneumococcal genome. Nevertheless, it is unlikely that DNA simply slides along the pilus towards the DNA uptake machinery on the cell surface, which is why it was speculated that ComGA could have dual functions and facilitate both extension and retraction (3). Interestingly, retractable T4P in enterotoxigenic *E. coli* and toxin-coregulated T4P in *Vibrio cholerae* also lack a retraction ATPase (38). The latter was recently shown to incorporate the minor pilin TcpB into the growing filament and thereby initiate pilus retraction (39). Whether a similar mechanism applies to the competence pilus in *S. pneumoniae* remains to be determined. Finally, it will be fundamental to understand the role of the pneumococcal minor pilins because these proteins are crucial for pilus assembly and function in other type T4P systems (40–44) and also play an important role in T2S systems (45–47).

Type IV Pilin Structures of Gram-Positive Bacteria

Many type IV pilins of Gram-negative bacteria have been studied from a structural point of view to understand the molecular details of T4P biogenesis. They typically share a common architecture and certain structural features (28). To date, there are three pilin structures from Gram-positive bacteria available, PilJ (12) and PilA1 (13) from *C. difficile* and ComGC from *S. pneumoniae* (16). All three structures show a typical pilin fold but great variability in their C termini. Interestingly, the minor pilin PilJ exhibits a dual-pilin-like fold, presumably the result of a gene duplication or fusion event (12). While PilA1 and ComGC lack cysteines, the three cysteine residues present in PilJ bind a structural zinc ion, suggesting that pilins in Gram-positive bacteria use different stabilization strategies than those previously known from counterparts in Gram-negative bacteria. The structure of soluble ComGC is the first pilin structure involved in the formation of a competence pilus in Gram-positive bacteria (16). It is exclusively α-helical, and its C-terminal head domain is much smaller than other previously described pilin structures.

SUMMARY

T4P are remarkably widespread in Gram-positive bacteria, and understanding their composition and biogenesis is an exciting area of increasing interest. While we are only beginning to explore different T4P in Gram-positive bacteria, it is clear that their overall architecture and their ability to assemble multiple subunits into long filamentous nanomachines are well conserved. A major outstanding question is how T4P in Gram-positive bacteria assemble across the thick layer of peptidoglycan and whether this process requires remodeling of the bacterial cell wall. Detailed genetic, molecular biology, and structural studies in combination with integrative imaging techniques are of great interest to help decipher the mechanisms of pilus formation and function in Gram-positive bacteria.

CITATION

Muschiol S, Aschtgen M-S, Nannapaneni P, Henriques-Normark B. 2019. Gram-positive type IV pili and competence. Microbiol Spectrum 7(1):PSIB-0011-2018.

REFERENCES

1. **Imam S, Chen Z, Roos DS, Pohlschröder M.** 2011. Identification of surprisingly diverse type IV pili, across a broad range of gram-positive bacteria. *PLoS One* **6**:e28919.
2. **Melville S, Craig L.** 2013. Type IV pili in Gram-positive bacteria. *Microbiol Mol Biol Rev* **77**:323–341.
3. **Piepenbrink KH, Sundberg EJ.** 2016. Motility and adhesion through type IV pili in Gram-positive bacteria. *Biochem Soc Trans* **44**:1659–1666.
4. **Tomich M, Planet PJ, Figurski DH.** 2007. The tad locus: postcards from the widespread colonization island. *Nat Rev Microbiol* **5**:363–375.
5. **Craig L, Pique ME, Tainer JA.** 2004. Type IV pilus structure and bacterial pathogenicity. *Nat Rev Microbiol* **2**:363–378.
6. **Berry JL, Pelicic V.** 2015. Exceptionally widespread nanomachines composed of type IV pilins: the prokaryotic Swiss Army knives. *FEMS Microbiol Rev* **39**:134–154.
7. **Chung YS, Breidt F, Dubnau D.** 1998. Cell surface localization and processing of the ComG proteins, required for DNA binding during transformation of *Bacillus subtilis*. *Mol Microbiol* **29**:905–913.

8. **Chung YS, Dubnau D.** 1995. ComC is required for the processing and translocation of comGC, a pilin-like competence protein of *Bacillus subtilis*. *Mol Microbiol* **15:**543–551.

9. **Chung YS, Dubnau D.** 1998. All seven comG open reading frames are required for DNA binding during transformation of competent *Bacillus subtilis*. *J Bacteriol* **180:**41–45.

10. **Chen I, Provvedi R, Dubnau D.** 2006. A macromolecular complex formed by a pilin-like protein in competent *Bacillus subtilis*. *J Biol Chem* **281:**21720–21727.

11. **Varga JJ, Nguyen V, O'Brien DK, Rodgers K, Walker RA, Melville SB.** 2006. Type IV pili-dependent gliding motility in the Gram-positive pathogen *Clostridium perfringens* and other clostridia. *Mol Microbiol* **62:**680–694.

12. **Piepenbrink KH, Maldarelli GA, de la Peña CF, Mulvey GL, Snyder GA, De Masi L, von Rosenvinge EC, Günther S, Armstrong GD, Donnenberg MS, Sundberg EJ.** 2014. Structure of *Clostridium difficile* PilJ exhibits unprecedented divergence from known type IV pilins. *J Biol Chem* **289:**4334–4345.

13. **Piepenbrink KH, Maldarelli GA, Martinez de la Peña CF, Dingle TC, Mulvey GL, Lee A, von Rosenvinge E, Armstrong GD, Donnenberg MS, Sundberg EJ.** 2015. Structural and evolutionary analyses show unique stabilization strategies in the type IV pili of *Clostridium difficile*. *Structure* **23:**385–396.

14. **Rakotoarivonina H, Jubelin G, Hebraud M, Gaillard-Martinie B, Forano E, Mosoni P.** 2002. Adhesion to cellulose of the Gram-positive bacterium *Ruminococcus albus* involves type IV pili. *Microbiology* **148:**1871–1880.

15. **Laurenceau R, Péhau-Arnaudet G, Baconnais S, Gault J, Malosse C, Dujeancourt A, Campo N, Chamot-Rooke J, Le Cam E, Claverys JP, Fronzes R.** 2013. A type IV pilus mediates DNA binding during natural transformation in *Streptococcus pneumoniae*. *PLoS Pathog* **9:**e1003473.

16. **Muschiol S, Erlendsson S, Aschtgen MS, Oliveira V, Schmieder P, de Lichtenberg C, Teilum K, Boesen T, Akbey U, Henriques-Normark B.** 2017. Structure of the competence pilus major pilin ComGC in *Streptococcus pneumoniae*. *J Biol Chem* **292:**14134–14146.

17. **Gurung I, Berry JL, Hall AMJ, Pelicic V.** 2017. Cloning-independent markerless gene editing in *Streptococcus sanguinis*: novel insights in type IV pilus biology. *Nucleic Acids Res* **45:**e40.

18. **Gurung I, Spielman I, Davies MR, Lala R, Gaustad P, Biais N, Pelicic V.** 2016. Functional analysis of an unusual type IV pilus in the Gram-positive *Streptococcus sanguinis*. *Mol Microbiol* **99:**380–392.

19. **Angelov A, Bergen P, Nadler F, Hornburg P, Lichev A, Übelacker M, Pachl F, Kuster B, Liebl W.** 2015. Novel Flp pilus biogenesis-dependent natural transformation. *Front Microbiol* **6:**84.

20. **Pegden RS, Larson MA, Grant RJ, Morrison M.** 1998. Adherence of the gram-positive bacterium *Ruminococcus albus* to cellulose and identification of a novel form of cellulose-binding protein which belongs to the Pil family of proteins. *J Bacteriol* **180:**5921–5927.

21. **Håvarstein LS, Coomaraswamy G, Morrison DA.** 1995. An unmodified heptadecapeptide pheromone induces competence for genetic transformation in *Streptococcus pneumoniae*. *Proc Natl Acad Sci U S A* **92:**11140–11144.

22. **Moreno-Gámez S, Sorg RA, Domenech A, Kjos M, Weissing FJ, van Doorn GS, Veening JW.** 2017. Quorum sensing integrates environmental cues, cell density and cell history to control bacterial competence. *Nat Commun* **8:**854.

23. **Johnsborg O, Eldholm V, Håvarstein LS.** 2007. Natural genetic transformation: prevalence, mechanisms and function. *Res Microbiol* **158:**767–778.

24. **Johnston C, Martin B, Fichant G, Polard P, Claverys JP.** 2014. Bacterial transformation: distribution, shared mechanisms and divergent control. *Nat Rev Microbiol* **12:**181–196.

25. **Lee MS, Morrison DA.** 1999. Identification of a new regulator in *Streptococcus pneumoniae* linking quorum sensing to competence for genetic transformation. *J Bacteriol* **181:**5004–5016.

26. **Peterson SN, Sung CK, Cline R, Desai BV, Snesrud EC, Luo P, Walling J, Li H, Mintz M, Tsegaye G, Burr PC, Do Y, Ahn S, Gilbert J, Fleischmann RD, Morrison DA.** 2004. Identification of competence pheromone responsive genes in *Streptococcus pneumoniae* by use of DNA microarrays. *Mol Microbiol* **51:**1051–1070.

27. **Briley K Jr, Dorsey-Oresto A, Prepiak P, Dias MJ, Mann JM, Dubnau D.** 2011. The secretion ATPase ComGA is required for the binding and transport of transforming DNA. *Mol Microbiol* **81:**818–830.

28. **Giltner CL, Nguyen Y, Burrows LL.** 2012. Type IV pilin proteins: versatile molecular modules. *Microbiol Mol Biol Rev* **76:**740–772.

29. **Sandkvist M.** 2001. Biology of type II secretion. *Mol Microbiol* **40:**271–283.

30. **Makarova KS, Koonin EV, Albers SV.** 2016. Diversity and evolution of type IV pili systems in Archaea. *Front Microbiol* **7:**667.

31. **Albers SV, Jarrell KF.** 2018. The archaellum: an update on the unique archaeal motility structure. *Trends Microbiol* **26:**351–362.

32. **Korotkov KV, Hol WG.** 2008. Structure of the GspK-GspI-GspJ complex from the enterotoxigenic *Escherichia coli* type 2 secretion system. *Nat Struct Mol Biol* **15:**462–468.

33. **Wolfgang M, Lauer P, Park HS, Brossay L, Hébert J, Koomey M.** 1998. PilT mutations lead to simultaneous defects in competence for natural transformation and twitching motility in piliated *Neisseria gonorrhoeae*. *Mol Microbiol* **29:**321–330.

34. **Obergfell KP, Seifert HS.** 2015. Mobile DNA in the pathogenic *Neisseria*. *Microbiol Spectr* **3:** MDNA3-0015-2014.

35. **Matthey N, Blokesch M.** 2016. The DNA-uptake process of naturally competent *Vibrio cholerae*. *Trends Microbiol* **24:**98–110.

36. **Ellison CK, Dalia TN, Vidal Ceballos A, Wang JC, Biais N, Brun YV, Dalia AB.** 2018. Retraction of DNA-bound type IV competence pili initiates DNA uptake during natural transformation in *Vibrio cholerae*. *Nat Microbiol* **3:**773–780.

37. **Cowley LA, Petersen FC, Junges R, Jimson D, Jimenez M, Morrison DA, Hanage WP.** 2018. Evolution via recombination: cell-to-cell contact facilitates larger recombination events in *Streptococcus pneumoniae*. *PLoS Genet* **14:** e1007410.

38. **Kolappan S, Ng D, Yang G, Harn T, Craig L.** 2015. Crystal structure of the minor pilin CofB, the initiator of CFA/III pilus assembly in enterotoxigenic *Escherichia coli*. *J Biol Chem* **290:**25805–25818.

39. **Ng D, Harn T, Altindal T, Kolappan S, Marles JM, Lala R, Spielman I, Gao Y, Hauke CA, Kovacikova G, Verjee Z, Taylor RK, Biais N, Craig L.** 2016. The *Vibrio cholerae* minor pilin TcpB initiates assembly and retraction of the toxin-coregulated pilus. *PLoS Pathog* **12:**e1006109.

40. **Giltner CL, Habash M, Burrows LL.** 2010. *Pseudomonas aeruginosa* minor pilins are incorporated into type IV pili. *J Mol Biol* **398:**444–461.

41. **Nguyen Y, Sugiman-Marangos S, Harvey H, Bell SD, Charlton CL, Junop MS, Burrows LL.** 2015. *Pseudomonas aeruginosa* minor pilins prime type IVa pilus assembly and promote surface display of the PilY1 adhesin. *J Biol Chem* **290:**601–611.

42. **Marko VA, Kilmury SLN, MacNeil LT, Burrows LL.** 2018. *Pseudomonas aeruginosa* type IV minor pilins and PilY1 regulate virulence by modulating FimS-AlgR activity. *PLoS Pathog* **14:**e1007074.

43. **Kuchma SL, Griffin EF, O'Toole GA.** 2012. Minor pilins of the type IV pilus system participate in the negative regulation of swarming motility. *J Bacteriol* **194:**5388–5403.

44. **Cehovin A, Simpson PJ, McDowell MA, Brown DR, Noschese R, Pallett M, Brady J, Baldwin GS, Lea SM, Matthews SJ, Pelicic V.** 2013. Specific DNA recognition mediated by a type IV pilin. *Proc Natl Acad Sci U S A* **110:**3065–3070.

45. **Cisneros DA, Bond PJ, Pugsley AP, Campos M, Francetic O.** 2012. Minor pseudopilin self-assembly primes type II secretion pseudopilus elongation. *EMBO J* **31:**1041–1053.

46. **Cisneros DA, Pehau-Arnaudet G, Francetic O.** 2012. Heterologous assembly of type IV pili by a type II secretion system reveals the role of minor pilins in assembly initiation. *Mol Microbiol* **86:**805–818.

47. **Burrows LL.** 2012. Prime time for minor subunits of the type II secretion and type IV pilus systems. *Mol Microbiol* **86:**765–769.

48. **Johnston C, Campo N, Bergé MJ, Polard P, Claverys JP.** 2014. *Streptococcus pneumoniae*, le transformiste. *Trends Microbiol* **22:**113–119.

49. **Diallo A, Foster HR, Gromek KA, Perry TN, Dujeancourt A, Krasteva PV, Gubellini F, Falbel TG, Burton BM, Fronzes R.** 2017. Bacterial transformation: ComFA is a DNA-dependent ATPase that forms complexes with ComFC and DprA. *Mol Microbiol* **105:**741–754.

The Remarkable Biomechanical Properties of the Type 1 Chaperone-Usher Pilus: A Structural and Molecular Perspective

12

MANUELA K. HOSPENTHAL[1,2] and GABRIEL WAKSMAN[1]

INTRODUCTION

Chaperone-usher (CU) pili are virulence factors displayed on a wide variety of Gram-negative bacterial pathogens (1), mediating bacterial attachment and biofilm formation (2). The two best-studied examples of CU pili are the type 1 and P pili of uropathogenic *Escherichia coli* (UPEC), which is the most important causative agent of urinary tract infections (3). We here summarize the steps of CU pilus biogenesis and highlight the most recent structural advances relating to type 1 pili that allow UPEC to thrive in the urinary tract.

BIOGENESIS OF CHAPERONE-USHER PILI

The individual building blocks required for type 1 and P pilus assembly are known as pilins (pilus subunits) and are encoded by the *fim* and *pap* operons, respectively (4). The majority of fully assembled CU pili adopt a composite architecture consisting of a thin tip fibrillum, attached to a long superhelical rod emanating from the outer membrane (Fig. 1a). The adhesin (FimH for

[1]Institute of Structural and Molecular Biology, University College London and Birkbeck, London WC1E 7HX, United Kingdom
[2]Institute of Molecular Biology and Biophysics, ETH Zürich, 8093 Zürich, Switzerland
Protein Secretion in Bacteria
Edited by Maria Sandkvist, Eric Cascales, and Peter J. Christie
© 2019 American Society for Microbiology, Washington, DC
doi:10.1128/microbiolspec.PSIB-0010-2018

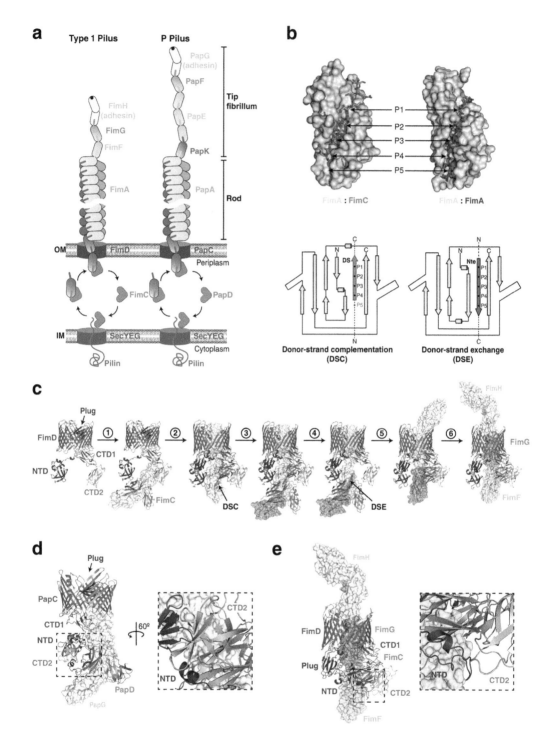

type 1 pili; PapG for P pili) is located at the distal end of the tip fibrillum and is composed of an N-terminal lectin domain, which is responsible for binding to specific host ligands, and a C-terminal pilin domain, which connects the adhesin to the remainder of the pilus (5, 6). The complete tip fibrillum is composed of further pilus subunits, namely, FimG and FimF in the case of type 1 pili or PapF, PapE, and PapK for P pili (Fig. 1a). The tip fibrillum of P pili is longer, as it contains 5 to 10 copies of PapE, while all other subunits are present as a single copy (7, 8). The largest section of CU pili, the rod, is composed of thousands of copies of a single pilin (FimA for type 1 pili; PapA for P pili) (2, 9, 10). All CU pilins are composed of C-terminally truncated and thus incomplete immunoglobulin (Ig)-like folds, lacking the 7th β-strand (Fig. 1b). Due to the missing strand, the pilins are unstable on their own and contain a long hydrophobic groove on their surface, composed of the so-called P1 to P5 hydrophobic pockets (6, 11). The pilins are stabilized by interacting with a dedicated chaperone (FimC for type 1 pili; PapD for P pili) through a process known as donor strand complementation (Fig. 1b). Here, the P1 to P4 pockets of the hydrophobic groove are

---

FIGURE 1 Architecture and assembly of chaperone-usher pili. (a) Type 1 (left) and P pili (right) are the two archetypal CU pili of UPEC. Pilins are transported through the inner membrane (IM) via the SecYEG machinery. Once in the periplasm, a dedicated chaperone (FimC for type 1 pili; PapD for P pili) helps to fold, stabilize, and transport individual pilins to the outer membrane (OM), where they are assembled into pili by the usher (FimD for type 1 pili; PapC for P pili). The largest section of CU pili is the rod, which is composed of thousands of copies of a single subunit (FimA for type 1 pili; PapA for P pili) arranged into a right-handed superhelical quaternary structure. On top of the rod, located at the pilus' distal end, is a thin and flexible tip fibrillum. The most important tip fibrillum subunit is the adhesin (FimH for type 1 pili; PapG for P pili), which is responsible for the interaction of CU pili with host cell receptors. The remainder of the tip fibrillum is formed by FimG and FimF for type 1 pili and PapF, PapE, and PapK for P pili. (b) Pilins are unstable on their own because they consist of C-terminally truncated incomplete Ig-like folds lacking the 7th β-strand. This creates a large hydrophobic groove on the subunit's surface. After their transport into the periplasm, the chaperone inserts its G1 β-strand into the hydrophobic groove, thereby completing and stabilizing its fold. This is known as donor strand complementation (DSC) (PDB code 4DWH [79]) (left side). The pilin's P1 to P4 pockets are occupied by the chaperone's P1 to P4 residues, while the P5 pocket remains empty. Once assembled into a pilus, the 10- to 20-residue-long N-terminal extension (Nte) of each subunit complements the preceding pilin's groove, stabilizing the structure and linking the subunits in the pilin polymer. This is referred to as donor strand exchange (DSE) (PDB code 5OH0 [10]) (right side). The Nte of FimA in the surface model on the right has been removed for clarity. A zip-in–zip-out mechanism is responsible for the transition from DSC to DSE, whereby the previously empty P5 pocket first becomes occupied by the incoming subunit's Nte, displacing the chaperone's complementing strand and subsequently allowing the Nte to fully occupy the pilin's P1 to P5 pockets. (c) In step 1, the chaperone-adhesin complex binds to the usher's NTD (PDB codes 3BWU [80], 1QUN [6], and 3OHN and 3RFZ [26]). In step 2, the plug relocates next to the periplasmically located NTD, while the chaperone-adhesin complex is transferred to the usher's CTDs, which interact with the adhesin's pilin domain. The adhesin's lectin domain begins to translocate through the usher pore (PDB code 3RFZ). In step 3, the next chaperone-pilin complex is recruited to the NTD and the Nte of this pilin is oriented towards the pilin domain of the adhesin (PDB codes 3RFZ and 3BWU). In step 4, the chaperone's donor strand is replaced by the Nte of the newly recruited pilin by the zip-in–zip-out mechanism. The displaced chaperone is recycled (PDB codes 3RFZ, 3BWU, and 4XOE [43]). In step 5, the chaperone-pilin complex is transferred to the CTDs and the adhesin continues to move up and out through the usher pore (PDB codes 3RFZ and 4J3O [56]). In step 6, the cycle is repeated and new pilins are incorporated into the growing pilus (PDB code 4J3O). The mechanism of translocation through the usher depicted is illustrated using both crystal and modeled structures of the CU pilus systems. (d and e) Two novel structures shed light on the chaperone-subunit handover mechanism from the NTD to the CTDs. Shown are the structures of PapCDG (PDB code 6CD2 [33]) in a preactivated state (d) and of FimDCFGH (PDB code 6E14 [34]) in an activated state (e), trapping conformations that show novel interactions between the NTD and CTD2, during chaperone-subunit handover. Dashed boxes and zoomed-in views highlight the NTD to CTD2 interactions.

complemented by residues of one of the chaperone's own β-strands, thereby completing the pilin's Ig-like fold, while the P5 pocket remains empty (6, 11–13). The chaperone-subunit complexes are then shuttled to the outer membrane-embedded usher (FimD for type 1 pili; PapC for P pili), where the pilus subunits are assembled into pili. During pilus assembly, pilins undergo a transition from a chaperone-stabilized binary complex to a stable polymer where the hydrophobic groove becomes complemented by a β-strand formed by the N-terminal extension (Nte) of the next subunit in assembly (14–16). This is termed donor strand exchange (DSE) (Fig. 1b).

The usher contains several distinct domains: the 24-stranded β-barrel pore, the N-terminal domain (NTD), two C-terminal domains (CTD1 and CTD2), and the plug (17, 18) (Fig. 1c). The steps of pilus biogenesis have been visualized by a series of structures and modeled states of the translocating usher of both the type 1 and P pilus systems (depicted in Fig. 1c). CU pili are assembled in a top-down manner starting with the adhesin, which primes the usher for pilus biogenesis when it is recruited to the periplasmic NTD (19–25). Next, the chaperone-adhesin complex is transferred to the higher-affinity CTDs, the plug is displaced from the usher pore into the periplasm, and the adhesin's lectin domain is translocated into the β-barrel channel (26, 27). This allows further chaperone-subunit complexes to be recruited to the NTD, bringing the Nte of the incoming subunit into close proximity with the hydrophobic groove of the preceding subunit (26). In a zip-in–zip-out mechanism, the P5 residue of the Nte first engages the previously empty P5 pocket before displacing the chaperone's donor strand entirely by sequentially invading the groove's P4, P3, P2, and P1 pockets (28, 29). The chaperone is recycled and further chaperone-subunit complexes continue to be incorporated into the growing pilus. The stochastic incorporation of PapH, the termination subunit of P pili, halts pilus biogenesis, as this subunit lacks the P5 pocket, making it

unable to undergo DSE (30). FimI also displays a closed P5 pocket once bound to the NTD of the usher and is likely the termination subunit of type 1 pili (31). For a more detailed description of pilus biogenesis, please refer to recent reviews on this topic (2, 8, 32).

Two recent structures are beginning to shed light onto the handover mechanism of chaperone-subunit complexes from the NTD to the CTDs. First, a crystal structure of the "pre-activated" PapC usher, in complex with the chaperone-adhesin complex (PapDG), was determined (33) (Fig. 1d). In this structure, the PapDG complex has been recruited to the NTD but has not yet been fully transferred to the CTDs, while the plug still occupies the usher's β-barrel pore. Interestingly, CTD2 has moved across to engage PapDG bound to the NTD, by forming contacts with both PapD and the NTD. Thus, this state potentially represents the moment prior to chaperone-adhesin handover from the NTD to the CTDs in the preactivated usher. The precise temporal order of plug displacement, transfer to the CTDs, and recruitment of the next chaperone-subunit complex remains to be determined. Second, a new conformation of the activated usher during chaperone-subunit handover was captured using cryo-electron microscopy (34) (Fig. 1e). This structure of the FimD usher in complex with the tip fibrillum (FimFGH) and the chaperone (FimC) shows the usher in the process of chaperone-subunit (FimCF) handover to the CTDs. During this transfer, the NTD remains bound to the chaperone-subunit complex as it swings across to engage CTD2. At this point, both the NTD and CTD2 are bound to the growing end of the pilus, which has been suggested to prevent the pilus fiber from diffusing away during pilus biogenesis. This recent structural information, for both the preactivated (PapCDG) and activated (FimDCFGH) ushers during chaperone-subunit transfer, raises interesting questions. While both structures revealed a novel interaction between the NTD and CTD2, the relative positions of the usher's various domains and the chaperone-subunit complex are distinct. Whether this is due to the stage of pilus biogenesis each structure represents (pre-activated versus activated) and/or whether differences between the type 1 and P pilus systems also play a role remains to be fully explored.

MAKING CONTACT: THE ROLE OF THE TYPE 1 PILUS ADHESIN

UPEC can cause disease by ascending up the urinary tract and colonizing host tissues in the bladder (cystitis) and the kidney (pyelonephritis) (3, 35). The urinary tract presents a unique challenge for UPEC organisms, as they periodically experience the shear forces resulting from urine flow, which is faster and more turbulent in the bladder and lower urinary tract than in the kidneys (36). The adhesins of type 1 and P pili differ with respect to their ligands and their modes of interaction. The FimH lectin domain interacts with mannosylated proteins expressed on the surface of bladder epithelial cells with a so-called "catch-bond" mechanism, whereas the weaker "slip-bond" interaction of the PapG lectin domain with galabiose-containing glycosphingolipids takes place primarily on the kidney epithelium (8, 37–39). UPEC's ability to travel up the urinary tract is, in part, thought to involve an increase in P pilus expression and a concomitant downregulation of type 1 pilus expression (40). We here focus on recent advances in our understanding of the FimH-mannose catch-bond interaction.

The mannose binding site is located at the tip of FimH's lectin domain (Fig. 2a) and consists of a negatively charged pocket surrounded by a hydrophobic ridge (6, 41). In the ligand-bound state, the loops surrounding the binding site tighten around the ligand, with the most substantial rearrangement occurring in the clamp loop (residues 8 to 16) (42, 43) (Fig. 2b). Due to FimH's

catch-bond mechanism, the affinity of its interaction with mannose increases when the bacterium experiences tensile mechanical force, as demonstrated in flow chamber (44, 45) and atomic force microscopy (46, 47) experiments. This allosteric mechanism depends on the relative orientation of FimH's lectin and pilin domains, which are connected by a linker (44) (Fig. 2a). FimH binds mannose with moderate affinity when the two domains are closely associated with each other and switches to a high-affinity binding mode when external forces (e.g., urine flow) separate the two domains. In a recent study, the binding kinetics of the FimH-mannose interaction were measured for donor strand-complemented full-length FimH (low-affinity state) and for the isolated lectin domain (a proxy for the high-affinity domain-separated state) (43). These measurements showed that in different *Escherichia coli* strains the low-affinity (domain-associated) state of FimH exhibits a dissociation constant in the micromolar range, whereas isolated lectin domains (high-affinity state) bound their ligands with ~3,300-fold-higher affinity in the low nanomolar range. This increase in affinity is due to a combination of a 30-fold-lower on-rate and >100,000-fold-lower off-rate (43).

The domain-separated high-affinity state is characterized by a rearrangement of the lectin domain's swing loop (residues 27 to 33), insertion loop (residues 112 to 118), and linker loop (residues 154 to 160), which form the interface with the pilin domain (42) (Fig. 2c). This conformational state has been observed in crystal structures of the isolated ligand-bound lectin domain (43, 48–53), FimH prior to DSE as observed in the FimCH (6, 41) and FimDCH complexes (26) (FimC keeps FimH in its domain-separated state), ligand-bound FimH complemented by a non-cognate donor strand peptide of FimF (43), and a FimH variant containing residues (A27V/V163A) that have been positively selected for pathogenicity among UPEC strains (54, 55). Structures of wild-type full-length FimH after DSE, either in the context of the remainder

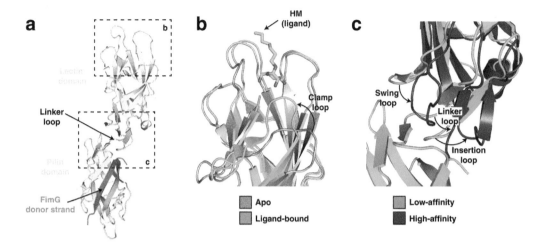

FIGURE 2 Structural rearrangements in FimH. (a) Ribbon diagram of full-length FimH (yellow), which is complemented by a donor strand peptide from FimG (orange) (PDB code 4XOE [43]). The lectin and pilin domains are labeled, and the dashed boxes highlight the ligand binding pocket (top) and the domain interface region (bottom) that are expanded in panels b and c. (b) Superposition of apo (gray) (PDB code 4XOD [43]) and ligand-bound (cyan) (PDB code 4XOE [43]) FimH, focusing on the ligand binding pocket. The ligand is *n*-heptyl α-d-mannoside (HM). An arrow indicates the structural rearrangement of the clamp loop. (c) Superposition of ligand-bound FimH (cyan) (PDB code 4XOE) and a ligand-bound construct of the FimH lectin domain only (purple) (PDB code 4XOC [43]), focusing on the lectin domain loops at the domain interface. The cyan structure is in a domain-associated (low-affinity) state, whereas the purple lectin domain-only structure represents the conformation of a domain-separated (high-affinity) state. Arrows indicate the rearrangements of the swing, linker, and insertion loops. The FimH pilin domain in all structures was stabilized by a FimG donor strand peptide.

of the tip fibrillum (FimCFGH [42] and FimDCFGH [34, 56]) or complemented by a FimG Nte peptide (43, 54), adopt the low-affinity domain-associated conformation.

Early crystal structures capturing both high- and low-affinity states led to the proposition of an allosteric mechanism where domain separation induced by external forces caused a structural rearrangement of the lectin domain loops at the domain interface (Fig. 2c). These changes would, in turn, cause the β-sandwich lectin domain to adopt a less twisted and more elongated conformation, releasing the autoinhibitory effect of the pilin domain and resulting in the ligand binding loops tightening around the ligand (42) (Fig. 2b). Additional crystal structures, together with small-angle X-ray scattering (SAXS) and ion mobility-mass spectrometry (IMMS) experiments supported by molecular dynamics (MD), have further expanded the catch-bond

mechanism. MD simulations showed that the relative conformational flexibility of the lectin and pilin domains is greater in the domain-separated state and that the overall conformation of the domain-associated state is fairly constrained (43, 54). Furthermore, the mode of ligand binding seemed to depend on the conformational state of FimH: in the domain-separated (high-affinity) state, the ligand bound tightly in a defined orientation (shown in Fig. 2b), whereas in the domain-associated (low-affinity) state, the ligand sampled several orientations (54). These simulations were in line with previous experiments that showed a high degree of conformational dynamics of the tyrosine gate (composed of two parallel tyrosines, Y48 and Y137) of the FimH binding pocket in the presence of mannosides (51, 57). On the other hand, crystal structures of both the high- and low-affinity states show the ligand bound in the same conformation, and the crystallographic B factors (temperature factor) of the bound ligand in these structures suggested that there is no difference in ligand binding mode (43). Furthermore, MD simulations suggested a high-energy barrier between the domain-associated and domain-separated states (43, 54), in agreement with ligand binding not leading to interface loop rearrangements and domain separation in crystal structures of donor strand-complemented full-length FimH (43). Nevertheless, insights from SAXS and IMMS experiments suggest that ligand binding may indeed start to shift the conformational equilibrium that exists in solution towards the domain-separated state (54).

This extraordinary catch-bond mechanism ensures that in the absence of urine flow, UPEC can overcome the strength of host cell binding to efficiently disseminate through the urinary tract using flagellar motility, while avoiding being flushed out during periods of urine flow (43, 45, 58, 59). In addition, the initial low-affinity interaction may prevent bacteria from engaging soluble mannose receptors, such as those on the Tamm-Horsfall protein, under low-shear conditions, which would constitute a nonproductive binding event (60).

CLINGING ON: THE IMPORTANCE OF THE PILUS ROD

The FimH catch-bond mechanism is not the only feature of type 1 pili that allows UPEC to biomechanically withstand the shear forces in the urinary tract. The rod, the largest CU pilus section, adopts a superhelical quaternary structure composed of 3 to 4 subunits (FimA for type 1 pili; PapA for P pili) per helical turn, with the most important interface (stacking interface) occurring between every n and $n + 3$ subunits (61–63) (Fig. 3a and b). This remarkable structure is able to uncoil in response to shear forces, by sequentially breaking the stack-to-stack interactions, thereby dissipating the forces experienced by the adhesin and enabling

UPEC to remain firmly attached (64, 65). The FimH catch-bond affinity switch and rod uncoiling mechanisms are functionally coupled, as the forces required for the two are similar (66, 67). In 2016, the first high-resolution cryo-electron microscopy structure of the P pilus revealed the molecular determinants that enable rod uncoiling (9). During urine flow, the DSE interactions that stabilize and link the pilins in the rod are extremely strong and will not break (68, 69) (Fig. 3c), whereas the largely polar interactions that mediate the subunit-subunit interactions of the rod's quaternary superhelical structure are much weaker and begin to break (9). This causes the rod to progressively uncoil, eventually adopting a head-to-tail configuration of pilins before recoiling once the external force has ceased. These biomechanical properties have been the subject of many atomic force microscopy and optical tweezer studies (36, 64–67, 70–77). Interestingly, such experiments showed that type 1 pilus rods require slightly higher forces to induce rod unwinding than do P pilus rods, prompting the suggestion that type 1 pili are better adapted to withstand the more turbulent flows of the lower urinary tract (36, 76). Several structures of the type 1 pilus rod have been determined (10, 77, 78) (Fig. 3), revealing a very similar overall architecture to the P pilus rod, except that the type 1 pilus rod lacks the so-called "staple" region (9) (Fig. 3c). It was suggested that the increased resistance against unwinding of type 1 pili could be explained, in part, by the larger stacking interface (10). The physiological relevance of these biomechanical properties was demonstrated when UPEC strains expressing type 1 pilus rods with weaker stack-to-stack interactions were significantly attenuated in their ability to cause intestinal colonization and bladder infection in mice (77).

Interestingly, the stability of type 1 pili seems to depend on the route of their assembly. Type 1 pili assembled *in vivo*, by the CU machinery, are significantly more stable

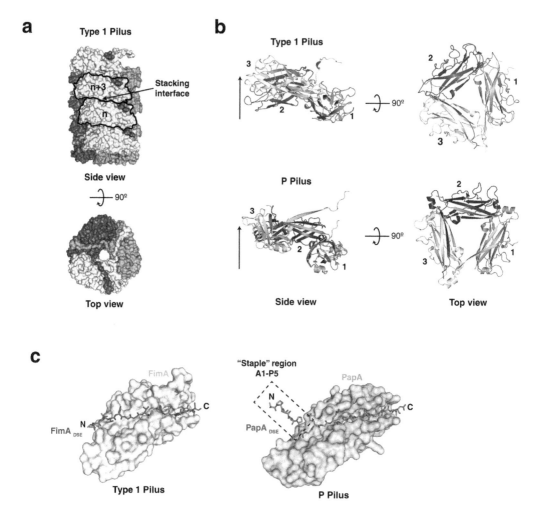

FIGURE 3 The structure of the rod. (a) Surface models showing the type 1 pilus rod structure (PDB code 5OH0 [10]) in a side view and a top view, which are 90° rotated with respect to each other. The Nte of the uppermost FimA molecule is removed in the top view for illustrative purposes. (b) Cartoon models showing three adjacent molecules or one "stack" of the type 1 pilus (blue) (PDB code 5OH0) and P pilus (green) (PDB code 5FLU [9]) rods. The left and right parts of this panel show one stack as a side view and a top view, respectively, rotated by 90°. Pilin subunits are arbitrarily numbered starting with the pilin at the bottom, which would be most proximal to the OM and last to be assembled, to show the nature of the right-handed superhelical structure. The arrow in the side view shows the upward trajectory of the subunits in the structure. (c) Surface representation of an individual FimA pilin subunit within the type 1 pilus rod (left, blue) (PDB code 5OH0) and a PapA pilin subunit within the P pilus rod (right, green) (PDB code 5FLU). The stick model shows the complementing donor strands, which originate from the Nte of the next subunit in assembly. The dashed box shows the staple region (residues 1 to 5) of the PapA Nte, a region not present in the type 1 pilus.

against dissociation and unfolding than pilus rods assembled from FimA alone *in vitro* (10). In fact, the unfolding rate constants differed by 3 to 4 orders of magnitude, raising the intriguing question of whether the usher, or indeed another factor, can guide and influence the assembly of an optimally stable structure *in vivo*. Future experiments will hopefully resolve this question and reveal the molecular details responsible.

CONCLUSIONS

In this brief review, we have summarized aspects of CU pilus structure, with a special focus on type 1 pili. The FimH catch-bond mechanism and the spring-like properties of the rod are crucial biomechanical adaptations that allow UPEC to maintain a foothold in the hostile environment of the urinary tract. These two features are functionally coupled, and perturbation of either results in less virulent bacteria.

ACKNOWLEDGMENTS

This work was funded by MRC grant 018434 to G.W.

CITATION

Hospenthal MK, Waksman G. 2019. The remarkable biomechanical properties of the type 1 chaperone-usher pilus: a structural and molecular perspective. Microbiol Spectrum 7(1):PSIB-0010-2018.

REFERENCES

1. **Thanassi DG, Saulino ET, Hultgren SJ.** 1998. The chaperone/usher pathway: a major terminal branch of the general secretory pathway. *Curr Opin Microbiol* **1:**223–231.

2. **Hospenthal MK, Costa TRD, Waksman G.** 2017. A comprehensive guide to pilus biogenesis in Gram-negative bacteria. *Nat Rev Microbiol* **15:**365–379.

3. **Flores-Mireles AL, Walker JN, Caparon M, Hultgren SJ.** 2015. Urinary tract infections: epidemiology, mechanisms of infection and treatment options. *Nat Rev Microbiol* **13:**269–284.

4. **Schwan WR.** 2011. Regulation of *fim* genes in uropathogenic *Escherichia coli*. *World J Clin Infect Dis* **1:**17–25.

5. **Dodson KW, Pinkner JS, Rose T, Magnusson G, Hultgren SJ, Waksman G.** 2001. Structural basis of the interaction of the pyelonephritic *E. coli* adhesin to its human kidney receptor. *Cell* **105:**733–743.

6. **Choudhury D, Thompson A, Stojanoff V, Langermann S, Pinkner J, Hultgren SJ, Knight SD.** 1999. X-ray structure of the FimC-FimH chaperone-adhesin complex from uropathogenic *Escherichia coli. Science* **285:**1061–1066.

7. **Kuehn MJ, Heuser J, Normark S, Hultgren SJ.** 1992. P pili in uropathogenic *E. coli* are composite fibres with distinct fibrillar adhesive tips. *Nature* **356:**252–255.

8. **Werneburg GT, Thanassi DG.** 2018. Pili assembled by the chaperone/usher pathway in *Escherichia coli* and *Salmonella. Ecosal Plus* **8:**1–37.

9. **Hospenthal MK, Redzej A, Dodson K, Ukleja M, Frenz B, Rodrigues C, Hultgren SJ, DiMaio F, Egelman EH, Waksman G.** 2016. Structure of a chaperone-usher pilus reveals the molecular basis of rod uncoiling. *Cell* **164:**269–278.

10. **Hospenthal MK, Zyla D, Costa TRD, Redzej A, Giese C, Lillington J, Glockshuber R, Waksman G.** 2017. The cryoelectron microscopy structure of the type 1 chaperone-usher pilus rod. *Structure* **25:**1829–1838.e4.

11. **Sauer FG, Fütterer K, Pinkner JS, Dodson KW, Hultgren SJ, Waksman G.** 1999. Structural basis of chaperone function and pilus biogenesis. *Science* **285:**1058–1061.

12. **Barnhart MM, Pinkner JS, Soto GE, Sauer FG, Langermann S, Waksman G, Frieden C, Hultgren SJ.** 2000. PapD-like chaperones provide the missing information for folding of pilin proteins. *Proc Natl Acad Sci U S A* **97:**7709–7714.

13. **Vetsch M, Puorger C, Spirig T, Grauschopf U, Weber-Ban EU, Glockshuber R.** 2004. Pilus chaperones represent a new type of protein-folding catalyst. *Nature* **431:**329–333.

14. **Sauer FG, Pinkner JS, Waksman G, Hultgren SJ.** 2002. Chaperone priming of pilus subunits facilitates a topological transition that drives fiber formation. *Cell* **111:**543–551.

15. **Zavialov AV, Berglund J, Pudney AF, Fooks LJ, Ibrahim TM, MacIntyre S, Knight SD.** 2003. Structure and biogenesis of the capsular F1 antigen from *Yersinia pestis*: preserved folding energy drives fiber formation. *Cell* **113:**587–596.

16. **Nishiyama M, Ishikawa T, Rechsteiner H, Glockshuber R.** 2008. Reconstitution of pilus assembly reveals a bacterial outer membrane catalyst. *Science* **320:**376–379.

17. **Remaut H, Tang C, Henderson NS, Pinkner JS, Wang T, Hultgren SJ, Thanassi DG, Waksman G, Li H.** 2008. Fiber formation across the bacterial outer membrane by the chaperone/usher pathway. *Cell* **133:**640–652.

18. **Huang Y, Smith BS, Chen LX, Baxter RHG, Deisenhofer J.** 2009. Insights into pilus assembly and secretion from the structure and functional characterization of usher PapC. *Proc Natl Acad Sci U S A* **106:**7403–7407.

19. **So SS, Thanassi DG.** 2006. Analysis of the requirements for pilus biogenesis at the outer membrane usher and the function of the usher C-terminus. *Mol Microbiol* **60**:364–375.

20. **Saulino ET, Thanassi DG, Pinkner JS, Hultgren SJ.** 1998. Ramifications of kinetic partitioning on usher-mediated pilus biogenesis. *EMBO J* **17**:2177–2185.

21. **Ng TW, Akman L, Osisami M, Thanassi DG.** 2004. The usher N terminus is the initial targeting site for chaperone-subunit complexes and participates in subsequent pilus biogenesis events. *J Bacteriol* **186**:5321–5331.

22. **Nishiyama M, Vetsch M, Puorger C, Jelesarov I, Glockshuber R.** 2003. Identification and characterization of the chaperone-subunit complex-binding domain from the type 1 pilus assembly platform FimD. *J Mol Biol* **330**:513–525.

23. **Henderson NS, Ng TW, Talukder I, Thanassi DG.** 2011. Function of the usher N-terminus in catalysing pilus assembly. *Mol Microbiol* **79**:954–967.

24. **Nishiyama M, Horst R, Eidam O, Herrmann T, Ignatov O, Vetsch M, Bettendorff P, Jelesarov I, Grütter MG, Wüthrich K, Glockshuber R, Capitani G.** 2005. Structural basis of chaperone-subunit complex recognition by the type 1 pilus assembly platform FimD. *EMBO J* **24**:2075–2086.

25. **Munera D, Hultgren S, Fernández LÁ.** 2007. Recognition of the N-terminal lectin domain of FimH adhesin by the usher FimD is required for type 1 pilus biogenesis. *Mol Microbiol* **64**:333–346.

26. **Phan G, Remaut H, Wang T, Allen WJ, Pirker KF, Lebedev A, Henderson NS, Geibel S, Volkan E, Yan J, Kunze MBA, Pinkner JS, Ford B, Kay CWM, Li H, Hultgren SJ, Thanassi DG, Waksman G.** 2011. Crystal structure of the FimD usher bound to its cognate FimC-FimH substrate. *Nature* **474**:49–53.

27. **Werneburg GT, Henderson NS, Portnoy EB, Sarowar S, Hultgren SJ, Li H, Thanassi DG.** 2015. The pilus usher controls protein interactions via domain masking and is functional as an oligomer. *Nat Struct Mol Biol* **22**:540–546.

28. **Remaut H, Rose RJ, Hannan TJ, Hultgren SJ, Radford SE, Ashcroft AE, Waksman G.** 2006. Donor-strand exchange in chaperone-assisted pilus assembly proceeds through a concerted beta strand displacement mechanism. *Mol Cell* **22**:831–842.

29. **Rose RJ, Welsh TS, Waksman G, Ashcroft AE, Radford SE, Paci E.** 2008. Donor-strand exchange in chaperone-assisted pilus assembly revealed in atomic detail by molecular dynamics. *J Mol Biol* **375**:908–919.

30. **Verger D, Miller E, Remaut H, Waksman G, Hultgren S.** 2006. Molecular mechanism of P pilus termination in uropathogenic *Escherichia coli*. *EMBO Rep* **7**:1228–1232.

31. **Bečárová Z.** 2015. *Mechanisms of FimI, the assembly termination subunit of the type 1 pili from uropathogenic* Escherichia coli. *PhD thesis*. ETH Zürich, Zürich, Switzerland.

32. **Waksman G.** 2017. Structural and molecular biology of a protein-polymerizing nanomachine for pilus biogenesis. *J Mol Biol* **429**:2654–2666.

33. **Omattage NS, Deng Z, Pinkner JS, Dodson KW, Almqvist F, Yuan P, Hultgren SJ.** 2018. Structural basis for usher activation and intramolecular subunit transfer in P pilus biogenesis in *Escherichia coli*. *Nat Microbiol* **3**:1362–1368.

34. **Du M, Yuan Z, Yu H, Henderson N, Sarowar S, Zhao G, Werneburg GT, Thanassi DG, Li H.** 2018. Handover mechanism of the growing pilus by the bacterial outer-membrane usher FimD. *Nature* **562**:444–447.

35. **Spaulding CN, Klein RD, Schreiber HL IV, Janetka JW, Hultgren SJ.** 2018. Precision antimicrobial therapeutics: the path of least resistance? *NPJ Biofilms Microbiomes* **4**:4.

36. **Andersson M, Uhlin BE, Fällman E.** 2007. The biomechanical properties of E. coli pili for urinary tract attachment reflect the host environment. *Biophys J* **93**:3008–3014.

37. **Roberts JA, Marklund BI, Ilver D, Haslam D, Kaack MB, Baskin G, Louis M, Möllby R, Winberg J, Normark S.** 1994. The Gal(alpha 1-4)Gal-specific tip adhesin of *Escherichia coli* P-fimbriae is needed for pyelonephritis to occur in the normal urinary tract. *Proc Natl Acad Sci U S A* **91**:11889–11893.

38. **Larsson A, Ohlsson J, Dodson KW, Hultgren SJ, Nilsson U, Kihlberg J.** 2003. Quantitative studies of the binding of the class II PapG adhesin from uropathogenic *Escherichia coli* to oligosaccharides. *Bioorg Med Chem* **11**:2255–2261.

39. **Hannan TJ, Totsika M, Mansfield KJ, Moore KH, Schembri MA, Hultgren SJ.** 2012. Host-pathogen checkpoints and population bottlenecks in persistent and intracellular uropathogenic *Escherichia coli* bladder infection. *FEMS Microbiol Rev* **36**:616–648.

40. **Schaeffer AJ, Schwan WR, Hultgren SJ, Duncan JL.** 1987. Relationship of type 1 pilus expression in *Escherichia coli* to ascending urinary tract infections in mice. *Infect Immun* **55**:373–380.

41. **Hung CS, Bouckaert J, Hung D, Pinkner J, Widberg C, DeFusco A, Auguste CG, Strouse**

R, Langermann S, Waksman G, Hultgren SJ. 2002. Structural basis of tropism of *Escherichia coli* to the bladder during urinary tract infection. *Mol Microbiol* **44:**903–915.

42. Le Trong I, Aprikian P, Kidd BA, Forero-Shelton M, Tchesnokova V, Rajagopal P, Rodriguez V, Interlandi G, Klevit R, Vogel V, Stenkamp RE, Sokurenko EV, Thomas WE. 2010. Structural basis for mechanical force regulation of the adhesin FimH via finger trap-like beta sheet twisting. *Cell* **141:**645–655.

43. Sauer MM, Jakob RP, Eras J, Baday S, Eriş D, Navarra G, Bernèche S, Ernst B, Maier T, Glockshuber R. 2016. Catch-bond mechanism of the bacterial adhesin FimH. *Nat Commun* **7:**10738.

44. Thomas WE, Trintchina E, Forero M, Vogel V, Sokurenko EV. 2002. Bacterial adhesion to target cells enhanced by shear force. *Cell* **109:**913–923.

45. Thomas WE, Nilsson LM, Forero M, Sokurenko EV, Vogel V. 2004. Shear-dependent 'stick-and-roll' adhesion of type 1 fimbriated *Escherichia coli*. *Mol Microbiol* **53:**1545–1557.

46. Yakovenko O, Sharma S, Forero M, Tchesnokova V, Aprikian P, Kidd B, Mach A, Vogel V, Sokurenko E, Thomas WE. 2008. FimH forms catch bonds that are enhanced by mechanical force due to allosteric regulation. *J Biol Chem* **283:**11596–11605.

47. Sokurenko EV, Vogel V, Thomas WE. 2008. Catch-bond mechanism of force-enhanced adhesion: counterintuitive, elusive, but ... widespread? *Cell Host Microbe* **4:**314–323.

48. Bouckaert J, Berglund J, Schembri M, De Genst E, Cools L, Wuhrer M, Hung C-S, Pinkner J, Slättegård R, Zavialov A, Choudhury D, Langermann S, Hultgren SJ, Wyns L, Klemm P, Oscarson S, Knight SD, De Greve H. 2005. Receptor binding studies disclose a novel class of high-affinity inhibitors of the *Escherichia coli* FimH adhesin. *Mol Microbiol* **55:**441–455.

49. Wellens A, Garofalo C, Nguyen H, Van Gerven N, Slättegård R, Hernalsteens J-P, Wyns L, Oscarson S, De Greve H, Hultgren S, Bouckaert J. 2008. Intervening with urinary tract infections using anti-adhesives based on the crystal structure of the FimH-oligomannose-3 complex. *PLoS One* **3:**e2040.

50. Han Z, Pinkner JS, Ford B, Obermann R, Nolan W, Wildman SA, Hobbs D, Ellenberger T, Cusumano CK, Hultgren SJ, Janetka JW. 2010. Structure-based drug design and optimization of mannoside bacterial FimH antagonists. *J Med Chem* **53:**4779–4792.

51. Wellens A, Lahmann M, Touaibia M, Vaucher J, Oscarson S, Roy R, Remaut H, Bouckaert J. 2012. The tyrosine gate as a potential entropic lever in the receptor-binding site of the bacterial adhesin FimH. *Biochemistry* **51:**4790–4799.

52. Brument S, Sivignon A, Dumych TI, Moreau N, Roos G, Guérardel Y, Chalopin T, Deniaud D, Bilyy RO, Darfeuille-Michaud A, Bouckaert J, Gouin SG. 2013. Thiazolylaminomannosides as potent antiadhesives of type 1 piliated *Escherichia coli* isolated from Crohn's disease patients. *J Med Chem* **56:**5395–5406.

53. Vanwetswinkel S, Volkov AN, Sterckx YGJ, Garcia-Pino A, Buts L, Vranken WF, Bouckaert J, Roy R, Wyns L, van Nuland NAJ. 2014. Study of the structural and dynamic effects in the FimH adhesin upon α-d-heptyl mannose binding. *J Med Chem* **57:**1416–1427.

54. Kalas V, Pinkner JS, Hannan TJ, Hibbing ME, Dodson KW, Holehouse AS, Zhang H, Tolia NH, Gross ML, Pappu RV, Janetka J, Hultgren SJ. 2017. Evolutionary fine-tuning of conformational ensembles in FimH during host-pathogen interactions. *Sci Adv* **3:**e1601944.

55. Chen SL, Hung CS, Pinkner JS, Walker JN, Cusumano CK, Li Z, Bouckaert J, Gordon JI, Hultgren SJ. 2009. Positive selection identifies an in vivo role for FimH during urinary tract infection in addition to mannose binding. *Proc Natl Acad Sci U S A* **106:**22439–22444.

56. Geibel S, Procko E, Hultgren SJ, Baker D, Waksman G. 2013. Structural and energetic basis of folded-protein transport by the FimD usher. *Nature* **496:**243–246.

57. Fiege B, Rabbani S, Preston RC, Jakob RP, Zihlmann P, Schwardt O, Jiang X, Maier T, Ernst B. 2015. The tyrosine gate of the bacterial lectin FimH: a conformational analysis by NMR spectroscopy and X-ray crystallography. *Chembiochem* **16:**1235–1246.

58. Wright KJ, Seed PC, Hultgren SJ. 2005. Uropathogenic *Escherichia coli* flagella aid in efficient urinary tract colonization. *Infect Immun* **73:**7657–7668.

59. Lane MC, Lockatell V, Monterosso G, Lamphier D, Weinert J, Hebel JR, Johnson DE, Mobley HLT. 2005. Role of motility in the colonization of uropathogenic *Escherichia coli* in the urinary tract. *Infect Immun* **73:**7644–7656.

60. Pak J, Pu Y, Zhang ZT, Hasty DL, Wu XR. 2001. Tamm-Horsfall protein binds to type 1 fimbriated *Escherichia coli* and prevents *E. coli* from binding to uroplakin Ia and Ib receptors. *J Biol Chem* **276:**9924–9930.

61. Bullitt E, Makowski L. 1995. Structural polymorphism of bacterial adhesion pili. *Nature* **373:**164–167.

62. Hahn E, Wild P, Hermanns U, Sebbel P, Glockshuber R, Häner M, Taschner N, Burkhard P, Aebi U, Müller SA. 2002. Exploring the 3D molecular architecture of *Escherichia coli* type 1 pili. *J Mol Biol* **323**:845–857.

63. Mu X-Q, Bullitt E. 2006. Structure and assembly of P-pili: a protruding hinge region used for assembly of a bacterial adhesion filament. *Proc Natl Acad Sci U S A* **103**:9861–9866.

64. Miller E, Garcia T, Hultgren S, Oberhauser AF. 2006. The mechanical properties of *E. coli* type 1 pili measured by atomic force microscopy techniques. *Biophys J* **91**:3848–3856.

65. Zakrisson J, Wiklund K, Axner O, Andersson M. 2012. Helix-like biopolymers can act as dampers of force for bacteria in flows. *Eur Biophys J* **41**:551–560.

66. Forero M, Yakovenko O, Sokurenko EV, Thomas WE, Vogel V. 2006. Uncoiling mechanics of *Escherichia coli* type I fimbriae are optimized for catch bonds. *PLoS Biol* **4**:e298–e299.

67. Zakrisson J, Wiklund K, Axner O, Andersson M. 2013. The shaft of the type 1 fimbriae regulates an external force to match the FimH catch bond. *Biophys J* **104**:2137–2148.

68. Puorger C, Eidam O, Capitani G, Erilov D, Grütter MG, Glockshuber R. 2008. Infinite kinetic stability against dissociation of supramolecular protein complexes through donor strand complementation. *Structure* **16**:631–642.

69. Alonso-Caballero A, Schönfelder J, Poly S, Corsetti F, De Sancho D, Artacho E, Perez-Jimenez R. 2018. Mechanical architecture and folding of *E. coli* type 1 pilus domains. *Nat Commun* **9**:2758.

70. Jass J, Schedin S, Fällman E, Ohlsson J, Nilsson UJ, Uhlin BE, Axner O. 2004. Physical properties of *Escherichia coli* P pili measured by optical tweezers. *Biophys J* **87**:4271–4283.

71. Fällman E, Schedin S, Jass J, Uhlin BE, Axner O. 2005. The unfolding of the P pili quaternary structure by stretching is reversible, not plastic. *EMBO Rep* **6**:52–56.

72. Andersson M, Fällman E, Uhlin BE, Axner O. 2006. A sticky chain model of the elongation and unfolding of *Escherichia coli* P pili under stress. *Biophys J* **90**:1521–1534.

73. Andersson M, Fällman E, Uhlin BE, Axner O. 2006. Dynamic force spectroscopy of *E. coli* P pili. *Biophys J* **91**:2717–2725.

74. Andersson M, Fällman E, Uhlin BE, Axner O. 2006. Force measuring optical tweezers system for long time measurements of P pili stability. Proceedings SPIE 6088, Imaging, Manipulation, and Analysis of Biomolecules, Cells, and Tissues IV, 608810 (21 February 2006).

75. Lugmaier RA, Schedin S, Kühner F, Benoit M. 2008. Dynamic restacking of *Escherichia coli* P-pili. *Eur Biophys J* **37**:111–120.

76. Andersson M, Axner O, Almqvist F, Uhlin BE, Fällman E. 2008. Physical properties of biopolymers assessed by optical tweezers: analysis of folding and refolding of bacterial pili. *Chemphyschem* **9**:221–235.

77. Spaulding CN, Schreiber HL IV, Zheng W, Dodson KW, Hazen JE, Conover MS, Wang F, Svenmarker P, Luna-Rico A, Francetic O, Andersson M, Hultgren S, Egelman EH. 2018. Functional role of the type 1 pilus rod structure in mediating host-pathogen interactions. *eLife* **7**:5145.

78. Habenstein B, Loquet A, Hwang S, Giller K, Vasa SK, Becker S, Habeck M, Lange A. 2015. Hybrid structure of the type 1 pilus of uropathogenic *Escherichia coli*. *Angew Chem Int Ed Engl* **54**:11691–11695.

79. Crespo MD, Puorger C, Schärer MA, Eidam O, Grütter MG, Capitani G, Glockshuber R. 2012. Quality control of disulfide bond formation in pilus subunits by the chaperone FimC. *Nat Chem Biol* **8**:707–713.

80. Eidam O, Dworkowski FSN, Glockshuber R, Grütter MG, Capitani G. 2008. Crystal structure of the ternary FimC-FimF(t)-FimD(N) complex indicates conserved pilus chaperone-subunit complex recognition by the usher FimD. *FEBS Lett* **582**:651–655.

Therapeutic Approaches Targeting the Assembly and Function of Chaperone-Usher Pili

13

JOHN J. PSONIS[1,2] and DAVID G. THANASSI[1,2]

INTRODUCTION

The chaperone-usher (CU) pathway is dedicated to the biogenesis of surface structures termed pili or fimbriae that play indispensable roles in the pathogenesis of a wide range of bacteria (1–4). Pili are hair-like fibers composed of multiple different subunit proteins. They are typically involved in adhesion, allowing bacteria to establish a foothold within the host. Following attachment, pili modulate host cell signaling pathways, promote or inhibit host cell invasion, and mediate bacterium-bacterium interactions leading to formation of community structures such as biofilms (5, 6). Gram-negative bacteria express multiple CU pili that contribute to their ability to colonize diverse environmental niches (1, 7–10). Pili thus function at the host-pathogen interface to both initiate and sustain infection and represent attractive therapeutic targets.

PILUS FUNCTION

The most extensively characterized CU pili are type 1 pili, found in members of the *Enterobacteriaceae*, and P pili, found in uropathogenic *Escherichia coli*

[1]Department of Molecular Genetics and Microbiology, School of Medicine, Stony Brook University, Stony Brook, NY 11794
[2]Center for Infectious Diseases, Stony Brook University, Stony Brook, NY 11794
Protein Secretion in Bacteria
Edited by Maria Sandkvist, Eric Cascales, and Peter J. Christie
© 2019 American Society for Microbiology, Washington, DC
doi:10.1128/ecosalplus.ESP-0033-2018

(UPEC). Both pili are key virulence factors for UPEC colonization of the urinary tract and the establishment of urinary tract infections (UTI) (Fig. 1). Type 1 pili bind to mannosylated proteins in the bladder, leading to cystitis, and P pili bind to di-galactose-containing moieties in kidney glycolipids, leading to pyelonephritis (11–13). Bacterial binding via type 1 pili also activates host cell pathways that lead to actin cytoskeletal rearrangements and subsequent bacterial invasion into the host cells via a zipper-like mechanism (14, 15). Type 1 pili contribute to the formation of extracellular biofilms (16),

as well as intracellular biofilm-like communities (IBCs) by UPEC during bladder infection (Fig. 1) (17). Bacteria within these IBCs are protected from antibiotics and immune surveillance (18, 19).

Type 1 and P pili expressed by UPEC are considered classical pili, which are heteropolymers of different protein subunits that form rigid, helical rods. Similarly, enterotoxigenic *E. coli* (ETEC) employs a large group of rigid pili, termed colonization factor antigen (CFA) or coli surface antigen (CS) pili, to adhere to the small intestine, facilitating toxin delivery into the gut lumen (20). Another

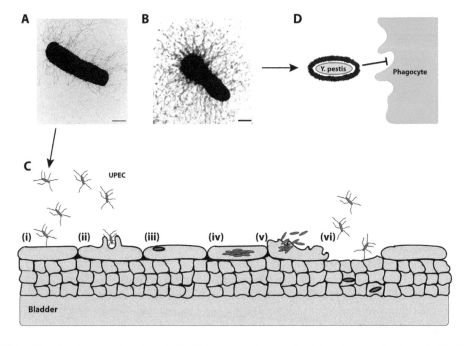

FIGURE 1 **Ultrastructure and function of CU pili. Electron micrographs of *E. coli* expressing type 1 pili (A) and *Y. pestis* expressing F1 capsule (B). Scale bars = 500 nm. (C) Cartoon for pilus-mediated bacterial interactions in the bladder. (i) Type 1-piliated UPEC binds to superficial umbrella cells that line the lumen of the bladder. (ii) Pilus-receptor interactions induce a signaling cascade that promotes internalization of adherent bacteria via a membrane zippering mechanism. (iii) Within bladder epithelial cells, UPEC are trafficked to membrane-bound, acidic compartments similar to lysosomes. (iv) In the superficial umbrella cells, UPEC break into the cytosol and rapidly multiply, forming intracellular biofilm-like communities. (v) Bladder cells containing large numbers of UPEC exfoliate, providing a mechanism for bacterial clearance by the flow of urine. (vi) This, however, leaves the underlying layers of immature bladder epithelial cells exposed. UPEC can invade these immature urothelial cells and persist in a quiescent stage in late endosome-like compartments, avoiding detection by immunosurveillance mechanisms. (D) The *Y. pestis* F1 capsule plays an antiphagocytic role by preventing opsonizing antibodies from binding to the bacterial surface, blocking Fc receptor phagocytosis. More generally, expression of the F1 capsule can mask bacterial adhesins and other surface structures, preventing interactions that lead to internalization into host cells.**

group of pili assembled by the CU pathway comprises thin, flexible fibers that in some cases form amorphous, capsular-like or "afimbrial" structures (3). Examples of these are the Afa/Dr pili (21–23), expressed by various pathogenic *E. coli* strains, and the F1 capsular antigen of *Yersinia pestis* (24, 25), which forms a dense coating around the bacteria and is involved in preventing uptake by macrophages (Fig. 1) (25, 26).

CU pili are remarkably adapted to colonization of specific environmental niches. To mediate colonization of the urinary tract, type 1 pili must be able to withstand the shear forces generated by the flow of urine. The FimH adhesin utilizes a catch bond mechanism to switch between low- and high-affinity binding conformations, facilitating migration (rolling) and receptor sampling in the absence of urinary flow and attachment (sticking) during periods of turbulence (27–29). The helical pilus rod exhibits properties of compliance and flexibility, which is also important for resistance to shear forces and allows bacteria to regain proximity to host cells after exposure to turbulence (30–32).

PILUS ASSEMBLY

The CU pathway harnesses protein-protein interactions to drive pilus fiber assembly and secretion in the absence of an external energy source such as ATP, which is not available in the bacterial periplasm (33, 34). Newly synthesized pilus subunits in the cytoplasm contain an N-terminal signal sequence that directs them to the SecYEG translocon in the inner membrane for translocation into the periplasm (Fig. 2). In the periplasm, the signal sequence is cleaved and the subunits undergo disulfide bond formation in a process catalyzed by the oxidoreductase DsbA (33, 35). The subunits then form binary complexes with chaperone proteins (FimC for type 1 pili and PapD for P pili). The chaperone recognizes only unfolded subunits that have already undergone disulfide bond formation. This

serves an important quality control role, ensuring that only oxidized, mechanically stable subunits are incorporated into the pilus (36–38). The chaperone donates a β-strand to complete the immunoglobulin-like fold of the subunits in a mechanism termed donor strand complementation (DSC) (39, 40). This process allows subunit folding and inhibits premature subunit-subunit interactions. In the absence of the chaperone, subunits misfold, aggregate, and are degraded by the DegP periplasmic protease (41, 42).

Periplasmic chaperone-subunit complexes interact with the outer membrane (OM) usher (FimD for type 1 pili and PapC for P pili) (Fig. 2), which catalyzes the exchange of chaperone-subunit for subunit-subunit interactions in a process termed donor strand exchange (DSE) (43–47). In DSE, the N-terminal extension of an incoming pilus subunit displaces the β-strand donated by the chaperone to release the chaperone and form a subunit-subunit interaction. This interaction is energetically favorable and initiates at a binding pocket on the subunit, termed the P5 pocket, which is left vacant by the chaperone donor strand (43–46). The correct ordering of subunits in the pilus fiber is determined by the differential affinities of chaperone-subunit complexes for the usher and the rate of DSE between different subunit-subunit pairs, as well as the periplasmic concentrations of different subunits (47–51). The usher thus promotes ordered polymerization of the pilus fiber and provides the channel for secretion of the pilus fiber to the cell surface.

The usher comprises a 24-stranded β-barrel channel domain, a plug domain that serves as a channel gate, an N-terminal periplasmic domain (NTD), and two C-terminal domains (CTD1 and CTD2) (52–55). In the resting (*apo*) usher, the plug domain occludes the channel pore and masks the usher C domains (56, 57). In the type 1 pilus system, the usher is activated by binding of a FimC-FimH chaperone-adhesin complex to the usher NTD (52). FimC-FimH binding results

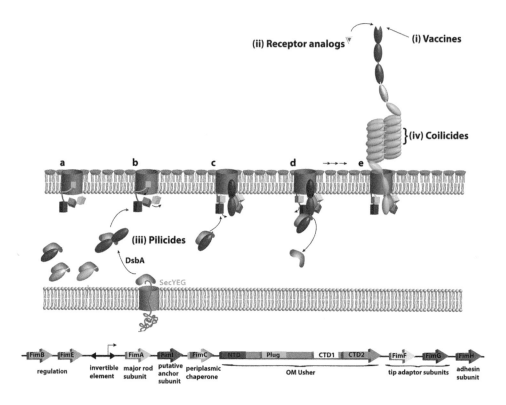

FIGURE 2 CU pilus assembly pathway and targets for therapeutic intervention. The *fim* gene cluster coding for type 1 pili, along with names and functions of encoded proteins, is shown at the bottom. Upon entering the periplasm via the SecYEG general secretory machinery, nascent pilus subunits form binary complexes with the pilus chaperone (FimC), which facilitates subunit folding by DSC, completing the Ig fold of the subunit's pilin domain. The adhesin subunit (FimH, red) is depicted with an additional N-terminal lectin domain, which contains the receptor-binding site. Chaperone-subunit complexes then interact with the OM usher (FimD), which comprises a β-barrel channel domain, a plug domain, an N-terminal periplasmic domain (NTD), and two C terminal domains (CTD1 and CTD2). (a) In the resting usher, the plug domain occludes the channel pore and masks the CTDs. (b and c) The usher is activated by binding of a FimC-FimH chaperone-adhesin complex to the usher NTD. This results in displacement of the plug from the channel and handoff of FimC-FimH to the usher CTDs, freeing the NTD to recruit the next chaperone-subunit complex (FimC-FimG). (d) The newly recruited complex bound to the NTD is oriented perfectly to undergo DSE with the previously recruited complex bound to the CTDs, forming the first link in the pilus fiber. The newly incorporated chaperone-subunit is then handed over from the NTD to the CTDs. (e) Repeated rounds of this process result in assembly and secretion of the pilus fiber. Different steps along this pathway are targets for antipilus therapeutics. (i) Vaccination using a full-length or truncated adhesin subunit inhibits pilus-mediated bacterial adhesion and pathogenesis. (ii) Small-molecule receptor analogs occupy the pilus adhesin binding site, preventing pili from adhering to host receptors. (iii) Pilicides inhibit pilus assembly via different mechanisms, such as interfering with chaperone-subunit or subunit-subunit interactions, interfering with binding of chaperone-subunit complexes to the usher, or inhibiting proper folding of the usher in the bacterial OM. (iv) Coilicides inhibit uncoiling and recoiling of the pilus rod, thus impairing resilience of the fibers during fluid flow.

in plug expulsion from the lumen of the usher channel, which also frees the CTDs (55, 57) (Fig. 2). FimC-FimH is then delivered from the usher NTD to the CTDs, in a handover process likely driven by differential affinity and direct domain-domain interactions (57–60). This frees the usher NTD for recruiting the next chaperone-subunit complex (FimC-FimG for type 1 pili) from the periplasm. The newly recruited complex bound to the usher

NTD is oriented perfectly to undergo DSE with the previously recruited complex bound at the usher CTDs (55) (Fig. 2). This displaces the chaperone from the subunit bound at the CTDs, forming the first link in the pilus fiber. The newly incorporated chaperone-subunit is then handed over from the usher NTD to the CTDs to reset the system for a new round of subunit recruitment and incorporation, which continues concomitantly with translocation of the nascent pilus fiber through the usher channel to the cell surface. The pilus fiber is thus assembled and secreted in a top-down manner, with the pilus rod adopting its final helical quaternary structure upon exiting the usher pore. The helical rod is stabilized by extensive polar interactions between the pilus rod subunits, providing a remarkable level of flexibility that is important for resistance to shear stress. Shear stress can disrupt these polar interactions, linearizing the pilus rod without breaking the strong hydrophobic interactions that mediate subunit polymerization (61–63). Each of these steps along the CU assembly pathway offers targets for the development of therapeutic inhibitors.

PILUS-DIRECTED THERAPEUTIC APPROACHES

The ever-increasing rate of antibiotic resistance among pathogenic bacteria is necessitating a hard look into alternative methods for treatment (64, 65). Antivirulence therapeutics that specifically target pilus function or assembly represent one such alternative approach to traditional antibiotics. In contrast to traditional antibiotics that nonspecifically interfere with essential biological processes, antivirulence therapeutics disrupt systems only required for bacterial pathogens to cause disease within the host, thus limiting detrimental side effects on commensal bacteria and the selective pressure that leads to antibiotic resistance (66–69).

The indispensable roles that CU pili play in bacterial pathogenesis make them attractive targets for directed therapeutic intervention. A number of approaches have been taken to develop antipilus therapeutics, including vaccines against pilus proteins, inhibitors of pilus-mediated adhesion, and small molecules that disrupt pilus biogenesis (Fig. 2). CU pili are prevalent among the *Enterobacteriaceae* and their structure, function, and mechanism of assembly are conserved, suggesting that pilus-targeting therapeutics may have activity against a broad range of bacterial pathogens.

Vaccination Strategies

Because of their important roles in bacterial virulence, pili have received considerable attention in vaccine development programs. Vaccination with whole pili has proven unsuccessful, mainly due to factors such as phase-variable expression and antigenic variation (70, 71). Moreover, in the case of monoadhesive pili such as type 1 and P pili, which have a single distal adhesin subunit, this approach may bias the immune system towards structural pilus subunits that are present in much higher copy than the adhesin, thus failing to inhibit pilus function (6). In contrast, vaccination with the adhesin subunit FimH (type 1 pili) or PapG (P pili) confers substantial protection against UPEC in both murine and primate infection models, without affecting the commensal *E. coli* in the gut (72–76). Vaccination with a truncated form of the mannose-resistant *Proteus*-like fimbriae (MR/P) tip adhesin MrpH, fused to FliC from *Salmonella enterica* serovar Typhimurium as an adjuvant, was also found to confer protection against UTI caused by *Proteus mirabilis* (77). Moreover, in a recent human clinical trial, oral administration of antibodies raised against the colonization factor I (CFA/I) pilus tip adhesin CfaE conferred substantial protection against ETEC colonization (78).

The catch-bond behavior of pilus adhesins poses some challenges to vaccination. In some cases, antibody binding to FimH stabilizes its high-affinity state and thereby enhances rather than inhibits binding of the adhesin to its receptor (79). Moreover, by shifting from the high- to the low-affinity state, FimH may shed bound antibodies (79). Novel approaches have been taken to overcome these issues. For example, a new type of antibody that binds to a single loop within the binding pocket of FimH can displace bound ligand. This parasteric antibody is potent not only in inhibiting but also in reversing bacterial adhesion, dissolving surface-adherent biofilms and conferring protection against cystitis in mice (80).

Small-Molecule Receptor Analogs

For type 1 pili, soluble receptor analogs, termed mannosides, are being developed as anti-adhesives by occupying the FimH receptor binding site. Mouse model studies have shown that these compounds can prophylactically prevent bacterial bladder colonization, with efficacy against established UTI as well as catheter-associated UTI (81–84). In a recent study, mannosides were found to selectively deplete intestinal UPEC reservoirs without altering the gut microbiota, which may have implications for reducing the rate of recurrent UTIs (85). Furthermore, by shifting the UPEC niche primarily to the extracellular milieu, mannosides may exhibit synergy with traditional antibiotics (86).

A variety of approaches are being taken to develop mannosides with improved or novel properties. FimH antagonist efficacy has traditionally been evaluated using a truncated FimH construct locked in a single conformation. New approaches taking into consideration the dynamic nature of FimH binding have led to the development of biphenyl mannosides that have excellent affinities for all physiologically relevant FimH conformations and exhibit increased potency compared to conventional FimH antagonists (87). Thiomannosides are reported to have improved metabolic stability and oral bioavailability (88). Galabiose-based soluble receptor analogs are also being developed to target P pilus adhesion (89). A similar strategy is being employed to develop receptor-

mimicking galactosides that target the F9 pilus adhesin FmlH. Lead FmlH antagonists significantly reduce bacterial burdens in the bladders and kidneys of infected mice (90). Thiazolylmannosides and heptylmannoside-based glycocompounds, a new class of FimH antagonists with greater stability at low pH, have shown efficacy against adherent-invasive *E. coli*, which plays a key role in the gut inflammation of patients with Crohn's disease (91–94). Finally, multivalent inhibitors that function as potent anti-adhesives by cross-linking bacteria have also been developed by coupling FimH antagonists on synthetic scaffolds (95–97).

Small-Molecule Pilicides

Another class of small-molecule CU pilus inhibitors, known as pilicides, inhibit the pilus assembly and secretion process. The original pilicides consist of molecules with a 2-pyridine scaffold (98, 99). These molecules bind to the periplasmic chaperone and interfere with chaperone-subunit interactions or the binding of chaperone-subunit complexes to the usher. These compounds have demonstrated efficacy against pilus-mediated adhesion and biofilm formation (99–101). New synthetic approaches have resulted in the development of new classes of pilicides with improved potency (102, 103). In a recent study, pilicide ec240 was found to disrupt assembly of type 1, P, and S pili, as well as flagellar motility. In addition to interfering with pilus assembly, ec240 induces *fimS*-mediated phase off variation, downregulating type 1 pilus gene expression. Treatment of UPEC with ec240 reduced biofilm formation and bacterial colonization in the mouse UTI model (104).

Computational screening for compounds with complementarity to the FimH P5 binding pocket led to identification of the small-molecule AL1, which inhibits pilus subunit polymerization by disrupting the DSE reaction between the FimH and FimG subunits (105). By disrupting type 1 pilus biogenesis, AL1 reduces biofilm formation and bacterial adhesion to human bladder epithelial cells. Another small-molecule compound, nitazoxanide (NTZ), was shown to inhibit biofilm formation by enteroaggregative *E. coli* by disrupting the assembly of AAF CU pili (106). Further analysis demonstrated that NTZ also inhibits type 1 and P pilus assembly via a novel mechanism of action, by interfering with proper folding of the usher protein in the OM (107).

Other Approaches

A novel approach to target bacterial adhesion is the use of coilicides, a recently developed class of antipilus inhibitors that act by impairing compliance of the CU pilus rod. In a proof-of-principle experiment, it was shown that purified PapD chaperone binds to uncoiled P pilus rods and prevents their recoiling, thus decreasing their ability to withstand shear forces caused by fluid flow (108). Similarly, polyclonal anti-PapA antibodies were found to reduce the elastic properties of P pili (109). Bivalent polyclonal antibodies have also been used to diminish the compliance of CFA/I and coli surface antigen 2 (CS2) pili, which play essential roles in ETEC pathogenesis. These antibodies, which recognize major pilin subunits, decrease pilus resilience during fluid flow by clamping together layers of the helical fiber or two individual pili, thereby increasing their stiffness and entangling them (110, 111). The salivary peptide histatin-5 was also found to bind to and stiffen CFA/I pili, inhibiting ETEC colonization in the gastrointestinal tract (112).

Another strategy that is being explored is the engineering of an avirulent asymptomatic bacteriuria strain, 83972, to synthesize a surface-located oligosaccharide P pilus receptor mimic. This strain can bind virulent P-piliated UPEC, impairing its adhesion to kidney epithelial cells (113, 114). In an alternate approach, a recombinant strain 83972 that expresses type 1 pili can interfere with urinary catheter biofilm formation by virulent enterococci (115).

CONCLUSIONS

Pili assembled by the CU pathway function as virulence factors for a range of Gram-negative pathogenic bacteria. Pili are attractive targets for therapeutic intervention, as they are required both for early stages of colonization in the host and maintenance of infection. Therapeutic agents that target CU pili such as vaccines, adhesin receptor analogs, and small molecules show promise in selectively disrupting host-pathogen interactions that are crucial for disease. Such agents offer alternatives to traditional antibiotics and a pathway forward to combat the rising threat of antibiotic resistance. Additional knowledge gained regarding the assembly, structure, and function of CU pili will provide new opportunities for the development of novel anti-infective therapeutics.

ACKNOWLEDGMENTS

J.J.P. is supported by Medical Scientist Training Program award T32GM008444 from the U.S. National Institutes of Health (NIH). Research in the Thanassi laboratory on the subject of this review is supported by NIH grants R01GM62987 and R21AI121639.

CITATION

EcoSal Plus 2019; doi:10.1128/ecosalplus.ESP-0033-2018.

REFERENCES

1. **Nuccio SP, Bäumler AJ.** 2007. Evolution of the chaperone/usher assembly pathway: fimbrial classification goes Greek. *Microbiol Mol Biol Rev* **71**:551–575.
2. **Zav'yalov V, Zavialov A, Zav'yalova G, Korpela T.** 2010. Adhesive organelles of Gram-negative pathogens assembled with the classical chaperone/usher machinery: structure and function from a clinical standpoint. *FEMS Microbiol Rev* **34**:317–378.
3. **Thanassi DG, Bliska JB, Christie PJ.** 2012. Surface organelles assembled by secretion systems of Gram-negative bacteria: diversity

in structure and function. *FEMS Microbiol Rev* **36:**1046–1082.

4. **Geibel S, Waksman G.** 2014. The molecular dissection of the chaperone-usher pathway. *Biochim Biophys Acta* **1843:**1559–1567.

5. **Chahales P, Thanassi DG.** 2015. Structure, function, and assembly of adhesive organelles by uropathogenic bacteria. *Microbiol Spectr* 3(5):UTI-0018-2013.

6. **Werneburg GT, Thanassi DG.** 2018. Pili assembled by the chaperone/usher pathway in *Escherichia coli* and *Salmonella. EcoSal Plus.*

7. **van der Velden AW, Bäumler AJ, Tsolis RM, Heffron F.** 1998. Multiple fimbrial adhesins are required for full virulence of *Salmonella typhimurium* in mice. *Infect Immun* **66:**2803–2808.

8. **Korea CG, Badouraly R, Prevost MC, Ghigo JM, Beloin C.** 2010. *Escherichia coli* K-12 possesses multiple cryptic but functional chaperone-usher fimbriae with distinct surface specificities. *Environ Microbiol* **12:**1957–1977.

9. **Felek S, Jeong JJ, Runco LM, Murray S, Thanassi DG, Krukonis ES.** 2011. Contributions of chaperone/usher systems to cell binding, biofilm formation and *Yersinia pestis* virulence. *Microbiology* **157:**805–818.

10. **Wurpel DJ, Beatson SA, Totsika M, Petty NK, Schembri MA.** 2013. Chaperone-usher fimbriae of *Escherichia coli. PLoS One* **8:**e52835.

11. **Roberts JA, Marklund BI, Ilver D, Haslam D, Kaack MB, Baskin G, Louis M, Möllby R, Winberg J, Normark S.** 1994. The Gal(alpha 1-4)Gal-specific tip adhesin of *Escherichia coli* P-fimbriae is needed for pyelonephritis to occur in the normal urinary tract. *Proc Natl Acad Sci U S A* **91:**11889–11893.

12. **Zhou G, Mo WJ, Sebbel P, Min G, Neubert TA, Glockshuber R, Wu XR, Sun TT, Kong XP.** 2001. Uroplakin Ia is the urothelial receptor for uropathogenic *Escherichia coli*: evidence from in vitro FimH binding. *J Cell Sci* **114:**4095–4103.

13. **Eto DS, Jones TA, Sundsbak JL, Mulvey MA.** 2007. Integrin-mediated host cell invasion by type 1-piliated uropathogenic *Escherichia coli. PLoS Pathog* **3:**e100.

14. **Mulvey MA, Lopez-Boado YS, Wilson CL, Roth R, Parks WC, Heuser J, Hultgren SJ.** 1998. Induction and evasion of host defenses by type 1-piliated uropathogenic *Escherichia coli. Science* **282:**1494–1497.

15. **Martinez JJ, Mulvey MA, Schilling JD, Pinkner JS, Hultgren SJ.** 2000. Type 1 pilus-mediated bacterial invasion of bladder epithelial cells. *EMBO J* **19:**2803–2812.

16. **Pratt LA, Kolter R.** 1998. Genetic analysis of *Escherichia coli* biofilm formation: roles of flagella, motility, chemotaxis and type I pili. *Mol Microbiol* **30:**285–293.

17. **Wright KJ, Seed PC, Hultgren SJ.** 2007. Development of intracellular bacterial communities of uropathogenic *Escherichia coli* depends on type 1 pili. *Cell Microbiol* **9:**2230–2241.

18. **Justice SS, Hung C, Theriot JA, Fletcher DA, Anderson GG, Footer MJ, Hultgren SJ.** 2004. Differentiation and developmental pathways of uropathogenic *Escherichia coli* in urinary tract pathogenesis. *Proc Natl Acad Sci U S A* **101:**1333–1338.

19. **Anderson GG, Palermo JJ, Schilling JD, Roth R, Heuser J, Hultgren SJ.** 2003. Intracellular bacterial biofilm-like pods in urinary tract infections. *Science* **301:**105–107.

20. **Gaastra W, Svennerholm AM.** 1996. Colonization factors of human enterotoxigenic *Escherichia coli* (ETEC). *Trends Microbiol* **4:**444–452.

21. **Labigne-Roussel A, Schmidt MA, Walz W, Falkow S.** 1985. Genetic organization of the afimbrial adhesin operon and nucleotide sequence from a uropathogenic *Escherichia coli* gene encoding an afimbrial adhesin. *J Bacteriol* **162:**1285–1292.

22. **Garcia MI, Gounon P, Courcoux P, Labigne A, Le Bouguenec C.** 1996. The afimbrial adhesive sheath encoded by the afa-3 gene cluster of pathogenic *Escherichia coli* is composed of two adhesins. *Mol Microbiol* **19:**683–693.

23. **Anderson KL, Billington J, Pettigrew D, Cota E, Simpson P, Roversi P, Chen HA, Urvil P, du Merle L, Barlow PN, Medof ME, Smith RA, Nowicki B, Le Bouguénec C, Lea SM, Matthews S.** 2004. An atomic resolution model for assembly, architecture, and function of the Dr adhesins. *Mol Cell* **15:**647–657.

24. **Titball RW, Howells AM, Oyston PC, Williamson ED.** 1997. Expression of the *Yersinia pestis* capsular antigen (F1 antigen) on the surface of an *aroA* mutant of *Salmonella typhimurium* induces high levels of protection against plague. *Infect Immun* **65:**1926–1930.

25. **Du Y, Rosqvist R, Forsberg A.** 2002. Role of fraction 1 antigen of *Yersinia pestis* in inhibition of phagocytosis. *Infect Immun* **70:**1453–1460.

26. **Liu F, Chen H, Galván EM, Lasaro MA, Schifferli DM.** 2006. Effects of Psa and F1 on the adhesive and invasive interactions of *Yersinia pestis* with human respiratory tract epithelial cells. *Infect Immun* **74:**5636–5644.

27. **Thomas WE, Nilsson LM, Forero M, Sokurenko EV, Vogel V.** 2004. Shear-dependent 'stick-and-roll' adhesion of type 1

fimbriated *Escherichia coli. Mol Microbiol* 53:1545–1557.

28. **Le Trong I, Aprikian P, Kidd BA, Forero-Shelton M, Tchesnokova V, Rajagopal P, Rodriguez V, Interlandi G, Klevit R, Vogel V, Stenkamp RE, Sokurenko EV, Thomas WE.** 2010. Structural basis for mechanical force regulation of the adhesin FimH via finger trap-like beta sheet twisting. *Cell* 141:645–655.

29. **Sauer MM, Jakob RP, Eras J, Baday S, Eriş D, Navarra G, Bernèche S, Ernst B, Maier T, Glockshuber R.** 2016. Catch-bond mechanism of the bacterial adhesin FimH. *Nat Commun* 7:10738.

30. **Jass J, Schedin S, Fällman E, Ohlsson J, Nilsson UJ, Uhlin BE, Axner O.** 2004. Physical properties of *Escherichia coli* P pili measured by optical tweezers. *Biophys J* 87:4271–4283.

31. **Fällman E, Schedin S, Jass J, Uhlin BE, Axner O.** 2005. The unfolding of the P pili quaternary structure by stretching is reversible, not plastic. *EMBO Rep* 6:52–56.

32. **Andersson M, Axner O, Almqvist F, Uhlin BE, Fällman E.** 2008. Physical properties of biopolymers assessed by optical tweezers: analysis of folding and refolding of bacterial pili. *Chemphyschem* 9:221–235.

33. **Jacob-Dubuisson F, Striker R, Hultgren SJ.** 1994. Chaperone-assisted self-assembly of pili independent of cellular energy. *J Biol Chem* 269:12447–12455.

34. **Thanassi DG, Stathopoulos C, Karkal A, Li H.** 2005. Protein secretion in the absence of ATP: the autotransporter, two-partner secretion and chaperone/usher pathways of gram-negative bacteria. *Mol Membr Biol* 22:63–72.

35. **Totsika M, Heras B, Wurpel DJ, Schembri MA.** 2009. Characterization of two homologous disulfide bond systems involved in virulence factor biogenesis in uropathogenic *Escherichia coli* CFT073. *J Bacteriol* 191:3901–3908.

36. **Crespo MD, Puorger C, Schärer MA, Eidam O, Grütter MG, Capitani G, Glockshuber R.** 2012. Quality control of disulfide bond formation in pilus subunits by the chaperone FimC. *Nat Chem Biol* 8:707–713.

37. **Pilipczuk J, Zalewska-Piątek B, Bruździak P, Czub J, Wieczór M, Olszewski M, Wanarska M, Nowicki B, Augustin-Nowacka D, Piątek R.** 2017. Role of the disulfide bond in stabilizing and folding of the fimbrial protein DraE from uropathogenic *Escherichia coli. J Biol Chem* 292:16136–16149.

38. **Alonso-Caballero A, Schönfelder J, Poly S, Corsetti F, De Sancho D, Artacho E, Perez-Jimenez R.** 2018. Mechanical architecture and folding of *E. coli* type 1 pilus domains. *Nat Commun* 9:2758.

39. **Sauer FG, Fütterer K, Pinkner JS, Dodson KW, Hultgren SJ, Waksman G.** 1999. Structural basis of chaperone function and pilus biogenesis. *Science* 285:1058–1061.

40. **Choudhury D, Thompson A, Stojanoff V, Langermann S, Pinkner J, Hultgren SJ, Knight SD.** 1999. X-ray structure of the FimC-FimH chaperone-adhesin complex from uropathogenic *Escherichia coli. Science* 285:1061–1066.

41. **Jones CH, Danese PN, Pinkner JS, Silhavy TJ, Hultgren SJ.** 1997. The chaperone-assisted membrane release and folding pathway is sensed by two signal transduction systems. *EMBO J* 16:6394–6406.

42. **Bann JG, Pinkner JS, Frieden C, Hultgren SJ.** 2004. Catalysis of protein folding by chaperones in pathogenic bacteria. *Proc Natl Acad Sci U S A* 101:17389–17393.

43. **Sauer FG, Pinkner JS, Waksman G, Hultgren SJ.** 2002. Chaperone priming of pilus subunits facilitates a topological transition that drives fiber formation. *Cell* 111:543–551.

44. **Zavialov AV, Berglund J, Pudney AF, Fooks LJ, Ibrahim TM, MacIntyre S, Knight SD.** 2003. Structure and biogenesis of the capsular F1 antigen from *Yersinia pestis*: preserved folding energy drives fiber formation. *Cell* 113:587–596.

45. **Zavialov AV, Tischenko VM, Fooks LJ, Brandsdal BO, Aqvist J, Zav'yalov VP, Macintyre S, Knight SD.** 2005. Resolving the energy paradox of chaperone/usher-mediated fibre assembly. *Biochem J* 389:685–694.

46. **Remaut H, Rose RJ, Hannan TJ, Hultgren SJ, Radford SE, Ashcroft AE, Waksman G.** 2006. Donor-strand exchange in chaperone-assisted pilus assembly proceeds through a concerted beta strand displacement mechanism. *Mol Cell* 22:831–842.

47. **Nishiyama M, Ishikawa T, Rechsteiner H, Glockshuber R.** 2008. Reconstitution of pilus assembly reveals a bacterial outer membrane catalyst. *Science* 320:376–379.

48. **Saulino ET, Thanassi DG, Pinkner JS, Hultgren SJ.** 1998. Ramifications of kinetic partitioning on usher-mediated pilus biogenesis. *EMBO J* 17:2177–2185.

49. **Nishiyama M, Glockshuber R.** 2010. The outer membrane usher guarantees the formation of functional pili by selectively catalyzing donor-strand exchange between subunits that are adjacent in the mature pilus. *J Mol Biol* 396:1–8.

50. **Li Q, Ng TW, Dodson KW, So SS, Bayle KM, Pinkner JS, Scarlata S, Hultgren SJ, Thanassi DG.** 2010. The differential affinity of the usher

for chaperone-subunit complexes is required for assembly of complete pili. *Mol Microbiol* **76:**159–172.

51. **Allen WJ, Phan G, Hultgren SJ, Waksman G.** 2013. Dissection of pilus tip assembly by the FimD usher monomer. *J Mol Biol* **425:**958–967.

52. **Nishiyama M, Horst R, Eidam O, Herrmann T, Ignatov O, Vetsch M, Bettendorff P, Jelesarov I, Grütter MG, Wüthrich K, Glockshuber R, Capitani G.** 2005. Structural basis of chaperone-subunit complex recognition by the type 1 pilus assembly platform FimD. *EMBO J* **24:**2075–2086.

53. **Remaut H, Tang C, Henderson NS, Pinkner JS, Wang T, Hultgren SJ, Thanassi DG, Waksman G, Li H.** 2008. Fiber formation across the bacterial outer membrane by the chaperone/usher pathway. *Cell* **133:**640–652.

54. **Huang Y, Smith BS, Chen LX, Baxter RH, Deisenhofer J.** 2009. Insights into pilus assembly and secretion from the structure and functional characterization of usher PapC. *Proc Natl Acad Sci U S A* **106:**7403–7407.

55. **Phan G, Remaut H, Wang T, Allen WJ, Pirker KF, Lebedev A, Henderson NS, Geibel S, Volkan E, Yan J, Kunze MB, Pinkner JS, Ford B, Kay CW, Li H, Hultgren SJ, Thanassi DG, Waksman G.** 2011. Crystal structure of the FimD usher bound to its cognate FimC-FimH substrate. *Nature* **474:**49–53.

56. **Mapingire OS, Henderson NS, Duret G, Thanassi DG, Delcour AH.** 2009. Modulating effects of the plug, helix, and N- and C-terminal domains on channel properties of the PapC usher. *J Biol Chem* **284:**36324–36333.

57. **Werneburg GT, Henderson NS, Portnoy EB, Sarowar S, Hultgren SJ, Li H, Thanassi DG.** 2015. The pilus usher controls protein interactions via domain masking and is functional as an oligomer. *Nat Struct Mol Biol* **22:**540–546.

58. **Volkan E, Ford BA, Pinkner JS, Dodson KW, Henderson NS, Thanassi DG, Waksman G, Hultgren SJ.** 2012. Domain activities of PapC usher reveal the mechanism of action of an *Escherichia coli* molecular machine. *Proc Natl Acad Sci U S A* **109:**9563–9568.

59. **Omattage NS, Deng Z, Pinkner JS, Dodson KW, Almqvist F, Yuan P, Hultgren SJ.** 2018. Structural basis for usher activation and intramolecular subunit transfer in P pilus biogenesis in *Escherichia coli*. *Nat Microbiol* **3:**1362–1368.

60. **Du M, Yuan Z, Yu H, Henderson N, Sarowar S, Zhao G, Werneburg GT, Thanassi DG, Li H.** 2018. Handover mechanism of the growing pilus by the bacterial outer-membrane usher FimD. *Nature* **562:**444–447.

61. **Hospenthal MK, Zyla D, Costa TRD, Redzej A, Giese C, Lillington J, Glockshuber R, Waksman G.** 2017. The cryoelectron microscopy structure of the type 1 chaperone-usher pilus rod. *Structure* **25:**1829–1838.e4.

62. **Hospenthal MK, Redzej A, Dodson K, Ukleja M, Frenz B, Rodrigues C, Hultgren SJ, DiMaio F, Egelman EH, Waksman G.** 2016. Structure of a chaperone-usher pilus reveals the molecular basis of rod uncoiling. *Cell* **164:**269–278.

63. **Spaulding CN, Schreiber HL IV, Zheng W, Dodson KW, Hazen JE, Conover MS, Wang F, Svenmarker P, Luna-Rico A, Francetic O, Andersson M, Hultgren S, Egelman EH.** 2018. Functional role of the type 1 pilus rod structure in mediating host-pathogen interactions. *eLife* **7:**e31662.

64. **Fauci AS, Morens DM.** 2012. The perpetual challenge of infectious diseases. *N Engl J Med* **366:**454–461.

65. **World Health Organization.** 2014. *Antimicrobial Resistance: Global Report on Surveillance.* WHO, Geneva, Switzerland. https://www.who.int/drugresistance/documents/surveillancereport/en/

66. **Cegelski L, Marshall GR, Eldridge GR, Hultgren SJ.** 2008. The biology and future prospects of antivirulence therapies. *Nat Rev Microbiol* **6:**17–27.

67. **Allen RC, Popat R, Diggle SP, Brown SP.** 2014. Targeting virulence: can we make evolution-proof drugs? *Nat Rev Microbiol* **12:**300–308.

68. **Zambelloni R, Marquez R, Roe AJ.** 2015. Development of antivirulence compounds: a biochemical review. *Chem Biol Drug Des* **85:**43–55.

69. **Paharik AE, Schreiber HL IV, Spaulding CN, Dodson KW, Hultgren SJ.** 2017. Narrowing the spectrum: the new frontier of precision antimicrobials. *Genome Med* **9:**110.

70. **Svennerholm AM, Tobias J.** 2008. Vaccines against enterotoxigenic *Escherichia coli.* *Expert Rev Vaccines* **7:**795–804.

71. **van der Woude MW, Bäumler AJ.** 2004. Phase and antigenic variation in bacteria. *Clin Microbiol Rev* **17:**581–611.

72. **Langermann S, Palaszynski S, Barnhart M, Auguste G, Pinkner JS, Burlein J, Barren P, Koenig S, Leath S, Jones CH, Hultgren SJ.** 1997. Prevention of mucosal *Escherichia coli* infection by FimH-adhesin-based systemic vaccination. *Science* **276:**607–611.

73. **Thankavel K, Madison B, Ikeda T, Malaviya R, Shah AH, Arumugam PM, Abraham SN.** 1997. Localization of a domain in the FimH adhesin of *Escherichia coli* type 1 fimbriae

capable of receptor recognition and use of a domain-specific antibody to confer protection against experimental urinary tract infection. *J Clin Invest* **100**:1123–1136.

74. **Langermann S, Möllby R, Burlein JE, Palaszynski SR, Auguste CG, DeFusco A, Strouse R, Schenerman MA, Hultgren SJ, Pinkner JS, Winberg J, Guldevall L, Söderhäll M, Ishikawa K, Normark S, Koenig S.** 2000. Vaccination with FimH adhesin protects cynomolgus monkeys from colonization and infection by uropathogenic *Escherichia coli*. *J Infect Dis* **181**:774–778.

75. **Roberts JA, Kaack MB, Baskin G, Chapman MR, Hunstad DA, Pinkner JS, Hultgren SJ.** 2004. Antibody responses and protection from pyelonephritis following vaccination with purified *Escherichia coli* PapDG protein. *J Urol* **171**:1682–1685.

76. **Poggio TV, La Torre JL, Scodeller EA.** 2006. Intranasal immunization with a recombinant truncated FimH adhesin adjuvanted with CpG oligodeoxynucleotides protects mice against uropathogenic *Escherichia coli* challenge. *Can J Microbiol* **52**:1093–1102.

77. **Bameri Z, Asadi Karam MR, Habibi M, Ehsani P, Bouzari S.** 2018. Determination immunogenic property of truncated MrpH. FliC as a vaccine candidate against urinary tract infections caused by *Proteus mirabilis*. *Microb Pathog* **114**:99–106.

78. **Savarino SJ, McKenzie R, Tribble DR, Porter CK, O'Dowd A, Cantrell JA, Sincock SA, Poole ST, DeNearing B, Woods CM, Kim H, Grahek SL, Brinkley C, Crabb JH, Bourgeois AL.** 2017. Prophylactic efficacy of hyperimmune bovine colostral antiadhesin antibodies against enterotoxigenic *Escherichia coli* diarrhea: a randomized, double-blind, placebo-controlled, phase 1 trial. *J Infect Dis* **216**:7–13.

79. **Tchesnokova V, Aprikian P, Kisiela D, Gowey S, Korotkova N, Thomas W, Sokurenko E.** 2011. Type 1 fimbrial adhesin FimH elicits an immune response that enhances cell adhesion of *Escherichia coli*. *Infect Immun* **79**:3895–3904.

80. **Kisiela DI, Avagyan H, Friend D, Jalan A, Gupta S, Interlandi G, Liu Y, Tchesnokova V, Rodriguez VB, Sumida JP, Strong RK, Wu XR, Thomas WE, Sokurenko EV.** 2015. Inhibition and reversal of microbial attachment by an antibody with parasteric activity against the FimH adhesin of uropathogenic *E. coli*. *PLoS Pathog* **11**:e1004857.

81. **Klein T, Abgottspon D, Wittwer M, Rabbani S, Herold J, Jiang X, Kleeb S, Lüthi C, Scharenberg M, Bezençon J, Gubler E, Pang L, Smiesko M, Cutting B, Schwardt O, Ernst B.** 2010. FimH antagonists for the oral treatment of urinary tract infections: from design and synthesis to in vitro and in vivo evaluation. *J Med Chem* **53**:8627–8641.

82. **Cusumano CK, Pinkner JS, Han Z, Greene SE, Ford BA, Crowley JR, Henderson JP, Janetka JW, Hultgren SJ.** 2011. Treatment and prevention of urinary tract infection with orally active FimH inhibitors. *Sci Transl Med* **3**:109ra115.

83. **Jiang X, Abgottspon D, Kleeb S, Rabbani S, Scharenberg M, Wittwer M, Haug M, Schwardt O, Ernst B.** 2012. Antiadhesion therapy for urinary tract infections—a balanced PK/PD profile proved to be key for success. *J Med Chem* **55**:4700–4713.

84. **Totsika M, Kostakioti M, Hannan TJ, Upton M, Beatson SA, Janetka JW, Hultgren SJ, Schembri MA.** 2013. A FimH inhibitor prevents acute bladder infection and treats chronic cystitis caused by multidrug-resistant uropathogenic *Escherichia coli* ST131. *J Infect Dis* **208**:921–928.

85. **Spaulding CN, Klein RD, Ruer S, Kau AL, Schreiber HL, Cusumano ZT, Dodson KW, Pinkner JS, Fremont DH, Janetka JW, Remaut H, Gordon JI, Hultgren SJ.** 2017. Selective depletion of uropathogenic *E. coli* from the gut by a FimH antagonist. *Nature* **546**:528–532.

86. **Guiton PS, Cusumano CK, Kline KA, Dodson KW, Han Z, Janetka JW, Henderson JP, Caparon MG, Hultgren SJ.** 2012. Combinatorial small-molecule therapy prevents uropathogenic *Escherichia coli* catheter-associated urinary tract infections in mice. *Antimicrob Agents Chemother* **56**:4738–4745.

87. **Mayer K, Eris D, Schwardt O, Sager CP, Rabbani S, Kleeb S, Ernst B.** 2017. Urinary tract infection: which conformation of the bacterial lectin FimH is therapeutically relevant? *J Med Chem* **60**:5646–5662.

88. **Sattigeri JA, Garg M, Bhateja P, Soni A, Rauf ARA, Gupta M, Deshmukh MS, Jain T, Alekar N, Barman TK, Jha P, Chaira T, Bambal RB, Upadhyay DJ, Nishi T.** 2018. Synthesis and evaluation of thiomannosides, potent and orally active FimH inhibitors. *Bioorg Med Chem Lett* **28**:2993–2997.

89. **Salminen A, Loimaranta V, Joosten JA, Khan AS, Hacker J, Pieters RJ, Finne J.** 2007. Inhibition of P-fimbriated *Escherichia coli* adhesion by multivalent galabiose derivatives studied by a live-bacteria application of surface plasmon resonance. *J Antimicrob Chemother* **60**:495–501.

90. **Kalas V, Hibbing ME, Maddirala AR, Chugani R, Pinkner JS, Mydock-McGrane LK, Conover**

MS, Janetka JW, Hultgren SJ. 2018. Structure-based discovery of glycomimetic FmlH ligands as inhibitors of bacterial adhesion during urinary tract infection. *Proc Natl Acad Sci U S A* **115**: E2819–E2828.

91. Brument S, Sivignon A, Dumych TI, Moreau N, Roos G, Guérardel Y, Chalopin T, Deniaud D, Bilyy RO, Darfeuille-Michaud A, Bouckaert J, Gouin SG. 2013. Thiazolylaminomannosides as potent antiadhesives of type 1 piliated *Escherichia coli* isolated from Crohn's disease patients. *J Med Chem* **56**:5395–5406.

92. Sivignon A, Yan X, Alvarez Dorta D, Bonnet R, Bouckaert J, Fleury E, Bernard J, Gouin SG, Darfeuille-Michaud A, Barnich N. 2015. Development of heptylmannoside-based glycoconjugate antiadhesive compounds against adherent-invasive *Escherichia coli* bacteria associated with Crohn's disease. *mBio* **6**:e01298-15.

93. Chalopin T, Alvarez Dorta D, Sivignon A, Caudan M, Dumych TI, Bilyy RO, Deniaud D, Barnich N, Bouckaert J, Gouin SG. 2016. Second generation of thiazolylmannosides, FimH antagonists for *E. coli*-induced Crohn's disease. *Org Biomol Chem* **14**:3913–3925.

94. Sivignon A, Bouckaert J, Bernard J, Gouin SG, Barnich N. 2017. The potential of FimH as a novel therapeutic target for the treatment of Crohn's disease. *Expert Opin Ther Targets* **21**:837–847.

95. Almant M, Moreau V, Kovensky J, Bouckaert J, Gouin SG. 2011. Clustering of *Escherichia coli* type-1 fimbrial adhesins by using multimeric heptyl α-d-mannoside probes with a carbohydrate core. *Chemistry* **17**:10029–10038.

96. Schierholt A, Hartmann M, Lindhorst TK. 2011. Bi- and trivalent glycopeptide mannopyranosides as inhibitors of type 1 fimbriae-mediated bacterial adhesion: variation of valency, aglycon and scaffolding. *Carbohydr Res* **346**:1519–1526.

97. Richards SJ, Jones MW, Hunaban M, Haddleton DM, Gibson MI. 2012. Probing bacterial-toxin inhibition with synthetic glycopolymers prepared by tandem post-polymerization modification: role of linker length and carbohydrate density. *Angew Chem Int Ed Engl* **51**:7812–7816.

98. Svensson A, Larsson A, Emtenäs H, Hedenström M, Fex T, Hultgren SJ, Pinkner JS, Almqvist F, Kihlberg J. 2001. Design and evaluation of pilicides: potential novel antibacterial agents directed against uropathogenic *Escherichia coli*. *Chembiochem* **2**:915–918.

99. Pinkner JS, Remaut H, Buelens F, Miller E, Aberg V, Pemberton N, Hedenström M, Larsson A, Seed P, Waksman G, Hultgren SJ, Almqvist F. 2006. Rationally designed small compounds inhibit pilus biogenesis in uropathogenic bacteria. *Proc Natl Acad Sci U S A* **103**:17897–17902.

100. Chorell E, Pinkner JS, Bengtsson C, Banchelin TS, Edvinsson S, Linusson A, Hultgren SJ, Almqvist F. 2012. Mapping pilicide anti-virulence effect in *Escherichia coli*, a comprehensive structure-activity study. *Bioorg Med Chem* **20**:3128–3142.

101. Piatek R, Zalewska-Piatek B, Dzierzbicka K, Makowiec S, Pilipczuk J, Szemiako K, Cyranka-Czaja A, Wojciechowski M. 2013. Pilicides inhibit the FGL chaperone/usher assisted biogenesis of the Dr fimbrial polyadhesin from uropathogenic *Escherichia coli*. *BMC Microbiol* **13**:131.

102. Chorell E, Pinkner JS, Phan G, Edvinsson S, Buelens F, Remaut H, Waksman G, Hultgren SJ, Almqvist F. 2010. Design and synthesis of C-2 substituted thiazolo and dihydrothiazolo ring-fused 2-pyridones: pilicides with increased antivirulence activity. *J Med Chem* **53**:5690–5695.

103. Dang HT, Chorell E, Uvell H, Pinkner JS, Hultgren SJ, Almqvist F. 2014. Syntheses and biological evaluation of 2-amino-3-acyl-tetrahydrobenzothiophene derivatives; antibacterial agents with antivirulence activity. *Org Biomol Chem* **12**:1942–1956.

104. Greene SE, Pinkner JS, Chorell E, Dodson KW, Shaffer CL, Conover MS, Livny J, Hadjifrangiskou M, Almqvist F, Hultgren SJ. 2014. Pilicide ec240 disrupts virulence circuits in uropathogenic *Escherichia coli*. *mBio* **5**: e02038-14.

105. Lo AW, Van de Water K, Gane PJ, Chan AW, Steadman D, Stevens K, Selwood DL, Waksman G, Remaut H. 2014. Suppression of type 1 pilus assembly in uropathogenic *Escherichia coli* by chemical inhibition of subunit polymerization. *J Antimicrob Chemother* **69**:1017–1026.

106. Shamir ER, Warthan M, Brown SP, Nataro JP, Guerrant RL, Hoffman PS. 2010. Nitazoxanide inhibits biofilm production and hemagglutination by enteroaggregative *Escherichia coli* strains by blocking assembly of AafA fimbriae. *Antimicrob Agents Chemother* **54**:1526–1533.

107. Chahales P, Hoffman PS, Thanassi DG. 2016. Nitazoxanide inhibits pilus biogenesis by interfering with folding of the usher protein in the outer membrane. *Antimicrob Agents Chemother* **60**:2028–2038.

108. Klinth JE, Pinkner JS, Hultgren SJ, Almqvist F, Uhlin BE, Axner O. 2012. Impairment of the biomechanical compliance of P pili: a novel means of inhibiting uropathogenic bacterial infections? *Eur Biophys J* **41**:285–295.

109. **Mortezaei N, Singh B, Bullitt E, Uhlin BE, Andersson M.** 2013. P-fimbriae in the presence of anti-PapA antibodies: new insight of antibodies action against pathogens. *Sci Rep* **3:**3393.

110. **Singh B, Mortezaei N, Uhlin BE, Savarino SJ, Bullitt E, Andersson M.** 2015. Antibody-mediated disruption of the mechanics of CS20 fimbriae of enterotoxigenic *Escherichia coli. Sci Rep* **5:**13678.

111. **Singh B, Mortezaei N, Savarino SJ, Uhlin BE, Bullitt E, Andersson M.** 2017. Antibodies damage the resilience of fimbriae, causing them to be stiff and tangled. *J Bacteriol* **199:** e00665-16.

112. **Brown JW, Badahdah A, Iticovici M, Vickers TJ, Alvarado DM, Helmerhorst EJ, Oppenheim FG, Mills JC, Ciorba MA, Fleckenstein JM, Bullitt E.** 2018. A role for salivary peptides in the innate defense against enterotoxigenic *Escherichia coli. J Infect Dis* **217:**1435–1441.

113. **Paton AW, Morona R, Paton JC.** 2000. A new biological agent for treatment of Shiga toxigenic *Escherichia coli* infections and dysentery in humans. *Nat Med* **6:**265–270.

114. **Watts RE, Tan CK, Ulett GC, Carey AJ, Totsika M, Idris A, Paton AW, Morona R, Paton JC, Schembri MA.** 2012. *Escherichia coli* 83972 expressing a P fimbriae oligosaccharide receptor mimic impairs adhesion of uropathogenic *E. coli. J Infect Dis* **206:**1242–1249.

115. **Trautner BW, Cevallos ME, Li H, Riosa S, Hull RA, Hull SI, Tweardy DJ, Darouiche RO.** 2008. Increased expression of type-1 fimbriae by nonpathogenic *Escherichia coli* 83972 results in an increased capacity for catheter adherence and bacterial interference. *J Infect Dis* **198:**899–906.

Curli Biogenesis: Bacterial Amyloid Assembly by the Type VIII Secretion Pathway

14

SUJEET BHOITE,[1] NANI VAN GERVEN,[2,3] MATTHEW R. CHAPMAN,[1] and HAN REMAUT[2,3]

Curli are extracellular proteinaceous fibers made by Gram-negative bacteria. Curli-specific genes (*csg*) are primarily found in *Proteobacteria* and *Bacteroidetes* (1–3). The main function of curli fibers is associated with a sedimentary lifestyle and multicellular behavior in biofilms, as they form scaffolds that provide adhesive and structural support to the community (4–8). In certain pathogenic bacteria, curli have also been implicated in host colonization, innate response activation, and cell invasion (9–13).

Curli appear as rigid, 2- to 5-nm-thick coiled fibers that entangle into dense congregates that surround the cell (Fig. 1A and B) (1, 7, 14). Curli fibers form via a nucleation-dependent self-assembly process, and they adopt an amyloid fold as their native structure (15, 16). Curli are the product of a dedicated secretion-assembly pathway known as the nucleation-precipitation pathway, or type VIII secretion system (T8SS) (3, 17, 18). In *Escherichia coli*, seven curli-specific genes (*csg*) are clustered in two divergently described operons, *csgBAC* and *csgDEFG* (17) (Fig. 1C), where *csgA* and *csgB* are structural components of the fiber (1, 16), *csgC* is a periplasmic chaperone (19), *csgE*, *csgF*, and *csgG* form the secretion-assembly machinery (20–22), and csgD is a transcriptional activator

[1]Department of Molecular, Cellular, and Developmental Biology, University of Michigan, Ann Arbor, MI 48109
[2]Structural Biology Brussels, Vrije Universiteit Brussel, 1050 Brussels, Belgium
[3]Structural and Molecular Microbiology, Structural Biology Research Center, VIB, 1050 Brussels, Belgium

Protein Secretion in Bacteria
Edited by Maria Sandkvist, Eric Cascales, and Peter J. Christie
© 2019 American Society for Microbiology, Washington, DC
doi:10.1128/ecosalplus.ESP-0037-2018

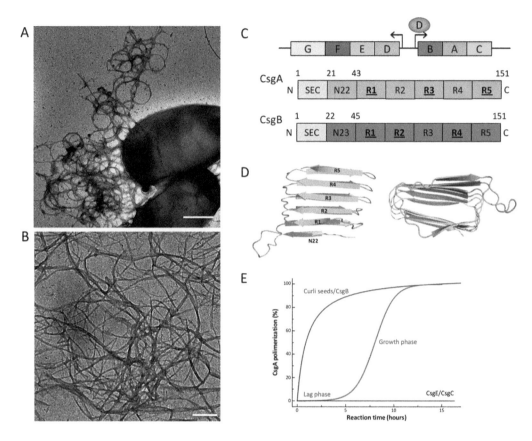

FIGURE 1 Curli composition and structure. (A and B) Transmission electron micrographs of individual *E. coli* cells producing curli fibers (A) and curli-like fibers grown *in vitro* from purified CsgA (B). Scale bars: 500 nm and 200 nm, respectively. **(C)** Schematic organization of the *csgDEFG* and *csgBAC* curli gene clusters and architecture of the curli subunits CsgA (blue) and CsgB (dark blue). Subunits comprise an N-terminal signal sequence (SEC) that is cleaved upon export into the periplasm. The mature proteins contain curlin pseudorepeat regions (N22, R1 to R5) that guide substrate specificity in the secretion pathway and form the amyloidogenic core of the curli subunits. Repeats that efficiently self-polymerize *in vitro* are underscored. **(D)** Theoretical model of CsgA predicted based on amino acid covariation analysis (42). The predictions point to a right- or left-handed β-helix made up from stacked curlin repeats (labeled R1 to R5). **(E)** Representation of typical *in vitro* CsgA polymerization profiles in the absence (red) or presence (blue) of preformed fibers or the CsgB nucleator. In the presence of CsgE (1:1 ratio) or CsgC (1:500 ratio), no CsgA polymerization is observed (black curve).

of the *csgBAC* operon (23). We review curli biogenesis and the T8SS using the *E. coli* components as a representative system.

THE SECRETION-ASSEMBLY MACHINERY

In Gram-negative bacteria, assembly of surface appendages requires the passage of two lipid bilayers, the cytoplasmic or inner mem-

brane and the outer membrane. The curli subunit proteins CsgA and CsgB cross the inner membrane via the Sec general secretory pathway, after which CsgE, CsgF, and CsgG orchestrate translocation of CsgA and CsgB across the outer membrane (20–22), where they assemble into curli polymers (21). CsgG is a 262-residue lipoprotein that forms the curli translocation channel in the outer membrane (20). It forms a nonameric complex with a 36-stranded β-barrel that inserts into the outer membrane and is connected to the periplasm via a cage-like vestibule with an inner diameter of approximately 35 Å (24, 25) (Fig. 2). The transmembrane and periplasmic

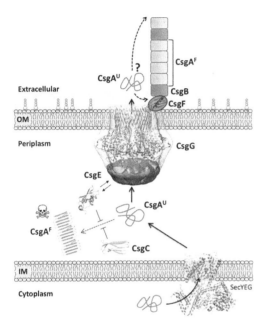

FIGURE 2 Integrated model for curli subunit secretion. Curli subunits enter the periplasm via the SecYEG translocon, from where they progress to the cell surface as unfolded polypeptides via the curli transporter CsgG. Premature folding and polymerization of CsgA in the periplasm (right dotted line) are inhibited by CsgE and CsgC. CsgE binds and targets subunits to the secretion channel, while CsgC provides a safeguard against runaway polymerization, likely by the binding and neutralization of early assembly intermediates and/or nascent fibers. CsgG forms a nonameric complex that acts as a peptide diffusion channel and cooperates with the periplasmic factor CsgE, which binds the channel and forms a capping structure to the secretion complex. Recruitment and (partial) enclosure of CsgA in the secretion complex are proposed to create an entropy gradient over the channel that favors CsgA's outward diffusion as an unfolded, soluble polypeptide. Once secreted, curli fiber formation and elongation are templated by CsgB, in a CsgF-dependent manner. CsgF is likely to be in contact or close proximity to the CsgG channel. The exact role of CsgF and whether fibers extend from the proximal or distal end (dashed arrows) are presently unknown. Abbreviations: IM, inner membrane; OM, outer membrane.

domains are separated by a 9-Å-wide channel constriction that would be compatible with the passage of unfolded polypeptides. Single-channel current recordings and *in vivo* bile salt sensitivity assays show a constitutively open channel, which can be gated in the presence of the periplasmic accessory factor CsgE (24). The latter is a soluble 107-residue protein that can oligomerize into a nonameric cap-like structure that binds the periplasmic entry to the CsgG secretion channel (24) (Fig. 2). Although at elevated concentrations CsgG can facilitate a nonselective leakage of periplasmic polypeptides, under native conditions secretion is specific for curli subunits and requires CsgE (22). *In vitro*, CsgE inhibits CsgA self-assembly when present at a 1:1 stoichiometric ratio, demonstrating that it can directly interact with curli subunits. The prevailing model is for CsgE to act as specificity factor by binding periplasmic CsgA and targeting it to the CsgG secretion channel. *In vitro* the CsgE:CsgG complex is in a reversible equilibrium (24), suggesting that CsgE may cycle between a periplasmic form and a CsgG-bound form. The nuclear magnetic resonance (NMR) structure of the CsgE monomer shows a compact alpha/beta sandwich protein with a flexible C-terminal tail (26) (Fig. 2). Two regions of CsgE involved in the CsgE-CsgA interaction were determined: a head comprising a positively charged patch centered around R47 and a stem comprising a negatively charged patch containing E31 and E85 (27). R47 was found to mediate an indispensable charge-charge interaction with CsgA, while mutations in the negatively charged neck region retain CsgA binding capacity but impact CsgE-dependent secretion of CsgA. The exact binding epitopes in CsgA are currently unknown, although the first pseudo-repeat in the curli subunits (in *E. coli* CsgA referred to as N22) is sufficient to target polypeptides for secretion through the curli pathway (20, 22). Secretion of nonnative sequences was found to have a size constraint in the case of folded fusion proteins, which is likely to reflect the width of the CsgG constriction (28).

The dimensions of the channel constriction suggest that the native substrate navigates the CsgG channel in an extended conformation (24, 25). In agreement with this, in strains with curli assembly defects, extracellular CsgA is found as an unfolded chain (16, 29). The transition of secreted CsgA to the amyloid state requires the activity of CsgF and the nucleator subunit CsgB (15, 21, 29–31).

CsgF is a 119-residue protein that is found to be surface exposed in a CsgG-dependent manner and is speculated to be in close proximity or direct contact with the CsgG channel (Fig. 2) (21). Structural data obtained by NMR spectroscopy revealed that the CsgF monomer consists of three independent elements: an N-terminal unstructured region, a 21-residue α-helix, and a C-terminal antiparallel β-sheet made of 4 strands (32). The absence of tertiary structure and the presence of exposed hydrophobic surfaces may correspond to putative interaction sides with CsgG or the secretion substrates CsgA and CsgB. Although CsgF appears to be dispensable for secretion of curli subunits, it is essential for curli fiber formation (21). When *csgF* is deleted, CsgB is no longer associated with the cell surface and is not able to exert its nucleating function. These observations implicate CsgF as a coupling factor between CsgA secretion and extracellular polymerization into curli fibers by coordinating or chaperoning the nucleating activity of the CsgB minor subunit (21).

A striking conundrum in curli assembly is the mode of protein translocation across the outer membrane. Protein transport is an energy-dependent process, in which the driving force is generally drawn from ATP or GTP hydrolysis, protein synthesis, or an electrochemical potential such as the proton motive force (33). However, the periplasmic space is devoid of known hydrolyzable energy carriers and the semiporous outer membrane does not support the buildup of solute or ion gradients. In Gram-negative secretion pathways, this hurdle is overcome by means of cell envelope-spanning complexes such as type I,

II, III, and IV secretion systems (34) or by drawing energy from substrate folding and assembly such as seen for the chaperone-usher assembly and type V secretion pathways (35, 36). Available structural data indicate that CsgG operates as a peptide diffusion channel (24, 25). The height of the CsgG constriction is compatible with the simultaneous binding of 4 or 5 residues along the length of an extended polypeptide chain, so that passage of the full protein implies stepwise Brownian diffusion along the length of the polypeptide chain. The rectifying force that ensures a net forward Brownian diffusion is presently unclear. The transition from the extended preassembly state to the amyloid state upon secretion and incorporation into curli fibers may provide such a driving force. However, in *csgB* deletion mutants, CsgA does not assemble into curli and can be found as disordered polypeptide in the cell medium, indicating that fiber assembly is not a prerequisite for secretion (21, 29). In an alternative hypothesis, secretion is driven by an entropy gradient across the outer membrane and the channel in a process that involves the secretion factor CsgE (24, 37). It is speculated that the local high concentration and conformational confinement of curli subunits in the CsgG vestibule would raise an entropic free-energy gradient over the translocation channel. Under physiological conditions, the entropy potential of CsgA capture at the translocation channel and the folding energy released from the disorder-order transition of subunits incorporating into surface-associated fibers may cooperate to ensure forward diffusion across the CsgG channel.

FIBER COMPONENTS AND SELF-ASSEMBLY

Curli fibers are linear, noncovalent polymers of the major and minor subunits CsgA and CsgB, respectively (1, 16). The major curlin subunit, CsgA, is a 151-amino-acid-long peptide with five imperfect repeat units, R1 to R5

(Fig. 1C). The repeating units in CsgA are predicted to form β-strand–loop–β-strand motifs that constitute the core of the β-sheet-rich curli amyloid fiber (3, 16, 38–42) (Fig. 1D). *In vitro*, CsgA amyloid formation can be monitored using the amyloid binding dye thioflavin T (ThT) (43). CsgA amyloid aggregation displays canonical sigmodal polymerization kinetics typical for amyloid proteins, with a lag phase, a growth phase, and a plateau phase (44) (Fig. 1E). The lag phase represents the time required for the buildup of enough primary nuclei to lead to the rapid incorporation of remaining free monomers into the nascent fibers during the growth phase (38, 44, 45). Recent *in vitro* studies suggest that the nucleation of CsgA into the amyloid template is a single-step process and does not pass through one or more oligomeric aggregated states as often seen in pathological amyloids (45). The phase transitions of the curli repeats from an unfolded to a folded β-strand–loop–β-strand motif, in the CsgA subunit itself or in a minimal oligomer such as a CsgA dimer, serve as the "nucleus" in this model (45). Once formed, CsgA fibers display distinctive structural and biophysical properties that are shared by all amyloids (16, 44, 45).

CsgA is secreted by the T8SS to the cell surface as an unstructured protein and remains in a nonamyloid form unless it interacts with the CsgB "nucleator" protein. CsgB shares 30% sequence identity with CsgA and is also predicted to form five imperfect β-strand–loop–β-strand repeats, R1 to R5 (31) (Fig. 1C). The C-terminal R5 repeat of CsgB has been shown to be necessary for anchoring CsgB to the cell surface (30, 31). While the precise mechanism of *in vivo* nucleation has not been elucidated, it is tempting to postulate that the R1 to R4 repeats of membrane-anchored CsgB assume a β-solenoid structure that provides a template for recognition by either R1 or R5 of CsgA (38, 41). The R1 and R5 repeats in CsgA are suggested to mediate interaction with CsgB, but also self-association by providing a template for secreted CsgA monomers to add on to the growing fiber tip (38, 46). A distinctive characteristic of curli assembly is that CsgA and CsgB can be independently secreted as unstructured subunits, which can then self-assemble into extracellular fibers when they contact each other on the cell surface (15, 29).

Interestingly, the R2, R3, and R4 repeating units of CsgA contain conserved Asp and Gly "gatekeeper" residues (47). The positions of these gatekeeper residues are conserved in most CsgA homologs, and certain gatekeeper residues modulate the amyloidogenic nature of CsgA by inhibiting its intrinsic aggregation propensity (47). The gatekeeper residues allow CsgA to remain in an unstructured form before being secreted to the cell surface and nucleated by CsgB (15, 30, 31, 38, 47). Interestingly, a CsgA mutant called CsgA* that lacks the gatekeeper residues polymerized *in vivo* in the absence of CsgB. However, the expression of CsgA* is toxic to cells, underlying the importance of controlled amyloid formation afforded by the T8SS system (47). In this respect, the T8SS evolved an intricate ability to maintain amyloidogenic proteins unstructured until secretion across the outer membrane.

PATHWAY CONTROL AND CHEMICAL INHIBITION

An intriguing issue in functional amyloid assembly pathways such as curli biogenesis is the avoidance of cytotoxic effects that are associated with amyloid formation in human and animal protein aggregation disorders. In addition to elaborate regulation at the transcriptional level (3, 17, 48–52), several adaptations within the pathway components ensure safe secretion of the amyloidogenic subunits and controlled fibril formation at the cell surface. A recent study found that unlike most pathological amyloid depositions, curli nucleation is a single-step process that does not involve intermediary oligomeric aggregates, which are often considered the more toxic species (45, 53). This suggests that curli subunits are evolutionarily optimized to have a sharp and direct transition from an intrinsically disordered conformation to the relatively inert amyloid conformation. In addition, by the provision of the major subunit CsgA and the minor subunit CsgB, the pathway has segregated efficient fiber elongation and fiber nucleation, providing spatiotemporal control over curli deposition. On top of that, the T8SS evolved a further safeguard to quell premature CsgA amyloid formation by having a dedicated periplasmic chaperone (19). CsgC serves as a potent chaperone-like protein to prevent the runaway polymerization of CsgA in the periplasmic space (19) (Fig. 2). *In vitro*, CsgC inhibits CsgA polymerization down to substoichiometric molar ratios of 1:500 (19). The exact mechanism by which CsgC inhibits CsgA aggregation is not completely understood. The observed substoichiometric molar ratio may suggest that CsgC transiently interacts with a soluble oligomeric pool of CsgA to prevent seed formation and subsequent polymerization of CsgA into amyloid fibers (19, 54). Another study, however, found that CsgC does not bind unfolded CsgA and instead acts at the growth poles of nuclei and curli fibers (45). Either way, a conserved patch of positively charged residues on the surface-exposed β-strand (β4-β5 edge) of CsgC appears to mediate an electrostatic interaction between CsgA and CsgC that drives the inhibitory effect (54).

Finally, because functional amyloids like curli are important for biofilm formation and also have been shown to be responsible for host colonization, virulence, and eliciting host immune response (45), chemical inhibitors of curli assembly have been developed (55, 56). A number of small-molecule inhibitors effectively inhibit curli formation at low micromolar to nanomolar concentrations, although these have yet to be evaluated in *in vivo* disease models. In recent years, at least 10 distinct bacterial amyloid systems have been described (9–13). Among these, the *Pseudomonas* Fap pathway is noticeable in encompassing a dedicated secretion-assembly pathway (57). Although nonhomologous to the type VIII secretion pathway, Fap assembly involves a dedicated outer membrane channel (FapF) and a major, polymerizing (FapC) and minor, nucleating (FapB) fiber subunit, reminiscent of curli assembly (58, 59).

ACKNOWLEDGMENTS

H.R. and N.V.G. acknowledge funding by VIB and ERC through consolidator grant BAS-SBBT (grant no. 649082). M.R.C. acknowledges generous support from NIH GM118651 and AI137535 and helpful discussions from members of the Chapman lab.

N.V.G. and H.R. declare to be named as inventors on three patent applications regarding the biotechnological use of curli components.

CITATION

EcoSal Plus 2019; doi:10.1128/ecosalplus.ESP-0037-2018.

REFERENCES

1. **Olsén A, Jonsson A, Normark S.** 1989. Fibronectin binding mediated by a novel class of surface organelles on *Escherichia coli*. *Nature* **338**:652–655.
2. **Dueholm MS, Albertsen M, Otzen D, Nielsen PH.** 2012. Curli functional amyloid systems are phylogenetically widespread and display large diversity in operon and protein structure. *PLoS One* **7**:e51274.
3. **Barnhart MM, Chapman MR.** 2006. Curli biogenesis and function. *Annu Rev Microbiol* **60**:131–147.
4. **Collinson SK, Doig PC, Doran JL, Clouthier S, Trust TJ, Kay WW.** 1993. Thin, aggregative fimbriae mediate binding of *Salmonella enteritidis* to fibronectin. *J Bacteriol* **175**:12–18.
5. **Römling U, Sierralta WD, Eriksson K, Normark S.** 1998. Multicellular and aggregative behaviour of *Salmonella typhimurium* strains is controlled by mutations in the agfD promoter. *Mol Microbiol* **28**:249–264.
6. **Kikuchi T, Mizunoe Y, Takade A, Naito S, Yoshida S.** 2005. Curli fibers are required for development of biofilm architecture in *Escherichia coli* K-12 and enhance bacterial adher-

ence to human uroepithelial cells. *Microbiol Immunol* **49**:875–884.

7. Hung C, Zhou Y, Pinkner JS, Dodson KW, Crowley JR, Heuser J, Chapman MR, Hadjifrangiskou M, Henderson JP, Hultgren SJ. 2013. *Escherichia coli* biofilms have an organized and complex extracellular matrix structure. *mBio* **4**:e00645-13.

8. Hufnagel DA, Depas WH, Chapman MR. 2015. The biology of the *Escherichia coli* extracellular matrix. *Microbiol Spectr* **3**:MB-0014-2014.

9. Herwald H, Mörgelin M, Olsén A, Rhen M, Dahlbäck B, Müller-Esterl W, Björck L. 1998. Activation of the contact-phase system on bacterial surfaces—a clue to serious complications in infectious diseases. *Nat Med* **4**:298–302.

10. Gophna U, Barlev M, Seijffers R, Oelschlager TA, Hacker J, Ron EZ. 2001. Curli fibers mediate internalization of *Escherichia coli* by eukaryotic cells. *Infect Immun* **69**:2659–2665.

11. Tükel C, Raffatellu M, Humphries AD, Wilson RP, Andrews-Polymenis HL, Gull T, Figueiredo JF, Wong MH, Michelsen KS, Akçelik M, Adams LG, Bäumler AJ. 2005. CsgA is a pathogen-associated molecular pattern of *Salmonella enterica* serotype Typhimurium that is recognized by Toll-like receptor 2. *Mol Microbiol* **58**:289–304.

12. Tükel C, Wilson RP, Nishimori JH, Pezeshki M, Chromy BA, Bäumler AJ. 2009. Responses to amyloids of microbial and host origin are mediated through Toll-like receptor 2. *Cell Host Microbe* **6**:45–53.

13. Van Gerven N, Van der Verren SE, Reiter DM, Remaut H. 2018. The role of functional amyloids in bacterial virulence. *J Mol Biol* **430**:3657–3684.

14. Serra DO, Klauck G, Hengge R. 2015. Vertical stratification of matrix production is essential for physical integrity and architecture of macrocolony biofilms of *Escherichia coli*. *Environ Microbiol* **17**:5073–5088.

15. Bian Z, Normark S. 1997. Nucleator function of CsgB for the assembly of adhesive surface organelles in *Escherichia coli*. *EMBO J* **16**:5827–5836.

16. Chapman MR, Robinson LS, Pinkner JS, Roth R, Heuser J, Hammar M, Normark S, Hultgren SJ. 2002. Role of *Escherichia coli* curli operons in directing amyloid fiber formation. *Science* **295**:851–855.

17. Hammar M, Arnqvist A, Bian Z, Olsen A, Normark S. 1995. Expression of two *csg* operons is required for production of fibronectin- and Congo red-binding curli polymers

in *Escherichia coli* K-12. *Mol Microbiol* **18**:661–670.

18. Van Gerven N, Klein RD, Hultgren SJ, Remaut H. 2015. Bacterial amyloid formation: structural insights into curli biogensis [*sic*]. *Trends Microbiol* **23**:693–706.

19. Evans ML, Chorell E, Taylor JD, Åden J, Götheson A, Li F, Koch M, Sefer L, Matthews SJ, Wittung-Stafshede P, Almqvist F, Chapman MR. 2015. The bacterial curli system possesses a potent and selective inhibitor of amyloid formation. *Mol Cell* **57**:445–455.

20. Robinson LS, Ashman EM, Hultgren SJ, Chapman MR. 2006. Secretion of curli fibre subunits is mediated by the outer membrane-localized CsgG protein. *Mol Microbiol* **59**:870–881.

21. Nenninger AA, Robinson LS, Hultgren SJ. 2009. Localized and efficient curli nucleation requires the chaperone-like amyloid assembly protein CsgF. *Proc Natl Acad Sci U S A* **106**:900–905.

22. Nenninger AA, Robinson LS, Hammer ND, Epstein EA, Badtke MP, Hultgren SJ, Chapman MR. 2011. CsgE is a curli secretion specificity factor that prevents amyloid fibre aggregation. *Mol Microbiol* **81**:486–499.

23. Prigent-Combaret C, Brombacher E, Vidal O, Ambert A, Lejeune P, Landini P, Dorel C. 2001. Complex regulatory network controls initial adhesion and biofilm formation in *Escherichia coli* via regulation of the *csgD* gene. *J Bacteriol* **183**:7213–7223.

24. Goyal P, Krasteva PV, Van Gerven N, Gubellini F, Van den Broeck I, Troupiotis-Tsaïlaki A, Jonckheere W, Péhau-Arnaudet G, Pinkner JS, Chapman MR, Hultgren SJ, Howorka S, Fronzes R, Remaut H. 2014. Structural and mechanistic insights into the bacterial amyloid secretion channel CsgG. *Nature* **516**:250–253.

25. Cao B, Zhao Y, Kou Y, Ni D, Zhang XC, Huang Y. 2014. Structure of the nonameric bacterial amyloid secretion channel. *Proc Natl Acad Sci U S A* **111**:E5439–E5444.

26. Shu Q, Krezel AM, Cusumano ZT, Pinkner JS, Klein R, Hultgren SJ, Frieden C. 2016. Solution NMR structure of CsgE: structural insights into a chaperone and regulator protein important for functional amyloid formation. *Proc Natl Acad Sci U S A* **113**:7130–7135.

27. Klein RD, Shu Q, Cusumano ZT, Nagamatsu K, Gualberto NC, Lynch AJL, Wu C, Wang W, Jain N, Pinkner JS, Amarasinghe GK, Hultgren SJ, Frieden C, Chapman MR. 2018. Structure-function analysis of the curli accessory protein CsgE defines surfaces essential

for coordinating amyloid fiber formation. *mBio* **9**:e01349-18.

28. **Van Gerven N, Goyal P, Vandenbussche G, De Kerpel M, Jonckheere W, De Greve H, Remaut H.** 2014. Secretion and functional display of fusion proteins through the curli biogenesis pathway. *Mol Microbiol* **91**:1022–1035.

29. **Hammar M, Bian Z, Normark S.** 1996. Nucleator-dependent intercellular assembly of adhesive curli organelles in *Escherichia coli. Proc Natl Acad Sci U S A* **93**:6562–6566.

30. **Hammer ND, McGuffie BA, Zhou Y, Badtke MP, Reinke AA, Brännström K, Gestwicki JE, Olofsson A, Almqvist F, Chapman MR.** 2012. The C-terminal repeating units of CsgB direct bacterial functional amyloid nucleation. *J Mol Biol* **422**:376–389.

31. **Hammer ND, Schmidt JC, Chapman MR.** 2007. The curli nucleator protein, CsgB, contains an amyloidogenic domain that directs CsgA polymerization. *Proc Natl Acad Sci U S A* **104**:12494–12499.

32. **Schubeis T, Spehr J, Viereck J, Köpping L, Nagaraj M, Ahmed M, Ritter C.** 2018. Structural and functional characterization of the curli adaptor protein CsgF. *FEBS Lett* **592**:1020–1029.

33. **Wickner W, Schekman R.** 2005. Protein translocation across biological membranes. *Science* **310**:1452–1456.

34. **Costa TRD, Felisberto-Rodrigues C, Meir A, Prevost MS, Redzej A, Trokter M, Waksman G.** 2015. Secretion systems in Gram-negative bacteria: structural and mechanistic insights. *Nat Rev Microbiol* **13**:343–359.

35. **Zavialov AV, Tischenko VM, Fooks LJ, Brandsdal BO, Aqvist J, Zav'yalov VP, Macintyre S, Knight SD.** 2005. Resolving the energy paradox of chaperone/usher-mediated fibre assembly. *Biochem J* **389**:685–694.

36. **Bernstein HD.** 2015. Looks can be deceiving: recent insights into the mechanism of protein secretion by the autotransporter pathway. *Mol Microbiol* **97**:205–215.

37. **Van den Broeck I, Goyal P, Remaut H.** 2015. Insights in peptide diffusion channels from the bacterial amyloid secretor CsgG. *Channels (Austin)* **9**:65–67.

38. **Wang X, Hammer ND, Chapman MR.** 2008. The molecular basis of functional bacterial amyloid polymerization and nucleation. *J Biol Chem* **283**:21530–21539.

39. **Debenedictis EP, Ma D, Keten S.** 2017. Structural predictions for curli amyloid fibril subunits CsgA and CsgB. *RSC Adv* **7**:48102–48112.

40. **Shewmaker F, McGlinchey RP, Thurber KR, McPhie P, Dyda F, Tycko R, Wickner RB.** 2009. The functional curli amyloid is not based on in-register parallel beta-sheet structure. *J Biol Chem* **284**:25065–25076.

41. **Louros NN, Bolas GMP, Tsiolaki PL, Hamodrakas SJ, Iconomidou VA.** 2016. Intrinsic aggregation propensity of the CsgB nucleator protein is crucial for curli fiber formation. *J Struct Biol* **195**:179–189.

42. **Tian P, Boomsma W, Wang Y, Otzen DE, Jensen MH, Lindorff-Larsen K.** 2015. Structure of a functional amyloid protein subunit computed using sequence variation. *J Am Chem Soc* **137**:22–25.

43. **Biancalana M, Koide S.** 2010. Molecular mechanism of thioflavin-T binding to amyloid fibrils. *Biochim Biophys Acta* **1804**:1405–1412.

44. **Arosio P, Knowles TPJ, Linse S.** 2015. On the lag phase in amyloid fibril formation. *Phys Chem Chem Phys* **17**:7606–7618.

45. **Sleutel M, Van den Broeck I, Van Gerven N, Feuillie C, Jonckheere W, Valotteau C, Dufrêne YF, Remaut H.** 2017. Nucleation and growth of a bacterial functional amyloid at single-fiber resolution. *Nat Chem Biol* **13**:902–908.

46. **Wang X, Chapman MR.** 2008. Sequence determinants of bacterial amyloid formation. *J Mol Biol* **380**:570–580.

47. **Wang X, Zhou Y, Ren J-J, Hammer ND, Chapman MR.** 2010. Gatekeeper residues in the major curlin subunit modulate bacterial amyloid fiber biogenesis. *Proc Natl Acad Sci U S A* **107**:163–168.

48. **Arnqvist A, Olsén A, Normark S.** 1994. σ^S-dependent growth-phase induction of the *csgBA* promoter in *Escherichia coli* can be achieved *in vivo* by σ70 in the absence of the nucleoid-associated protein H-NS. *Mol Microbiol* **13**:1021–1032.

49. **Brown PK, Dozois CM, Nickerson CA, Zuppardo A, Terlonge J, Curtiss R III.** 2001. MlrA, a novel regulator of curli (AgF) and extracellular matrix synthesis by *Escherichia coli* and *Salmonella enterica* serovar Typhimurium. *Mol Microbiol* **41**:349–363.

50. **Gerstel U, Park C, Römling U.** 2003. Complex regulation of *csgD* promoter activity by global regulatory proteins. *Mol Microbiol* **49**:639–654.

51. **Gerstel U, Römling U.** 2003. The *csgD* promoter, a control unit for biofilm formation in *Salmonella typhimurium. Res Microbiol* **154**:659–667.

52. **Olsén A, Arnqvist A, Hammar M, Sukupolvi S, Normark S.** 1993. The RpoS sigma factor relieves H-NS-mediated transcriptional repression of *csgA*, the subunit gene of fibronectin-binding curli in *Escherichia coli. Mol Microbiol* **7**:523–536.

53. **Chiti F, Dobson CM.** 2017. Protein misfolding, amyloid formation, and human disease: a summary of progress over the last decade. *Annu Rev Biochem* **86:**27–68.

54. **Taylor JD, Hawthorne WJ, Lo J, Dear A, Jain N, Meisl G, Andreasen M, Fletcher C, Koch M, Darvill N, Scull N, Escalera-Maurer A, Sefer L, Wenman R, Lambert S, Jean J, Xu Y, Turner B, Kazarian SG, Chapman MR, Bubeck D, de Simone A, Knowles TPJ, Matthews SJ.** 2016. Electrostatically-guided inhibition of curli amyloid nucleation by the CsgC-like family of chaperones. *Sci Rep* **6:**24656.

55. **Cegelski L, Pinkner JS, Hammer ND, Cusumano CK, Hung CS, Chorell E, Aberg V, Walker JN, Seed PC, Almqvist F, Chapman MR, Hultgren SJ.** 2009. Small-molecule inhibitors target *Escherichia coli* amyloid biogenesis and biofilm formation. *Nat Chem Biol* **5:**913–919.

56. **Andersson EK, Bengtsson C, Evans ML, Chorell E, Sellstedt M, Lindgren AEG,** **Hufnagel DA, Bhattacharya M, Tessier PM, Wittung-Stafshede P, Almqvist F, Chapman MR.** 2013. Modulation of curli assembly and pellicle biofilm formation by chemical and protein chaperones. *Chem Biol* **20:**1245–1254.

57. **Dueholm MS, Petersen SV, Sønderkær M, Larsen P, Christiansen G, Hein KL, Enghild JJ, Nielsen JL, Nielsen KL, Nielsen PH, Otzen DE.** 2010. Functional amyloid in *Pseudomonas*. *Mol Microbiol* **77:**1009–1020.

58. **Dueholm MS, Søndergaard MT, Nilsson M, Christiansen G, Stensballe A, Overgaard MT, Givskov M, Tolker-Nielsen T, Otzen DE, Nielsen PH.** 2013. Expression of Fap amyloids in *Pseudomonas aeruginosa, P. fluorescens*, and *P. putida* results in aggregation and increased biofilm formation. *Microbiologyopen* **2:**365–382.

59. **Rouse SL, Stylianou F, Wu HYG, Berry JL, Sewell L, Morgan RML, Sauerwein AC, Matthews S.** 2018. The FapF amyloid secretion transporter possesses an atypical asymmetric coiled coil. *J Mol Biol* **430:**3863–3871.

Sortases, Surface Proteins, and Their Roles in *Staphylococcus aureus* Disease and Vaccine Development

15

OLAF SCHNEEWIND[1] and DOMINIQUE MISSIAKAS[1]

INTRODUCTION

Prior to bacterial genome sequencing and the genetic analysis of pathogenesis, microbiologists identified molecules on microbial surfaces and studied their role in disease processes (1). The ultimate goal of this research was the identification of molecular formulations inciting antibody responses in vaccine recipients that prevented disease yet would otherwise not cause harm (2). Oswald Avery's discovery of the pneumococcus capsule and the demonstration that capsular polysaccharide vaccine protects against pneumococcal pneumonia represent an important paradigm (3, 4). Another was Rebecca Lancefield's characterization of M protein as the determinant of type-specific immunity against *Streptococcus pyogenes*, the causative agent of streptococcal pharyngitis and rheumatic fever (2). Lancefield and Sjöquist required proteases or peptidoglycan (murein) hydrolases, but not membrane detergents, to solubilize surface proteins of Gram-positive bacteria (2, 5, 6). The underlying reason for this biochemical phenomenon is that surface proteins are covalently linked to peptidoglycan at their C-terminal ends (7, 8).

Whole-genome sequencing enabled bioinformatic studies providing rapid answers about the universality of genetic traits among pathogens or about

[1]Department of Microbiology, University of Chicago, Chicago, IL 60637

Protein Secretion in Bacteria
Edited by Maria Sandkvist, Eric Cascales, and Peter J. Christie
© 2019 American Society for Microbiology, Washington, DC
doi:10.1128/microbiolspec.PSIB-0004-2018

sequence variation in response to host adaptive immune (antibody) responses (9). While bioinformatic analyses have had tremendous impact on supporting or refuting hypotheses about surface proteins in Gram-positive bacteria, experimental work represents the bedrock for hypothesis testing and for the alignment of arguments supporting bacterial vaccine development.

STAPHYLOCOCCAL SORTASES AND THEIR SURFACE PROTEIN SUBSTRATES

Surface proteins of *S. aureus* are amide linked to the pentaglycine cross bridge of the bacterial cell wall via their C-terminal threonine (T) residue (8). Precursors of staphylococcal surface proteins are synthesized in the bacterial cytoplasm with N-terminal signal peptides for Sec-mediated secretion and C-terminal LPXTG motif sorting signals that promote cell wall anchoring (Fig. 1A) (10). Sortase A, a type II membrane protein (N-terminal membrane anchor), cleaves the LPXTG motif of the sorting signal between its T and glycine (G) residues to form a thioester-linked acyl enzyme intermediate with its active-site cysteine thiol (11, 12) (Fig. 1B). The acyl enzyme is relieved by the nucleophilic attack of the amino group of the pentaglycine cross bridge within lipid II, the precursor to peptidoglycan biosynthesis (13, 14) (Fig. 1B). Surface protein linked to lipid II is subsequently incorporated into the cell wall envelope via the transglycosylation and transpeptidation reactions of bacterial cell wall synthesis (15–18) (Fig. 1B). *S. aureus srtA* (sortase A) mutants cannot assemble surface proteins into the cell wall envelope (19). The mechanism of action of *S. aureus* sortase A was validated for *Listeria monocytogenes* and *Bacillus anthracis* (20–22) and is considered to be universal in Gram-positive bacteria (23).

Genome sequences of all clinical *S. aureus* isolates harbor two sortase genes, *srtA* and *srtB*; however, the number of surface protein genes is variable (Table 1) (24–26). Sortase A

substrates bear the LPXTG motif sorting signal at the C-terminal end (Table 1) (27). Sortase B cleaves the NPQTN sorting signal of IsdC (iron-regulated surface determinant C), a protein that is linked to the cell wall when staphylococci are grown under iron starvation conditions, as occurs during host invasion (28). Several sortase A substrates have been described as microbial surface components recognizing adherence matrix molecules (MSCRAMMs) (29). These include ClfA, ClfB, Cna, FnbpA, FnbpB, and presumably also Pls, SraP, SasG, SrdC, and SdrD, although the identity of surface protein ligands in the latter group of proteins remains unclear (Table 1). Each MSCRAMM represents a mosaic of modular domains (30, 31). A surface-exposed, N-terminal A domain is generally endowed with ligand-binding activity. Repeat structural modules allow MSCRAMMs to span the thick peptidoglycan layer of staphylococci (30, 31). ClfA, ClfB, SrdC, SdrD, SdrE, Pls, and SraP each encompass extensively glycosylated serine-aspartate (SD) repeat domains (32–34) (Table 1).

The *srtB* and *isdC* genes are located in the *isd* locus, which also encodes sortase A-anchored products IsdA and IsdB, the membrane transporter IsdEF, and the cytoplasmic protein IsdG (35). The structural gene for sortase A-anchored IsdH is located outside of the *isd* locus (36). IsdB and IsdH function as hemophores to remove heme iron from hemoglobin and haptoglobin when hemoproteins are released from lysed host cells (36–39). IsdH competes with macrophage receptor CD163, the host recycling system for free hemoglobin, for the capture of heme from haptoglobin-hemoglobin (40). Bound heme iron is transferred from the NEAT (near-iron transporter) domains of IsdB or IsdH to the NEAT domain of IsdA for subsequent passage across the cell wall to IsdC and IsdEF-mediated import across the membrane (35). IsdG and its paralog IsdI cleave the tetrapyrrole ring of heme iron to liberate iron as a bacterial nutrient and enzyme cofactor (37, 41, 42). The sortase B-

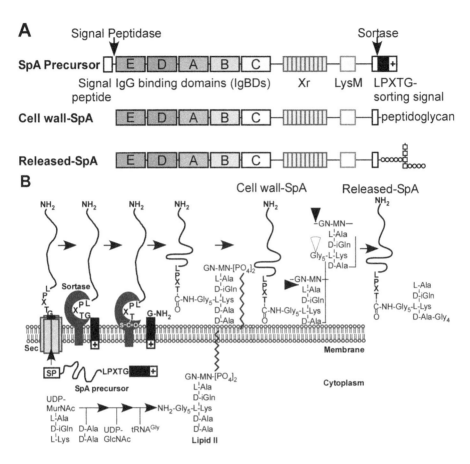

FIGURE 1 **Sortase-mediated anchoring to the cell wall envelope of *Staphylococcus aureus* using SpA as a model substrate. (A) Drawing to illustrate the primary structure of the SpA precursor with its N-terminal signal peptide and signal peptidase cleavage site, the five immunoglobulin binding domains (IgBDs), region X (Xr) LysM domain, and C-terminal LPXTG motif sorting signal with cleavage site for sortase A. Cell wall SpA is linked to peptidoglycan via an amide bond between the carboxyl group of the C-terminal threonine and the amino group of the pentaglycine cross bridge. Released SpA is liberated from the cell wall envelope via the action of several murein hydrolases. (B) Drawing to illustrate *S. aureus* secretion of SpA precursor, sortase-mediated cleavage of SpA precursor and acyl enzyme formation, resolution of the acyl enzyme by lipid II to generate SpA linked to lipid II, incorporation of SpA into the cell wall via the transpeptidation and transglycosylation reaction, and release of SpA from the cell wall envelope by murein hydrolases. Released SpA bears the overall structure L-Ala–D-iGln–L-Lys(SpA-LPET-Gly$_5$)–D-AlaGly$_4$.**

IsdC acyl enzyme intermediate is resolved by the nucleophilic attack of assembled peptidoglycan instead of lipid II (43). This mechanism ensures that IsdC is attached to peptidoglycan in the vicinity to the IsdEF membrane transporter, whereas IsdA and IsdB are deposited across the peptidoglycan layer (44).

SORTASES AND SURFACE PROTEIN CONTRIBUTIONS TO *S. AUREUS* COLONIZATION AND DISEASE PATHOGENESIS

S. aureus srtA mutants cannot colonize the nasopharynx and gastrointestinal tract of

TABLE 1 *Staphylococcus aureus* cell wall-anchored surface proteins[a]

Protein	Name(s)	GenBank accession number	aa[b]	Ligand(s)[c]	YSIRK motif[d]	Sorting motif[e]	Reference(s)
Sortase A-anchored proteins							
Adenosine synthase A	AdsA (SasH)	ABD22278.1	772	Adenosine and dAdo synthesis	No	LPKTG	106, 108
Clumping factor A	ClfA	ABD20644.1	933	Fibrinogen (γ chain) factor I	Yes	LPDTG	144, 145
Clumping factor B	ClfB	ABD21326.1	899	Fibrinogen (α chain), cytokeratins 8 and 10, loricrin	Yes	LPETG	97–102
Collagen adhesin	Cna	BAF45800.1	1,183	CollagenC1q	No	LPKTG	146, 147
Factor affecting methicillin resistance in Triton X-100 B	FmtB (SasB)	ATC68490.1	2,478	Unknown	Yes	LPDTG	148
Fibronectin binding protein A	FnbpA	ABD21634.1	1,018	Fibronectin, fibrinogen (γ chain), elastin	Yes	LPETG	30
Fibronectin binding protein B	FnbpB	ABD22827.1	940	Fibronectin, fibrinogen (α chain), elastin	Yes	LPETG	30
Iron-regulated surface determinant A	IsdA (SasE)	ABD21627.1	350	Heme transferred from IsdB/H	No	LPKTG	35
Iron-regulated surface determinant B	IsdB (SasJ)	ABD21843.1	645	Hemoglobin heme	Yes	LPQTG	36–39
Iron-regulated surface determinant H	IsdH (SasI/HarA)	ABD20516.1	895	Haptoglobin-hemoglobin heme	Yes	LPKTG	36–40
Plasmin-sensitive surface protein	Pls	AAD09131.1	1,637	Unknown	Yes	LPDTG	149, 150
S. aureus surface protein C	SasC	ABD21355.1	2,186	Promotes intercellular adhesion	Yes	LPNTG	151
S. aureus surface protein D	SasD	ABD21427.1	241	Unknown	No	LPAAG	
S. aureus surface protein F	SasF	ABD21199.1	635	Unknown	No	LPKAG	
S. aureus surface protein G	SasG	BAU36055.1	1,115	Unknown	Yes	LPKTG	
S. aureus surface protein K	SasK	ADC38744.1	211	Unknown	No	LPKTG	
Serine aspartic repeat protein C	SdrC	ABD21592.1	947	β-Neurexin, homophylic bonds	Yes	LPETG	152, 153
Serine aspartic repeat protein D	SdrD	ABD20874.1	1,381	Desmoglein 1	Yes	LPETG	154
Serine aspartic repeat protein E	SdrE	ABD22410.1	1,154	Factor H	Yes	LPETG	155
S. aureus protein A	SpA	ABD22331.1	508	Immunoglobulin (Fcγ, Fab V_H3)	Yes	LPETG	70, 71, 156, 157
Serine-rich adhesin for platelets	SraP (SasA)	ABD21900.1	2,271	Salivary agglutinin (gp340)	Possibly	LPDTG	34, 158
Sortase B-anchored protein							
Iron-regulated surface determinant C	IsdC	ABD20415.1	227	Heme transferred from IsdA	No	NPQTN	28

[a]The numbers of cell wall-anchored surface proteins vary among strains of *S. aureus* (26). For example, in strain *Staphylococcus aureus* subsp. *aureus* USA300_FPR3757, genes for Cna, SasK, and Pls are missing; the presence of stop codons results in truncated FmtB (SasB), SasC, and SasG products.

[b]aa, protein length in amino acids.

[c]Molecular component(s) recognized and bound by protein, or molecules synthesized in case of AdsA.

[d]Consensus motif found in some signal sequences which presumably accounts for secretion of proteins at the cross walls (62).

[e]Consensus motif recognized by sortases and present in C-terminal cell wall sorting signal.

mice (45, 46). Further, staphylococcal *srtA* mutants cannot form abscess lesions or survive in mouse tissues (19, 47). Following intravenous *S. aureus* inoculation to precipitate lethal bacteremia in mice or guinea pigs, *srtA* mutants are avirulent and cannot cause disease (48, 49). In the mouse skin abscess lesion and pneumonia models, *S. aureus srtA* mutants display smaller reductions in virulence. We attribute the smaller phenotypic defects to the models' requirements for large bacterial inocula and α-hemolysin secretion (50–52). *S. aureus srtB* mutants exhibit small but significant reductions in virulence in the mouse renal abscess, bloodstream, and infectious arthritis models; these defects are additive with those of sortase A mutants (53).

Cheng and coworkers (47) isolated *S. aureus* Newman mutants with insertional lesions in any one gene encoding LPXTG motif surface proteins. Unlike *srtA* variants, all mutants retained the ability to cause renal abscess lesions and lethal bacteremia in mice (47, 48). However, loss of *spa* (staphylococcal protein A), *isdA*, and *isdB* resulted in significant reductions in the number of abscess lesions (47). Mutations in the genes for clumping factor A (*clfA*) or adenosine synthase A (*adsA*) caused significant delays in time to death in the murine model for *S. aureus* bacteremia (48). When analyzed with human nasal epithelial cells, cotton rats, or mice as models for *S. aureus* colonization, *srtA* mutants have been shown to be unable to colonize the nasopharynx and gastrointestinal tract (54–56). In these models, clumping factor B (ClfB) and IsdA stand out as key contributors to *S. aureus* colonization (55, 57, 58). Thus, compared to any other virulence gene, *srtA* mutations exhibit the largest reduction in the ability of *S. aureus* to colonize and invade its hosts. Further, the sortase substrates AdsA, ClfA, ClfB, IsdA, IsdB, and SpA make important, nonredundant contributions towards colonization, invasion of host tissues, or the establishment of abscess lesions.

Staphylococcal Protein A

All clinical *S. aureus* isolates harbor the *spa* gene, which generates a precursor comprised of an N-terminal YSIRK/GXXS signal peptide, followed by 4 or 5 immunoglobulin binding domains (IgBDs), the region X repeats (Xr), LysM domain, and LPXTG sorting signal (23, 59, 60) (Fig. 1). SpA precursors enter the secretory pathway at septal membranes via their YSIRK/GXXS signal peptide (61–63). Once SpA is deposited into the cross wall, septal peptidoglycan is split and the cross wall assumes one-half of the spherical surface of *S. aureus* cells (61, 63). Staphylococci divide perpendicular to previous cell division planes, resulting in rapid SpA distribution over the entire bacterial surface (61). During cell division, dedicated murein hydrolases release SpA molecules from the peptidoglycan (64, 65). SpA linked to cell wall peptide fragments is thereby released into host tissues (66) (Fig. 1). Released SpA activates V_H3 idiotype B cell receptors (BCRs) and promotes IgG and IgM secretion in activated plasmablasts (67, 68) (Fig. 2A). When displayed in the bacterial envelope, SpA binds to Fcγ, i.e., the effector domain of IgG, and protects staphylococci from opsonophagocytic killing by immune cells (49, 69) (Fig. 2A). The five IgBDs of SpA each bind to Fcγ of human (IgG1, IgG2, and IgG4) and mouse (IgG1, IgG2a to -c, and IgG3) IgG (70, 71) (Fig. 2B). Each IgBD also binds V_H3 heavy chains of human and mouse immunoglobulin, including IgM (BCRs), IgG, IgE, IgD, and IgA (49, 68, 69, 72, 73) (Fig. 2B). Thus, released SpA functions as a B cell superantigen that promotes systemic production of V_H3-clonal IgG and IgM antibodies that do not recognize staphylococcal antigens, thereby preventing the development of pathogen-specific antibodies and the establishment of protective immunity (49, 67, 68). In spite of the B cell superantigen activity of SpA, *S. aureus* colonization and invasive disease in humans are associated with the development of antibody responses against some staphylococcal anti-

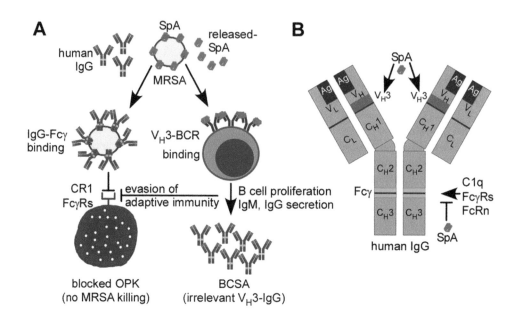

FIGURE 2 Biological functions of staphylococcal protein A (SpA). (A) *Staphylococcus aureus* and its antibiotic-resistant isolates (MRSA) harbor SpA in the cell wall envelope or released into the extracellular milieu (released SpA). Cell wall SpA binds Fcγ of human and animal IgG (green segment within blue IgG) and blocks the effector functions of antibodies, thereby preventing opsonophagocytic killing (OPK) of MRSA by immune cells through interference with complement (CR1) and Fcγ receptors (FcγRs). Released SpA cross-links V_H3-clonal B cell receptors (V_H3-BCR on the surface of B cells), triggering B cell proliferation and secretion of V_H3-clonal IgM and IgG (pink segments within blue IgG) without antigen specificity for *S. aureus*. This B cell superantigen activity (BCSA) of SpA produces irrelevant V_H3-clonal IgG and prevents the establishment of protective immunity against *S. aureus*. (B) Drawing to illustrate the primary structure of human IgG with variable (V_L and V_H) and conserved (C_L, C_H1, C_H2, and C_H3) light (L) and heavy (H) chains, their antigen-binding paratope (Ag), V_H3, and Fcγ domains. SpA binding sites at V_H3 heavy chains and Fcγ are in pink and green, respectively.

gens, predominantly serum IgG4 (74–76). These antibodies are, however, not protective and cannot promote opsonophagocytic killing because they are captured by cell wall-anchored SpA (71, 77–80).

Clumping factors A and B

Vascular damage triggers blood coagulation, a process whereby soluble fibrinogen, a 340-kDa dimer of trimers (α, β, and γ chains), is converted to insoluble fibrin following cleavage of fibrinopeptides A and B from the α and γ chains by thrombin; the prothrombinase complex Va/Xa is responsible for the conversion of prothrombin (PT) to active thrombin (81–83). The hemostatic system also immobilizes microbial invaders for

destruction by the immune system (84). However, this does not occur with *S. aureus*. All clinical *S. aureus* isolates clot human or animal blood even in the presence of coagulation inhibitors (85). Coagulation is promoted by secreted coagulase (Coa) and von Willebrand factor binding protein (vWbp) bound to PT (86). Coa-PT and vWbp-PT complexes cleave the A and B fibrinopeptides of fibrinogen but do not cut any of the other thrombin substrates (FV, FVIII, FXI, FXIII, protein C, antithrombin, and plasmin) (87). ClfA triggers *S. aureus* agglutination by binding to the C-terminal end of the fibrinogen γ chain (residues 395 to 411), effectively capping and tethering Coa-PT- and vWbp-PT-polymerized fibrin cables to the staphylococcal surface (48). ClfA, the prototypical MSCRAMM, is comprised of an N-terminal A domain with N1, N2, and N3 subdomains, an EF-hand like calcium binding module, and the SD repeat domain with 154 tandem seryl-aspartyl repeats (88). The N2 and N3 domains of ClfA (residues 229 to 545) assume immunoglobulin-like folds and bind their fibrin/fibrinogen ligand via the "dock, lock, and latch" mechanism (89–93). This interaction prevents further binding between fibrin/fibrinogen and the platelet integrin $\alpha_{IIb}\beta_3$ (94, 95). Thus, in addition to binding fibrinogen, ClfA functions as an inhibitor of platelet-fibrin clots. ClfB, which is also conserved among *S. aureus* isolates, represents a homolog of ClfA. The A domains of the two proteins are 26% identical (96), and both proteins use YSIRK/GXXS signal peptides, glycosylated SD repeats, and LPXTG motif sequences as topogenic elements (32, 62). ClfB binds to several host proteins, including the Aα chain of fibrinogen (97, 98), cytokeratin 8 (99), cytokeratin 10 (100, 101), and loricrin (102) (Table 1). These mammalian proteins harbor a motif sequence, GSSGXG, that represents the binding site for ClfB (103) and contributes to *S. aureus* colonization of the nasopharynx of mice (102).

Adenosine Synthase A

S. aureus abscess lesions are composed of a bacterial nidus, the staphylococcal abscess community (SAC), encased within a pseudo-capsule of fibrin and surrounded by layers of immune cells (86, 104). In spite of large numbers of infiltrated neutrophils, mice are unable to eliminate staphylococci from abscess lesions and eventually succumb to the persistent infection (47). Although neutrophils use NETosis (extracellular DNA) to entangle staphylococci, NETs are degraded by staphylococcal nuclease and thereby fail to exert bactericidal activities (105). Nuclease digestion of NETs releases 5′ and 3′ monophosphate nucleotides that are converted by *S. aureus* AdsA into deoxyadenosine (dAdo) (106). AdsA-mediated dAdo production triggers caspase-3-induced apoptosis of mouse and human macrophages and prevents phagocyte entry into the SAC (106). Human equilibrative nucleoside transporter 1 is responsible for the uptake of dAdo in phagocytes (107). Conversion of dAdo to dAMP is catalyzed by deoxycytidine kinase and adenosine kinase, and the subsequent formation of dATP triggers caspase-3-induced cell death (107). AdsA also converts adenosine nucleosides and nucleotides released during host cell lysis into adenosine, which binds adenosine receptors and triggers host immune suppression during bloodstream infection (108, 109).

USING SORTASES AND SURFACE PROTEINS FOR VACCINE DEVELOPMENT

The contribution of sortases towards *S. aureus* colonization and invasive disease provoked interest in surface proteins as vaccine antigens. Purified recombinant ClfA (A domain) generates antibodies that neutralize ClfA binding to fibrin(ogen) and provide partial protection against lethal bloodstream infection and infectious arthritis in mice (110). Anti-ClfA mouse hybridoma antibody or its cloned humanized variant tefibazumab binds to the ClfA N3 domain, inhibits fibrinogen binding (111, 112), and provides partial protection against lethal bloodstream infection in mice (113). Administration of clinical-grade tefibazumab was safe in healthy human volunteers and in patients with methicillin-resistant *S. aureus* (MRSA) bacteremia but could not improve the clinical outcomes for these patients (114). Using ClfA-immunized VelocImmune mice, MedImmune investigators isolated monoclonal antibody 11H10, with inhibitory activity for ClfA binding to fibrinogen (115). Human 11H10 IgG1 promotes MRSA opsonophagocytic killing with differentiated HL-60 neutrophils (115) and increases the survival of mice with lethal MRSA bloodstream infection (116, 117). MedImmune seeks to develop 11H10 IgG1 in conjunction with monoclonal antibody against α-hemolysin to improve the outcome of patients with ventilator-associated pneumonia and other invasive diseases (115). Pfizer developed SA4Ag, a multicomponent vaccine composed of ClfA, capsular polysaccharide type 5 and 8 conjugates, and manganese transporter C (118). SA4Ag is currently undergoing clinical efficacy evaluation in patients with instrumented posterior spinal fusion to protect against *S. aureus* surgical site and bloodstream infections (119).

Purified IsdB elicits antibodies that block heme iron scavenging and provide partial protection against *S. aureus* bacteremia in preclinical models (120–122). IsdB-specific antibodies may also promote opsonophagocytosis of *S. aureus* (121, 123). In a phase 3 clinical trial, IsdB (V710) immunization did not protect thoracic surgery patients from *S. aureus* surgical site infections (124). V710 immunization increased the risk for fatal *S. aureus* bacteremia 5-fold over the control cohort; the molecular basis for this safety concern is not known (124).

Humans and mice cannot generate antibodies against the IgBDs of SpA; however, SpA variants, engineered to exhibit reduced immunoglobulin binding, elicit SpA-neutralizing antibody responses (73). Animals with SpA-neutralizing antibodies exhibit dramatic increases in pathogen-specific antibody responses during colonization or invasive disease (46, 49, 69, 73). In fact, the corresponding SpA vaccine can protect against *S. aureus* colonization, renal abscess formation, and lethal bloodstream infection (46, 49, 69, 73). Similarly, SpA-neutralizing monoclonal antibody protects against *S. aureus* colonization and invasive disease in mice (125, 126). SpA vaccines have not yet been subjected to clinical testing.

SORTASE INHIBITORS

The complete transpeptidation reaction that is carried out by sortases can be recapitulated *in vitro* (12, 14, 127). However, most screens for sortase inhibitors have been conducted with assays measuring SrtA cleavage of LPXTG peptide (128). These inhibitors are generally not active *in vivo*, suggesting that in the envelope of *S. aureus*, sortase A may predominantly exist as an acyl enzyme (129). Other inhibitors can block sortase A activity *in vivo*, and such compounds abolish surface protein anchoring to the cell wall envelope of *S. aureus* and protect animals against lethal bloodstream infection (130, 131). Of note, sortase inhibitors may be useful for the prevention of *S. aureus* disease, as they can be expected to block colonization and invasion. Owing to the fact that the compounds cannot kill *S. aureus*, sortase inhibitors are unlikely to exhibit a therapeutic effect in individuals with active infectious disease (130).

SORTASES IN OTHER PATHOGENIC MICROBES

Gram-positive bacteria often harbor homologs of staphylococcal sortase A or class A sortases; only some microbes express sortase B homologs or class B sortases (132, 133). Based on structural features and substrate specificity, sortase homologs have been classified into six distinct classes, A to F (134). Among bacterial pathogens, *Corynebacterium diphtheriae* and *Bacillus anthracis* harbor class C sortase genes, which are clustered with surface protein genes containing LPXTG- and motif-specific sorting signals (135, 136). These genes encode pilus components that include adhesin and pilin subunits. Class C sortases link adhesin and pilin subunits together to construct a pilus (135–139). Class C sortases cleave the LPXTG motif of pilins to form acyl enzyme intermediates that are relieved by the nucleophilic attack of the ε-amino group of a conserved lysine (K) residue within the pilin motif of an incoming subunit (140–142). Pilin protomers are joined progressively to the pilus base; a housekeeping sortase terminates polymerization by transferring the whole structure to the peptidoglycan

(141, 143). For additional information on the different classes of sortases and their distribution among various phyla, the reader is referred to a recently published review (134).

In conclusion, sortases are ubiquitous in Gram-positive bacteria, anchoring proteins and pili to peptidoglycan via a conserved transpeptidation mechanism. Sortase-mediated attachment of virulence factors in *S. aureus* has stimulated searches for sortase inhibitors and protective antigens. These strategies may lead to the development of drugs that can prevent hospital-acquired infections or to protective vaccines that can prevent *S. aureus* colonization and/or invasive diseases.

ACKNOWLEDGMENTS

We thank laboratory members past and present for their contributions to the field of *S. aureus* sortases and surface proteins.

Work on *Staphylococcus aureus* in our laboratories is supported by grants AI038897, AI052474, and AI110937 from the National Institute of Allergy and Infectious Diseases.

We declare conflicts of interest as inventors of patents under commercial license for *S. aureus* vaccine development. We declare no further competing financial interests.

CITATION

Schneewind O, Missiakas D. 2019. Sortases, surface proteins, and their roles in *Staphylococcus aureus* disease and vaccine development. Microbiol Spectrum 7(1):PSIB-0004-2018.

REFERENCES

1. **Lancefield RC.** 1928. The antigenic complex of *Streptococcus hemolyticus*. I. Demonstration of a type-specific substance in extracts of *Streptococcus hemolyticus*. *J Exp Med* **47:**91–103.
2. **Lancefield RC.** 1962. Current knowledge of type-specific M antigens of group A streptococci. *J Immunol* **89:**307–313.
3. **Avery OT.** 1915. A further study on the biologic classification of pneumococci. *J Exp Med* **22:** 804–819.
4. **MacLeod CM, Hodges RG, Heidelberger M, Bernhard WG.** 1945. Prevention of pneumococcal pneumonia by immunization with specific capsular polysaccharides. *J Exp Med* **82:**445–465.
5. **Sjöquist J, Meloun B, Hjelm H.** 1972. Protein A isolated from *Staphylococcus aureus* after digestion with lysostaphin. *Eur J Biochem* **29:**572–578.
6. **Fischetti VA.** 1989. Streptococcal M protein: molecular design and biological behavior. *Clin Microbiol Rev* **2:**285–314.
7. **Schneewind O, Fowler A, Faull KF.** 1995. Structure of the cell wall anchor of surface proteins in *Staphylococcus aureus*. *Science* **268:**103–106.
8. **Marraffini LA, Dedent AC, Schneewind O.** 2006. Sortases and the art of anchoring proteins to the envelopes of gram-positive bacteria. *Microbiol Mol Biol Rev* **70:**192–221.
9. **Musser JM, Shelburne SA III.** 2009. A decade of molecular pathogenomic analysis of group A *Streptococcus*. *J Clin Invest* **119:**2455–2463.
10. **Schneewind O, Model P, Fischetti VA.** 1992. Sorting of protein A to the staphylococcal cell wall. *Cell* **70:**267–281.
11. **Mazmanian SK, Liu G, Ton-That H, Schneewind O.** 1999. *Staphylococcus aureus* sortase, an enzyme that anchors surface proteins to the cell wall. *Science* **285:**760–763.
12. **Ton-That H, Liu G, Mazmanian SK, Faull KF, Schneewind O.** 1999. Purification and characterization of sortase, the transpeptidase that cleaves surface proteins of *Staphylococcus aureus* at the LPXTG motif. *Proc Natl Acad Sci USA* **96:**12424–12429.
13. **Perry AM, Ton-That H, Mazmanian SK, Schneewind O.** 2002. Anchoring of surface proteins to the cell wall of *Staphylococcus aureus*. III. Lipid II is an *in vivo* peptidoglycan substrate for sortase-catalyzed surface protein anchoring. *J Biol Chem* **277:**16241–16248.
14. **Ton-That H, Mazmanian SK, Faull KF, Schneewind O.** 2000. Anchoring of surface proteins to the cell wall of *Staphylococcus aureus*. Sortase catalyzed in vitro transpeptidation reaction using LPXTG peptide and NH(2)-Gly(3) substrates. *J Biol Chem* **275:** 9876–9881.
15. **Ton-That H, Labischinski H, Berger-Bächi B, Schneewind O.** 1998. Anchor structure of staphylococcal surface proteins. III. Role of the FemA, FemB, and FemX factors in anchoring surface proteins to the bacterial cell wall. *J Biol Chem* **273:**29143–29149.

16. **Ton-That H, Schneewind O.** 1999. Anchor structure of staphylococcal surface proteins. IV. Inhibitors of the cell wall sorting reaction. *J Biol Chem* **274:**24316–24320.

17. **Ton-That H, Faull KF, Schneewind O.** 1997. Anchor structure of staphylococcal surface proteins. A branched peptide that links the carboxyl terminus of proteins to the cell wall. *J Biol Chem* **272:**22285–22292.

18. **Navarre WW, Ton-That H, Faull KF, Schneewind O.** 1998. Anchor structure of staphylococcal surface proteins. II. COOH-terminal structure of muramidase and amidase-solubilized surface protein. *J Biol Chem* **273:** 29135–29142.

19. **Mazmanian SK, Liu G, Jensen ER, Lenoy E, Schneewind O.** 2000. *Staphylococcus aureus* sortase mutants defective in the display of surface proteins and in the pathogenesis of animal infections. *Proc Natl Acad Sci U S A* **97:**5510–5515.

20. **Dhar G, Faull KF, Schneewind O.** 2000. Anchor structure of cell wall surface proteins in *Listeria monocytogenes. Biochemistry* **39:**3725–3733.

21. **Bierne H, Mazmanian SK, Trost M, Pucciarelli MG, Liu G, Dehoux P, Jänsch L, Garcia-del Portillo F, Schneewind O, Cossart P, European Listeria Genome Consortium.** 2002. Inactivation of the *srtA* gene in *Listeria monocytogenes* inhibits anchoring of surface proteins and affects virulence. *Mol Microbiol* **43:**869–881.

22. **Gaspar AH, Marraffini LA, Glass EM, Debord KL, Ton-That H, Schneewind O.** 2005. *Bacillus anthracis* sortase A (SrtA) anchors LPXTG motif-containing surface proteins to the cell wall envelope. *J Bacteriol* **187:**4646–4655.

23. **Mazmanian SK, Ton-That H, Schneewind O.** 2001. Sortase-catalysed anchoring of surface proteins to the cell wall of *Staphylococcus aureus. Mol Microbiol* **40:**1049–1057.

24. **Baba T, Bae T, Schneewind O, Takeuchi F, Hiramatsu K.** 2008. Genome sequence of *Staphylococcus aureus* strain Newman and comparative analysis of staphylococcal genomes: polymorphism and evolution of two major pathogenicity islands. *J Bacteriol* **190:**300–310.

25. **Kuroda M, Ohta T, Uchiyama I, Baba T, Yuzawa H, Kobayashi I, Cui L, Oguchi A, Aoki K, Nagai Y, Lian J, Ito T, Kanamori M, Matsumaru H, Maruyama A, Murakami H, Hosoyama A, Mizutani-Ui Y, Takahashi NK, Sawano T, Inoue R, Kaito C, Sekimizu K, Hirakawa H, Kuhara S, Goto S, Yabuzaki J, Kanehisa M, Yamashita A, Oshima K, Furuya K, Yoshino C, Shiba T, Hattori M, Ogasawara N, Hayashi H, Hiramatsu K.** 2001. Whole genome sequencing of meticillin-resistant *Staphylococcus aureus. Lancet* **357:**1225–1240.

26. **McCarthy AJ, Lindsay JA.** 2010. Genetic variation in *Staphylococcus aureus* surface and immune evasion genes is lineage associated: implications for vaccine design and host-pathogen interactions. *BMC Microbiol* **10:**173.

27. **Schneewind O, Mihaylova-Petkov D, Model P.** 1993. Cell wall sorting signals in surface proteins of gram-positive bacteria. *EMBO J* **12:**4803–4811.

28. **Mazmanian SK, Ton-That H, Su K, Schneewind O.** 2002. An iron-regulated sortase anchors a class of surface protein during *Staphylococcus aureus* pathogenesis. *Proc Natl Acad Sci U S A* **99:**2293–2298.

29. **Patti JM, Allen BL, McGavin MJ, Höök M.** 1994. MSCRAMM-mediated adherence of microorganisms to host tissues. *Annu Rev Microbiol* **48:**585–617.

30. **Foster TJ.** 2016. The remarkably multifunctional fibronectin binding proteins of *Staphylococcus aureus. Eur J Clin Microbiol Infect Dis* **35:**1923–1931.

31. **Foster TJ, Geoghegan JA, Ganesh VK, Höök M.** 2014. Adhesion, invasion and evasion: the many functions of the surface proteins of *Staphylococcus aureus. Nat Rev Microbiol* **12:**49–62.

32. **Thomer L, Becker S, Emolo C, Quach A, Kim HK, Rauch S, Anderson M, Leblanc JF, Schneewind O, Faull KF, Missiakas D.** 2014. *N*-Acetylglucosaminylation of serine-aspartate repeat proteins promotes *Staphylococcus aureus* bloodstream infection. *J Biol Chem* **289:**3478–3486.

33. **Bleiziffer I, Eikmeier J, Pohlentz G, McAulay K, Xia G, Hussain M, Peschel A, Foster S, Peters G, Heilmann C.** 2017. The plasmin-sensitive protein Pls in methicillin-resistant *Staphylococcus aureus* (MRSA) is a glycoprotein. *PLoS Pathog* **13:**e1006110.

34. **Siboo IR, Chambers HF, Sullam PM.** 2005. Role of SraP, a serine-rich surface protein of *Staphylococcus aureus*, in binding to human platelets. *Infect Immun* **73:**2273–2280.

35. **Mazmanian SK, Skaar EP, Gaspar AH, Humayun M, Gornicki P, Jelenska J, Joachmiak A, Missiakas DM, Schneewind O.** 2003. Passage of heme-iron across the envelope of *Staphylococcus aureus. Science* **299:**906–909.

36. **Dryla A, Gelbmann D, von Gabain A, Nagy E.** 2003. Identification of a novel iron regulated staphylococcal surface protein with haptoglobin-haemoglobin binding activity. *Mol Microbiol* **49:**37–53.

37. **Skaar EP, Humayun M, Bae T, DeBord KL, Schneewind O.** 2004. Iron-source preference of *Staphylococcus aureus* infections. *Science* **305:**1626–1628.

38. **Pishchany G, Sheldon JR, Dickson CF, Alam MT, Read TD, Gell DA, Heinrichs DE, Skaar EP.** 2014. IsdB-dependent hemoglobin binding is required for acquisition of heme by *Staphylococcus aureus*. *J Infect Dis* **209:**1764–1772.

39. **Choby JE, Skaar EP.** 2016. Heme synthesis and acquisition in bacterial pathogens. *J Mol Biol* **428:**3408–3428.

40. **Sæderup KL, Stødkilde K, Graversen JH, Dickson CF, Etzerodt A, Hansen SW, Fago A, Gell D, Andersen CB, Moestrup SK.** 2016. The *Staphylococcus aureus* protein IsdH inhibits host hemoglobin scavenging to promote heme acquisition by the pathogen. *J Biol Chem* **291:**23989–23998.

41. **Skaar EP, Gaspar AH, Schneewind O.** 2004. IsdG and IsdI, heme-degrading enzymes in the cytoplasm of *Staphylococcus aureus*. *J Biol Chem* **279:**436–443.

42. **Reniere ML, Ukpabi GN, Harry SR, Stec DF, Krull R, Wright DW, Bachmann BO, Murphy ME, Skaar EP.** 2010. The IsdG-family of haem oxygenases degrades haem to a novel chromophore. *Mol Microbiol* **75:**1529–1538.

43. **Marraffini LA, Schneewind O.** 2005. Anchor structure of staphylococcal surface proteins. V. Anchor structure of the sortase B substrate IsdC. *J Biol Chem* **280:**16263–16271.

44. **Maresso AW, Schneewind O.** 2006. Iron acquisition and transport in *Staphylococcus aureus*. *Biometals* **19:**193–203.

45. **Kiser KB, Cantey-Kiser JM, Lee JC.** 1999. Development and characterization of a *Staphylococcus aureus* nasal colonization model in mice. *Infect Immun* **67:**5001–5006.

46. **Sun Y, Emolo C, Holtfreter S, Wiles S, Kreiswirth B, Missiakas D, Schneewind O.** 2018. Staphylococcal protein A contributes to persistent colonization of mice with *Staphylococcus aureus*. *J Bacteriol* **200:**e00735-17.

47. **Cheng AG, Kim HK, Burts ML, Krausz T, Schneewind O, Missiakas DM.** 2009. Genetic requirements for *Staphylococcus aureus* abscess formation and persistence in host tissues. *FASEB J* **23:**3393–3404.

48. **McAdow M, Kim HK, Dedent AC, Hendrickx APA, Schneewind O, Missiakas DM.** 2011. Preventing *Staphylococcus aureus* sepsis through the inhibition of its agglutination in blood. *PLoS Pathog* **7:**e1002307.

49. **Kim HK, Falugi F, Thomer L, Missiakas DM, Schneewind O.** 2015. Protein A suppresses immune responses during *Staphylococcus*

aureus bloodstream infection in guinea pigs. *mBio* **6:**e02369-14.

50. **Bubeck Wardenburg J, Patel RJ, Schneewind O.** 2007. Surface proteins and exotoxins are required for the pathogenesis of *Staphylococcus aureus* pneumonia. *Infect Immun* **75:**1040–1044.

51. **Bubeck Wardenburg J, Schneewind O.** 2008. Vaccine protection against *Staphylococcus aureus* pneumonia. *J Exp Med* **205:**287–294.

52. **Kennedy AD, Bubeck Wardenburg J, Gardner DJ, Long D, Whitney AR, Braughton KR, Schneewind O, DeLeo FR.** 2010. Targeting of alpha-hemolysin by active or passive immunization decreases severity of USA300 skin infection in a mouse model. *J Infect Dis* **202:**1050–1058.

53. **Jonsson IM, Mazmanian SK, Schneewind O, Bremell T, Tarkowski A.** 2003. The role of *Staphylococcus aureus* sortase A and sortase B in murine arthritis. *Microbes Infect* **5:**775–780.

54. **Corrigan RM, Miajlovic H, Foster TJ.** 2009. Surface proteins that promote adherence of *Staphylococcus aureus* to human desquamated nasal epithelial cells. *BMC Microbiol* **9:**22.

55. **Schaffer AC, Solinga RM, Cocchiaro J, Portoles M, Kiser KB, Risley A, Randall SM, Valtulina V, Speziale P, Walsh E, Foster T, Lee JC.** 2006. Immunization with *Staphylococcus aureus* clumping factor B, a major determinant in nasal carriage, reduces nasal colonization in a murine model. *Infect Immun* **74:**2145–2153.

56. **Misawa Y, Kelley KA, Wang X, Wang L, Park WB, Birtel J, Saslowsky D, Lee JC.** 2015. *Staphylococcus aureus* colonization of the mouse gastrointestinal tract is modulated by wall teichoic acid, capsule, and surface proteins. *PLoS Pathog* **11:**e1005061.

57. **Clarke SR, Brummell KJ, Horsburgh MJ, McDowell PW, Mohamad SA, Stapleton MR, Acevedo J, Read RC, Day NP, Peacock SJ, Mond JJ, Kokai-Kun JF, Foster SJ.** 2006. Identification of *in vivo*-expressed antigens of *Staphylococcus aureus* and their use in vaccinations for protection against nasal carriage. *J Infect Dis* **193:**1098–1108.

58. **Wertheim HF, Walsh E, Choudhurry R, Melles DC, Boelens HA, Miajlovic H, Verbrugh HA, Foster T, van Belkum A.** 2008. Key role for clumping factor B in *Staphylococcus aureus* nasal colonization of humans. *PLoS Med* **5:**e17.

59. **Forsgren A.** 1970. Significance of protein A production by staphylococci. *Infect Immun* **2:**672–673.

60. **Votintseva AA, Fung R, Miller RR, Knox K, Godwin H, Wyllie DH, Bowden R, Crook DW, Walker AS.** 2014. Prevalence of *Staphylococcus aureus* protein A (*spa*) mutants in the

community and hospitals in Oxfordshire. *BMC Microbiol* **14**:63.

61. **DeDent AC, McAdow M, Schneewind O.** 2007. Distribution of protein A on the surface of *Staphylococcus aureus*. *J Bacteriol* **189**:4473–4484.

62. **DeDent A, Bae T, Missiakas DM, Schneewind O.** 2008. Signal peptides direct surface proteins to two distinct envelope locations of *Staphylococcus aureus*. *EMBO J* **27**:2656–2668.

63. **Yu W, Missiakas D, Schneewind O.** 2018. Septal secretion of protein A in *Staphylococcus aureus* requires SecA and lipoteichoic acid synthesis. *eLife* **7**:e34092.

64. **Frankel MB, Hendrickx AP, Missiakas DM, Schneewind O.** 2011. LytN, a murein hydrolase in the cross-wall compartment of *Staphylococcus aureus*, is involved in proper bacterial growth and envelope assembly. *J Biol Chem* **286**:32593–32605.

65. **Frankel MB, Schneewind O.** 2012. Determinants of murein hydrolase targeting to cross-wall of *Staphylococcus aureus* peptidoglycan. *J Biol Chem* **287**:10460–10471.

66. **Becker S, Frankel MB, Schneewind O, Missiakas D.** 2014. Release of protein A from the cell wall of *Staphylococcus aureus*. *Proc Natl Acad Sci U S A* **111**:1574–1579.

67. **Kim HK, Falugi F, Missiakas DM, Schneewind O.** 2016. Peptidoglycan-linked protein A promotes T cell-dependent antibody expansion during *Staphylococcus aureus* infection. *Proc Natl Acad Sci U S A* **113**:5718–5723.

68. **Pauli NT, Kim HK, Falugi F, Huang M, Dulac J, Henry Dunand C, Zheng NY, Kaur K, Andrews SF, Huang Y, DeDent A, Frank KM, Charnot-Katsikas A, Schneewind O, Wilson PC.** 2014. *Staphylococcus aureus* infection induces protein A-mediated immune evasion in humans. *J Exp Med* **211**:2331–2339.

69. **Falugi F, Kim HK, Missiakas DM, Schneewind O.** 2013. Role of protein A in the evasion of host adaptive immune responses by *Staphylococcus aureus*. *mBio* **4**:e00575-13.

70. **Forsgren A, Sjöquist J.** 1966. "Protein A" from *S. aureus*. I. Pseudo-immune reaction with human gamma-globulin. *J Immunol* **97**:822–827.

71. **Forsgren A.** 1968. Protein A from *Staphylococcus aureus*. VI. Reaction with subunits from guinea pig γ-1- and γ-2-globulin. *J Immunol* **100**:927–930.

72. **Sasso EH, Silverman GJ, Mannik M.** 1989. Human IgM molecules that bind staphylococcal protein A contain VHIII H chains. *J Immunol* **142**:2778–2783.

73. **Kim HK, Cheng AG, Kim H-Y, Missiakas DM, Schneewind O.** 2010. Nontoxigenic pro-tein A vaccine for methicillin-resistant *Staphylococcus aureus* infections in mice. *J Exp Med* **207**:1863–1870.

74. **Verkaik NJ, Lebon A, de Vogel CP, Hooijkaas H, Verbrugh HA, Jaddoe VW, Hofman A, Moll HA, van Belkum A, van Wamel WJ.** 2010. Induction of antibodies by *Staphylococcus aureus* nasal colonization in young children. *Clin Microbiol Infect* **16**:1312–1317.

75. **Swierstra J, Debets S, de Vogel C, Lemmens-den Toom N, Verkaik N, Ramdani-Bouguessa N, Jonkman MF, van Dijl JM, Fahal A, van Belkum A, van Wamel W.** 2015. IgG4 subclass-specific responses to *Staphylococcus aureus* antigens shed new light on host-pathogen inter-action. *Infect Immun* **83**:492–501.

76. **Holtfreter S, Jursa-Kulesza J, Masiuk H, Verkaik NJ, de Vogel C, Kolata J, Nowosiad M, Steil L, van Wamel W, van Belkum A, Völker U, Giedrys-Kalemba S, Bröker BM.** 2011. Antibody responses in furunculosis patients vaccinated with autologous formalin-killed *Staphylococcus aureus*. *Eur J Clin Microbiol Infect Dis* **30**:707–717.

77. **Kluytmans J, van Belkum A, Verbrugh H.** 1997. Nasal carriage of *Staphylococcus aureus*: epidemiology, underlying mechanisms, and associated risks. *Clin Microbiol Rev* **10**:505–520.

78. **Weinstein HJ.** 1959. The relation between the nasal-staphylococcal-carrier state and the in-cidence of postoperative complications. *N Engl J Med* **260**:1303–1308.

79. **Wertheim HF, Melles DC, Vos MC, van Leeuwen W, van Belkum A, Verbrugh HA, Nouwen JL.** 2005. The role of nasal carriage in *Staphylococcus aureus* infections. *Lancet Infect Dis* **5**:751–762.

80. **Forsgren A, Quie PG.** 1974. Effects of staphylococcal protein A on heat labile opsonins. *J Immunol* **112**:1177–1180.

81. **Adams RL, Bird RJ.** 2009. Review article: coagulation cascade and therapeutics update: relevance to nephrology. Part 1: overview of coagulation, thrombophilias and history of anticoagulants. *Nephrology (Carlton)* **14**:462–470.

82. **Doolittle RF.** 2003. Structural basis of the fibrinogen-fibrin transformation: contributions from X-ray crystallography. *Blood Rev* **17**:33–41.

83. **Ware S, Donahue JP, Hawiger J, Anderson WF.** 1999. Structure of the fibrinogen gamma-chain integrin binding and factor XIIIa cross-linking sites obtained through carrier protein driven crystallization. *Protein Sci* **8**:2663–2671.

84. **Levi M, Keller TT, van Gorp E, ten Cate H.** 2003. Infection and inflammation and the coagulation system. *Cardiovasc Res* **60**:26–39.

85. **Much H.** 1908. Über eine Vorstufe des Fibrinfermentes in Kulturen von *Staphylokokkus aureus. Biochem Z* **14:**143–155.

86. **Cheng AG, McAdow M, Kim HK, Bae T, Missiakas DM, Schneewind O.** 2010. Contribution of coagulases towards *Staphylococcus aureus* disease and protective immunity. *PLoS Pathog* **6:**e1001036.

87. **Thomer L, Schneewind O, Missiakas D.** 2016. Pathogenesis of *Staphylococcus aureus* bloodstream infections. *Annu Rev Pathol* **11:**343–364.

88. **O'Connell DP, Nanavaty T, McDevitt D, Gurusiddappa S, Höök M, Foster TJ.** 1998. The fibrinogen-binding MSCRAMM (clumping factor) of *Staphylococcus aureus* has a Ca2+-dependent inhibitory site. *J Biol Chem* **273:**6821–6829.

89. **Strong DD, Laudano AP, Hawiger J, Doolittle RF.** 1982. Isolation, characterization, and synthesis of peptides from human fibrinogen that block the staphylococcal clumping reaction and construction of a synthetic clumping particle. *Biochemistry* **21:**1414–1420.

90. **Ganesh VK, Rivera JJ, Smeds E, Ko Y-P, Bowden MG, Wann ER, Gurusiddappa S, Fitzgerald JR, Höök M.** 2008. A structural model of the *Staphylococcus aureus* ClfA-fibrinogen interaction opens new avenues for the design of anti-staphylococcal therapeutics. *PLoS Pathog* **4:**e1000226.

91. **Ponnuraj K, Bowden MG, Davis S, Gurusiddappa S, Moore D, Choe D, Xu Y, Höök M, Narayana SV.** 2003. A "dock, lock, and latch" structural model for a staphylococcal adhesin binding to fibrinogen. *Cell* **115:**217–228.

92. **Bowden MG, Heuck AP, Ponnuraj K, Kolosova E, Choe D, Gurusiddappa S, Narayana SV, Johnson AE, Höök M.** 2008. Evidence for the "dock, lock, and latch" ligand binding mechanism of the staphylococcal microbial surface component recognizing adhesive matrix molecules (MSCRAMM) SdrG. *J Biol Chem* **283:**638–647.

93. **Flick MJ, Du X, Prasad JM, Raghu H, Palumbo JS, Smeds E, Höök M, Degen JL.** 2013. Genetic elimination of the binding motif on fibrinogen for the *S. aureus* virulence factor ClfA improves host survival in septicemia. *Blood* **121:**1783–1794.

94. **O'Brien L, Kerrigan SW, Kaw G, Hogan M, Penadés J, Litt D, Fitzgerald DJ, Foster TJ, Cox D.** 2002. Multiple mechanisms for the activation of human platelet aggregation by *Staphylococcus aureus:* roles for the clumping factors ClfA and ClfB, the serine-aspartate repeat protein SdrE and protein A. *Mol Microbiol* **44:**1033–1044.

95. **Loughman A, Fitzgerald JR, Brennan MP, Higgins J, Downer R, Cox D, Foster TJ.** 2005. Roles for fibrinogen, immunoglobulin and complement in platelet activation promoted by *Staphylococcus aureus* clumping factor A. *Mol Microbiol* **57:**804–818.

96. **Ní Eidhin D, Perkins S, Francois P, Vaudaux P, Höök M, Foster TJ.** 1998. Clumping factor B (ClfB), a new surface-located fibrinogen-binding adhesin of *Staphylococcus aureus. Mol Microbiol* **30:**245–257.

97. **Walsh EJ, Miajlovic H, Gorkun OV, Foster TJ.** 2008. Identification of the *Staphylococcus aureus* MSCRAMM clumping factor B (ClfB) binding site in the alphaC-domain of human fibrinogen. *Microbiology* **154:**550–558.

98. **Perkins S, Walsh EJ, Deivanayagam CC, Narayana SV, Foster TJ, Höök M.** 2001. Structural organization of the fibrinogen-binding region of the clumping factor B MSCRAMM of *Staphylococcus aureus. J Biol Chem* **276:**44721–44728.

99. **Haim M, Trost A, Maier CJ, Achatz G, Feichtner S, Hintner H, Bauer JW, Onder K.** 2010. Cytokeratin 8 interacts with clumping factor B: a new possible virulence factor target. *Microbiology* **156:**3710–3721.

100. **Walsh EJ, O'Brien LM, Liang X, Höök M, Foster TJ.** 2004. Clumping factor B, a fibrinogen-binding MSCRAMM (microbial surface components recognizing adhesive matrix molecules) adhesin of *Staphylococcus aureus*, also binds to the tail region of type I cytokeratin 10. *J Biol Chem* **279:**50691–50699.

101. **O'Brien LM, Walsh EJ, Massey RC, Peacock SJ, Foster TJ.** 2002. *Staphylococcus aureus* clumping factor B (ClfB) promotes adherence to human type I cytokeratin 10: implications for nasal colonization. *Cell Microbiol* **4:**759–770.

102. **Mulcahy ME, Geoghegan JA, Monk IR, O'Keeffe KM, Walsh EJ, Foster TJ, McLoughlin RM.** 2012. Nasal colonisation by *Staphylococcus aureus* depends upon clumping factor B binding to the squamous epithelial cell envelope protein loricrin. *PLoS Pathog* **8:**e1003092.

103. **Ganesh VK, Barbu EM, Deivanayagam CC, Le B, Anderson AS, Matsuka YV, Lin SL, Foster TJ, Narayana SV, Höök M.** 2011. Structural and biochemical characterization of *Staphylococcus aureus* clumping factor B/ligand interactions. *J Biol Chem* **286:**25963–25972.

104. **Cheng AG, DeDent AC, Schneewind O, Missiakas D.** 2011. A play in four acts: *Staphylococcus aureus* abscess formation. *Trends Microbiol* **19:**225–232.

105. **Berends ET, Horswill AR, Haste NM, Monestier M, Nizet V, von Köckritz-Blickwede M.** 2010. Nuclease expression by *Staphylococcus aureus* facilitates escape from neutrophil extracellular traps. *J Innate Immun* **2**:576–586.

106. **Thammavongsa V, Missiakas DM, Schneewind O.** 2013. *Staphylococcus aureus* conversion of neutrophil extracellular traps into deoxyadenosine promotes immune cell death. *Science* **342**:863–866.

107. **Winstel V, Missiakas D, Schneewind O.** 2018. *Staphylococcus aureus* targets the purine salvage pathway to kill phagocytes. *Proc Natl Acad Sci U S A* **115**:6846–6851.

108. **Thammavongsa V, Kern JW, Missiakas DM, Schneewind O.** 2009. *Staphylococcus aureus* synthesizes adenosine to escape host immune responses. *J Exp Med* **206**:2417–2427.

109. **Thammavongsa V, Schneewind O, Missiakas DM.** 2011. Enzymatic properties of *Staphylococcus aureus* adenosine synthase (AdsA). *BMC Biochem* **12**:56.

110. **Josefsson E, Higgins J, Foster TJ, Tarkowski A.** 2008. Fibrinogen binding sites P336 and Y338 of clumping factor A are crucial for *Staphylococcus aureus* virulence. *PLoS One* **3**:e2206.

111. **Domanski PJ, Patel PR, Bayer AS, Zhang L, Hall AE, Syribeys PJ, Gorovits EL, Bryant D, Vernachio JH, Hutchins JT, Patti JM.** 2005. Characterization of a humanized monoclonal antibody recognizing clumping factor A expressed by *Staphylococcus aureus*. *Infect Immun* **73**:5229–5232.

112. **Ganesh VK, Liang X, Geoghegan JA, Cohen ALV, Venugopalan N, Foster TJ, Höök M.** 2016. Lessons from the crystal structure of the *S. aureus* surface protein clumping factor A in complex with tefibazumab, an inhibiting monoclonal antibody. *EBioMedicine* **13**:328–338.

113. **Hall AE, Domanski PJ, Patel PR, Vernachio JH, Syribeys PJ, Gorovits EL, Johnson MA, Ross JM, Hutchins JT, Patti JM.** 2003. Characterization of a protective monoclonal antibody recognizing *Staphylococcus aureus* MSCRAMM protein clumping factor A. *Infect Immun* **71**:6864–6870.

114. **Weems JJ Jr, Steinberg JP, Filler S, Baddley JW, Corey GR, Sampathkumar P, Winston L, John JF, Kubin CJ, Talwani R, Moore T, Patti JM, Hetherington S, Texter M, Wenzel E, Kelley VA, Fowler VG Jr.** 2006. Phase II, randomized, double-blind, multicenter study comparing the safety and pharmacokinetics of tefibazumab to placebo for treatment of *Staphylococcus aureus* bacteremia. *Antimicrob Agents Chemother* **50**:2751–2755.

115. **Tkaczyk C, Hamilton MM, Sadowska A, Shi Y, Chang CS, Chowdhury P, Buonapane R, Xiao X, Warrener P, Mediavilla J, Kreiswirth B, Suzich J, Stover CK, Sellman BR.** 2016. Targeting alpha toxin and ClfA with a multi-mechanistic monoclonal-antibody-based approach for prophylaxis of serious *Staphylococcus aureus* disease. *mBio* **7**:e00528-16.

116. **Tkaczyk C, Hua L, Varkey R, Shi Y, Dettinger L, Woods R, Barnes A, MacGill RS, Wilson S, Chowdhury P, Stover CK, Sellman BR.** 2012. Identification of anti-alpha toxin monoclonal antibodies that reduce the severity of *Staphylococcus aureus* dermonecrosis and exhibit a correlation between affinity and potency. *Clin Vaccine Immunol* **19**:377–385.

117. **Tkaczyk C, Kasturirangan S, Minola A, Jones-Nelson O, Gunter V, Shi YY, Rosenthal K, Aleti V, Semenova E, Warrener P, Tabor D, Stover CK, Corti D, Rainey G, Sellman BR.** 2017. Multimechanistic monoclonal antibodies (MAbs) targeting *Staphylococcus aureus* alpha-toxin and clumping factor A: activity and efficacy comparisons of a MAb combination and an engineered bispecific antibody approach. *Antimicrob Agents Chemother* **61**:e00629-17.

118. **Creech CB, Frenck RWJ Jr, Sheldon EA, Seiden DJ, Kankam MK, Zito ET, Girgenti D, Severs JM, Immermann FW, McNeil LK, Cooper D, Jansen KU, Gruber W, Eiden J, Anderson AS, Baber J.** 2017. Safety, tolerability, and immunogenicity of a single dose 4-antigen or 3-antigen *Staphylococcus aureus* vaccine in healthy older adults: results of a randomised trial. *Vaccine* **35**:385–394.

119. **Scully IL, Liberator PA, Jansen KU, Anderson AS.** 2014. Covering all the bases: preclinical development of an effective *Staphylococcus aureus* vaccine. *Front Immunol* **5**:109.

120. **Stranger-Jones YK, Bae T, Schneewind O.** 2006. Vaccine assembly from surface proteins of *Staphylococcus aureus*. *Proc Natl Acad Sci U S A* **103**:16942–16947.

121. **Kuklin NA, Clark DJ, Secore S, Cook J, Cope LD, McNeely T, Noble L, Brown MJ, Zorman JK, Wang XM, Pancari G, Fan H, Isett K, Burgess B, Bryan J, Brownlow M, George H, Meinz M, Liddell ME, Kelly R, Schultz L, Montgomery D, Onishi J, Losada M, Martin M, Ebert T, Tan CY, Schofield TL, Nagy E, Meineke A, Joyce JG, Kurtz MB, Caulfield MJ, Jansen KU, McClements W, Anderson AS.** 2006. A novel *Staphylococcus aureus* vaccine: iron surface determinant B induces rapid antibody responses in rhesus macaques and specific increased survival in a murine *S. aureus* sepsis model. *Infect Immun* **74**:2215–2223.

122. Kim HK, DeDent A, Cheng AG, McAdow M, Bagnoli F, Missiakas DM, Schneewind O. 2010. IsdA and IsdB antibodies protect mice against *Staphylococcus aureus* abscess formation and lethal challenge. *Vaccine* **28:**6382–6392.

123. Brown M, Kowalski R, Zorman J, Wang XM, Towne V, Zhao Q, Secore S, Finnefrock AC, Ebert T, Pancari G, Isett K, Zhang Y, Anderson AS, Montgomery D, Cope L, McNeely T. 2009. Selection and characterization of murine monoclonal antibodies to *Staphylococcus aureus* iron-regulated surface determinant B with functional activity *in vitro* and *in vivo*. *Clin Vaccine Immunol* **16:**1095–1104.

124. Fowler VG, Allen KB, Moreira ED, Moustafa M, Isgro F, Boucher HW, Corey GR, Carmeli Y, Betts R, Hartzel JS, Chan IS, McNeely TB, Kartsonis NA, Guris D, Onorato MT, Smugar SS, DiNubile MJ, Sobanjo-ter Meulen A. 2013. Effect of an investigational vaccine for preventing *Staphylococcus aureus* infections after cardiothoracic surgery: a randomized trial. *JAMA* **309:**1368–1378.

125. Kim HK, Emolo C, DeDent AC, Falugi F, Missiakas DM, Schneewind O. 2012. Protein A-specific monoclonal antibodies and prevention of *Staphylococcus aureus* disease in mice. *Infect Immun* **80:**3460–3470.

126. Thammavongsa V, Rauch S, Kim HK, Missiakas DM, Schneewind O. 2015. Protein A-neutralizing monoclonal antibody protects neonatal mice against *Staphylococcus aureus*. *Vaccine* **33:**523–526.

127. Ton-That H, Mazmanian SK, Alksne L, Schneewind O. 2002. Anchoring of surface proteins to the cell wall of *Staphylococcus aureus*. Cysteine 184 and histidine 120 of sortase form a thiolate-imidazolium ion pair for catalysis. *J Biol Chem* **277:**7447–7452.

128. Maresso AW, Wu R, Kern JW, Zhang R, Janik D, Missiakas DM, Duban ME, Joachimiak A, Schneewind O. 2007. Activation of inhibitors by sortase triggers irreversible modification of the active site. *J Biol Chem* **282:**23129–23139.

129. Maresso AW, Schneewind O. 2008. Sortase as a target of anti-infective therapy. *Pharmacol Rev* **60:**128–141.

130. Zhang J, Liu H, Zhu K, Gong S, Dramsi S, Wang YT, Li J, Chen F, Zhang R, Zhou L, Lan L, Jiang H, Schneewind O, Luo C, Yang CG. 2014. Antiinfective therapy with a small molecule inhibitor of *Staphylococcus aureus* sortase. *Proc Natl Acad Sci U S A* **111:**13517–13522.

131. Oh KB, Nam KW, Ahn H, Shin J, Kim S, Mar W. 2010. Therapeutic effect of (Z)-3-(2,5-dimethoxyphenyl)-2-(4-methoxyphenyl) acrylonitrile (DMMA) against *Staphylococcus aureus* infection in a murine model. *Biochem Biophys Res Commun* **396:**440–444.

132. Pallen MJ, Lam AC, Antonio M, Dunbar K. 2001. An embarrassment of sortases—a richness of substrates? *Trends Microbiol* **9:**97–102.

133. Hendrickx AP, Budzik JM, Oh SY, Schneewind O. 2011. Architects at the bacterial surface—sortases and the assembly of pili with isopeptide bonds. *Nat Rev Microbiol* **9:**166–176.

134. Jacobitz AW, Kattke MD, Wereszczynski J, Clubb RT. 2017. Sortase transpeptidases: structural biology and catalytic mechanism. *Adv Protein Chem Struct Biol* **109:**223–264.

135. Ton-That H, Schneewind O. 2003. Assembly of pili on the surface of *Corynebacterium diphtheriae*. *Mol Microbiol* **50:**1429–1438.

136. Mandlik A, Swierczynski A, Das A, Ton-That H. 2007. *Corynebacterium diphtheriae* employs specific minor pilins to target human pharyngeal epithelial cells. *Mol Microbiol* **64:**111–124.

137. Budzik JM, Marraffini LA, Souda P, Whitelegge JP, Faull KF, Schneewind O. 2008. Amide bonds assemble pili on the surface of bacilli. *Proc Natl Acad Sci U S A* **105:**10215–10220.

138. Budzik JM, Oh SY, Schneewind O. 2008. Cell wall anchor structure of BcpA pili in *Bacillus anthracis*. *J Biol Chem* **283:**36676–36686.

139. Budzik JM, Oh SY, Schneewind O. 2009. Sortase D forms the covalent bond that links BcpB to the tip of *Bacillus cereus* pili. *J Biol Chem* **284:**12989–12997.

140. Ton-That H, Marraffini LA, Schneewind O. 2004. Sortases and pilin elements involved in pilus assembly of *Corynebacterium diphtheriae*. *Mol Microbiol* **53:**251–261.

141. Swaminathan A, Mandlik A, Swierczynski A, Gaspar A, Das A, Ton-That H. 2007. Housekeeping sortase facilitates the cell wall anchoring of pilus polymers in *Corynebacterium diphtheriae*. *Mol Microbiol* **66:**961–974.

142. Mandlik A, Das A, Ton-That H. 2008. The molecular switch that activates the cell wall anchoring step of pilus assembly in gram-positive bacteria. *Proc Natl Acad Sci U S A* **105:**14147–14152.

143. Chang C, Amer BR, Osipiuk J, McConnell SA, Huang IH, Hsieh V, Fu J, Nguyen HH, Muroski J, Flores E, Ogorzalek Loo RR, Loo JA, Putkey JA, Joachimiak A, Das A, Clubb RT, Ton-That H. 2018. In vitro reconstitution of sortase-catalyzed pilus polymerization reveals structural elements involved in pilin cross-linking. *Proc Natl Acad Sci U S A* **115:**E5477–E5486.

144. **McDevitt D, Francois P, Vaudaux P, Foster TJ.** 1994. Molecular characterization of the clumping factor (fibrinogen receptor) of *Staphylococcus aureus*. *Mol Microbiol* **11**:237–248.

145. **Hair PS, Ward MD, Semmes OJ, Foster TJ, Cunnion KM.** 2008. *Staphylococcus aureus* clumping factor A binds to complement regulator factor I and increases factor I cleavage of C3b. *J Infect Dis* **198**:125–133.

146. **Zong Y, Xu Y, Liang X, Keene DR, Höök A, Gurusiddappa S, Höök M, Narayana SV.** 2005. A 'Collagen Hug' model for *Staphylococcus aureus* CNA binding to collagen. *EMBO J* **24**:4224–4236.

147. **Kang M, Ko YP, Liang X, Ross CL, Liu Q, Murray BE, Höök M.** 2013. Collagen-binding microbial surface components recognizing adhesive matrix molecule (MSCRAMM) of Gram-positive bacteria inhibit complement activation via the classical pathway. *J Biol Chem* **288**:20520–20531.

148. **Komatsuzawa H, Sugai M, Ohta K, Fujiwara T, Nakashima S, Suzuki J, Lee CY, Suginaka H.** 1997. Cloning and characterization of the fmt gene which affects the methicillin resistance level and autolysis in the presence of Triton X-100 in methicillin-resistant *Staphylococcus aureus*. *Antimicrob Agents Chemother* **41**:2355–2361.

149. **Kuusela P, Hildén P, Savolainen K, Vuento M, Lyytikäinen O, Vuopio-Varkila J.** 1994. Rapid detection of methicillin-resistant *Staphylococcus aureus* strains not identified by slide agglutination tests. *J Clin Microbiol* **32**:143–147.

150. **Kuusela P, Saksela O.** 1990. Binding and activation of plasminogen at the surface of *Staphylococcus aureus*. Increase in affinity after conversion to the Lys form of the ligand. *Eur J Biochem* **193**:759–765.

151. **Schroeder K, Jularic M, Horsburgh SM, Hirschhausen N, Neumann C, Bertling A, Schulte A, Foster S, Kehrel BE, Peters G, Heilmann C.** 2009. Molecular characterization of a novel *Staphylococcus aureus* surface protein (SasC) involved in cell aggregation and biofilm accumulation. *PLoS One* **4**:e7567.

152. **Barbu EM, Ganesh VK, Gurusiddappa S, Mackenzie RC, Foster TJ, Sudhof TC, Höök M.** 2010. β-Neurexin is a ligand for the *Staphylococcus aureus* MSCRAMM SdrC. *PLoS Pathog* **6**:e1000726.

153. **Feuillie C, Formosa-Dague C, Hays LM, Vervaeck O, Derclaye S, Brennan MP, Foster TJ, Geoghegan JA, Dufrêne YF.** 2017. Molecular interactions and inhibition of the staphylococcal biofilm-forming protein SdrC. *Proc Natl Acad Sci U S A* **114**:3738–3743.

154. **Askarian F, Ajayi C, Hanssen AM, van Sorge NM, Pettersen I, Diep DB, Sollid JU, Johannessen M.** 2016. The interaction between *Staphylococcus aureus* SdrD and desmoglein 1 is important for adhesion to host cells. *Sci Rep* **6**:22134.

155. **Zhang Y, Wu M, Hang T, Wang C, Yang Y, Pan W, Zang J, Zhang M, Zhang X.** 2017. *Staphylococcus aureus* SdrE captures complement factor H's C-terminus via a novel 'close, dock, lock and latch' mechanism for complement evasion. *Biochem J* **474**:1619–1631.

156. **Uhlén M, Guss B, Nilsson B, Gatenbeck S, Philipson L, Lindberg M.** 1984. Complete sequence of the staphylococcal gene encoding protein A. A gene evolved through multiple duplications. *J Biol Chem* **259**:1695–1702.

157. **Graille M, Stura EA, Corper AL, Sutton BJ, Taussig MJ, Charbonnier JB, Silverman GJ.** 2000. Crystal structure of a *Staphylococcus aureus* protein A domain complexed with the Fab fragment of a human IgM antibody: structural basis for recognition of B-cell receptors and superantigen activity. *Proc Natl Acad Sci U S A* **97**:5399–5404.

158. **Kukita K, Kawada-Matsuo M, Oho T, Nagatomo M, Oogai Y, Hashimoto M, Suda Y, Tanaka T, Komatsuzawa H.** 2013. *Staphylococcus aureus* SasA is responsible for binding to the salivary agglutinin gp340, derived from human saliva. *Infect Immun* **81**:1870–1879.

Architecture and Assembly of Periplasmic Flagellum

16

YUNJIE CHANG[1,2] and JUN LIU[1,2]

The flagellum is a major organelle for motility in many bacterial species. It confers locomotion and is often associated with virulence of bacterial pathogens. Flagella from different species share a conserved core but also exhibit profound variations in flagellar structure, flagellar number, and placement (1, 2), resulting in distinct flagella that appear to be adapted to the specific environments that the bacteria encounter. While many bacteria possess multiple peritrichous flagella, such as those found in *Escherichia coli* and *Salmonella enterica*, other bacteria, such as *Vibrio* spp. and *Pseudomonas aeruginosa*, normally have a single flagellum at one cell pole (Fig. 1). Spirochetes uniquely assemble flagella that are embedded in periplasmic space between their inner and outer membranes, thus called periplasmic flagella (3). Although the flagella of *E. coli* and *Salmonella* have been extensively studied for several decades, periplasmic flagella are less understood, despite their profound impact on the distinctive morphology and motility of spirochetes. In this chapter, many aspects of periplasmic flagella are discussed, with particular focus on their structure and assembly.

[1]Department of Microbial Pathogenesis, Yale University School of Medicine, New Haven, CT 06536
[2]Microbial Sciences Institute, Yale University, West Haven, CT 06516
Protein Secretion in Bacteria
Edited by Maria Sandkvist, Eric Cascales, and Peter J. Christie
© 2019 American Society for Microbiology, Washington, DC
doi:10.1128/microbiolspec.PSIB-0030-2019

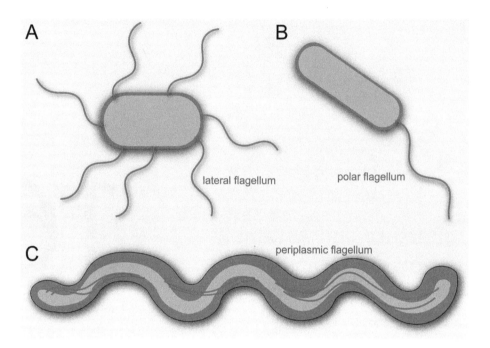

FIGURE 1 Distinctive placement of bacterial flagellum. (A) Bacteria with flagella distributed all over the cell (e.g., *Escherichia coli*) are peritrichous. (B) Monotrichous bacteria, such as *Vibrio cholerae*, *Pseudomonas aeruginosa*, and *Caulobacter crescentus*, have a single flagellum present at one end of the cell. (C) Spirochetes, including species of *Borrelia*, *Treponema*, and *Leptospira*, possess specialized flagella located within the periplasmic space. The rotation of the periplasmic flagella allows the bacterium to swim forward in a corkscrew-like motion.

SPIROCHETES ARE A DISTINCTIVE GROUP OF BACTERIA OF SIGNIFICANT IMPORTANCE IN HUMAN HEALTH

Spirochetes cause several major diseases in humans, such as Lyme disease (*Borrelia burgdorferi*), syphilis (*Treponema pallidum*), leptospirosis (*Leptospira interrogans*), and periodontitis (*Treponema* spp.). Lyme disease is the most commonly reported tick-borne illness in the United States, and the incidence is growing rapidly. The number of patients diagnosed with Lyme disease each year in the United States is approximately 300,000. The disease is caused by *B. burgdorferi* and related organisms, and it is transmitted to humans through the bite of infected *Ixodes* ticks (4). Syphilis is a common sexually transmitted disease in many areas of the world. Leptospirosis is the most common waterborne zoonosis worldwide.

MOTILITY IS ESSENTIAL FOR SPIROCHETES TO INFECT AND DISSEMINATE IN MAMMALIAN HOSTS

Spirochetal motility is unique, as the entire bacterium is involved in translocation without the involvement of external appendages. The motility is driven by periplasmic flagella, and rotation of the flagella causes a serpentine movement, allowing the organism to very efficiently bore its way through viscous media or tissue (3). To complete the host-vector life cycle, *B. burgdorferi* is able to adapt to divergent host environments and also evade the defense of its mammalian reservoir (4).

Several studies provide direct evidence that the unique motility and chemotaxis of *B. burgdorferi* are essential for the establishment of infection in mammals and the completion of its enzootic cycle (5–8).

PERIPLASMIC FLAGELLA ARE NECESSARY FOR THE FLAT-WAVE MORPHOLOGY AND DISTINCTIVE MOTILITY OF *B. BURGDORFERI*

B. burgdorferi possesses 7 to 11 periplasmic flagella that are inserted at each cell pole and wrap around the cell cylinder to produce the spirochete's distinctive flat-wave morphology. Periplasmic flagella are crucial not only for motility but also for the overall shape of *B. burgdorferi*, as mutant cells lacking flagella are nonmotile and exhibit a rod-shaped morphology (9). Similar to the external flagella found in the model organisms *E. coli* and *S. enterica*, periplasmic flagella are composed of the flagellar motor, the hook, and the filament. The flagellar motor is a rotary motor that anchors the flagellum to the inner membrane. The motors of the periplasmic flagella are noticeably larger than those of external flagella (Fig. 2). Importantly, the periplasmic flagellar motor possesses a spirochete-specific "collar" (Fig. 2) (10–13). The motor can be further divided into two parts: the rotor and the stator. The rotation of the motor is driven by the torque generated by the stator-rotor interaction, utilizing energy generated by the flow of protons through the stator channel. The rotor is composed of the MS ring, the C ring, and the rod. The MS ring is the base of the rotor, and it is formed by multiple copies of FliF. The C ring is located in the cytoplasm and is also known as the switch complex. It consists of the proteins FliG, FliM, and FliN and controls the direction of flagellar rotation. The rod serves as a drive shaft and consists of multiple different proteins (FlgB, FlgC, FlgF, and FlgG). The hook of the periplasmic flagellum is located in periplasmic space, in contrast to the externally localized hook in *E. coli* and *S. enterica*. A recent study indicates that the hook proteins are cross-linked by a covalent bond, an unusual property necessary for transmission of high rotational torque from the motor to the filament (14). The filament is the longest component of the periplasmic flagella. Multiple filaments arising from both poles form flat ribbons that wrap around the spirochete cell body in a right-handed fashion (15). The flagellar type III secretion system (fT3SS), which is embedded in the flagellar motor, is responsible for the transport and assembly of the protein components of the rod, the hook, and the filament (16).

CHARACTERIZATION OF THE UNIQUE PERIPLASMIC STRUCTURE OF SPIROCHETAL FLAGELLA

The periplasmic flagella possess a unique spirochete-specific collar that has not been found in other bacterial flagella reported to date (1, 2). The collar in the *B. burgdorferi* periplasmic flagellar motor is ~71 nm in diameter and ~24 nm in height, presumably composed of many different proteins. However, there is limited information regarding its structure, function, and protein components. Recently, a hypothetical membrane protein, FlbB, was identified as a candidate for involvement in collar assembly (10). In addition, the novel tetratricopeptide repeat protein BB0236 was also proposed to contribute to collar assembly (17). Mutants deficient in either FlbB or BB0236 are nonmotile, and their periplasmic flagella lack the collar, its associated proteins (including FliL), and the stator (Fig. 3). These findings provide direct evidence that the collar is indeed an important (as well as unique) component of periplasmic flagella (Fig. 2). Although additional unknown proteins are likely involved in collar assembly, it is evident that the periplasmic collar provides a static framework promoting the recruitment and stable association of stator units, which could, in turn, facilitate the

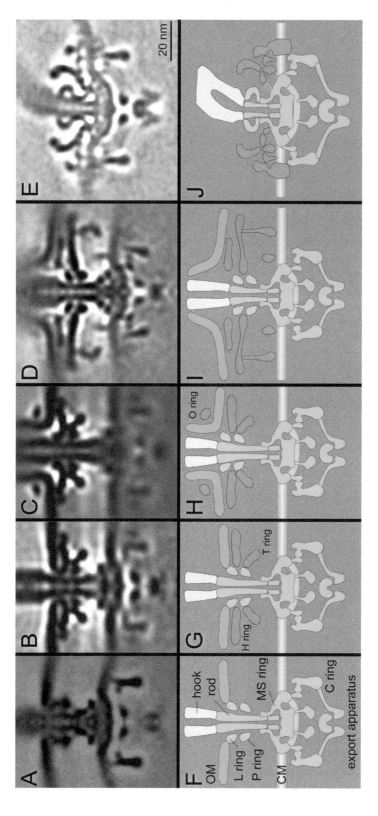

FIGURE 2 Comparison of motor structures from *E. coli, Vibrio, H. pylori,* and *Borrelia*. (A) A central section of an *E. coli* flagellar motor. (B) A central section from a nonsheathed *Vibrio* flagellar motor. (C) A central section from a sheathed *Vibrio* flagellar motor. (D) A central section from a sheathed flagellar motor of *H. pylori*. (E) A central section from a *Borrelia* flagellar motor. (F to J) Schematic models derived from the central sections shown in panels A to E, respectively. Adapted from prior publications (16, 50, 51), with permission.

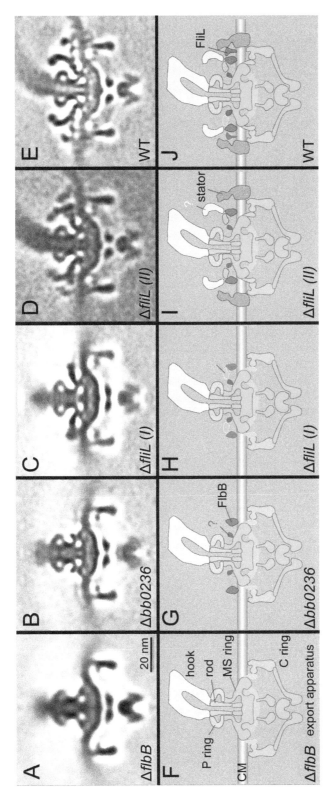

FIGURE 3 Characterization of the unique features in periplasmic flagella, as examined through mutational analysis. (A) Central section from a mutant lacking FlbB. (B) Central section from a mutant lacking BB0236. (C) Central section from a class average of a mutant lacking FliL. (D) Central section from another class average of a mutant lacking FliL. (E) A central section from wild-type flagellar motor. (F to J) Schematic models derived from panels A to E, respectively. Adapted from a prior publication (17), with permission.

higher torques to rotate the periplasmic flagellum. The rotation of the flagella within the confinements of the periplasm enables the spirochete to bore its way through complex, viscous environments in vertebrate and tick tissues.

STATOR-ROTOR INTERACTION

Powered by the electrochemical gradient across the cytoplasmic membrane, the flagellar motor can rotate the filament at high speed. It is believed that the flagellar rotation is mediated by the interaction between the cytoplasmic loop region of MotA and the C-terminal domain of FliG in the C ring. However, there is limited structural information on stator-rotor interaction in model systems *E. coli* and *S. enterica*, largely because the stator is poorly resolved due to its dynamic nature and low occupancy (18–20). In contrast, the *in situ* flagellar motor structures of *B. burgdorferi* and other spirochetes determined by cryo-electron tomography (cryo-ET) reveal more detailed information regarding the stators and their interactions with the C ring (1, 12, 13, 21–23). The presence of the collar in spirochetes is likely essential for the better visualization of the stator and its interaction with the rotor, because the collar provides a stable framework to recruit and stabilize the stators.

CheY-P BINDING AND FLAGELLAR SWITCHING OF ROTATIONAL DIRECTION

The flagellar motor in many bacterial species can rotate in both counterclockwise (CCW) and clockwise (CW) directions to achieve swimming towards attractants or away from repellents. The rotation direction is controlled by a sophisticated chemotactic system. In the signaling pathway, CheY is phosphorylated by CheA kinase; then the phosphorylated CheY binds to the FliM protein in the C ring and induces conformational changes

that alter the stator-rotation interaction and cause switching (24). Studies with *E. coli* of the correlation between the CW rotation and the intracellular level of phosphorylated CheY in individual cells indicated that binding and switching are highly cooperative (25). The switching spreads from one or more nucleation points on the C ring, a phenomenon referred to as conformational spread (26). Recent experiments revealed that the flagellar motor can adapt to varied levels of phosphorylated CheY by increasing the content of FliM (27). Additional experiments suggested that it is not CheY-P binding but rather the direction of motor rotation that has the largest effect on remodeling of the FliM (28). It was suggested that there are ~34 molecules of FliM in a motor with exclusively CW rotation and ~44 molecules in a motor with CCW rotation. These *E. coli* studies also indicate that motors with even more FliM molecules may exist. It is unclear how the C ring can accommodate such a large change and if similar C-ring modifications also occur in spirochetes.

Because periplasmic flagellar motors are located at the two cell poles, it was hypothesized that spirochetal motors rotate asymmetrically at one end relative to the other during a run (29, 30). CheX is the only CheY-P phosphatase identified in the *B. burgdorferi* genome. A *cheX* mutant constantly flexes and is not able to run or reverse (31), while both *cheA2* and *cheY3* mutants constantly run in one direction (30, 32). A comparison of the motor structures from two different motions (flex and run) will likely shed new light upon the mechanisms underlying CheY-P binding and the switching of rotational direction.

FLAGELLAR ASSEMBLY

The bacterial flagellum is built from the inside out, from proximal to distal structures, in a temporally and spatially regulated fashion. Detailed insights into the flagellar assembly have been well established for *E.*

coli and *S. enterica* (33, 34). In these organisms, multiple copies of FliF form the MS ring (35), which serves as the initial base for flagellar assembly, structural maturation, and function. The MS ring also serves as a scaffold for the assembly of the C ring. FliG proteins directly associate with the cytoplasmic face of the MS ring and form the FliG ring (36, 37). FliM and FliN proteins form a stable complex with a stoichiometry of 1:4 (38, 39). The FliM-FliN$_4$ complex binds to the FliG ring to form the completed C ring. The export apparatus, which is assembled inside the MS ring and the C ring, is responsible for exporting flagellar axial protein components from the cytoplasm to the distal end of the nascent flagellar apparatus. FliE proteins are likely assembled first and form a junction between the MS ring and the rod to overcome their symmetry mismatch. Then multiple copies of FlgB, -C, -F, and -G form the rod, FlgI proteins form the P ring, and FliH proteins form the L ring. The FlgD cap assembles at the rod tip to support the assembly of the hook. After the hook assembly, the filament cap (FliD) is formed to support the assembly of the filament (40). In *B. burgdorferi*, a series of genetic mutations were introduced to arrest its assembly, and the assembling process was then imaged by cryo-ET (16).

FLAGELLAR EXPORT APPARATUS AND ITS EVOLUTIONARILY RELATED INJECTISOME

The fT3SS consists of five integral membrane proteins (FlhA, FlhB, FliP, FliQ, and FliR) and three soluble proteins (FliH, FliI, and FliJ) and is located at the center of the cytoplasmic face of the MS ring. The ATP complex promotes the export process by binding and delivering substrates to the export apparatus (41, 42). FliI is an ATPase and shows structural similarity with the α and β subunits of the F_0F_1-ATP synthase (43); it exhibits its full ATPase activity when it self-assembles into a homohexamer (44).

FliH, FliI, and FliJ coordinately deliver a chaperone-substrate complex to the export gate by binding to the docking platform of the fT3SS for substrate export. FliP, FliQ, and FliR form an export gate complex with helical symmetry (45).

fT3SSs in different bacterial species are highly conserved. In addition, they are evolutionarily related to virulence T3SSs (vT3SSs). The evolutionary relationship between the flagellum and the injectisome has garnered significant debate. The latest phylogenomic and comparative analyses of fT3SSs and vT3SSs suggest that the vT3SS derived from a flagellar ancestor. The loss of flagellum-specific genes led to an eventual loss in the motility function, but this system presumably kept the ability to secrete proteins (46).

The overall organization of the fT3SS machine in periplasmic flagella shares many features similar to those observed in the vT3SS machine (47–49) (Fig. 4). However, the ATPase complex of the periplasmic flagella is noticeably different from those observed in the injectisome (Fig. 4). There are 23 spokes and one hub in the ATPase complex of the *B. burgdorferi* periplasmic flagella. Only 6 spokes and one hub were observed in *Salmonella* injectisome, presumably optimizing for substrate recruitment and export. In contrast, the ATPase complex in the periplasmic flagella not only facilitates substrate recruitment and secretion but also supports the integrity of the C ring, which undergoes rotation and switches rotational direction between CW and CCW.

OUTLOOK AND PERSPECTIVE

Although the structure and functions of the bacterial flagellum have been studied for several decades, many important questions remain to be addressed. For example, how does the stator couple proton gradient to generate the torque? How does the C ring change its conformation to generate rotation or switch rotational direction? How

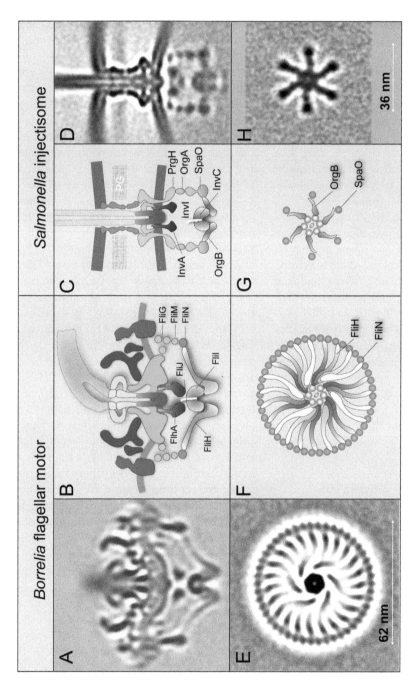

FIGURE 4 Comparison of the fT3SS from *B. burgdorferi* and the vT3SS from *Salmonella*. (A) A central section from the *B. burgdorferi* motor. (B) The fT3SS in the spirochete motor includes the ATPase complex (orange) and the export apparatus (purple) underneath the MS ring. (C and D) The vT3SS from the *Salmonella* injectisome is modeled in a similar color scheme. The difference between the two T3SSs is striking in a comparison of the cross sections of their ATPase complexes. Note that the C ring from the *B. burgdorferi* motor is a continuous ring with ~46 copies of FliN tetramer. There are 23 visible FliN spokes (E and F). There are six pods in the *Salmonella* injectisome. Only six spokes of the FliH homolog OrgB connect the ATPase complex to the SpaO molecules that compose the pod of the injectisome. Adapted from a prior publication (52), with permission.

does the proton gradient power the export apparatus and facilitate protein transport across inner membrane? Periplasmic flagella inspire additional questions: how do the periplasmic flagella coordinate their rotation from two cell poles, and how does the spirochete-specific collar assemble? Given that periplasmic flagella play critical roles in many bacterial pathogens, it will be important to understand not only the conserved structure and function among bacterial flagella but also the specific features that distinguish periplasmic flagella from others. Emerging techniques such as cryo-ET will be increasingly valuable to address many of these fundamental questions in periplasmic flagellum.

ACKNOWLEDGMENTS

The work in the Liu laboratory was supported by grants GM107629 and R01AI087946 from the National Institutes of Health.

CITATION

Chang Y, Liu J. 2019. Architecture and assembly of periplasmic flagellum. Microbiol Spectrum 7(4):PSIB-0030-2019.

REFERENCES

1. Chen S, Beeby M, Murphy GE, Leadbetter JR, Hendrixson DR, Briegel A, Li Z, Shi J, Tocheva EI, Müller A, Dobro MJ, Jensen GJ. 2011. Structural diversity of bacterial flagellar motors. *EMBO J* **30:**2972–2981.

2. Zhao X, Norris SJ, Liu J. 2014. Molecular architecture of the bacterial flagellar motor in cells. *Biochemistry* **53:**4323–4333.

3. Charon NW, Cockburn A, Li C, Liu J, Miller KA, Miller MR, Motaleb MA, Wolgemuth CW. 2012. The unique paradigm of spirochete motility and chemotaxis. *Annu Rev Microbiol* **66:**349–370.

4. Radolf JD, Caimano MJ, Stevenson B, Hu LT. 2012. Of ticks, mice and men: understanding the dual-host lifestyle of Lyme disease spirochaetes. *Nat Rev Microbiol* **10:**87–99.

5. Sultan SZ, Sekar P, Zhao X, Manne A, Liu J, Wooten RM, Motaleb MA. 2015. Motor rotation is essential for the formation of the periplasmic flagellar ribbon, cellular morphology, and *Borrelia burgdorferi* persistence within *Ixodes scapularis* tick and murine hosts. *Infect Immun* **83:**1765–1777.

6. Motaleb MA, Liu J, Wooten RM. 2015. Spirochetal motility and chemotaxis in the natural enzootic cycle and development of Lyme disease. *Curr Opin Microbiol* **28:**106–113.

7. Li C, Xu H, Zhang K, Liang FT. 2010. Inactivation of a putative flagellar motor switch protein FliG1 prevents *Borrelia burgdorferi* from swimming in highly viscous media and blocks its infectivity. *Mol Microbiol* **75:**1563–1576.

8. Sultan SZ, Manne A, Stewart PE, Bestor A, Rosa PA, Charon NW, Motaleb MA. 2013. Motility is crucial for the infectious life cycle of *Borrelia burgdorferi. Infect Immun* **81:**2012–2021.

9. Motaleb MA, Corum L, Bono JL, Elias AF, Rosa P, Samuels DS, Charon NW. 2000. *Borrelia burgdorferi* periplasmic flagella have both skeletal and motility functions. *Proc Natl Acad Sci U S A* **97:**10899–10904.

10. Moon KH, Zhao X, Manne A, Wang J, Yu Z, Liu J, Motaleb MA. 2016. Spirochetes flagellar collar protein FlbB has astounding effects in orientation of periplasmic flagella, bacterial shape, motility, and assembly of motors in *Borrelia burgdorferi. Mol Microbiol* **102:**336–348.

11. Kudryashev M, Cyrklaff M, Baumeister W, Simon MM, Wallich R, Frischknecht F. 2009. Comparative cryo-electron tomography of pathogenic Lyme disease spirochetes. *Mol Microbiol* **71:**1415–1434.

12. Liu J, Lin T, Botkin DJ, McCrum E, Winkler H, Norris SJ. 2009. Intact flagellar motor of *Borrelia burgdorferi* revealed by cryo-electron tomography: evidence for stator ring curvature and rotor/C-ring assembly flexion. *J Bacteriol* **191:**5026–5036.

13. Murphy GE, Leadbetter JR, Jensen GJ. 2006. In situ structure of the complete *Treponema primitia* flagellar motor. *Nature* **442:**1062–1064.

14. Miller MR, Miller KA, Bian J, James ME, Zhang S, Lynch MJ, Callery PS, Hettick JM, Cockburn A, Liu J, Li C, Crane BR, Charon NW. 2016. Spirochaete flagella hook proteins self-catalyse a lysinoalanine covalent crosslink for motility. *Nat Microbiol* **1:**16134.

15. Charon NW, Goldstein SF, Marko M, Hsieh C, Gebhardt LL, Motaleb MA, Wolgemuth CW, Limberger RJ, Rowe N. 2009. The flat-ribbon configuration of the periplasmic flagella of *Borrelia burgdorferi* and its relation-

ship to motility and morphology. *J Bacteriol* **191:**600–607.

16. Zhao X, Zhang K, Boquoi T, Hu B, Motaleb MA, Miller KA, James ME, Charon NW, Manson MD, Norris SJ, Li C, Liu J. 2013. Cryoelectron tomography reveals the sequential assembly of bacterial flagella in *Borrelia burgdorferi*. *Proc Natl Acad Sci U S A* **110:**14390–14395.

17. Moon KH, Zhao X, Xu H, Liu J, Motaleb MA. 2018. A tetratricopeptide repeat domain protein has profound effects on assembly of periplasmic flagella, morphology and motility of the Lyme disease spirochete *Borrelia burgdorferi*. *Mol Microbiol* **110:**634–647.

18. Leake MC, Chandler JH, Wadhams GH, Bai F, Berry RM, Armitage JP. 2006. Stoichiometry and turnover in single, functioning membrane protein complexes. *Nature* **443:**355–358.

19. Fukuoka H, Wada T, Kojima S, Ishijima A, Homma M. 2009. Sodium-dependent dynamic assembly of membrane complexes in sodium-driven flagellar motors. *Mol Microbiol* **71:**825–835.

20. Paulick A, Koerdt A, Lassak J, Huntley S, Wilms I, Narberhaus F, Thormann KM. 2009. Two different stator systems drive a single polar flagellum in *Shewanella oneidensis* MR-1. *Mol Microbiol* **71:**836–850.

21. Liu J, Howell JK, Bradley SD, Zheng Y, Zhou ZH, Norris SJ. 2010. Cellular architecture of *Treponema pallidum*: novel flagellum, periplasmic cone, and cell envelope as revealed by cryo electron tomography. *J Mol Biol* **403:**546–561.

22. Raddi G, Morado DR, Yan J, Haake DA, Yang XF, Liu J. 2012. Three-dimensional structures of pathogenic and saprophytic *Leptospira* species revealed by cryo-electron tomography. *J Bacteriol* **194:**1299–1306.

23. Kudryashev M, Cyrklaff M, Wallich R, Baumeister W, Frischknecht F. 2010. Distinct in situ structures of the *Borrelia* flagellar motor. *J Struct Biol* **169:**54–61.

24. Welch M, Oosawa K, Aizawa S, Eisenbach M. 1993. Phosphorylation-dependent binding of a signal molecule to the flagellar switch of bacteria. *Proc Natl Acad Sci U S A* **90:**8787–8791.

25. Cluzel P, Surette M, Leibler S. 2000. An ultrasensitive bacterial motor revealed by monitoring signaling proteins in single cells. *Science* **287:**1652–1655.

26. Bai F, Branch RW, Nicolau DV Jr, Pilizota T, Steel BC, Maini PK, Berry RM. 2010. Conformational spread as a mechanism for cooperativity in the bacterial flagellar switch. *Science* **327:**685–689.

27. Yuan J, Branch RW, Hosu BG, Berg HC. 2012. Adaptation at the output of the chemotaxis signalling pathway. *Nature* **484:**233–236.

28. Lele PP, Branch RW, Nathan VSJ, Berg HC. 2012. Mechanism for adaptive remodeling of the bacterial flagellar switch. *Proc Natl Acad Sci U S A* **109:**20018–20022.

29. Charon NW, Goldstein SF. 2002. Genetics of motility and chemotaxis of a fascinating group of bacteria: the spirochetes. *Annu Rev Genet* **36:**47–73.

30. Li C, Bakker RG, Motaleb MA, Sartakova ML, Cabello FC, Charon NW. 2002. Asymmetrical flagellar rotation in *Borrelia burgdorferi* non-chemotactic mutants. *Proc Natl Acad Sci U S A* **99:**6169–6174.

31. Motaleb MA, Miller MR, Li C, Bakker RG, Goldstein SF, Silversmith RE, Bourret RB, Charon NW. 2005. CheX is a phosphorylated CheY phosphatase essential for *Borrelia burgdorferi* chemotaxis. *J Bacteriol* **187:**7963–7969.

32. Motaleb MA, Sultan SZ, Miller MR, Li C, Charon NW. 2011. CheY3 of *Borrelia burgdorferi* is the key response regulator essential for chemotaxis and forms a long-lived phosphorylated intermediate. *J Bacteriol* **193:**3332–3341.

33. Chevance FF, Hughes KT. 2008. Coordinating assembly of a bacterial macromolecular machine. *Nat Rev Microbiol* **6:**455–465.

34. Macnab RM. 2003. How bacteria assemble flagella. *Annu Rev Microbiol* **57:**77–100.

35. Suzuki H, Yonekura K, Namba K. 2004. Structure of the rotor of the bacterial flagellar motor revealed by electron cryomicroscopy and single-particle image analysis. *J Mol Biol* **337:**105–113.

36. Minamino T, Imada K, Kinoshita M, Nakamura S, Morimoto YV, Namba K. 2011. Structural insight into the rotational switching mechanism of the bacterial flagellar motor. *PLoS Biol* **9:**e1000616.

37. Lee LK, Ginsburg MA, Crovace C, Donohoe M, Stock D. 2010. Structure of the torque ring of the flagellar motor and the molecular basis for rotational switching. *Nature* **466:**996–1000.

38. Delalez NJ, Berry RM, Armitage JP. 2014. Stoichiometry and turnover of the bacterial flagellar switch protein FliN. *mBio* **5:**e01216-14.

39. Brown PN, Mathews MAA, Joss LA, Hill CP, Blair DF. 2005. Crystal structure of the flagellar rotor protein FliN from *Thermotoga maritima*. *J Bacteriol* **187:**2890–2902.

40. Zhang K, Qin Z, Chang Y, Liu J, Malkowski MG, Shipa S, Li L, Qiu W, Zhang J-R, Li C. 2019. Analysis of a flagellar filament cap mutant reveals that HtrA serine protease degrades unfolded flagellin protein in the periplasm of *Borrelia burgdorferi*. *Mol Microbiol* **111:**1652–1670.

41. **Fraser GM, González-Pedrajo B, Tame JR, Macnab RM.** 2003. Interactions of FliJ with the *Salmonella* type III flagellar export apparatus. *J Bacteriol* **185:**5546–5554.

42. **Minamino T, Imada K.** 2015. The bacterial flagellar motor and its structural diversity. *Trends Microbiol* **23:**267–274.

43. **Ibuki T, Imada K, Minamino T, Kato T, Miyata T, Namba K.** 2011. Common architecture of the flagellar type III protein export apparatus and F- and V-type ATPases. *Nat Struct Mol Biol* **18:**277–282.

44. **Imada K, Minamino T, Tahara A, Namba K.** 2007. Structural similarity between the flagellar type III ATPase FliI and F1-ATPase subunits. *Proc Natl Acad Sci U S A* **104:**485–490.

45. **Kuhlen L, Abrusci P, Johnson S, Gault J, Deme J, Caesar J, Dietsche T, Mebrhatu MT, Ganief T, Macek B, Wagner S, Robinson CV, Lea SM.** 2018. Structure of the core of the type III secretion system export apparatus. *Nat Struct Mol Biol* **25:**583–590.

46. **Abby SS, Rocha EP.** 2012. The non-flagellar type III secretion system evolved from the bacterial flagellum and diversified into host-cell adapted systems. *PLoS Genet* **8:**e1002983.

47. **Hu B, Lara-Tejero M, Kong Q, Galan JE, Liu J.** 2017. In situ molecular architecture of the *Salmonella* type III secretion machine. *Cell* **168:**1065–1074.e1010.

48. **Hu B, Morado DR, Margolin W, Rohde JR, Arizmendi O, Picking WL, Picking WD, Liu J.** 2015. Visualization of the type III secretion sorting platform of *Shigella flexneri*. *Proc Natl Acad Sci U S A* **112:**1047–1052.

49. **Kawamoto A, Morimoto YV, Miyata T, Minamino T, Hughes KT, Kato T, Namba K.** 2013. Common and distinct structural features of *Salmonella* injectisome and flagellar basal body. *Sci Rep* **3:**3369.

50. **Zhu S, Nishikino T, Hu B, Kojima S, Homma M, Liu J.** 2017. Molecular architecture of the sheathed polar flagellum in *Vibrio alginolyticus*. *Proc Natl Acad Sci U S A* **114:**10966–10971.

51. **Qin Z, Lin WT, Zhu S, Franco AT, Liu J.** 2017. Imaging the motility and chemotaxis machineries in *Helicobacter pylori* by cryo-electron tomography. *J Bacteriol* **199:**e00695-16.

52. **Qin Z, Tu J, Lin T, Norris SJ, Li C, Motaleb MA, Liu J.** 2018. Cryo-electron tomography of periplasmic flagella in *Borrelia burgdorferi* reveals a distinct cytoplasmic ATPase complex. *PLoS Biol* **16:**e3000050.

Outer Membrane Vesicle-Host Cell Interactions

17

JESSICA D. CECIL,[1,*] NATALIE SIRISAENGTAKSIN,[2,*]
NEIL M. O'BRIEN-SIMPSON,[1] and ANNE MARIE KRACHLER[2]

INTRODUCTION TO OUTER MEMBRANE VESICLES

Outer membrane vesicles (OMVs) are nanosized, spherical proteoliposomes. They are secreted via vesiculation of the outer membrane by Gram-negative bacteria as part of the normal growth process (1). OMVs play diverse roles in intracellular communication, microbial virulence, and modulation of the host immune response (2).

The surface of OMVs is composed of a phospholipid bilayer with an outer layer of lipopolysaccharide (LPS), outer membrane proteins, and receptors. Internally, OMVs possess a thin layer of peptidoglycan and contain periplasmic proteins as well as nucleic acids (2–6) (Fig. 1). Specific components may selectively be enriched or depleted from OMVs, suggesting that vesiculation is a deliberate and regulated process (6, 7). Many theories on the mechanism of vesiculation exist, including the accumulation of envelope components, increased membrane curvature, or reduction in lipoprotein to peptidoglycan cross-links (2, 8–11). In all cases, vesiculation requires the outer membrane to

[1]Oral Health Cooperative Research Centre, Melbourne Dental School, Bio21 Institute, The University of Melbourne, Melbourne, Victoria 3052, Australia
[2]Department of Microbiology and Molecular Genetics, McGovern Medical School, The University of Texas Health Science Center at Houston, Houston, TX 77030
[*]These authors contributed equally.

Protein Secretion in Bacteria
Edited by Maria Sandkvist, Eric Cascales, and Peter J. Christie
© 2019 American Society for Microbiology, Washington, DC
doi:10.1128/microbiolspec.PSIB-0001-2018

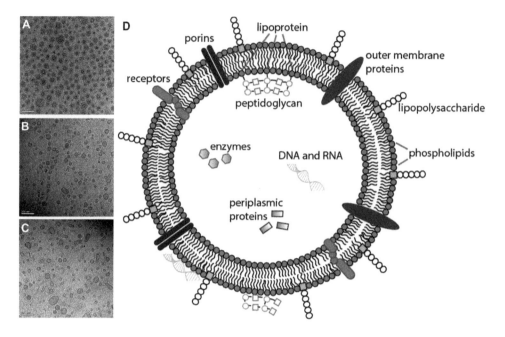

FIGURE 1 Structure and composition of bacterial OMVs. (A to C) Examples of purified OMVs isolated from *Porphyromonas gingivalis* (A), *Treponema denticola* (B), and *Tannerella forsythia* (C). OMVs were purified using an optiprep gradient and visualized using cryo-transmission electron microscopy as previously described (70). Scale bars, 200 nm. (D) Typical composition of bacterial OMVs.

separate from the underlying peptidoglycan layer and outwards budding until a vesicle can detach from the bacterial surface. The exception is the recently described mechanism of vesicle formation by "explosive" cell lysis, which is initiated by a prophage endolysin (12).

Variations in temperature, growth medium, growth phase, and many other factors can quantitatively and qualitatively influence vesiculation and OMV composition. For example, OMVs secreted by biofilms, as opposed to planktonic bacteria, are smaller, more gelatinous, and produced in larger quantities (13). Vesiculation in *Helicobacter pylori* biofilms is enhanced by the addition of serum to the growth medium (14). OMVs produced during stationary growth have different physiochemical properties than those produced during exponential phase, including differences in protein and lipid composition, a higher buoyant density, and higher negative charge (15, 16). Vesiculation

also increases in response to nutrient restriction and exposure to chemical stressors and during infection (10, 17–20). This implies that vesiculation is an envelope stress response promoting bacterial survival (21–23). Mutations leading to increased vesiculation enhance the pathogenic potential of bacteria, despite the associated increase in metabolic burden (24). Stress-inducing conditions are often used to stimulate vesiculation, although the composition of such preparations is altered by the stress (25). Alterations in amounts of lipoprotein, LPS, and other pathogen-associated molecular patterns contained in such OMVs likely trigger different immunological responses, and this should be considered when using them to study host-microbe interactions.

The sensitivity of OMVs to altered growth conditions as well as the inherent variability between OMVs of different species and strains demands precise and standardized methods of growth and OMV isolation. Additionally, biochemical and immunological characterization of OMVs is greatly complicated by their nanosize, which precludes many established methods of quantification and analysis. These obstacles complicate OMV research and make comparisons and conclusions regarding OMV-host cell interactions difficult. We and others have reported methods to separate immunogenic cell debris from OMVs and combined this separation with enumeration of the OMVs via flow cytometry and nanoparticle tracking techniques to allow quantitative comparison of OMV-host cell interactions (26–28).

Techniques used to characterize and visualize OMV-host cell interactions were recently reviewed elsewhere (29). Here we discuss the role of OMVs in microbe-host interactions, with particular emphasis on two areas that have seen major recent advances: OMVs in microbe-host interactions in the oral cavity and the same in the gastrointestinal tract.

OMV-HOST CELL INTERACTIONS IN THE ORAL CAVITY

Chronic periodontitis is an inflammatory, polymicrobial disease promoting the progressive destruction of bone and ligament tissue supporting the teeth (30). Destruction of these support structures leads to tooth mobility and loss (30, 31). Chronic periodontitis is associated with an increased risk of cardiovascular disease, adverse pregnancy outcomes, respiratory infections, and rheumatoid arthritis (32, 33).

The subgingival plaque is home to multispecies biofilms, and its development is strongly associated with the onset of chronic periodontitis. These biofilms are protected within the periodontal pocket, the space around a tooth left behind by degraded bone and tissue. Although more than 700 bacterial species make up this structured biofilm (34), only a handful of them are associated with disease progression (35). Increased concentrations of the Gram-negative, anaerobic,

proteolytic bacteria *Porphyromonas gingivalis*, *Treponema denticola*, and *Tannerella forsythia* are strongly associated with symptoms of chronic periodontitis (36–38), and all three species secrete OMVs (Fig. 1).

Role of OMVs in Plaque Formation

P. gingivalis is the most widely studied bacterial species in the periodontal disease field and a major contributor to chronic periodontitis. Implantation of *P. gingivalis* was sufficient to induce periodontitis in nonhuman primates, supporting its role as a keystone pathogen in the disease (39). Its success as an oral pathogen is largely attributed to its arsenal of pathogenicity factors, many of which are secreted from the cell in OMVs (6, 40). The most prominent of these are gingipains, secreted cysteine proteases (41). *P. gingivalis* OMVs are enriched in gingipains compared to the bacterial surface (6, 42). Gingipains promote *P. gingivalis* cell spread throughout periodontal tissue, by promoting the destruction of supportive bone and tissue within the oral cavity, and host cell invasion (43–45).

P. gingivalis-derived OMVs exhibit a functional flexibility that allows for the elimination of competitors and promotion of bacteria advantageous to *P. gingivalis* (40). They are also capable of inhibiting and dispersing competing biofilms, such as those composed of *Streptococcus gordonii*, in a gingipain-dependent manner (40). *P. gingivalis* OMVs have a strong tendency to form aggregates, both between themselves and between themselves and OMVs of other microbes, which facilitates the coaggregation of nonaggregating species, such as *T. denticola*, *Eubacterium saburreum*, and *Capnocytophaga ochracea* (4, 9). *T. forsythia* also benefits from the secretion of gingipain-containing OMVs, which enhance the attachment of whole-cell *T. forsythia* to epithelial cells (46). These studies suggest that *P. gingivalis* OMVs influence the bacterial composition of the periodontal plaque. *T. forsythia* secretes OMVs that may also encourage and strengthen subgingival biofilms. They contain both the sialidase SiaHI and the β-*N*-acetylglucosaminidase HexA, which are thought to be involved in biofilm formation (7, 47, 48).

OMV Interactions with Gingival Tissues

Nanoparticles 10 to 100 nm in diameter penetrate the extracellular matrix protecting host cells (49). Therefore, it has been proposed that periodontal OMVs act as a novel secretion system that delivers virulence factors deep into host tissues, eliminating the need for direct bacterial contact (50). The OMVs' ability to adhere to and fuse with host cells provides a mechanism of cellular entry for pathogenicity factors (51, 52).

P. gingivalis secretes OMVs that are internalized by both gingival epithelial and endothelial cells (53–55) (Fig. 2). Multiple internalization pathways for these OMVs have been proposed. These include caveolin-dependent endocytosis and a fimbria-dependent pathway which relies on lipid raft-mediated endocytosis (56). The exact manner of endocytosis depends upon vesicle size (56). Once internalized, *P. gingivalis* OMVs survive briefly within endocytic organelles before being sorted into lysosomal compartments and degraded (53, 57). Despite the fast turnover of *P. gingivalis* OMVs within host cells, their entry still causes functional impairment of epithelial cells by degrading signaling molecules required for cellular migration (58). *P. gingivalis* OMVs also disrupt oral squamous epithelial cell monolayers by inducing gingipain-dependent cell detachment (59) (Fig. 2). Likewise, *T. denticola*

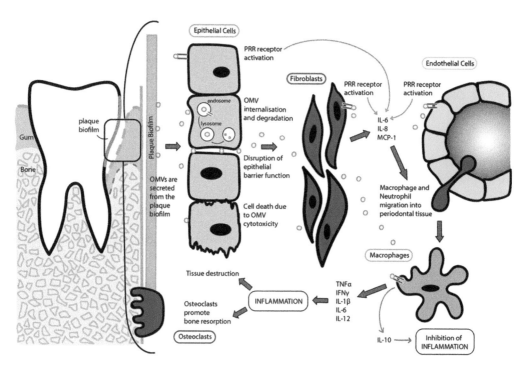

FIGURE 2 A summary of OMV interactions within gingival tissues. Bacteria accrete on the tooth's surface and form a bacterial (plaque) biofilm that is adjacent to the epithelial cells in the gingival (gum) tissue. OMVs secreted from bacteria in this plaque biofilm bind to and penetrate into the mucosal tissue and generate host cell interactions and responses. These responses culminate in a chronic inflammatory response resulting in osteoclast activation which, in turn, promotes bone resorption and eventual tooth loss.

secretes dentilisin-containing OMVs that disrupt tight junctions in epithelial monolayers. The protease activity of dentilisin results in the degradation of intercellular adhesion proteins, which facilitates bacterial penetration of underlying tissues (50).

A factor contributing to advanced periodontal tissue destruction is host cell death, reported to affect both epithelial cells (60) and fibroblasts (61) in gingival biopsy samples of periodontitis patients. *T. denticola* outer membranes and purified outer membrane proteins Msp and chymotrypsin-like proteinase are highly cytotoxic to periodontal ligament epithelial cells due to their pore-forming activity (62) (Fig. 2). Lipooligosaccharide (LOS) on *T. denticola* OMVs is highly toxic to gingival epithelial cells (63). Additionally, *P. gingivalis* OMVs hinder the proliferation of fibroblasts and endothelial cells and suppress angiogenesis *in vitro*, contributing to inhibited wound repair in periodontal tissues (64). Interestingly, low concentrations of *T. forsythia* OMVs promote cell survival in human gingival fibroblasts over short periods (7). *P. gingivalis* OMVs protect endothelial cells by reducing eNOS expression, an indicator of oxidative damage and metabolic dysfunction (65). In some cases, cell death is preceded by autophagy, a highly regulated process leading to degradation of damaged organelles and cytosolic products (66, 67). Autophagy is an important mechanism during periodontal inflammation (68) and is stimulated by peptidoglycan contained within bacterial OMVs (69).

Manipulation of the Host Immune Response by OMVs

OMVs from *P. gingivalis*, *T. denticola*, and *T. forsythia* interact intimately with mucosal epithelial cells, connective tissue fibroblasts, endothelial cells, and innate immune cells to facilitate and dysregulate inflammation within gingival tissue (26). These OMVs activate pattern recognition receptors (PRRs) in gingival epithelial cells, resulting in cell activation, cytokine secretion, or apoptotic cell death

(Fig. 2). OMVs interact with macrophages through PRRs to induce the secretion of both proinflammatory and anti-inflammatory cytokines that dysregulate chronic inflammation (70). The effects of proinflammatory cytokines on periodontitis include the activation of neutrophils, T and B lymphocytes, macrophages, natural killer cells, and osteoclasts. This promotes connective tissue destruction and alveolar bone resorption, all clinical hallmarks of chronic periodontitis (Fig. 2) (71–73).

Human periodontal ligament fibroblasts express significantly higher levels of interleukin 6 (IL-6), IL-8, and monocyte chemoattractant protein 1 (MCP-1) when exposed to *T. forsythia* OMVs than in response to *T. forsythia* bacteria (7). IL-8 and MCP-1 are chemoattractants that induce the migration of neutrophils and monocytes, respectively, to the site of inflammation (71, 74) (Fig. 2). Once recruited to tissues, monocytes are stimulated by *P. gingivalis* OMVs to induce foam cell formation (75), release inducible nitric oxide synthase and nitric oxide (NO) (76), and secrete significant amounts of cytokines (Fig. 2).

P. gingivalis OMVs induce IL-8 secretion from human gingival fibroblasts (77), and LOS on *T. denticola* OMVs induces strong proinflammatory responses from gingival fibroblasts, including secretion of IL-6, IL-8, MCP-1, prostaglandin E, matrix metalloproteinase 3, and NO (63). The continuous secretion of cytokines from host tissues promotes periodontal tissue destruction and leads to the recruitment of innate and adaptive immune cells to the site of infection.

P. gingivalis OMVs also inhibit the surface expression of human leukocyte antigen—antigen D-related (HLA-DR) molecules on human umbilical cord vascular endothelial cells, limiting major histocompatibility complex class II-induced active immunity (54). Gingipains contained within the OMVs compromise the protective action of human serum through the degradation of IgG, IgM, and complement factor C3 (78, 79). Further, *P. gingivalis* OMVs mediate LPS tolerance in monocyte/macrophage cell lines, limiting pro-inflammatory responses. Prior exposure to *P. gingivalis* OMVs greatly inhibits tumor necrosis factor alpha and IL-1β secretion in response to either *Escherichia coli* LPS or whole-cell, live *P. gingivalis* (74, 80). Likewise, *T. denticola* LOS and the outer membrane protein Msp can induce macrophage tolerance to further stimulation with intact bacteria (81) (Fig. 2).

Tolerance to LPS assists the host by minimizing inflammatory damage induced by high OMV/bacterial concentrations and prolonged or repeated exposure; however, it also benefits bacterial survival by inhibiting bacterial clearance. *P. gingivalis* OMVs also manipulate adaptive immunity and elicit humoral immune responses. Intranasal immunization with *P. gingivalis* OMVs induces *P. gingivalis*-specific IgG and IgA antibodies in blood as well as mucosal IgA in saliva (3, 82). The structural and functional stability of *P. gingivalis* OMVs, combined with their ability to induce mucosal immunity, makes them a promising candidate for future immunization studies (83).

OMV-HOST CELL INTERACTIONS IN THE GASTROINTESTINAL TRACT

The human intestinal mucosa is one of the largest interfaces mediating host-microbe interactions. The human gut alone plays host to more than 500 species of bacteria, which fall largely into four major phyla: *Bacteroidetes*, *Firmicutes*, *Actinobacteria*, and *Proteobacteria* (84, 85). These microbial populations play diverse and critical roles in intestinal homeostasis and overall human health. Certain types of bacteria may benefit the host, as they produce metabolites that allow energy recovery and aid in nutrient absorption (86). Commensal bacteria may also positively promote differentiation and proliferation of the intestinal epithelial cell layer (87). However, disruption in the composition of the gut microflora may contribute

to many different pathologies, including inflammatory bowel diseases, obesity, cancer, diabetes, and neurological disorders, among others (88–92). Although the intestinal epithelium communicates with the gut microbiota, the two entities are physically separated by a mucosal barrier (93). As in other environments, gut bacteria generate OMVs, which they use as a means of bacterium-host communication at the mucosal interface (Fig. 3). The intestinal milieu and mucus layer have been reported to stimulate vesiculation (20).

Role of OMVs in Intercellular Trafficking

Many toxins once classically viewed as secreted proteins have recently been shown to be associated with and trafficked by OMVs. These include, most prominently, Shiga toxins of enterohemorrhagic *E. coli* (EHEC) (94, 95). EHEC is a foodborne pathogen causing hemorrhagic colitis and hemolytic-uremic syndrome, a life-threatening complication of EHEC infection that may result in kidney failure (96). OMV-associated Shiga toxin is sufficient to cause hemolytic-uremic syndrome in a mouse model (97). EHEC vesicles traffic accessory toxins, including hemolysin and cytolethal distending toxin V, into host cells (98, 99). Exposure to the gastrointestinal milieu, including conditions of low pH and the presence of mucin and antimicrobial peptides, stimulates the release of OMVs from EHEC and other enteric pathogens (20, 100). A majority of cholera toxin (CTx) produced by *Vibrio cholerae* is also secreted in OMV-associated form, rather than as soluble protein as previously thought (101). Since CTx is contained within the vesicle lumen, it is taken up in a manner independent of host glycoproteins such as Lewis X glycan, which acts as a receptor for soluble CTx (102, 103). The finding that many toxins are vesicle associated significantly impacts our understanding of their entry kinetics, intracellular trafficking, and intoxication of host cells (99). Vesicle-associated LPS modulates OMV entry kinetics (28), and OMV-associated LPS is the main

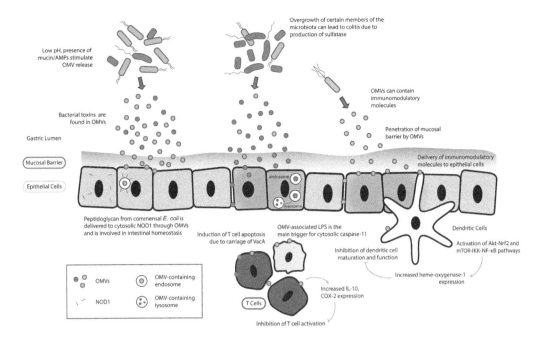

FIGURE 3 **Summary of OMV interactions with the gastrointestinal tract. Environmental pressures, such as low pH, mucin, or peptides or overgrowth of certain bacteria, result in the release of OMVs in the gastric lumen. These OMVs are able to penetrate the mucosal barrier and via different mechanisms adhere to and interact with the underlying epithelial cell and immune cells, inducing homeostasis or pathology.**

trigger for cytosolic caspase 11, which contributes to innate immune responses during EHEC infection (104).

Subversion of Gastrointestinal Immunity by OMVs

Helicobacter pylori is a pathogen that colonizes the upper gastrointestinal tract of around half the human population, although carriage remains asymptomatic in 80% of patients. However, persistent infection induces chronic inflammation of the gastric mucosa, which can lead to gastric ulceration and cancer (105). *H. pylori* releases cytotoxic proteins, including the vacuolating cytotoxin VacA, as OMV cargo (106, 107). These OMVs contribute to the carcinogenic potential of *H. pylori* infections (108), although the exact mechanisms remain elusive. OMVs are small enough to penetrate the mucosal barrier and deliver immunomodulatory molecules to the

gastric epithelium (Fig. 3). This interaction between host cells and bacterial proteins dysregulates both proinflammatory and anti-inflammatory processes, contributing to overall pathogenesis. For example, *H. pylori*-derived OMVs induce T cell apoptosis, partially due to carriage of VacA (109). At lower concentrations, they inhibit T cell activation in a cyclo-ogygenase 2-dependent manner, by increasing the expression of the anti-inflammatory cytokine IL-10 (110). *H. pylori* OMVs also subvert the activation of dendritic cells, by modulating both the Akt-Nrf2 and mTOR–IκB kinase (IKK)–NF-κB signaling axes (Fig. 3). This induces the expression of heme oxygenase 1, which regulates dendritic cell maturation and function (111).

OMVs from commensals play a potent role in mediating anti-inflammatory responses and microbial immune tolerance (Fig. 3). Peptidoglycan from commensal *E. coli* is delivered to cytosolic nucleotide-binding oligomerization domain-containing protein 1 (NOD1) through OMVs, and this process is involved in intestinal homeostasis (112). In the absence of these mechanisms, the host is more susceptible to inflammatory disease, such as colitis (113). On the other hand, overgrowth of certain members of the microbiota, for example, *Bacteroides thetaiotaomicron*, can lead to colitis due to their production of sulfatase, which permits access of OMVs to immune cells and proinflammatory responses (114).

OMVs as a Platform for Vaccine Development

Acute diarrheal diseases remain a major cause of mortality worldwide and particularly affect infants and the elderly. Despite intense efforts, vaccines that both efficiently target enteric pathogens and are effective in populations in regions of endemicity have remained elusive. However, the finding that OMVs play a crucial role in the pathogenesis of enteric infections has brought about a renewed interest and directed a new wave of research in this area. OMVs have been tested as antigens for vaccines that target pathogenic *E. coli* (115), *Salmonella* (116), *Vibrio cholerae* (117, 118), and combinations of pathogens (119). OMVs carry a complex mixture of biologically active molecules, including LPS, proteins, and phospholipids. This means that they stimulate more protective immune responses than purified proteins and elicit more robust protection than recombinant protein vaccines (120, 121). They may also offer prolonged protection compared to that of protein-based vaccines (122).

OMVs derived from *Salmonella enterica* serovar Enteritidis provided intranasally or intraperitoneally elicited both robust humoral and mucosal immune responses and were protective against *S.* Enteritidis infection (123). Engineered strains overexpressing the small RNA MicA displayed hypervesiculation, with OMVs enriched in outer membrane porins. These provoked robust Th1- and Th17-

type immune responses and were protective against lethal challenge with *Salmonella* (124). In another study, a modified strain expressing penta-acetylated lipid A was used, which produced OMVs that retained their immunogenicity against enterotoxigenic *E. coli*, but with reduced reactogenicity (120).

OUTLOOK AND FUTURE RESEARCH

Many advancements have been made in the field of OMV research over the past few years, and we have significantly furthered our understanding of OMV-host cell interactions, including in previously understudied niches, such as the oral cavity. An emerging field is the investigation of OMVs secreted by the microbiota and the contributions of these to gut homeostasis. Recent studies indicate an extensive metabolic link between the intestinal microbiota and host, and OMVs serve as a shuttle of small molecules between bacteria and the host (125). Another exciting area is the investigation of immune responses triggered by OMVs. A detailed understanding of these interactions will allow us to tailor OMV-based vaccines and improve our ability to target pathogens which have so far evaded vaccination efforts. These developments ensure that the investigation of OMV-host interactions will stay a busy and rewarding field in years to come.

ACKNOWLEDGMENTS

We apologize to all researchers in this exciting field whose studies we had to omit due to space limitations. We encourage the reader to further explore the literature to discover the abundance of fantastic research in this area.

We thank the O'Brien-Simpson and Krachler labs for helpful comments and critical reading of drafts of the manuscript.

Our research is funded by a UTSystems STAR award and the NIH (grant R01 AI132354) and by the Australian Government, Department of Industry, Innovation and Science, the Australian Dental Research Foundation (ADRF), and National Health and Medical Research Council (NHMRC) project grant APP1101935.

CITATION

Cecil JD, Sirisaengtaksin N, O'Brien-Simpson NM, Krachler AM. 2019. Outer membrane vesicle-host cell interactions. Microbiol Spectrum 7(1):PSIB-0001-2018.

REFERENCES

1. **Beveridge TJ.** 1999. Structures of gram-negative cell walls and their derived membrane vesicles. *J Bacteriol* **181:**4725–4733.
2. **Kuehn MJ, Kesty NC.** 2005. Bacterial outer membrane vesicles and the host-pathogen interaction. *Genes Dev* **19:**2645–2655.
3. **Nakao R, Hasegawa H, Ochiai K, Takashiba S, Ainai A, Ohnishi M, Watanabe H, Senpuku H.** 2011. Outer membrane vesicles of *Porphyromonas gingivalis* elicit a mucosal immune response. *PLoS One* **6:**e26163.
4. **Grenier D, Mayrand D.** 1987. Functional characterization of extracellular vesicles produced by *Bacteroides gingivalis*. *Infect Immun* **55:**111–117.
5. **Haurat MF, Aduse-Opoku J, Rangarajan M, Dorobantu L, Gray MR, Curtis MA, Feldman MF.** 2011. Selective sorting of cargo proteins into bacterial membrane vesicles. *J Biol Chem* **286:**1269–1276.
6. **Veith PD, Chen Y-Y, Gorasia DG, Chen D, Glew MD, O'Brien-Simpson NM, Cecil JD, Holden JA, Reynolds EC.** 2014. *Porphyromonas gingivalis* outer membrane vesicles exclusively contain outer membrane and periplasmic proteins and carry a cargo enriched with virulence factors. *J Proteome Res* **13:**2420–2432.
7. **Friedrich V, Gruber C, Nimeth I, Pabinger S, Sekot G, Posch G, Altmann F, Messner P, Andrukhov O, Schäffer C.** 2015. Outer membrane vesicles of *Tannerella forsythia*: biogenesis, composition, and virulence. *Mol Oral Microbiol* **30:**451–473.
8. **Zhou L, Srisatjaluk R, Justus DE, Doyle RJ.** 1998. On the origin of membrane vesicles in gram-negative bacteria. *FEMS Microbiol Lett* **163:**223–228.
9. **Grenier D.** 2013. *Porphyromonas gingivalis* outer membrane vesicles mediate coaggregation and piggybacking of *Treponema denticola* and *Lachnoanaerobaculum saburreum*. *Int J Dent* **2013:**305476.
10. **McBroom AJ, Kuehn MJ.** 2007. Release of outer membrane vesicles by Gram-negative bacteria is a novel envelope stress response. *Mol Microbiol* **63:**545–558.
11. **Roier S, Zingl FG, Cakar F, Durakovic S, Kohl P, Eichmann TO, Klug L, Gadermaier B, Weinzerl K, Prassl R, Lass A, Daum G, Reidl J, Feldman MF, Schild S.** 2016. A novel mechanism for the biogenesis of outer membrane vesicles in Gram-negative bacteria. *Nat Commun* **7:**10515.
12. **Turnbull L, Toyofuku M, Hynen AL, Kurosawa M, Pessi G, Petty NK, Osvath SR, Cárcamo-Oyarce G, Gloag ES, Shimoni R, Omasits U, Ito S, Yap X, Monahan LG, Cavaliere R, Ahrens CH, Charles IG, Nomura N, Eberl L, Whitchurch CB.** 2016. Explosive cell lysis as a mechanism for the biogenesis of bacterial membrane vesicles and biofilms. *Nat Commun* **7:**11220.
13. **Schooling SR, Beveridge TJ.** 2006. Membrane vesicles: an overlooked component of the matrices of biofilms. *J Bacteriol* **188:**5945–5957.
14. **Yonezawa H, Osaki T, Kurata S, Fukuda M, Kawakami H, Ochiai K, Hanawa T, Kamiya S.** 2009. Outer membrane vesicles of *Helicobacter pylori* TK1402 are involved in biofilm formation. *BMC Microbiol* **9:**197.
15. **Tashiro Y, Ichikawa S, Shimizu M, Toyofuku M, Takaya N, Nakajima-Kambe T, Uchiyama H, Nomura N.** 2010. Variation of physiochemical properties and cell association activity of membrane vesicles with growth phase in *Pseudomonas aeruginosa*. *Appl Environ Microbiol* **76:**3732–3739.
16. **McCaig WD, Koller A, Thanassi DG.** 2013. Production of outer membrane vesicles and outer membrane tubes by *Francisella novicida*. *J Bacteriol* **195:**1120–1132.
17. **Ellis TN, Kuehn MJ.** 2010. Virulence and immunomodulatory roles of bacterial outer membrane vesicles. *Microbiol Mol Biol Rev* **74:**81–94.
18. **McKee AS, McDermid AS, Baskerville A, Dowsett AB, Ellwood DC, Marsh PD.** 1986. Effect of hemin on the physiology and virulence of *Bacteroides gingivalis* W50. *Infect Immun* **52:**349–355.
19. **Smalley JW, Birss AJ, McKee AS, Marsh PD.** 1991. Haemin-restriction influences haemin-binding, haemagglutination and protease activity of cells and extracellular membrane vesicles of *Porphyromonas gingivalis* W50. *FEMS Microbiol Lett* **69:**63–67.
20. **Bauwens A, Kunsmann L, Marejková M, Zhang W, Karch H, Bielaszewska M, Mellmann A.**

2017. Intrahost milieu modulates production of outer membrane vesicles, vesicle-associated Shiga toxin 2a and cytotoxicity in *Escherichia coli* O157:H7 and O104:H4. *Environ Microbiol Rep* **9**:626–634.

21. **Macdonald IA, Kuehn MJ.** 2013. Stress-induced outer membrane vesicle production by *Pseudomonas aeruginosa*. *J Bacteriol* **195**:2971–2981.

22. **Marisa Heredia R, Sabrina Boeris P, Sebastián Liffourrena A, Fernanda Bergero M, Alberto López G, Inés Lucchesi G.** 2016. Release of outer membrane vesicles in *Pseudomonas putida* as a response to stress caused by cationic surfactants. *Microbiology* **162**:813–822.

23. **Kulp A, Kuehn MJ.** 2010. Biological functions and biogenesis of secreted bacterial outer membrane vesicles. *Annu Rev Microbiol* **64**:163–184.

24. **Song T, Mika F, Lindmark B, Liu Z, Schild S, Bishop A, Zhu J, Camilli A, Johansson J, Vogel J, Wai SN.** 2008. A new *Vibrio cholerae* sRNA modulates colonization and affects release of outer membrane vesicles. *Mol Microbiol* **70**:100–111.

25. **Collins BS.** 2011. Gram-negative outer membrane vesicles in vaccine development. *Discov Med* **12**:7–15.

26. **Cecil JD, O'Brien-Simpson NM, Lenzo JC, Holden JA, Chen YY, Singleton W, Gause KT, Yan Y, Caruso F, Reynolds EC.** 2016. Differential responses of pattern recognition receptors to outer membrane vesicles of three periodontal pathogens. *PLoS One* **11**:e0151967.

27. **Wieser A, Storz E, Liegl G, Peter A, Pritsch M, Shock J, Wai SN, Schubert S.** 2014. Efficient quantification and characterization of bacterial outer membrane derived nano-particles with flow cytometric analysis. *Int J Med Microbiol* **304**:1032–1037.

28. **O'Donoghue EJ, Sirisaengtaksin N, Browning DF, Bielska E, Hadis M, Fernandez-Trillo F, Alderwick L, Jabbari S, Krachler AM.** 2017. Lipopolysaccharide structure impacts the entry kinetics of bacterial outer membrane vesicles into host cells. *PLoS Pathog* **13**:e1006760.

29. **O'Donoghue EJ, Krachler AM.** 2016. Mechanisms of outer membrane vesicle entry into host cells. *Cell Microbiol* **18**:1508–1517.

30. **Oliver RC, Brown LJ.** 1993. Periodontal diseases and tooth loss. *Periodontol 2000* **2**:117–127.

31. **Shaddox LM, Walker CB.** 2010. Treating chronic periodontitis: current status, challenges, and future directions. *Clin Cosmet Investig Dent* **2**:79–91.

32. **Kim J, Amar S.** 2006. Periodontal disease and systemic conditions: a bidirectional relationship. *Odontology* **94**:10–21.

33. **Linden GJ, Lyons A, Scannapieco FA.** 2013. Periodontal systemic associations: review of the evidence. *J Periodontol* **84**(Suppl):S8–S19.

34. **Aas JA, Paster BJ, Stokes LN, Olsen I, Dewhirst FE.** 2005. Defining the normal bacterial flora of the oral cavity. *J Clin Microbiol* **43**:5721–5732.

35. **Paster BJ, Olsen I, Aas JA, Dewhirst FE.** 2006. The breadth of bacterial diversity in the human periodontal pocket and other oral sites. *Periodontol 2000* **42**:80–87.

36. **Socransky SS, Haffajee AD, Cugini MA, Smith C, Kent RL Jr.** 1998. Microbial complexes in subgingival plaque. *J Clin Periodontol* **25**:134–144.

37. **Haffajee AD, Cugini MA, Tanner A, Pollack RP, Smith C, Kent RL Jr, Socransky SS.** 1998. Subgingival microbiota in healthy, well-maintained elder and periodontitis subjects. *J Clin Periodontol* **25**:346–353.

38. **Byrne SJ, Dashper SG, Darby IB, Adams GG, Hoffmann B, Reynolds EC.** 2009. Progression of chronic periodontitis can be predicted by the levels of *Porphyromonas gingivalis* and *Treponema denticola* in subgingival plaque. *Oral Microbiol Immunol* **24**:469–477.

39. **Holt SC, Ebersole J, Felton J, Brunsvold M, Kornman KS.** 1988. Implantation of *Bacteroides gingivalis* in nonhuman primates initiates progression of periodontitis. *Science* **239**:55–57.

40. **Ho M-H, Chen C-H, Goodwin JS, Wang B-Y, Xie H.** 2015. Functional advantages of *Porphyromonas gingivalis* vesicles. *PLoS One* **10**:e0123448.

41. **Sheets SM, Robles-Price AG, McKenzie RM, Casiano CA, Fletcher HM.** 2008. Gingipain-dependent interactions with the host are important for survival of *Porphyromonas gingivalis*. *Front Biosci* **13**:3215–3238.

42. **O'Brien-Simpson NM, Veith PD, Dashper SG, Reynolds EC.** 2003. *Porphyromonas gingivalis* gingipains: the molecular teeth of a microbial vampire. *Curr Protein Pept Sci* **4**:409–426.

43. **Li N, Collyer CA.** 2011. Gingipains from *Porphyromonas gingivalis*—complex domain structures confer diverse functions. *Eur J Microbiol Immunol* **1**:41–58.

44. **Pathirana RD, O'Brien-Simpson NM, Brammar GC, Slakeski N, Reynolds EC.** 2007. Kgp and RgpB, but not RgpA, are important for *Porphyromonas gingivalis* virulence in the murine periodontitis model. *Infect Immun* **75**:1436–1442.

45. **Pathirana RD, O'Brien-Simpson NM, Reynolds EC.** 2010. Host immune responses to *Porphyromonas gingivalis* antigens. *Periodontol 2000* **52**:218–237.

46. **Inagaki S, Onishi S, Kuramitsu HK, Sharma A.** 2006. *Porphyromonas gingivalis* vesicles

enhance attachment, and the leucine-rich repeat BspA protein is required for invasion of epithelial cells by "*Tannerella forsythia.*" *Infect Immun* **74:**5023–5028.

47. **Yoshimura M, Ohara N, Kondo Y, Shoji M, Okano S, Nakano Y, Abiko Y, Nakayama K.** 2008. Proteome analysis of *Porphyromonas gingivalis* cells placed in a subcutaneous chamber of mice. *Oral Microbiol Immunol* **23:**413–418.

48. **Veith PD, Chen YY, Chen D, O'Brien-Simpson NM, Cecil JD, Holden JA, Lenzo JC, Reynolds EC.** 2015. *Tannerella forsythia* outer membrane vesicles are enriched with substrates of the type IX secretion system and TonB-dependent receptors. *J Proteome Res* **14:**5355–5366.

49. **Dane KY, Nembrini C, Tomei AA, Eby JK, O'Neil CP, Velluto D, Swartz MA, Inverardi L, Hubbell JA.** 2011. Nano-sized drug-loaded micelles deliver payload to lymph node immune cells and prolong allograft survival. *J Control Release* **156:**154–160.

50. **Chi B, Qi M, Kuramitsu HK.** 2003. Role of dentilisin in *Treponema denticola* epithelial cell layer penetration. *Res Microbiol* **154:**637–643.

51. **Galka F, Wai SN, Kusch H, Engelmann S, Hecker M, Schmeck B, Hippenstiel S, Uhlin BE, Steinert M.** 2008. Proteomic characterization of the whole secretome of *Legionella pneumophila* and functional analysis of outer membrane vesicles. *Infect Immun* **76:**1825–1836.

52. **Bomberger JM, Maceachran DP, Coutermarsh BA, Ye S, O'Toole GA, Stanton BA.** 2009. Long-distance delivery of bacterial virulence factors by *Pseudomonas aeruginosa* outer membrane vesicles. *PLoS Pathog* **5:**e1000382.

53. **Furuta N, Tsuda K, Omori H, Yoshimori T, Yoshimura F, Amano A.** 2009. *Porphyromonas gingivalis* outer membrane vesicles enter human epithelial cells via an endocytic pathway and are sorted to lysosomal compartments. *Infect Immun* **77:**4187–4196.

54. **Srisatjaluk R, Kotwal GJ, Hunt LA, Justus DE.** 2002. Modulation of gamma interferon-induced major histocompatibility complex class II gene expression by *Porphyromonas gingivalis* membrane vesicles. *Infect Immun* **70:**1185–1192.

55. **Tsuda K, Amano A, Umebayashi K, Inaba H, Nakagawa I, Nakanishi Y, Yoshimori T.** 2005. Molecular dissection of internalization of *Porphyromonas gingivalis* by cells using fluorescent beads coated with bacterial membrane vesicle. *Cell Struct Funct* **30:**81–91.

56. **Gui MJ, Dashper SG, Slakeski N, Chen YY, Reynolds EC.** 2016. Spheres of influence: *Porphyromonas gingivalis* outer membrane vesicles. *Mol Oral Microbiol* **31:**365–378.

57. **Amano A, Kuboniwa M, Takeuchi H.** 2014. Transcellular invasive mechanisms of *Porphyromonas gingivalis* in host-parasite interactions. *J Oral Biosci* **56:**58–62.

58. **Furuta N, Takeuchi H, Amano A.** 2009. Entry of *Porphyromonas gingivalis* outer membrane vesicles into epithelial cells causes cellular functional impairment. *Infect Immun* **77:**4761–4770.

59. **Nakao R, Takashiba S, Kosono S, Yoshida M, Watanabe H, Ohnishi M, Senpuku H.** 2014. Effect of *Porphyromonas gingivalis* outer membrane vesicles on gingipain-mediated detachment of cultured oral epithelial cells and immune responses. *Microbes Infect* **16:**6–16.

60. **Jarnbring F, Somogyi E, Dalton J, Gustafsson A, Klinge B.** 2002. Quantitative assessment of apoptotic and proliferative gingival keratinocytes in oral and sulcular epithelium in patients with gingivitis and periodontitis. *J Clin Periodontol* **29:**1065–1071.

61. **Arce RM, Tamayo O, Cortés A.** 2007. Apoptosis of gingival fibroblasts in periodontitis. *Colomb Med* **38:**197–209.

62. **Fenno JC, Hannam PM, Leung WK, Tamura M, Uitto VJ, McBride BC.** 1998. Cytopathic effects of the major surface protein and the chymotrypsinlike protease of *Treponema denticola*. *Infect Immun* **66:**1869–1877.

63. **Tanabe S, Bodet C, Grenier D.** 2008. *Treponema denticola* lipooligosaccharide activates gingival fibroblasts and upregulates inflammatory mediator production. *J Cell Physiol* **216:**727–731.

64. **Bartruff JB, Yukna RA, Layman DL.** 2005. Outer membrane vesicles from *Porphyromonas gingivalis* affect the growth and function of cultured human gingival fibroblasts and umbilical vein endothelial cells. *J Periodontol* **76:**972–979.

65. **Jia Y, Guo B, Yang W, Zhao Q, Jia W, Wu Y.** 2015. Rho kinase mediates *Porphyromonas gingivalis* outer membrane vesicle-induced suppression of endothelial nitric oxide synthase through ERK1/2 and p38 MAPK. *Arch Oral Biol* **60:**488–495.

66. **Levine B, Mizushima N, Virgin HW.** 2011. Autophagy in immunity and inflammation. *Nature* **469:**323–335.

67. **Kroemer G, Mariño G, Levine B.** 2010. Autophagy and the integrated stress response. *Mol Cell* **40:**280–293.

68. **Bullon P, Cordero MD, Quiles JL, Ramirez-Tortosa MC, Gonzalez-Alonso A, Alfonsi S, García-Marín R, de Miguel M, Battino M.** 2012. Autophagy in periodontitis patients and gingival fibroblasts: unraveling the link between

chronic diseases and inflammation. *BMC Med* **10**:122.

69. **Irving AT, Mimuro H, Kufer TA, Lo C, Wheeler R, Turner LJ, Thomas BJ, Malosse C, Gantier MP, Casillas LN, Votta BJ, Bertin J, Boneca IG, Sasakawa C, Philpott DJ, Ferrero RL, Kaparakis-Liaskos M.** 2014. The immune receptor NOD1 and kinase RIP2 interact with bacterial peptidoglycan on early endosomes to promote autophagy and inflammatory signaling. *Cell Host Microbe* **15**:623–635.

70. **Cecil JD, O'Brien-Simpson NM, Lenzo JC, Holden JA, Singleton W, Perez-Gonzalez A, Mansell A, Reynolds EC.** 2017. Outer membrane vesicles prime and activate macrophage inflammasomes and cytokine secretion *in vitro* and *in vivo*. *Front Immunol* **8**:1017.

71. **Silva TA, Garlet GP, Fukada SY, Silva JS, Cunha FQ.** 2007. Chemokines in oral inflammatory diseases: apical periodontitis and periodontal disease. *J Dent Res* **86**:306–319.

72. **Stashenko P, Teles R, D'Souza R.** 1998. Periapical inflammatory responses and their modulation. *Crit Rev Oral Biol Med* **9**:498–521.

73. **Graunaite I, Lodiene G, Maciulskiene V.** 2011. Pathogenesis of apical periodontitis: a literature review. *J Oral Maxillofac Res* **2**(4):e1.

74. **Waller T, Kesper L, Hirschfeld J, Dommisch H, Kölpin J, Oldenburg J, Uebele J, Hoerauf A, Deschner J, Jepsen S, Bekeredjian-Ding I.** 2016. *Porphyromonas gingivalis* outer membrane vesicles induce selective tumor necrosis factor tolerance in a Toll-like receptor 4 and mTOR-dependent manner. *Infect Immun* **84**:1194–1204.

75. **Qi M, Miyakawa H, Kuramitsu HK.** 2003. *Porphyromonas gingivalis* induces murine macrophage foam cell formation. *Microb Pathog* **35**:259–267.

76. **Imayoshi R, Cho T, Kaminishi H.** 2011. NO production in RAW264 cells stimulated with *Porphyromonas gingivalis* extracellular vesicles. *Oral Dis* **17**:83–89.

77. **Hijiya T, Shibata Y, Hayakawa M, Abiko Y.** 2010. A monoclonal antibody against *fimA* type II *Porphyromonas gingivalis* inhibits IL-8 production in human gingival fibroblasts. *Hybridoma (Larchmt)* **29**:201–204.

78. **Potempa J, Mikolajczyk-Pawlinska J, Brassell D, Nelson D, Thøgersen IB, Enghild JJ, Travis J.** 1998. Comparative properties of two cysteine proteinases (gingipains R), the products of two related but individual genes of *Porphyromonas gingivalis*. *J Biol Chem* **273**:21648–21657.

79. **Grenier D.** 1992. Inactivation of human serum bactericidal activity by a trypsinlike protease isolated from *Porphyromonas gingivalis*. *Infect Immun* **60**:1854–1857.

80. **Duncan L, Yoshioka M, Chandad F, Grenier D.** 2004. Loss of lipopolysaccharide receptor CD14 from the surface of human macrophage-like cells mediated by *Porphyromonas gingivalis* outer membrane vesicles. *Microb Pathog* **36**:319–325.

81. **Nussbaum G, Ben-Adi S, Genzler T, Sela M, Rosen G.** 2009. Involvement of Toll-like receptors 2 and 4 in the innate immune response to *Treponema denticola* and its outer sheath components. *Infect Immun* **77**:3939–3947.

82. **Kesavalu L, Ebersole JL, Machen RL, Holt SC.** 1992. *Porphyromonas gingivalis* virulence in mice: induction of immunity to bacterial components. *Infect Immun* **60**:1455–1464.

83. **Nakao R, Hasegawa H, Dongying B, Ohnishi M, Senpuku H.** 2016. Assessment of outer membrane vesicles of periodontopathic bacterium *Porphyromonas gingivalis* as possible mucosal immunogen. *Vaccine* **34**:4626–4634.

84. **Human Microbiome Project Consortium.** 2012. Structure, function and diversity of the healthy human microbiome. *Nature* **486**:207–214.

85. **Faust K, Sathirapongsasuti JF, Izard J, Segata N, Gevers D, Raes J, Huttenhower C.** 2012. Microbial co-occurrence relationships in the human microbiome. *PLOS Comput Biol* **8**:e1002606.

86. **O'Hara AM, Shanahan F.** 2006. The gut flora as a forgotten organ. *EMBO Rep* **7**:688–693.

87. **Davenport AP, Nunez DJ, Hall JA, Kaumann AJ, Brown MJ.** 1989. Autoradiographical localization of binding sites for porcine [125I] endothelin-1 in humans, pigs, and rats: functional relevance in humans. *J Cardiovasc Pharmacol* **13**(Suppl 5):S166–S170.

88. **Zeng MY, Inohara N, Nuñez G.** 2017. Mechanisms of inflammation-driven bacterial dysbiosis in the gut. *Mucosal Immunol* **10**:18–26.

89. **Manichanh C, Borruel N, Casellas F, Guarner F.** 2012. The gut microbiota in IBD. *Nat Rev Gastroenterol Hepatol* **9**:599–608.

90. **Turnbaugh PJ, Ley RE, Mahowald MA, Magrini V, Mardis ER, Gordon JI.** 2006. An obesity-associated gut microbiome with increased capacity for energy harvest. *Nature* **444**:1027–1031.

91. **Brennan CA, Garrett WS.** 2016. Gut microbiota, inflammation, and colorectal cancer. *Annu Rev Microbiol* **70**:395–411.

92. **Naseer MI, Bibi F, Alqahtani MH, Chaudhary AG, Azhar EI, Kamal MA, Yasir M.** 2014. Role of gut microbiota in obesity, type 2 diabetes and Alzheimer's disease. *CNS Neurol Disord Drug Targets* **13**:305–311.

93. **Johansson ME, Phillipson M, Petersson J, Velcich A, Holm L, Hansson GC.** 2008. The

inner of the two Muc2 mucin-dependent mucus layers in colon is devoid of bacteria. *Proc Natl Acad Sci U S A* **105:**15064–15069.

94. **Kolling GL, Matthews KR.** 1999. Export of virulence genes and Shiga toxin by membrane vesicles of *Escherichia coli* O157:H7. *Appl Environ Microbiol* **65:**1843–1848.

95. **Kunsmann L, Rüter C, Bauwens A, Greune L, Glüder M, Kemper B, Fruth A, Wai SN, He X, Lloubes R, Schmidt MA, Dobrindt U, Mellmann A, Karch H, Bielaszewska M.** 2015. Virulence from vesicles: novel mechanisms of host cell injury by *Escherichia coli* O104:H4 outbreak strain. *Sci Rep* **5:**13252.

96. **Mayer CL, Leibowitz CS, Kurosawa S, Stearns-Kurosawa DJ.** 2012. Shiga toxins and the pathophysiology of hemolytic uremic syndrome in humans and animals. *Toxins (Basel)* **4:**1261–1287.

97. **Kim SH, Lee YH, Lee SH, Lee SR, Huh JW, Kim SU, Chang KT.** 2011. Mouse model for hemolytic uremic syndrome induced by outer membrane vesicles of *Escherichia coli* O157:H7. *FEMS Immunol Med Microbiol* **63:**427–434.

98. **Bielaszewska M, Rüter C, Kunsmann L, Greune L, Bauwens A, Zhang W, Kuczius T, Kim KS, Mellmann A, Schmidt MA, Karch H.** 2013. Enterohemorrhagic *Escherichia coli* hemolysin employs outer membrane vesicles to target mitochondria and cause endothelial and epithelial apoptosis. *PLoS Pathog* **9:**e1003797.

99. **Bielaszewska M, Rüter C, Bauwens A, Greune L, Jarosch KA, Steil D, Zhang W, He X, Lloubes R, Fruth A, Kim KS, Schmidt MA, Dobrindt U, Mellmann A, Karch H.** 2017. Host cell interactions of outer membrane vesicle-associated virulence factors of enterohemorrhagic *Escherichia coli* O157: intracellular delivery, trafficking and mechanisms of cell injury. *PLoS Pathog* **13:**e1006159.

100. **Elmi A, Dorey A, Watson E, Jagatia H, Inglis NF, Gundogdu O, Bajaj-Elliott M, Wren BW, Smith DGE, Dorrell N.** 2018. The bile salt sodium taurocholate induces *Campylobacter jejuni* outer membrane vesicle production and increases OMV-associated proteolytic activity. *Cell Microbiol* **20:**e12814.

101. **Chatterjee D, Chaudhuri K.** 2011. Association of cholera toxin with *Vibrio cholerae* outer membrane vesicles which are internalized by human intestinal epithelial cells. *FEBS Lett* **585:**1357–1362.

102. **Cervin J, Wands AM, Casselbrant A, Wu H, Krishnamurthy S, Cvjetkovic A, Estelius J, Dedic B, Sethi A, Wallom KL, Riise R, Bäckström M, Wallenius V, Platt FM, Lebens M, Teneberg S, Fändriks L, Kohler JJ, Yrlid U.** 2018. GM1 ganglioside-independent intoxication by cholera toxin. *PLoS Pathog* **14:**e1006862.

103. **Rasti ES, Schappert ML, Brown AC.** 2018. Association of *Vibrio cholerae* 569B outer membrane vesicles with host cells occurs in a GM1-independent manner. *Cell Microbiol* **20:** e12828.

104. **Vanaja SK, Russo AJ, Behl B, Banerjee I, Yankova M, Deshmukh SD, Rathinam VAK.** 2016. Bacterial outer membrane vesicles mediate cytosolic localization of LPS and caspase-11 activation. *Cell* **165:**1106–1119.

105. **Amieva M, Peek RM Jr.** 2016. Pathobiology of *Helicobacter pylori*-induced gastric cancer. *Gastroenterology* **150:**64–78.

106. **Fiocca R, Necchi V, Sommi P, Ricci V, Telford J, Cover TL, Solcia E.** 1999. Release of *Helicobacter pylori* vacuolating cytotoxin by both a specific secretion pathway and budding of outer membrane vesicles. Uptake of released toxin and vesicles by gastric epithelium. *J Pathol* **188:**220–226.

107. **Keenan J, Day T, Neal S, Cook B, Perez-Perez G, Allardyce R, Bagshaw P.** 2000. A role for the bacterial outer membrane in the pathogenesis of *Helicobacter pylori* infection. *FEMS Microbiol Lett* **182:**259–264.

108. **Chitcholtan K, Hampton MB, Keenan JI.** 2008. Outer membrane vesicles enhance the carcinogenic potential of *Helicobacter pylori*. *Carcinogenesis* **29:**2400–2405.

109. **Winter J, Letley D, Rhead J, Atherton J, Robinson K.** 2014. *Helicobacter pylori* membrane vesicles stimulate innate pro- and anti-inflammatory responses and induce apoptosis in Jurkat T cells. *Infect Immun* **82:**1372–1381.

110. **Hock BD, McKenzie JL, Keenan JI.** 2017. *Helicobacter pylori* outer membrane vesicles inhibit human T cell responses via induction of monocyte COX-2 expression. *Pathog Dis* **75:** ftx034.

111. **Ko SH, Rho DJ, Jeon JI, Kim YJ, Woo HA, Kim N, Kim JM.** 2016. Crude preparations of *Helicobacter pylori* outer membrane vesicles induce upregulation of heme oxygenase-1 via activating Akt-Nrf2 and mTOR–IκB kinase–NF-κB pathways in dendritic cells. *Infect Immun* **84:**2162–2174.

112. **Cañas MA, Fàbrega MJ, Giménez R, Badia J, Baldomà L.** 2018. Outer membrane vesicles from probiotic and commensal *Escherichia coli* activate NOD1-mediated immune responses in intestinal epithelial cells. *Front Microbiol* **9:**498.

113. **Fàbrega MJ, Rodríguez-Nogales A, Garrido-Mesa J, Algieri F, Badía J, Giménez R, Gálvez J, Baldomà L.** 2017. Intestinal anti-inflammatory effects of outer membrane vesicles from *Esche-*

richia coli Nissle 1917 in DSS-experimental colitis in mice. *Front Microbiol* 8:1274.

114. **Hickey CA, Kuhn KA, Donermeyer DL, Porter NT, Jin C, Cameron EA, Jung H, Kaiko GE, Wegorzewska M, Malvin NP, Glowacki RW, Hansson GC, Allen PM, Martens EC, Stappenbeck TS.** 2015. Colitogenic *Bacteroides thetaiotaomicron* antigens access host immune cells in a sulfatase-dependent manner via outer membrane vesicles. *Cell Host Microbe* 17:672–680.

115. **Roy K, Hamilton DJ, Munson GP, Fleckenstein JM.** 2011. Outer membrane vesicles induce immune responses to virulence proteins and protect against colonization by enterotoxigenic *Escherichia coli*. *Clin Vaccine Immunol* 18:1803–1808.

116. **De Benedetto G, Alfini R, Cescutti P, Caboni M, Lanzilao L, Necchi F, Saul A, MacLennan CA, Rondini S, Micoli F.** 2017. Characterization of O-antigen delivered by generalized modules for membrane antigens (GMMA) vaccine candidates against nontyphoidal *Salmonella*. *Vaccine* 35:419–426.

117. **Adriani R, Mousavi Gargari SL, Nazarian S, Sarvary S, Noroozi N.** 2018. Immunogenicity of *Vibrio cholerae* outer membrane vesicles secreted at various environmental conditions. *Vaccine* 36:322–330.

118. **Sinha R, Howlader DR, Ta A, Mitra S, Das S, Koley H.** 2017. Retinoic acid pre-treatment down regulates *V. cholerae* outer membrane vesicles induced acute inflammation and enhances mucosal immunity. *Vaccine* 35:3534–3547.

119. **Leitner DR, Lichtenegger S, Temel P, Zingl FG, Ratzberger D, Roier S, Schild-Prüfert K, Feichter S, Reidl J, Schild S.** 2015. A combined vaccine approach against *Vibrio cholerae*

and ETEC based on outer membrane vesicles. *Front Microbiol* 6:823.

120. **Hays MP, Houben D, Yang Y, Luirink J, Hardwidge PR.** 2018. Immunization with Skp delivered on outer membrane vesicles protects mice against enterotoxigenic *Escherichia coli* challenge. *Front Cell Infect Microbiol* 8:132.

121. **Wang S, Gao J, Wang Z.** 2018. Outer membrane vesicles for vaccination and targeted drug delivery. *Wiley Interdisc Rev Nanomed Nanobiotechnol* 2018:e1523.

122. **Schager AE, Dominguez-Medina CC, Necchi F, Micoli F, Goh YS, Goodall M, Flores-Langarica A, Bobat S, Cook CNL, Arcuri M, Marini A, King LDW, Morris FC, Anderson G, Toellner KM, Henderson IR, López-Macías C, MacLennan CA, Cunningham AF.** 2018. IgG responses to porins and lipopolysaccharide within an outer membrane-based vaccine against nontyphoidal *Salmonella* develop at discordant rates. *mBio* 9:e02379-17.

123. **Liu Q, Yi J, Liang K, Zhang X, Liu Q.** 2017. Outer membrane vesicles derived from *Salmonella* Enteritidis protect against the virulent wild-type strain infection in a mouse model. *J Microbiol Biotechnol* 27:1519–1528.

124. **Choi HI, Kim M, Jeon J, Han JK, Kim KS.** 2017. Overexpression of MicA induces production of OmpC-enriched outer membrane vesicles that protect against *Salmonella* challenge. *Biochem Biophys Res Commun* 490:991–996.

125. **Bryant WA, Stentz R, Le Gall G, Sternberg MJE, Carding SR, Wilhelm T.** 2017. *In silico* analysis of the small molecule content of outer membrane vesicles produced by *Bacteroides thetaiotaomicron* indicates an extensive metabolic link between microbe and host. *Front Microbiol* 8:2440.

Type I Secretion Systems— One Mechanism for All?

18

OLIVIA SPITZ,[1] ISABELLE N. ERENBURG,[1] TOBIAS BEER,[1]
KERSTIN KANONENBERG,[1] I. BARRY HOLLAND,[2] and LUTZ SCHMITT[1]

INTRODUCTION

Gram-negative bacteria are equipped with at least seven dedicated secretion systems that mediate the export of proteins beyond the outer membrane (1, 2). These are called type 1 to 6 and type 9 secretion systems (T1SS to T6SS and T9SS). Among those, T3SS, T4SS, and T6SS are even capable of delivering their cargo directly into the cytosol of the host cell. In this minireview, we place the major emphasis on the hemolysin A (HlyA) secretion system in *Escherichia coli*. This is by far the most studied and illustrates very well the largely conserved, essential features of T1SS. Interestingly, however, an important mechanistic variation in the translocation of some of the unusually extended giant RTX proteins—adhesins—was discovered recently (3) and is also discussed.

T1SS substrates are usually defined by the presence of several blocks of nonapeptide-binding sequences with the consensus GGxGxDxUx (4, 5), where x can be any amino acid and U is a large hydrophobic amino acid. The exceptions are the SiiE-like adhesins (Fig. 1) (6). These nonapeptides gave rise to the abbreviation RTX (repeats in toxins), the name for the family.

[1]Institute of Biochemistry, Heinrich Heine University Düsseldorf, Düsseldorf, Germany
[2]Institute of Genetics and Microbiology, University of Paris-Sud, Orsay, France
Protein Secretion in Bacteria
Edited by Maria Sandkvist, Eric Cascales, and Peter J. Christie
© 2019 American Society for Microbiology, Washington, DC
doi:10.1128/microbiolspec.PSIB-0003-2018

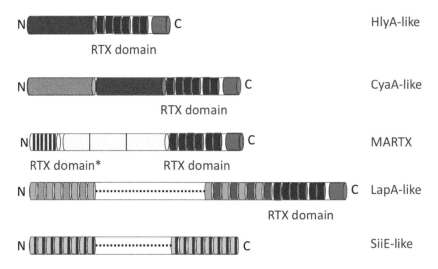

FIGURE 1 Architecture of substrates of T1SS. The primary structure of a canonical substrate of a T1SS is shown as white cylinder with the N and C termini labeled by "N" and "C," respectively. The secretion sequence (approximately 50 to 100 amino acids depending on the substrate) at the C terminus is in red, the GG repeats forming the classic RTX domain are in blue (six GG repeats as in the case of HlyA have been chosen as an example), and the functional, N-terminal domain is in brown. However, the number and types of architectures of this functional domain have increased in recent years. HlyA-like proteins contain only one domain with dedicated activity (pore-forming activity in the case of HlyA), while, for example, CyaA-like proteins contain two domains, which possess an adenylate cyclase (light brown) and a pore-forming (brown) activity in the case of CyaA. A third class are MARTX proteins (exemplified here by a MARTX protein from *V. cholerae*). The effector domains (yellow and separated by black vertical lines) that are autocatalytically excised after secretion are flanked by an N-terminal RTX-like domain (marked as RTX domain*) and a C-terminal RTX domain. The C-terminal domain corresponds to the canonical sequence, while the conserved aspartate is missing in the N-terminal one. Another architecture is present in LapA-like adhesins (or bacterial transglutaminase-like cysteine proteinases) that contain multiple, different domains. In the case of LapA, two different colors indicate two different domains. However, the number of different domains is not restricted to two. Additionally, the double-alanine motif in the N termini of LapA-like RTX adhesins is not shown. Finally, SiiE-like adhesins contain multiple identical domains, such as the 53 copies of the BIg domain in the case of SiiE (6, 71). The vertical blue line indicates that the GG repeats are integrated within the Ig-like domains and do not form a separate RTX domain. Please note that the drawing of the functional domains is not to scale.

These motifs, also called GG repeats, specifically bind Ca^{2+} (see below) and are implicated in posttranslation folding. The RTX domain (Fig. 1) is located N terminal to the secretion signal at the extreme C terminus.

Like T1SS substrates, the very large and widespread group of peptide bacteriocins in Gram-negative bacteria (7–9) also require an ABC transporter, a membrane fusion protein (MFP), and an outer membrane (OM) protein for secretion. However, these antimicrobials lack RTX repeats, have a cleavable N-terminal secretion sequence instead of the "classical" C-terminal signal, and have

a quite distinctive translocation mechanism (10, 11): alternating access rather than extrusion through an OM "tunnel." In view of these properties, we decided not to include them in this minireview. The interested reader is directed to references 7 to 11.

The first molecular identification of a T1SS, the secretion machinery of the pore-forming toxin HlyA from *E. coli* (12), was made in the 1980s and 1990s (13–15) with the demonstration that two inner membrane proteins, an ABC transporter and an MFP, encoded together with the toxin in the same operon were required for secretion. A fourth gene in the *hly* operon encoded an acyltransferase, HlyC (16), catalyzing the posttranslational modification of two internal lysine residues (17–19). This modification, with fatty acids ranging from C_{14} to C_{17} in length, requires acyl carrier protein (20) and is essential for HlyA to form a pore in the host membrane. Thus, only the acylated, toxic form of hemolysin should be called HlyA, while the nonacylated form should correctly be called pro-HlyA. Such acylation conferring toxicity is observed not only in HlyA but also in other hemolysins, leukotoxins, and cytolysins that are members of the RTX family (4). The recently published crystal structure of a homologue of the *E. coli* HlyC (21) allows a more detailed understanding of how acylation is installed. Notably, however, acylation is not required for secretion into the extracellular space. On the other hand, the proteins encoded in the *hly* operon are not sufficient for secretion of (pro-)HlyA. The OM component of the translocon is TolC. The TolC protein is encoded elsewhere in the chromosome and was first described for a related T1SS by Wandersman and Delepelaire (22).

Equally important, the work by the laboratories of Koronakis, Holland, and Goebel demonstrated that substrate secretion by the T1SS is a one-step process, i.e., directly from the cytosol into the extracellular space without any periplasmic intermediate. Furthermore, these data established that the entire process was Sec independent, relying on a novel C-terminal secretion signal (5, 23–31).

However, the view that T1SS is mediated by the one-step translocation of proteins has been challenged. Recently, so-called periplasmic intermediates for a proposed two-step secretion process were described for the adhesins LapA and IBA (3). The exciting results identified a "retention module" (RM) at the N terminus that anchors the adhesion to the cell surface by stalling further translocation. This leaves a stalled short stub in the periplasm, apparently stuck in TolC, and a fully translocated, functional adhesin in the extracellular space. When conditions change, for example, in the case of LapA, to conditions unfavorable for biofilm formation, proteolysis removes the RM and releases the adhesin. Therefore, the secretion of the adhesin, as the authors described it, occurs in two steps. Our interpretation is that this is an exciting and important variation of the T1SS but that translocation is still effectively one step, and therefore, we prefer to call the adhesin-TolC-RM complex a pseudoperiplasmic intermediate.

Here we summarize our current knowledge of the molecular processes that underlie the T1SS and focus on the molecular events that result in secretion of substrates that harbor a C-terminal secretion sequence.

THE SUBSTRATES OF THE T1SS

The N-terminal domain of an RTX protein like HlyA contains one functional domain, the HlyA pore-forming toxin. CyaA from *Bordetella pertussis* (32) harbors an HlyA-like toxin but also an adenylate cyclase that, following translocation into the cytosol of a host cell, manipulates cAMP levels. More complex architectures are present in MARTX (multifunctional autoprocessing repeats in toxins), LapA, and SiiE-like proteins. MARTX proteins are of enormous size (approximately 500 to 900 kDa). This protein family is encoded in a chromosomal island in human pathogens such as *Vibrio cholerae* (33, 34). The extreme N-terminal part of these proteins is composed of an RTX-like domain that, however, lacks the conserved aspartate residue that normally coordinates the Ca^{2+} ion. Spaced between this domain and the RTX domain near the C terminus are effector proteins that are autoprocessed, posttranslocationally, to release a cocktail of different effectors into a host cell (35). However, little is still known about the mechanism of secretion of MARTX proteins. The RTX adhesins, LapA from *Pseudomonas fluorescens*, and SiiE, the RTX-like protein from *Salmonella*, are even larger, reaching up to 1.5 MDa (3, 36, 37). In LapA

(3, 38), the functional domain contains a varying number of domains that mediate adhesin functions. Strikingly, the SiiE-like adhesins deviate from the canonical architecture of RTX proteins, particularly with respect to calcium binding (39). Ca^{2+} binding sites are distributed virtually throughout the entire molecule, which is composed of 53 bacterial immunoglobulin-like (BIg) domains constituting the functional domain—the adhesin (Fig. 1). Ca^{2+} type I sites (three aspartate residues) fulfill the role of RTX repeats in secretion and are positioned at all the interfaces between two BIg domains (6). On the other hand, the translocon is composed of the familiar tripartite complex; translocation depends on a C-terminal secretion sequence (40) inferred to be extruded first (39).

A C-terminal secretion signal remains as a signature characteristic of RTX substrates. Signals appear to be conserved only within groups of related proteins, with no evidence of widespread conservation as far as we are aware. For the hemolysin group, competitive hypotheses have postulated a specific linear code, a structural code, or a combination of the two, but the question remains unresolved (see the extensive discussion in reference 41).

RTX Motifs and Ca^{2+} Promote Extracellular Folding of Substrates

A bioinformational approach based on the presence of the GG repeats revealed more than 1,000 putative RTX proteins in approximately 250 bacterial species (4). Since that study was published in 2010, the number of putative RTX proteins is necessarily much larger today, given the enormous number of genomes now sequenced. However, only the compilation of Linhartová et al. (4) is currently available. The number of identified RTX repeats ranged from below 10 to more than 40, with a slight tendency of the number of repeats to correlate with molecular weight. Additionally, more than 90% of the putative RTX proteins displayed an isoelectric point below 5.0, suggesting that electrophoretic

mobility (42) might be important for the secretion process.

Structural studies of the alkaline protease from *Pseudomonas aeruginosa* (43) and other substrates of the T1SS (6, 44, 45) confirmed that the nonapeptide repeats bind Ca^{2+} ions. Two GG repeats coordinate one Ca^{2+} ion by interaction of the side chain of the aspartate residue and the carbonyl oxygens of the amino acids forming the repeat. This architecture creates a right-handed, so-called β-roll motif (Fig. 2). Functional *in vitro* studies demonstrated that Ca^{2+} ions are a strict requirement for folding. In other words, in the absence of Ca^{2+} ions, substrates, such as HlyA from *E. coli* (46–49) or CyaA from *Bordetella pertussis* (50–52), remain unfolded or in a molten globular state (53). Subsequent studies revealed that the dissociation constant of Ca^{2+} ions from the RTX domain is in the high micromolar range. The concentration of free Ca^{2+} ions in the bacterial cytosol is strictly regulated and remains in the high nanomolar range (100 to 300 nM in *E. coli*) (54). Secretion of RTX proteins therefore presumably occurs in the unfolded state. This hypothesis was indeed experimentally verified by the fusion of maltose binding protein to a C-terminal fragment of HlyA that only harbored the secretion signal and three of the six GG repeats (55). Given that the extracellular concentration of Ca^{2+} ions is normally in the millimolar range (54), this suggests that RTX repeats immediately bind Ca^{2+} upon exit from the bacterium. Elegant *in vitro* studies with CyaA have also demonstrated that binding of Ca^{2+} ions to the RTX domain induces immediate folding of the entire protein (46–48), suggesting that Ca^{2+} ions act as a chemical foldase.

CURRENT WORKING MODEL FOR CLASSIC T1SS

The process of secreting a substrate by a T1SS starts at the ribosome. However, only after the extreme C terminus of the substrate

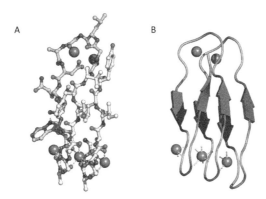

FIGURE 2 **Structure of GG repeats of alkaline protease (PDB entry 1KAP) from _P. aeruginosa_ in its Ca²⁺-bound state, resulting in the classic β-roll motif. (A) The five Ca²⁺ ions are shown as blue spheres. For simplicity, only the first three GG repeats are shown in ball-and-stick representation. The carbon atoms of GG repeat one are in gray, the carbon atoms of the second GG repeat in green, and the ones of the third repeat in yellow. The interactions of repeat one with the bound Ca²⁺ ion are indicated by gray dashed lines, and the interaction of the third repeat with the bound Ca²⁺ ions is in yellow. As it is evident, one Ca²⁺ ion is coordinated by repeat _n_ and repeat _n_ + 2. (B) RTX domain of alkaline protease from _P. aeruginosa_ in cartoon representation. The orientation is identical to that in panel A, and the gray and yellow dashed lines indicate the interactions.**

containing the secretion signal (Fig. 1), around 50 to 100 amino acids, has been synthesized will secretion be initiated, since all information necessary and sufficient for secretion is encoded in the secretion signal. Bearing in mind that the sizes of T1SS proteins range from 20 kDa up to 1,500 kDa, two obvious questions arise: why do substrates of T1SS not aggregate and precipitate prior to secretion, and why are these proteins not immediately degraded by cytosolic proteases? Unfortunately, we do not yet have answers to these important questions.

In the second step, the unfolded substrate interacts with both of the two membrane proteins of the inner membrane, the ABC transporter and the MFP (56, 57). Based on cross-linking studies with the HlyA system, these two proteins were shown to form a stable complex in the inner membrane, a dimer of the ABC transporter and a trimer of the MFP (57). However, the remarkable similarity of the T1SS translocon to tripartite drug efflux pumps, such as the AcrB-AcrA-TolC system from _E. coli_, in which there is a 2:6 stoichiometry (ABC:MFP) (58), makes it most likely that the T1SS MFP is also a hexamer. Nevertheless, further research should be undertaken to resolve this obvious discrepancy. Deletion studies by the Koronakis laboratory showed that the cytoplasmic domain of the MFP is required to recruit the OM component, TolC in the case of the HlyA machinery (56). However, the engagement occurred only in the presence of the substrate, indicating that docking of HlyA with the inner membrane complex transmits a signal to the periplasmic domain of HlyD that results in the formation of a transient HlyB-HlyD-TolC complex, a so-called "channel-tunnel" bridging the entire distance from the cytosol to the extracellular space across the periplasm and two membranes. The timing of these events also explains why deletion or inactivation of one of the three translocon components completely abolishes secretion without the appearance of a periplasmic intermediate. Finally, biophysical studies with the isolated nucleotide binding domain of HlyB, the ABC transporter of the HlyA secretion machinery, demonstrated an interaction with the substrate in the low micromolar range that required the secretion signal (59).

As soon as the outer membrane protein is engaged and a continuous channel tunnel has formed, the substrate enters the translocation pathway (Fig. 3A). For HlyA, it was experimentally demonstrated that secretion is directional, with the C terminus extended first onto the cell surface (60). Furthermore, the entire process proceeds with a secretion rate of 16 amino acids per transporter per second (61). At this stage, Ca^{2+} ions must bind to the RTX motifs and induce folding as soon as the substrate appears at the cell surface

(Fig. 3A). This should prevent backsliding of the entire protein. Interestingly, reducing the external Ca^{2+} concentration below the dissociation constant of the ion from the RTX motif did not reduce the secretion rate in the HlyA system (61). This clearly demonstrates that the secretion rate is independent of Ca^{2+} and that Ca^{2+}-induced folding does not represent a driving force for secretion. A seemingly different scenario was observed for the much larger adenylate cyclase toxin (CyaA) from *B. pertussis* (62). A Ca^{2+} concentration of 2 mM in the media accelerated the efficiency of secretion. However, even when the Ca^{2+}

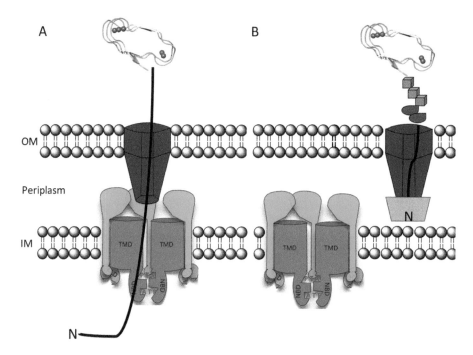

FIGURE 3 Schematic summary of the classic T1SS-mediated substrate secretion (A) and the recently discovered secretion mechanism for some RTX adhesins in which secretion stalls just before completion, creating a so-called two-step process with a pseudoperiplasmic intermediate (B). The ABC transporter and the MFP are shown in blue and green, respectively, and the OM protein is in maroon. (A) The unfolded substrate is secreted with its C terminus first. At the cell surface, Ca^{2+} ions (blue spheres) bind to the GG repeats and induce folding, which results in formation of the β-roll (indicated in cartoon representation). (B) In the case of adhesins such as IBA or LapA, the N-terminal domain starts folding prior to or during secretion, which plugs the translocon (indicated by the light brown polygon) and tethers the entire substrate at the cell surface within the OM component of the translocon of the T1SS. The brown cubes and distorted ellipse represent folded domains of the substrate. This scheme clearly demonstrates that the classic T1SS disassembles only after the entire substrate is translocated, while in two-step T1SS disassembly earlier, e.g., when the N-terminal plug domain has not passed the OM. For further details, see the text. IM, inner membrane; NBD, nucleotide binding domain; TMD, transmembrane domain.

concentration was reduced to 0.1 mM (which does not allow folding of CyaA), 50% of produced CyaA still reached the cell surface. These differences might be due to the diverging sizes of the two RTX proteins or details of the architecture of the RTX domain. Thus, further experimental approaches and analysis of additional T1SS substrates are required to completely understand the molecular mechanisms of secretion, the influence of Ca^{2+} ion folding and secretion (if any), and the molecular signals that regulate substrate translocation across two membranes in one step for this group of classic T1SS substrates.

RTX ADHESINS—NEW KIDS ON THE BLOCK

RTX proteins include not only toxins but also lipases, S-layer proteins, MARTX, and adhesins, which are extremely large in size. Recently, the structure of an ice-binding adhesin (IBA) of the marine Gram-negative bacterium *Marinomonas primoryensis* (molecular mass, 1.5 MDa) was determined and the putative mechanism of translocation modeled (37). IBA contains the hallmarks of substrates of a T1SS, an RTX domain and a C-terminal secretion sequence. N terminal to the RTX domain, three additional domains are located, namely, peptide-, sugar-, and ice-binding domains. While interactions of the peptide- and sugar-binding domains with surface receptors of other microorganisms allow formation of mixed aggregates of microorganisms, the ice-binding domain anchors *M. primoryensis* to ice in seas, lakes, or rivers. Surprisingly, and in contrast to the classical RTX proteins described earlier that are directly secreted into the extracellular space, IBA is translocated but then retained on the cell surface (Fig. 3B). Guo et al. (37) identified a conserved region (homologous to RM in LapA described above) at the extreme N terminus of IBA that they proposed could plug the channel-tunnel of the T1SS. Based on their structural analysis, Guo et al. proposed

that this N-terminal region forms two domains: a proximal sequence that folds and a distal region that is sufficiently unfolded to traverse the TolC homologue into the outer membrane. This prevents further translocation and retains the adhesin at the cell surface. In 2018, exciting new data concerning the LapA adhesin from *Pseudomonas fluorescens* (3; for a summary, see reference 38) provided direct experimental evidence for this model in a comprehensive multidisciplinary study. The 160-residue RM was shown to be essential to tether the adhesin to the translocator and thus to the cell surface (Fig. 3B). Moreover, the RM consists of two domains, folded and unfolded, with the former specifically cleaved by a dedicated protease, LapG, to release the adhesin. In another exciting twist, LapG is normally inactivated by binding to its membrane receptor, LapD. Binding is controlled by c-di-GMP to favor binding under conditions suitable for biofilm formation (63). Finally, we note that both of these studies (for IBA and LapA) confirm the directionality of translocation, C terminal first, for T1SS secretion.

Finally, it must be stressed that while IBA and LapA are anchored to the surface by stalling translocation, other strategies are used to retain adhesins at the cell surface. SiiE from *Salmonella enterica* contains a putative coiled-coil motif that facilitates immobilization of the entire protein on the surface of the cell envelope, which is controlled by SiiA and SiiB (36). Thus, continued efforts are needed to see whether additional mechanisms and modifications of the classic T1SS exist that are used by Gram-negative bacteria to cope with the demands of their ecological niches.

SUMMARY AND OUTLOOK

An enormous amount of data on T1SS has been gathered since the discovery of the first system. The amount of available structural information on the components of the translocon machinery is increasing constantly.

These components include the OM protein TolC (64), a closely related homologue of HlyC (21), isolated domains of the ABC transporter HlyB (65–67) and other ABC transporters (68), a soluble fragment of the MFP HlyD (69), and an entire structure of an ABC transporter (70) of a putative T1SS with unknown substrate from *Aquifex aeolicus*. This article is unable to cover all aspects of type I secretion; however, it provides a broad summary of the accumulating data on functional aspects of the secretion process. The review does not engage with the possibilities of the T1SS in biotechnological applications (for a recent summary, see reference 41) that go well beyond basic research and would allow large-scale protein production and isolation via protein secretion.

However, we are still some distance from a systematic understanding of the T1SS since the nature of molecular signals and intramolecular communication within this nanomachinery remains unclear. In summary, there are still many open questions that have to be addressed and many more fascinating variations and novel insights to be discovered for the T1SS in Gram-negative bacteria.

ACKNOWLEDGMENTS

We apologize to all our colleagues whose work is not cited due to space limitations. We thank all current and former members of the group for fruitful discussions.

Research on hemolysin A and the hemolysin A secretion machinery is funded by the MOI III graduate school under project name Molecules of Infection and the DFG through CRC 1208 under project name Identity and Dynamics of Membrane Systems—From Molecules to Cellular Functions (project A01 to L.S.).

CITATION

Spitz O, Erenburg IN, Beer T, Kanonenberg K, Holland IB, Schmitt L. 2019. Type I secretion systems—one mechanism for all? Microbiol Spectrum 7(2):PSIB-0003-2018.

REFERENCES

1. **Costa TR, Felisberto-Rodrigues C, Meir A, Prevost MS, Redzej A, Trokter M, Waksman G.** 2015. Secretion systems in Gram-negative bacteria: structural and mechanistic insights. *Nat Rev Microbiol* **13**:343–359.
2. **Veith PD, Glew MD, Gorasia DG, Reynolds EC.** 2017. Type IX secretion: the generation of bacterial cell surface coatings involved in virulence, gliding motility and the degradation of complex biopolymers. *Mol Microbiol* **106**:35–53.
3. **Smith TJ, Font ME, Kelly CM, Sondermann H, O'Toole GA.** 2018. An N-terminal retention module anchors the giant adhesin LapA of *Pseudomonas fluorescens* at the cell surface: a novel sub-family of type I secretion systems. *J Bacteriol* **200**:e00734-17.
4. **Linhartová I, Bumba L, Mašín J, Basler M, Osička R, Kamanová J, Procházková K, Adkins I, Hejnová-Holubová J, Sadílková L, Morová J, Sebo P.** 2010. RTX proteins: a highly diverse family secreted by a common mechanism. *FEMS Microbiol Rev* **34**:1076–1112.
5. **Felmlee T, Welch RA.** 1988. Alterations of amino acid repeats in the *Escherichia coli* hemolysin affect cytolytic activity and secretion. *Proc Natl Acad Sci U S A* **85**:5269–5273.
6. **Griessl MH, Schmid B, Kassler K, Braunsmann C, Ritter R, Barlag B, Stierhof YD, Sturm KU, Danzer C, Wagner C, Schäffer TE, Sticht H, Hensel M, Muller YA.** 2013. Structural insight into the giant Ca^{2+}-binding adhesin SiiE: implications for the adhesion of *Salmonella enterica* to polarized epithelial cells. *Structure* **21**:741–752.
7. **Håvarstein LS, Diep DB, Nes IF.** 1995. A family of bacteriocin ABC transporters carry out proteolytic processing of their substrates concomitant with export. *Mol Microbiol* **16**:229–240.
8. **Håvarstein LS, Holo H, Nes IF.** 1994. The leader peptide of colicin V shares consensus sequences with leader peptides that are common among peptide bacteriocins produced by gram-positive bacteria. *Microbiology* **140**:2383–2389.
9. **Michiels J, Dirix G, Vanderleyden J, Xi C.** 2001. Processing and export of peptide pheromones and bacteriocins in Gram-negative bacteria. *Trends Microbiol* **9**:164–168.
10. **Choudhury HG, Tong Z, Mathavan I, Li Y, Iwata S, Zirah S, Rebuffat S, van Veen HW, Beis K.** 2014. Structure of an antibacterial peptide ATP-binding cassette transporter in a novel outward occluded state. *Proc Natl Acad Sci U S A* **111**:9145–9150.

11. **Husada F, Bountra K, Tassis K, de Boer M, Romano M, Rebuffat S, Beis K, Cordes T.** 2018. Conformational dynamics of the ABC transporter McjD seen by single-molecule FRET. *EMBO J* **37:**e100056.

12. **Felmlee T, Pellett S, Welch RA.** 1985. Nucleotide sequence of an *Escherichia coli* chromosomal hemolysin. *J Bacteriol* **163:**94–105.

13. **Härtlein M, Schiessl S, Wagner W, Rdest U, Kreft J, Goebel W.** 1983. Transport of hemolysin by *Escherichia coli*. *J Cell Biochem* **22:**87–97.

14. **Noegel A, Rdest U, Springer W, Goebel W.** 1979. Plasmid cistrons controlling synthesis and excretion of the exotoxin alpha-haemolysin of *Escherichia coli*. *Mol Gen Genet* **175:**343–350.

15. **Springer W, Goebel W.** 1980. Synthesis and secretion of hemolysin by *Escherichia coli*. *J Bacteriol* **144:**53–59.

16. **Nicaud JM, Mackman N, Gray L, Holland IB.** 1985. Characterisation of HlyC and mechanism of activation and secretion of haemolysin from *E. coli* 2001. *FEBS Lett* **187:**339–344.

17. **Stanley P, Hyland C, Koronakis V, Hughes C.** 1999. An ordered reaction mechanism for bacterial toxin acylation by the specialized acyltransferase HlyC: formation of a ternary complex with acylACP and protoxin substrates. *Mol Microbiol* **34:**887–901.

18. **Stanley P, Packman LC, Koronakis V, Hughes C.** 1994. Fatty acylation of two internal lysine residues required for the toxic activity of *Escherichia coli* hemolysin. *Science* **266:**1992–1996.

19. **Trent MS, Worsham LM, Ernst-Fonberg ML.** 1998. The biochemistry of hemolysin toxin activation: characterization of HlyC, an internal protein acyltransferase. *Biochemistry* **37:**4644–4652.

20. **Issartel JP, Koronakis V, Hughes C.** 1991. Activation of *Escherichia coli* prohaemolysin to the mature toxin by acyl carrier protein-dependent fatty acylation. *Nature* **351:**759–761.

21. **Greene NP, Crow A, Hughes C, Koronakis V.** 2015. Structure of a bacterial toxin-activating acyltransferase. *Proc Natl Acad Sci U S A* **112:** E3058–E3066.

22. **Wandersman C, Delepelaire P.** 1990. TolC, an *Escherichia coli* outer membrane protein required for hemolysin secretion. *Proc Natl Acad Sci U S A* **87:**4776–4780.

23. **Chervaux C, Sauvonnet N, Le Clainche A, Kenny B, Hung AL, Broome-Smith JK, Holland IB.** 1995. Secretion of active beta-lactamase to the medium mediated by the *Escherichia coli* haemolysin transport pathway. *Mol Gen Genet* **249:**237–245.

24. **Gentschev I, Hess J, Goebel W.** 1990. Change in the cellular localization of alkaline phosphatase by alteration of its carboxy-terminal sequence. *Mol Gen Genet* **222:**211–216.

25. **Gray L, Baker K, Kenny B, Mackman N, Haigh R, Holland IB.** 1989. A novel C-terminal signal sequence targets *Escherichia coli* haemolysin directly to the medium. *J Cell Sci Suppl* **11:**45–57.

26. **Gray L, Mackman N, Nicaud JM, Holland IB.** 1986. The carboxy-terminal region of haemolysin 2001 is required for secretion of the toxin from *Escherichia coli*. *Mol Gen Genet* **205:**127–133.

27. **Koronakis V, Koronakis E, Hughes C.** 1989. Isolation and analysis of the C-terminal signal directing export of *Escherichia coli* hemolysin protein across both bacterial membranes. *EMBO J* **8:**595–605.

28. **Mackman N, Baker K, Gray L, Haigh R, Nicaud JM, Holland IB.** 1987. Release of a chimeric protein into the medium from *Escherichia coli* using the C-terminal secretion signal of haemolysin. *EMBO J* **6:**2835–2841.

29. **Mackman N, Holland IB.** 1984. Functional characterization of a cloned haemolysin determinant from *E. coli* of human origin, encoding information for the secretion of a 107K polypeptide. *Mol Gen Genet* **196:**129–134.

30. **Mackman N, Nicaud JM, Gray L, Holland IB.** 1985. Genetical and functional organisation of the *Escherichia coli* haemolysin determinant 2001. *Mol Gen Genet* **201:**282–288.

31. **Mackman N, Nicaud JM, Gray L, Holland IB.** 1985. Identification of polypeptides required for the export of haemolysin 2001 from *E. coli*. *Mol Gen Genet* **201:**529–536.

32. **Goyard S, Sebo P, D'Andria O, Ladant D, Ullmann A.** 1993. *Bordetella pertussis* adenylate cyclase: a toxin with multiple talents. *Zentralbl Bakteriol* **278:**326–333.

33. **Satchell KJ.** 2011. Structure and function of MARTX toxins and other large repetitive RTX proteins. *Annu Rev Microbiol* **65:**71–90.

34. **Satchell KJF.** 2015. Multifunctional-autoprocessing repeats-in-toxin (MARTX) toxins of vibrios. *Microbiol Spectr* **3**(3):VE-0002-2014.

35. **Kim BS, Gavin HE, Satchell KJ.** 2015. Distinct roles of the repeat-containing regions and effector domains of the *Vibrio vulnificus* multifunctional-autoprocessing repeats-in-toxin (MARTX) toxin. *mBio* **6:**e00324-15.

36. **Barlag B, Hensel M.** 2015. The giant adhesin SiiE of *Salmonella enterica*. *Molecules* **20:**1134–1150.

37. **Guo S, Stevens CA, Vance TDR, Olijve LLC, Graham LA, Campbell RL, Yazdi SR, Escobedo C, Bar-Dolev M, Yashunsky V, Braslavsky I,**

Langelaan DN, Smith SP, Allingham JS, Voets IK, Davies PL. 2017. Structure of a 1.5-MDa adhesin that binds its Antarctic bacterium to diatoms and ice. *Sci Adv* **3**:e1701440.

38. Smith TJ, Sondermann H, O'Toole GA. 2018. Type 1 does the two-step: type 1 secretion substrates with a functional periplasmic intermediate. *J Bacteriol* **200**:e00168-18.

39. Peters B, Stein J, Klingl S, Sander N, Sandmann A, Taccardi N, Sticht H, Gerlach RG, Muller YA, Hensel M. 2017. Structural and functional dissection reveals distinct roles of Ca2+-binding sites in the giant adhesin SiiE of *Salmonella enterica*. *PLoS Pathog* **13**: e1006418.

40. Wagner C, Polke M, Gerlach RG, Linke D, Stierhof YD, Schwarz H, Hensel M. 2011. Functional dissection of SiiE, a giant non-fimbrial adhesin of *Salmonella enterica*. *Cell Microbiol* **13**:1286–1301.

41. Holland IB, Peherstorfer S, Kanonenberg K, Lenders M, Reimann S, Schmitt L. 2016. Type I protein secretion—deceptively simple yet with a wide range of mechanistic variability across the family. *EcoSal Plus* **7**:ESP-0019-2015.

42. Cao G, Kuhn A, Dalbey RE. 1995. The translocation of negatively charged residues across the membrane is driven by the electrochemical potential: evidence for an electrophoresis-like membrane transfer mechanism. *EMBO J* **14**:866–875.

43. Baumann U, Wu S, Flaherty KM, McKay DB. 1993. Three-dimensional structure of the alkaline protease of *Pseudomonas aeruginosa*: a two-domain protein with a calcium binding parallel beta roll motif. *EMBO J* **12**:3357–3364.

44. Baumann U, Bauer M, Létoffé S, Delepelaire P, Wandersman C. 1995. Crystal structure of a complex between *Serratia marcescens* metalloprotease and an inhibitor from *Erwinia chrysanthemi*. *J Mol Biol* **248**:653–661.

45. Meier R, Drepper T, Svensson V, Jaeger KE, Baumann U. 2007. A calcium-gated lid and a large beta-roll sandwich are revealed by the crystal structure of extracellular lipase from *Serratia marcescens*. *J Biol Chem* **282**:31477–31483.

46. Ostolaza H, Soloaga A, Goñi FM. 1995. The binding of divalent cations to *Escherichia coli* alpha-haemolysin. *Eur J Biochem* **228**:39–44.

47. Sánchez-Magraner L, Viguera AR, García-Pacios M, Garcillán MP, Arrondo JL, de la Cruz F, Goñi FM, Ostolaza H. 2007. The calcium-binding C-terminal domain of *Escherichia coli* alpha-hemolysin is a major determinant in the surface-active properties of the protein. *J Biol Chem* **282**:11827–11835.

48. Soloaga A, Ramírez JM, Goñi FM. 1998. Reversible denaturation, self-aggregation, and membrane activity of *Escherichia coli* alpha-hemolysin, a protein stable in 6 M urea. *Biochemistry* **37**:6387–6393.

49. Thomas S, Bakkes PJ, Smits SH, Schmitt L. 2014. Equilibrium folding of pro-HlyA from *Escherichia coli* reveals a stable calcium ion dependent folding intermediate. *Biochim Biophys Acta* **1844**:1500–1510.

50. Blenner MA, Shur O, Szilvay GR, Cropek DM, Banta S. 2010. Calcium-induced folding of a beta roll motif requires C-terminal entropic stabilization. *J Mol Biol* **400**:244–256.

51. Chenal A, Guijarro JI, Raynal B, Delepierre M, Ladant D. 2009. RTX calcium binding motifs are intrinsically disordered in the absence of calcium: implication for protein secretion. *J Biol Chem* **284**:1781–1789.

52. Sotomayor Pérez AC, Karst JC, Davi M, Guijarro JI, Ladant D, Chenal A. 2010. Characterization of the regions involved in the calcium-induced folding of the intrinsically disordered RTX motifs from the *Bordetella pertussis* adenylate cyclase toxin. *J Mol Biol* **397**:534–549.

53. Zhang L, Conway JF, Thibodeau PH. 2012. Calcium-induced folding and stabilization of the *Pseudomonas aeruginosa* alkaline protease. *J Biol Chem* **287**:4311–4322.

54. Jones HE, Holland IB, Baker HL, Campbell AK. 1999. Slow changes in cytosolic free Ca2+ in *Escherichia coli* highlight two putative influx mechanisms in response to changes in extracellular calcium. *Cell Calcium* **25**:265–274.

55. Bakkes PJ, Jenewein S, Smits SH, Holland IB, Schmitt L. 2010. The rate of folding dictates substrate secretion by the *Escherichia coli* hemolysin type 1 secretion system. *J Biol Chem* **285**:40573–40580.

56. Balakrishnan L, Hughes C, Koronakis V. 2001. Substrate-triggered recruitment of the TolC channel-tunnel during type I export of hemolysin by *Escherichia coli*. *J Mol Biol* **313**:501–510.

57. Thanabalu T, Koronakis E, Hughes C, Koronakis V. 1998. Substrate-induced assembly of a contiguous channel for protein export from *E. coli*: reversible bridging of an inner-membrane translocase to an outer membrane exit pore. *EMBO J* **17**:6487–6496.

58. Du D, Wang Z, James NR, Voss JE, Klimont E, Ohene-Agyei T, Venter H, Chiu W, Luisi BF. 2014. Structure of the AcrAB-TolC multi-drug efflux pump. *Nature* **509**:512–515.

59. Benabdelhak H, Kiontke S, Horn C, Ernst R, Blight MA, Holland IB, Schmitt L. 2003. A

specific interaction between the NBD of the ABC-transporter HlyB and a C-terminal fragment of its transport substrate haemolysin A. *J Mol Biol* **327**:1169–1179.

60. **Lenders MHH, Weidtkamp-Peters S, Kleinschrodt D, Jaeger K-E, Smits SHJ, Schmitt L.** 2015. Directionality of substrate translocation of the hemolysin A type I secretion system. *Sci Rep* **5**:12470.

61. **Lenders MH, Beer T, Smits SH, Schmitt L.** 2016. In vivo quantification of the secretion rates of the hemolysin A type I secretion system. *Sci Rep* **6**:33275.

62. **Bumba L, Masin J, Macek P, Wald T, Motlova L, Bibova I, Klimova N, Bednarova L, Veverka V, Kachala M, Svergun DI, Barinka C, Sebo P.** 2016. Calcium-driven folding of RTX domain β-rolls ratchets translocation of RTX proteins through type I secretion ducts. *Mol Cell* **62**:47–62.

63. **Monds RD, Newell PD, Gross RH, O'Toole GA.** 2007. Phosphate-dependent modulation of c-di-GMP levels regulates *Pseudomonas fluorescens* Pf0-1 biofilm formation by controlling secretion of the adhesin LapA. *Mol Microbiol* **63**:656–679.

64. **Koronakis V, Sharff A, Koronakis E, Luisi B, Hughes C.** 2000. Crystal structure of the bacterial membrane protein TolC central to multidrug efflux and protein export. *Nature* **405**:914–919.

65. **Lecher J, Schwarz CK, Stoldt M, Smits SH, Willbold D, Schmitt L.** 2012. An RTX transporter tethers its unfolded substrate during secretion via a unique N-terminal domain. *Structure* **20**:1778–1787.

66. **Zaitseva J, Jenewein S, Jumpertz T, Holland IB, Schmitt L.** 2005. H662 is the linchpin of ATP hydrolysis in the nucleotide-binding domain of the ABC transporter HlyB. *EMBO J* **24**:1901–1910.

67. **Zaitseva J, Oswald C, Jumpertz T, Jenewein S, Wiedenmann A, Holland IB, Schmitt L.** 2006. A structural analysis of asymmetry required for catalytic activity of an ABC-ATPase domain dimer. *EMBO J* **25**:3432–3443.

68. **Murata D, Okano H, Angkawidjaja C, Akutsu M, Tanaka SI, Kitahara K, Yoshizawa T, Matsumura H, Kado Y, Mizohata E, Inoue T, Sano S, Koga Y, Kanaya S, Takano K.** 2017. Structural basis for the *Serratia marcescens* lipase secretion system: crystal structures of the membrane fusion protein and nucleotide-binding domain. *Biochemistry* **56**:6281–6291.

69. **Kim JS, Song S, Lee M, Lee S, Lee K, Ha NC.** 2016. Crystal structure of a soluble fragment of the membrane fusion protein HlyD in a type i secretion system of Gram-negative bacteria. *Structure* **24**:477–485.

70. **Morgan JLW, Acheson JF, Zimmer J.** 2017. Structure of a type-1 secretion system ABC transporter. *Structure* **25**:522–529.

71. **Gerlach RG, Jäckel D, Stecher B, Wagner C, Lupas A, Hardt WD, Hensel M.** 2007. *Salmonella* pathogenicity island 4 encodes a giant non-fimbrial adhesin and the cognate type 1 secretion system. *Cell Microbiol* **9**:1834–1850.

Architecture, Function, and Substrates of the Type II Secretion System

19

KONSTANTIN V. KOROTKOV[1] and MARIA SANDKVIST[2]

INTRODUCTION

The type II secretion system (T2SS) is one of several extracellular secretion systems in Gram-negative bacteria. While highly prevalent in gamma- and betaproteobacteria, the T2SS is also recognized to a lesser extent in members of the delta and alpha classes (1, 2). It is known for its prolific protease secretion activity. In addition, the T2SS mediates extracellular delivery of a variety of toxins, lipases, and enzymes that break down complex carbohydrates, thus conferring a survival advantage to pathogenic as well as environmental species (2–4). The T2SS is not restricted to extracellular pathogens, such as *Acinetobacter baumannii*, *Escherichia coli*, *Klebsiella pneumoniae*, *Pseudomonas aeruginosa*, and *Vibrio cholerae*; it is also present and contributes to growth of intracellular pathogens, including *Legionella pneumophila*, which replicates in aquatic amoebae, alveolar macrophages, and epithelial cells (5–7). The obligate intracellular pathogen *Chlamydia trachomatis* also depends on T2SS components for extracellular secretion; however, its T2SS is atypical, as some components are missing or are too different from homologs in other species to be identified using BLAST algorithms (8, 9).

[1]Department of Molecular and Cellular Biochemistry, University of Kentucky, Lexington, KY 40506
[2]Department of Microbiology and Immunology, University of Michigan Medical School, Ann Arbor, MI 48109

Protein Secretion in Bacteria
Edited by Maria Sandkvist, Eric Cascales, and Peter J. Christie
© 2019 American Society for Microbiology, Washington, DC
doi:10.1128/ecosalplus.ESP-0034-2018

With 12 to 15 different components distributed in the cytoplasm, cytoplasmic membrane (CM), and outer membrane (OM), the large multiprotein T2SS spans the entire Gram-negative cell envelope (Fig. 1A). While many of the T2SS constituents are structurally and functionally related to those of type IV pilus systems (10), some of the components are unique to the T2SS and are therefore likely to have a specific role in the secretion process. Energy through the hydrolysis of ATP is provided by GspE, a cytoplasmic hexameric ATPase that interacts with the cytoplasmic domains of GspL and GspF, two CM components (Fig. 2) (11–21). GspL, in turn, forms a tight complex with GspM, a structural homolog (22–27). The CM complex also consists of GspC (Fig. 2), which reaches into the periplasmic space, making contact with the secretin that forms the OM conduit, consisting of 15 copies of GspD (Fig. 3A) (28–38). A gene for GspN, a fifth CM component, is present in the T2SS operons of a subset of Gram-negative species; however, its removal has often no discernible effect on secretion and its function remains unknown (39, 40). Interestingly, *Xanthomonas campestris* lacks GspC. Instead, it expresses GspN, which may substitute for GspC (41). In addition, the function of some T2SSs is supported by the CM proteins GspA and GspB, which contribute to GspD assembly and transport to the OM, possibly by increasing the pore size of the peptidoglycan or anchoring it to this structural meshwork (42–44). Finally, GspG forms a periplasmic pseudopilus that extends from the CM and is likely capped by the minor pseudopilins GspH, GspI, GspJ, and GspK, components that initiate the formation of the pseudopilus (Fig. 2) (45–50). Prior to assembly of the pseudopilus, which involves extracting the pseudopilins from the CM and polymerizing them into short helical pilus-like fibers, they are N-terminally cleaved and methylated by the prepilin peptidase GspO (PilD) (51–53).

Proteins to be secreted by the T2SS are initially produced as precursors with N-terminal signal peptides that are removed following translocation across the CM by the Sec or TAT pathways (54–57). While tethered to the CM or following release to the periplasmic compartment, they then undergo folding and, in some cases, oligomerization into larger complexes (58–65). Figure 1B shows examples of T2SS substrates with known structures. Many T2SS substrates, particularly proteases, are also produced with a removable propeptide, in addition to the signal peptide, that functions as an intramolecular chaperone and/or inhibitor (64, 66–68). Other T2SS substrates require dedicated, often CM-tethered, chaperones that assist in the folding and/or engagement with the T2SS prior to OM translocation (see the crystal structure of the *Burkholderia glumae* lipase in complex with a soluble form of its chaperone in Fig. 1B) (69–74).

Here we discuss the latest findings relating to the T2SS substrates, the structure and assembly of the secretin, and the mechanism of secretion, focusing on the role of the pseudopilus.

TRANSPORT AND FUNCTION OF T2SS SUBSTRATES

The secretion of exoproteins by the T2SS is considered a two-step process, where the two steps—transport across the CM and OM—can be genetically and physically separated (61, 62). All T2SS substrates have to be exposed to the periplasmic compartment to be recognized by the T2SS. Some enter the T2SS as soluble periplasmic intermediates, while others are extracted directly from the CM. Examples of the latter include the prolipoproteins pullulanase and SslE produced by *Klebsiella pneumoniae* and enteropathogenic *Escherichia coli* (EPEC), respectively, which are expressed with signal peptides that contain a lipobox with a conserved cysteine (75–77). The cysteine is acylated and the signal peptide is removed. The lipidated cysteine remains with the

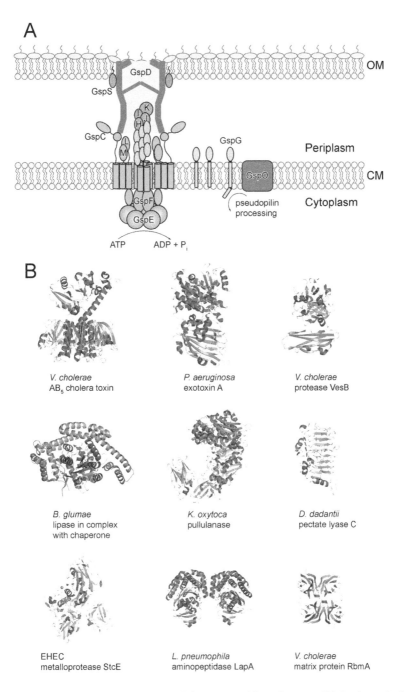

FIGURE 1 Overview of the general architecture of the T2SS and its substrates. (A) A schematic diagram of topology and location of the conserved core components of the T2SS. The accessory components GspN, GspA, and GspB are not shown. **(B)** A selection of the T2SS substrates of variable functions. Protein toxins include *V. cholerae* AB$_5$ cholera toxin (139) and *P. aeruginosa* exotoxin A (140). Hydrolytic enzymes include *V. cholerae* VesB (68), *B. glumae* lipase in complex with chaperone (shown in purple) (71), *K. oxytoca* pullulanase (77), *D. dadantii* pectate lyase C (141), EHEC metalloprotease StcE (142), and *L. pneumophila* aminopeptidase LapA (91). *V. cholerae* biofilm matrix protein RbmA is a scaffolding protein (143, 144).

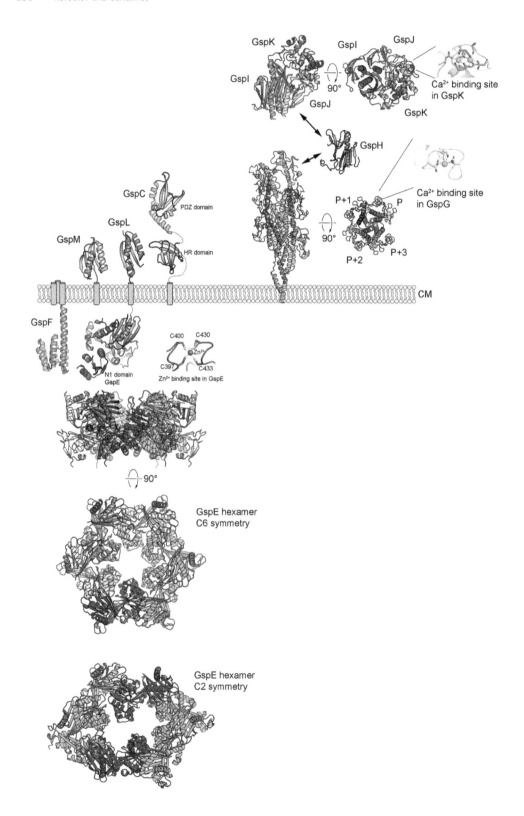

mature protein and is further modified by an *N*-acyltransferase prior to engagement with the T2SS (78). In contrast to many other, soluble T2SS substrates, which are released from the cells following OM transfer, these lipoproteins remain primarily associated with the bacterial cell surface (76, 79). Presumably they are retained with the OM through their lipidated N termini, because a pullulanase variant produced with a typical Sec signal peptide is solubly released following OM translocation (62). Another example of surface retention includes the cell association of heat-labile enterotoxin (LT) produced by enterotoxigenic *Escherichia coli* (ETEC), which binds via the B subunit oligomer to lipopolysaccharide in a Kdo core-dependent manner (80, 81). Although the B subunits of LT and cholera toxin are nearly identical, cholera toxin does not remain associated with the *V. cholerae* surface because its Kdo core is phosphorylated, thus preventing the binding. A third form of surface retention is typified by the pectin lyase PnlH, which is anchored in the *Dickeya dadantii* OM by a noncleavable TAT signal peptide (82). *V. cholerae* provides yet another means by which T2SS substrates associate with the cell surface. This involves the production of proteins with a C-terminally located tripartite motif, GlyGly-CTERM, which consists of residues rich in glycines and serines followed by a stretch of hydrophobic amino acids and positively charged residues (83, 84). A recent study has

shown that the GlyGly-CTERM domain of one of these proteins, VesB, is cleaved off in the CM by an intramembrane protease, rhombosortase, and the newly generated C terminus is capped with a glycerol-phosphoethanolamine unit that may be acylated (84). This is followed by OM translocation and surface localization of VesB. Expression of VesB without its GlyGly-CTERM domain results in the release of VesB to the extracellular compartment, signifying the importance of the GlyGly-CTERM extension and C-terminal modification for VesB surface retention. While the above-listed methods exemplify ways to retain T2SS substrates on the bacterial surface, these proteins can also be found in various amounts in culture supernatants. This is due to release of OM vesicles, the formation of micelles, or removal of the proteins from the cell surface by extracellular proteases (79, 84).

The contribution of the T2SS to environmental growth and virulence of human, animal, and plant pathogens is apparent when one considers the secreted proteins and their activities (Table 1). Devastating diseases such as cholera and childhood diarrhea caused by ETEC are mediated by cholera toxin and LT and result in severe dehydration and even death when not treated (85, 86). By inducing watery diarrhea, the enterotoxins aid in the spread and transmission of these diseases. Another means by which T2SS substrates benefit bacteria is through nutrient acquisition in the eukaryotic host and environment. The release and generation of nutrients are

---

FIGURE 2 Structures of the T2SS assembly platform and pseudopilus components. The ATPase is hexameric *V. cholerae* GspE with C6 and C2 symmetries (20). A close-up view shows the Zn^{2+} binding site, which is required for the function of GspE (14, 145). Inner membrane components include the cytoplasmic domain of *V. cholerae* GspF (19), cytoplasmic domain of GspL in complex with N1 domain of *V. cholerae* GspE (16), periplasmic domain of *V. parahaemolyticus* GspL (26), periplasmic domain of *V. cholerae* GspM (25), the homology region (HR) domain of ETEC GspC (32), and the PDZ domain of *V. cholerae* GspC (29). The structure of periplasmic domain of *P. aeruginosa* GspL (XcpY) has been recently published (146). Regarding pseudopilus components, in the *K. oxytoca* GspG pseudopilus model based on the cryo-EM reconstruction (50), a close-up view shows the Ca^{2+} binding site of *K. oxytoca* GspG, *V. cholerae* minor pseudopilin GspH (47), and the trimeric complex of ETEC GspK-GspI-GspJ (48), and a close-up view shows a double-Ca^{2+} binding site of GspK. The structure of a homologous XcpX-XcpV-XcpW complex from *P. aeruginosa* has been recently reported (147).

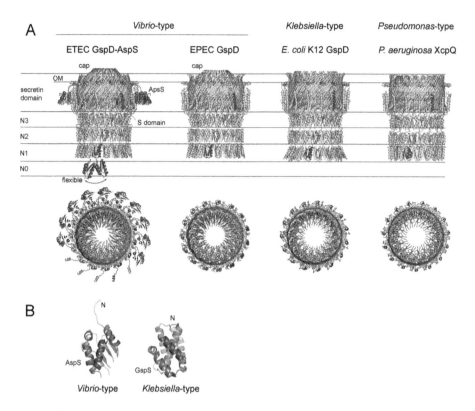

FIGURE 3 Structures of the T2SS secretins and pilotins. (A) The side and top views of ETEC GspD-AspS complex (37), EPEC GspD (36), *E. coli* K-12 GspD (34), and *P. aeruginosa* GspD (35). A single secretin protomer is highlighted, with N1, N2, and N3 domains in shades of blue, the secretin domain in green, and the S domain in magenta. Several AspS protomers (brown) were omitted to clearly show the location of the S domain. The cap subdomain in the *Vibrio*-type secretins is highlighted in orange. The N0 domains (purple) were not resolved in the available cryo-EM reconstructions due to flexibility. Instead, its approximate location is indicated (148). Note that the N1-N2 domains of EPEC GspD (36) and the N1 domain of *P. aeruginosa* GspD (35) have been placed as rigid fit models. (B) Structures of pilotins in complex with the secretin S domains (magenta). Structures of *Vibrio*-type ETEC AspS (37) and *Klebsiella*-type *D. dadantii* GspS (116) are shown.

the result of the action of pore-forming toxins such as aerolysin, which causes osmotic host cell lysis, and a range of enzymes, including proteases, lipases, DNases, and carbohydrate-degrading enzymes that digest host components and tissue (40, 63, 87–92). For example, plant cell wall-degrading enzymes, such as cellulases, pectinases, and pectate lyases, promote growth of phytopathogenic bacteria, resulting in crop losses (55, 87, 93). *L. pneumophila* aminopeptidase LapA is another recently identified example of a T2SS protein that contributes to nutrient acquisition by

generating amino acids and thus supporting intracellular growth of *L. pneumophila* in amoebae (91). Degradative enzymes also aid pathogens in gaining access to new niches. Enzymes that target mucins, which cover and protect cells of the digestive tract, generate direct access to the epithelial cell surface for colonization of enteric pathogens and facilitate delivery of toxins (94–96). Recent work has also recognized the contribution of many proteases and other effectors in immune evasion. *Burkholderia cenocepacia* ZmpA and ZmpB are examples of proteases that cleave

TABLE 1 Examples of T2SS substrates

Protein(s)	Type(s)	Activity(ies)	Reference(s)
Toxins	Enterotoxin (cholera toxin, *E. coli* heat-labile enterotoxin)	ADP-ribosylation of Gsα subunit leading to increased adenylate cyclase activity and raising cAMP levels, which activates protein kinase A, followed by phosphorylation of the CFTR channel. This leads to efflux of chloride ions and water release into the intestinal lumen and consequent secretory diarrhea.	85, 86
	Exotoxin A	ADP-ribosylation of elongation factor 2, inhibition of protein synthesis in host cells	149
	Pore-forming toxin (aerolysin, cytolysin)	Host cell membrane depolarization and lysis	63, 150–152
Proteases	Metalloprotease, serine protease, cysteine protease, aminopeptidase	Cleavage of proteins or peptides, breakdown of host extracellular matrix, tissue damage, detachment from host cells, nutrient acquisition, evasion of host defense system, translocation to new niche	73, 90, 91, 96, 97, 99–101, 103, 149, 153–156
Lipid-modifying enzymes	Lipase, phospholipase, glycerophospholipid cholesterol acyltransferase	Breakdown of lipids to fatty acids and glycerol, nutrient acquisition, translocation to new niche; breakdown of phospholipids and destabilization of host cell membranes	40, 70, 73, 88, 91, 149, 157, 158
Carbohydrate-active enzymes	Chitinase, amylase, cellulase, pullulanase, xylanase, pectinase, pectin methylesterase, pectate lyase, levansucrase	Breakdown of polysaccharides such as chitin, cellulose, pectin, amylose, pullulan, and xylan; targeting of O-GlcNAcylated proteins in the host; nutrient acquisition; depolymerization of plant cell wall; wilting; soft rot	55, 87, 93, 157, 159, 160
Phosphatases	Alkaline phosphatase, acid phosphatase	Dephosphorylation, phosphate acquisition, phosphate solubilization	161–163
Nucleic acid-targeting enzymes	DNase, RNase	Hydrolysis of DNA and RNA; generation of nutrients, including carbon, nitrogen, and phosphate, that support bacterial growth; evasion of neutrophil extracellular traps	92, 164, 165
Metal reductase	C-type cytochromes	Reduction of insoluble metal oxides; electron transport; anaerobic respiration	166, 167
Others	Chitin-binding protein, collagen-like protein, biofilm-associated proteins	Adherence, biofilm formation	105, 106, 168, 169

antimicrobial peptides, and the *L. pneumophila* T2SS reduces the output of cytokines and chemokines during infection, in part due to secretion and activity of the metalloprotease ProA (97–99). Another secreted metalloprotease, StcE, protects enterohemor-rhagic *Escherichia coli* (EHEC) from host defense mechanisms, including complement-mediated killing, by cleaving C1 esterase inhibitor and neutrophil-associated proteins, while the *A. baumannii* metalloprotease CpaA interferes with blood coagulation by inactivating factor XII, which may result in the escape from clearance by intravascular clots and dissemination of *A. baumannii* (73, 100–103). In the environment, bacterial life frequently occurs as matrix-encased biofilm. The major component is often polysaccharides, but the matrix also contains specific proteins secreted by the T2SS that contribute to the formation, architecture, and stability of the biofilm (76, 104–109).

SECRETIN STRUCTURE AND ASSEMBLY

The secretin of the T2SS forms the OM channel for the passage of secreted proteins. Topologically, it contains an N-terminal N0 domain, homologous repeat N1 to N3 domains, and a C-terminal secretin core domain (Fig. 3A). The majority of T2SS secretins have an additional C-terminal S domain required for interaction with pilotins, which delivers them to the OM (see below and Fig. 3B). Early electron microscopy (EM) studies revealed the general architecture of secretins as a multimeric channel with an open periplasmic chamber, a closed central gate, and a top chamber (31, 110–113). The crystal structures of the N0-N1-N2 domains of ETEC GspD (30) allowed modeling of the N0-N1-N2-N3 domains in the cryo-EM structure of *V. cholerae* GspD (31). However, only the recent progress in cryo-EM methodology enabled investigators to solve the full-length secretin structures of GspD from *E. coli* K-12 and *V. cholerae* (34), *P. aeruginosa* (35), and EPEC (36) and of the GspD-pilotin complex from ETEC (37) (Fig. 3A). The general structural features are well conserved between secretins of the *Vibrio* type, the *Klebsiella* type (114), and the more sequence-divergent *Pseudomonas* type (Fig. 3A) (35). The secretins form cylindrical structures ~150 Å in diameter and ~195 Å in length that display 15-fold symmetry, in contrast with previously reported 12-fold symmetry, a discrepancy likely caused by the limited resolution of the earlier studies. The major

difference between the secretins is the presence of an extracellular gate (cap) of unknown function in the *Vibrio*-type secretins (Fig. 3A) (34, 36).

In the secretin monomer, the domains are arranged in line that is tilted at ~30° (Fig. 3A). The N0 domain is not modeled in the available reconstructions, as it was disordered, although weak smeared density was observed in some two-dimensional (2D) class averages (34). Together, the secretin core domains form a double-barrel structure that includes inner and outer barrels. Each secretin monomer contributes 4 shorter beta-sheets to the inner barrel and 4 longer beta-sheets to the outer barrel. The β-hairpin extensions from the inner barrel form the periplasmic gate in the secretin channel. A mechanism of periplasmic gate opening during secretion has been suggested based on a partially open structure of a G453A mutant of *V. cholerae* GspD, which showed an upward rotation of the β-hairpins (34). This is consistent with a recent structure of a homologous type III secretion system secretin in the open form, which revealed that the β-hairpins of the periplasmic gate are in upward position (115). The aromatic and aliphatic residues on the top of the outer barrel contribute to the hydrophobic belt that creates the transmembrane region. The S domain provides overall stability to the secretin oligomer by interacting with an adjacent protomer. The N3 domains form a ring below the secretin core domains with which they make extensive contacts. The interactions between N1-N2 and N2-N3 domain rings, on the other hand, are limited to the linkers and several specific contacts, which led to lower resolution of 3D reconstructions in this region.

The majority of T2SSs contain a *Klebsiella*-type pilotin, a fatty-acylated protein which has an α-helical structure that provides a hydrophobic groove for interactions with the second α-helix from the S domain of the secretin (Fig. 3B) (116). In contrast, *Vibrio*-

type secretins utilize an alternative, structurally unrelated pilotin, although this pilotin fulfills the same function (114, 117). The structure of the *Vibrio*-type pilotin in complex with the secretin adopts an open conformation compared to the apo-pilotin to accommodate the α-helix from the S domain (Fig. 3B) (37). While pilotins direct secretins to the OM via the lipoprotein sorting system (118), the mechanism by which the secretin protomers acquire the assembled state and insert into the OM independently of the BAM complex is less clear (119, 120). However, a recent study that subjected the N3 domain of the *Klebsiella oxytoca* GspD to mutational analysis underscores the importance of the N3 domain in the early steps of secretin assembly (121). The results suggest that the N3 domain provides stability to the prepore, a prerequisite for OM insertion and pore formation of the secretin.

PSEUDOPILUS AND ASSEMBLY PLATFORM

Topologically, all pseudopilins consist of an N-terminal hydrophobic helix that extends into variable C-terminal globular domains. The major pseudopilin GspG contains a conserved Ca^{2+} binding site (Fig. 2) (50, 122). Disruption of Ca^{2+} binding by either amino acid substitutions or Ca^{2+} ion removal with chelating agents affects the stability of the pseudopilin monomer, pseudopilus assembly, and substrate secretion (50, 122). Interestingly, the minor pseudopilin GspK also contains a Ca^{2+} binding site (Fig. 2); however, the functional significance of this feature has yet to be determined (48). The minor pseudopilins GspK, GspI, and GspJ form a quasihelical trimeric complex that is thought to be located at the pseudopilus tip (Fig. 2) (48). It has been demonstrated that this complex serves as a priming site for pseudopilus assembly (123). The minor pseudopilin GspH possibly acts as an adapter between the GspK-GspI-GspJ tip and the

poly-GspG fiber (47). The most recent model of the *K. oxytoca* GspG pseudopilus based on a cryo-EM reconstruction revealed a right-handed helical fiber with a 10-Å rise (50). The N-terminal hydrophobic helices of the GspG subunits arrange within the core of the pseudopilus, with the C-terminal domains and the Ca^{2+} binding sites located at the surface (Fig. 2). Interestingly, the N-terminal hydrophobic helix is connected to the C-terminal domain by an extended linker, a feature distinct from the type IV pilus.

While a number of GspE homologs from the type IV pili systems have been structurally characterized in hexameric form (124–126), the structure of hexameric GspE has been elusive. Employing a "scaffold" Hcp1 fusion strategy allowed visualization of GspE in two hexameric conformations: a symmetrical C6 form and an extended C2 form (Fig. 2) (20). These conformations may reflect structural transitions in GspE during ATP hydrolysis and transfer of mechanical energy to support pseudopilus assembly. The details of this process are not completely understood, although it is believed to involve GspF, GspL, and GpsM, which have all been shown to interact with the pseudopilus (13, 18, 127, 128). While the structures of the extramembrane domains of GspC, GspF, GspL, and GspM have been solved (Fig. 2), the structure of the assembly platform formed by these components has not yet been reported.

IMPLICATIONS FOR MECHANISM

Despite the progress in understanding the structure-function relationship of the various components of the T2SS, the mechanism by which this important secretion system transports both soluble and lipidated proteins across the OM remains poorly understood. As folding of the T2SS substrates is a prerequisite for engagement with the T2SS, a secretion signal is thought to be formed in the folded structures; however, the structures of T2SS substrates greatly vary and a common, general secretion signal has yet to be identified. The protein-protein interaction domain PDZ of GspC has been suggested to recognize and recruit the T2SS substrates once they arrive in the periplasmic compartment, yet interactions between the T2SS substrates and GspD, GspH-GspI-GspJ-GspK (which forms the tip of the pseudopilus), and the CM proteins GspL and GspM have also been demonstrated (129–134). These interactions are, for the most part, consistent with two prevailing models for driving proteins through the secretin pore: the piston machinery and the Archimedes screw (135). In the piston model the pseudopilus tip supposedly pushes the substrates through the secretin in a linear fashion, although this mechanism cannot fully account for the required retraction of the pseudopilus and recharging of substrate, as the T2SS lacks a retraction ATPase (48, 129, 132, 135, 136). In the Archimedes screw model, the rotary motion via interactions with the poly-GspG shaft of the pseudopilus threads the T2SS substrates out through the secretin pore; however, this model requires a continuous degradation and replenishment of GspG (137, 138). Dedicated removal of GspG-bound calcium and subsequent destabilization of GspG may result in degradation (50), but given the rigid structure of the double-barrel secretin domain, the dimensions of the pseudopilus, and the mounting evidence for the rotation of the pseudopilus itself, perhaps a composite model should be considered, in which the rotary motion of the pseudopilus drives the secretion of the substrates that are pushed out by the pseudopilus tip.

CONCLUSION

In conclusion, while structural information is now available for most of the T2SS components and many T2SS substrates and the general architecture of the T2SS is understood, there are still multiple unanswered questions about the precise stoichiometry of this secretion complex, the detailed mecha-

nism of pseudopilus assembly and possibly disassembly, and the molecular basis for substrate recognition. Before long, however, answers to these quandaries are expected to be revealed, as the field is progressing rapidly due to technology advances.

ACKNOWLEDGMENTS

We thank Iain Hay and Trevor Lithgow for the models of EPEC GspD and *P. aeruginosa* XcpQ secretins.

This work was supported in part by grants R01AI127085 from National Institute of Allergy and Infectious Diseases (to M.S.) and W81XWH-18-1-0587 from the Department of Defense (to M.S.).

CITATION

EcoSal Plus 2019; doi:10.1128/ecosalplus. ESP-0034-2018.

REFERENCES

1. **Abby SS, Cury J, Guglielmini J, Néron B, Touchon M, Rocha EP.** 2016. Identification of protein secretion systems in bacterial genomes. *Sci Rep* **6**:23080.
2. **Cianciotto NP, White RC.** 2017. Expanding role of type II secretion in bacterial pathogenesis and beyond. *Infect Immun* **85**:e00014-17.
3. **Korotkov KV, Sandkvist M, Hol WGJ.** 2012. The type II secretion system: biogenesis, molecular architecture and mechanism. *Nat Rev Microbiol* **10**:336–351.
4. **Sikora AE.** 2013. Proteins secreted via the type II secretion system: smart strategies of *Vibrio cholerae* to maintain fitness in different ecological niches. *PLoS Pathog* **9**:e1003126.
5. **Liles MR, Edelstein PH, Cianciotto NP.** 1999. The prepilin peptidase is required for protein secretion by and the virulence of the intracellular pathogen *Legionella pneumophila*. *Mol Microbiol* **31**:959–970.
6. **Hales LM, Shuman HA.** 1999. *Legionella pneumophila* contains a type II general secretion pathway required for growth in amoebae as well as for secretion of the Msp protease. *Infect Immun* **67**:3662–3666.
7. **Rossier O, Cianciotto NP.** 2001. Type II protein secretion is a subset of the PilD-

dependent processes that facilitate intracellular infection by *Legionella pneumophila*. *Infect Immun* 69:2092–2098.

8. **Nguyen BD, Valdivia RH.** 2012. Virulence determinants in the obligate intracellular pathogen *Chlamydia trachomatis* revealed by forward genetic approaches. *Proc Natl Acad Sci U S A* 109:1263–1268.

9. **Snavely EA, Kokes M, Dunn JD, Saka HA, Nguyen BD, Bastidas RJ, McCafferty DG, Valdivia RH.** 2014. Reassessing the role of the secreted protease CPAF in *Chlamydia trachomatis* infection through genetic approaches. *Pathog Dis* 71:336–351.

10. **McCallum M, Burrows LL, Howell PL.** 2018. The dynamic structures of the type IV pilus. *Microbiol Spectr* 7:PSIB-0006-2018.

11. **Sandkvist M, Bagdasarian M, Howard SP, DiRita VJ.** 1995. Interaction between the autokinase EpsE and EpsL in the cytoplasmic membrane is required for extracellular secretion in *Vibrio cholerae*. *EMBO J* 14:1664–1673.

12. **Sandkvist M, Keith JM, Bagdasarian M, Howard SP.** 2000. Two regions of EpsL involved in species-specific protein-protein interactions with EpsE and EpsM of the general secretion pathway in *Vibrio cholerae*. *J Bacteriol* 182:742–748.

13. **Py B, Loiseau L, Barras F.** 2001. An inner membrane platform in the type II secretion machinery of Gram-negative bacteria. *EMBO Rep* 2:244–248.

14. **Robien MA, Krumm BE, Sandkvist M, Hol WGJ.** 2003. Crystal structure of the extracellular protein secretion NTPase EpsE of *Vibrio cholerae*. *J Mol Biol* 333:657–674.

15. **Camberg JL, Sandkvist M.** 2005. Molecular analysis of the *Vibrio cholerae* type II secretion ATPase EpsE. *J Bacteriol* 187:249–256.

16. **Abendroth J, Murphy P, Sandkvist M, Bagdasarian M, Hol WGJ.** 2005. The X-ray structure of the type II secretion system complex formed by the N-terminal domain of EpsE and the cytoplasmic domain of EpsL of *Vibrio cholerae*. *J Mol Biol* 348:845–855.

17. **Camberg JL, Johnson TL, Patrick M, Abendroth J, Hol WGJ, Sandkvist M.** 2007. Synergistic stimulation of EpsE ATP hydrolysis by EpsL and acidic phospholipids. *EMBO J* 26:19–27.

18. **Arts J, de Groot A, Ball G, Durand E, El Khattabi M, Filloux A, Tommassen J, Koster M.** 2007. Interaction domains in the *Pseudomonas aeruginosa* type II secretory apparatus component XcpS (GspF). *Microbiology* 153:1582–1592.

19. **Abendroth J, Mitchell DD, Korotkov KV, Johnson TL, Kreger A, Sandkvist M, Hol WGJ.** 2009. The three-dimensional structure of the cytoplasmic domains of EpsF from the type 2 secretion system of *Vibrio cholerae*. *J Struct Biol* 166:303–315.

20. **Lu C, Turley S, Marionni ST, Park YJ, Lee KK, Patrick M, Shah R, Sandkvist M, Bush MF, Hol WGJ.** 2013. Hexamers of the type II secretion ATPase GspE from *Vibrio cholerae* with increased ATPase activity. *Structure* 21:1707–1717.

21. **Lu C, Korotkov KV, Hol WGJ.** 2014. Crystal structure of the full-length ATPase GspE from the *Vibrio vulnificus* type II secretion system in complex with the cytoplasmic domain of GspL. *J Struct Biol* 187:223–235.

22. **Michel G, Bleves S, Ball G, Lazdunski A, Filloux A.** 1998. Mutual stabilization of the XcpZ and XcpY components of the secretory apparatus in *Pseudomonas aeruginosa*. *Microbiology* 144:3379–3386.

23. **Sandkvist M, Hough LP, Bagdasarian MM, Bagdasarian M.** 1999. Direct interaction of the EpsL and EpsM proteins of the general secretion apparatus in *Vibrio cholerae*. *J Bacteriol* 181:3129–3135.

24. **Robert V, Hayes F, Lazdunski A, Michel GP.** 2002. Identification of XcpZ domains required for assembly of the secreton of *Pseudomonas aeruginosa*. *J Bacteriol* 184:1779–1782.

25. **Abendroth J, Rice AE, McLuskey K, Bagdasarian M, Hol WGJ.** 2004. The crystal structure of the periplasmic domain of the type II secretion system protein EpsM from *Vibrio cholerae*: the simplest version of the ferredoxin fold. *J Mol Biol* 338:585–596.

26. **Abendroth J, Kreger AC, Hol WGJ.** 2009. The dimer formed by the periplasmic domain of EpsL from the type 2 secretion system of *Vibrio parahaemolyticus*. *J Struct Biol* 168:313–322.

27. **Lallemand M, Login FH, Guschinskaya N, Pineau C, Effantin G, Robert X, Shevchik VE.** 2013. Dynamic interplay between the periplasmic and transmembrane domains of GspL and GspM in the type II secretion system. *PLoS One* 8:e79562.

28. **Lybarger SR, Johnson TL, Gray MD, Sikora AE, Sandkvist M.** 2009. Docking and assembly of the type II secretion complex of *Vibrio cholerae*. *J Bacteriol* 191:3149–3161.

29. **Korotkov KV, Krumm B, Bagdasarian M, Hol WGJ.** 2006. Structural and functional studies of EpsC, a crucial component of the type 2 secretion system from *Vibrio cholerae*. *J Mol Biol* 363:311–321.

30. **Korotkov KV, Pardon E, Steyaert J, Hol WGJ.** 2009. Crystal structure of the N-terminal domain of the secretin GspD from ETEC determined with the assistance of a nanobody. *Structure* 17:255–265.

31. **Reichow SL, Korotkov KV, Hol WGJ, Gonen T.** 2010. Structure of the cholera toxin secretion channel in its closed state. *Nat Struct Mol Biol* 17:1226–1232.

32. **Korotkov KV, Johnson TL, Jobling MG, Pruneda J, Pardon E, Héroux A, Turley S, Steyaert J, Holmes RK, Sandkvist M, Hol WGJ.** 2011. Structural and functional studies on the interaction of GspC and GspD in the type II secretion system. *PLoS Pathog* 7:e1002228.

33. **Wang X, Pineau C, Gu S, Guschinskaya N, Pickersgill RW, Shevchik VE.** 2012. Cysteine scanning mutagenesis and disulfide mapping analysis of arrangement of GspC and GspD protomers within the type 2 secretion system. *J Biol Chem* 287:19082–19093.

34. **Yan Z, Yin M, Xu D, Zhu Y, Li X.** 2017. Structural insights into the secretin translocation channel in the type II secretion system. *Nat Struct Mol Biol* 24:177–183.

35. **Hay ID, Belousoff MJ, Lithgow T.** 2017. Structural basis of type 2 secretion system engagement between the inner and outer bacterial membranes. *mBio* 8:e01344-17.

36. **Hay ID, Belousoff MJ, Dunstan RA, Bamert RS, Lithgow T.** 2018. Structure and membrane topography of the *Vibrio*-type secretin complex from the type 2 secretion system of enteropathogenic *Escherichia coli. J Bacteriol* 200:e00521-17.

37. **Yin M, Yan Z, Li X.** 2018. Structural insight into the assembly of the type II secretion system pilotin-secretin complex from enterotoxigenic *Escherichia coli. Nat Microbiol* 3:581–587.

38. **Majewski DD, Worrall LJ, Strynadka NC.** 2018. Secretins revealed: structural insights into the giant gated outer membrane portals of bacteria. *Curr Opin Struct Biol* 51:61–72.

39. **Possot OM, Vignon G, Bomchil N, Ebel F, Pugsley AP.** 2000. Multiple interactions between pullulanase secreton components involved in stabilization and cytoplasmic membrane association of PulE. *J Bacteriol* 182:2142–2152.

40. **Johnson TL, Waack U, Smith S, Mobley H, Sandkvist M.** 2016. *Acinetobacter baumannii* is dependent on the type II secretion system and its substrate LipA for lipid utilization and *in vivo* fitness. *J Bacteriol* 198:711–719.

41. **Lee HM, Wang KC, Liu YL, Yew HY, Chen LY, Leu WM, Chen DC, Hu NT.** 2000. Association of the cytoplasmic membrane protein XpsN with the outer membrane protein XpsD in the type II protein secretion apparatus of *Xanthomonas campestris* pv. campestris. *J Bacteriol* 182:1549–1557.

42. **Li G, Miller A, Bull H, Howard SP.** 2011. Assembly of the type II secretion system: identification of ExeA residues critical for peptidoglycan binding and secretin multimerization. *J Bacteriol* 193:197–204.

43. **Strozen TG, Stanley H, Gu Y, Boyd J, Bagdasarian M, Sandkvist M, Howard SP.** 2011. Involvement of the GspAB complex in assembly of the type II secretion system secretin of *Aeromonas* and *Vibrio* species. *J Bacteriol* 193:2322–2331.

44. **Vanderlinde EM, Strozen TG, Hernández SB, Cava F, Howard SP.** 2017. Alterations in peptidoglycan cross-linking suppress the secretin assembly defect caused by mutation of GspA in the type II secretion system. *J Bacteriol* 199:e00617-17.

45. **Sauvonnet N, Vignon G, Pugsley AP, Gounon P.** 2000. Pilus formation and protein secretion by the same machinery in *Escherichia coli. EMBO J* 19:2221–2228.

46. **Durand E, Bernadac A, Ball G, Lazdunski A, Sturgis JN, Filloux A.** 2003. Type II protein secretion in *Pseudomonas aeruginosa*: the pseudopilus is a multifibrillar and adhesive structure. *J Bacteriol* 185:2749–2758.

47. **Yanez ME, Korotkov KV, Abendroth J, Hol WGJ.** 2008. Structure of the minor pseudopilin EpsH from the type 2 secretion system of *Vibrio cholerae. J Mol Biol* 377:91–103.

48. **Korotkov KV, Hol WGJ.** 2008. Structure of the GspK-GspI-GspJ complex from the enterotoxigenic *Escherichia coli* type 2 secretion system. *Nat Struct Mol Biol* 15:462–468.

49. **Douzi B, Durand E, Bernard C, Alphonse S, Cambillau C, Filloux A, Tegoni M, Voulhoux R.** 2009. The XcpV/GspI pseudopilin has a central role in the assembly of a quaternary complex within the T2SS pseudopilus. *J Biol Chem* 284:34580–34589.

50. **López-Castilla A, Thomassin JL, Bardiaux B, Zheng W, Nivaskumar M, Yu X, Nilges M, Egelman EH, Izadi-Pruneyre N, Francetic O.** 2017. Structure of the calcium-dependent type 2 secretion pseudopilus. *Nat Microbiol* 2:1686–1695.

51. **Pugsley AP, Dupuy B.** 1992. An enzyme with type IV prepilin peptidase activity is required to process components of the general extracellular protein secretion pathway of *Klebsiella oxytoca. Mol Microbiol* 6:751–760.

52. **Pugsley AP.** 1993. Processing and methylation of PulG, a pilin-like component of the general

secretory pathway of *Klebsiella oxytoca*. *Mol Microbiol* **9**:295–308.

53. **Nunn DN, Lory S.** 1993. Cleavage, methylation, and localization of the *Pseudomonas aeruginosa* export proteins XcpT, -U, -V, and -W. *J Bacteriol* **175**:4375–4382.

54. **Pugsley AP, Kornacker MG, Poquet I.** 1991. The general protein-export pathway is directly required for extracellular pullulanase secretion in *Escherichia coli* K12. *Mol Microbiol* **5**:343–352.

55. **He SY, Schoedel C, Chatterjee AK, Collmer A.** 1991. Extracellular secretion of pectate lyase by the *Erwinia chrysanthemi* out pathway is dependent upon Sec-mediated export across the inner membrane. *J Bacteriol* **173**:4310–4317.

56. **Voulhoux R, Ball G, Ize B, Vasil ML, Lazdunski A, Wu LF, Filloux A.** 2001. Involvement of the twin-arginine translocation system in protein secretion via the type II pathway. *EMBO J* **20**:6735–6741.

57. **Ball G, Antelmann H, Imbert PR, Gimenez MR, Voulhoux R, Ize B.** 2016. Contribution of the twin arginine translocation system to the exoproteome of *Pseudomonas aeruginosa*. *Sci Rep* **6**:27675.

58. **Hirst TR, Holmgren J.** 1987. Transient entry of enterotoxin subunits into the periplasm occurs during their secretion from *Vibrio cholerae*. *J Bacteriol* **169**:1037–1045.

59. **Hirst TR, Holmgren J.** 1987. Conformation of protein secreted across bacterial outer membranes: a study of enterotoxin translocation from *Vibrio cholerae*. *Proc Natl Acad Sci U S A* **84**:7418–7422.

60. **Hardy SJ, Holmgren J, Johansson S, Sanchez J, Hirst TR.** 1988. Coordinated assembly of multisubunit proteins: oligomerization of bacterial enterotoxins *in vivo* and *in vitro*. *Proc Natl Acad Sci U S A* **85**:7109–7113.

61. **Pugsley AP.** 1992. Translocation of a folded protein across the outer membrane in *Escherichia coli*. *Proc Natl Acad Sci U S A* **89**:12058–12062.

62. **Poquet I, Faucher D, Pugsley AP.** 1993. Stable periplasmic secretion intermediate in the general secretory pathway of *Escherichia coli*. *EMBO J* **12**:271–278.

63. **Hardie KR, Schulze A, Parker MW, Buckley JT.** 1995. *Vibrio* spp. secrete proaerolysin as a folded dimer without the need for disulphide bond formation. *Mol Microbiol* **17**:1035–1044.

64. **Braun P, Tommassen J, Filloux A.** 1996. Role of the propeptide in folding and secretion of elastase of *Pseudomonas aeruginosa*. *Mol Microbiol* **19**:297–306.

65. **Voulhoux R, Taupiac MP, Czjzek M, Beaumelle B, Filloux A.** 2000. Influence of

deletions within domain II of exotoxin A on its extracellular secretion from *Pseudomonas aeruginosa*. *J Bacteriol* **182**:4051–4058.

66. **Häse CC, Finkelstein RA.** 1991. Cloning and nucleotide sequence of the *Vibrio cholerae* hemagglutinin/protease (HA/protease) gene and construction of an HA/protease-negative strain. *J Bacteriol* **173**:3311–3317.

67. **McIver KS, Kessler E, Olson JC, Ohman DE.** 1995. The elastase propeptide functions as an intramolecular chaperone required for elastase activity and secretion in *Pseudomonas aeruginosa*. *Mol Microbiol* **18**:877–889.

68. **Gadwal S, Korotkov KV, Delarosa JR, Hol WGJ, Sandkvist M.** 2014. Functional and structural characterization of *Vibrio cholerae* extracellular serine protease B, VesB. *J Biol Chem* **289**:8288–8298.

69. **Hobson AH, Buckley CM, Aamand JL, Jørgensen ST, Diderichsen B, McConnell DJ.** 1993. Activation of a bacterial lipase by its chaperone. *Proc Natl Acad Sci U S A* **90**:5682–5686.

70. **Martínez A, Ostrovsky P, Nunn DN.** 1999. LipC, a second lipase of *Pseudomonas aeruginosa*, is LipB and Xcp dependent and is transcriptionally regulated by pilus biogenesis components. *Mol Microbiol* **34**:317–326.

71. **Pauwels K, Lustig A, Wyns L, Tommassen J, Savvides SN, Van Gelder P.** 2006. Structure of a membrane-based steric chaperone in complex with its lipase substrate. *Nat Struct Mol Biol* **13**:374–375.

72. **Coulthurst SJ, Lilley KS, Hedley PE, Liu H, Toth IK, Salmond GP.** 2008. DsbA plays a critical and multifaceted role in the production of secreted virulence factors by the phytopathogen *Erwinia carotovora* subsp. *atroseptica*. *J Biol Chem* **283**:23739–23753.

73. **Harding CM, Kinsella RL, Palmer LD, Skaar EP, Feldman MF.** 2016. Medically relevant *Acinetobacter* species require a type II secretion system and specific membrane-associated chaperones for the export of multiple substrates and full virulence. *PLoS Pathog* **12**:e1005391.

74. **Kinsella RL, Lopez J, Palmer LD, Salinas ND, Skaar EP, Tolia NH, Feldman MF.** 2017. Defining the interaction of the protease CpaA with its type II secretion chaperone CpaB and its contribution to virulence in *Acinetobacter* species. *J Biol Chem* **292**:19628–19638.

75. **Pugsley AP, Chapon C, Schwartz M.** 1986. Extracellular pullulanase of *Klebsiella pneumoniae* is a lipoprotein. *J Bacteriol* **166**:1083–1088.

76. **Baldi DL, Higginson EE, Hocking DM, Praszkier J, Cavaliere R, James CE, Bennett-**

Wood V, Azzopardi KI, Turnbull L, Lithgow T, Robins-Browne RM, Whitchurch CB, Tauschek M. 2012. The type II secretion system and its ubiquitous lipoprotein substrate, SslE, are required for biofilm formation and virulence of enteropathogenic *Escherichia coli*. *Infect Immun* 80:2042–2052.

77. East A, Mechaly AE, Huysmans GHM, Bernarde C, Tello-Manigne D, Nadeau N, Pugsley AP, Buschiazzo A, Alzari PM, Bond PJ, Francetic O. 2016. Structural basis of pullulanase membrane binding and secretion revealed by X-ray crystallography, molecular dynamics and biochemical analysis. *Structure* 24:92–104.

78. Zückert WR. 2014. Secretion of bacterial lipoproteins: through the cytoplasmic membrane, the periplasm and beyond. *Biochim Biophys Acta* 1843:1509–1516.

79. d'Enfert C, Chapon C, Pugsley AP. 1987. Export and secretion of the lipoprotein pullulanase by *Klebsiella pneumoniae*. *Mol Microbiol* 1:107–116.

80. Horstman AL, Kuehn MJ. 2002. Bacterial surface association of heat-labile enterotoxin through lipopolysaccharide after secretion via the general secretory pathway. *J Biol Chem* 277:32538–32545.

81. Horstman AL, Bauman SJ, Kuehn MJ. 2004. Lipopolysaccharide 3-deoxy-ᴅ-manno-octulosonic acid (Kdo) core determines bacterial association of secreted toxins. *J Biol Chem* 279:8070–8075.

82. Ferrandez Y, Condemine G. 2008. Novel mechanism of outer membrane targeting of proteins in Gram-negative bacteria. *Mol Microbiol* 69:1349–1357.

83. Haft DH, Varghese N. 2011. GlyGly-CTERM and rhombosortase: a C-terminal protein processing signal in a many-to-one pairing with a rhomboid family intramembrane serine protease. *PLoS One* 6:e28886.

84. Gadwal S, Johnson TL, Remmer H, Sandkvist M. 2018. C-terminal processing of GlyGly-CTERM containing proteins by rhombosortase in *Vibrio cholerae*. *PLoS Pathog* 14:e1007341.

85. Sandkvist M, Michel LO, Hough LP, Morales VM, Bagdasarian M, Koomey M, DiRita VJ, Bagdasarian M. 1997. General secretion pathway (*eps*) genes required for toxin secretion and outer membrane biogenesis in *Vibrio cholerae*. *J Bacteriol* 179:6994–7003.

86. Tauschek M, Gorrell RJ, Strugnell RA, Robins-Browne RM. 2002. Identification of a protein secretory pathway for the secretion of heat-labile enterotoxin by an enterotoxi-

genic strain of *Escherichia coli*. *Proc Natl Acad Sci U S A* 99:7066–7071.

87. Lindeberg M, Collmer A. 1992. Analysis of eight *out* genes in a cluster required for pectic enzyme secretion by *Erwinia chrysanthemi*: sequence comparison with secretion genes from other gram-negative bacteria. *J Bacteriol* 174:7385–7397.

88. Kagami Y, Ratliff M, Surber M, Martinez A, Nunn DN. 1998. Type II protein secretion by *Pseudomonas aeruginosa*: genetic suppression of a conditional mutation in the pilin-like component XcpT by the cytoplasmic component XcpR. *Mol Microbiol* 27:221–233.

89. Sikora AE, Lybarger SR, Sandkvist M. 2007. Compromised outer membrane integrity in *Vibrio cholerae* type II secretion mutants. *J Bacteriol* 189:8484–8495.

90. Park BR, Zielke RA, Wierzbicki IH, Mitchell KC, Withey JH, Sikora AE. 2015. A metalloprotease secreted by the type II secretion system links *Vibrio cholerae* with collagen. *J Bacteriol* 197:1051–1064.

91. White RC, Gunderson FF, Tyson JY, Richardson KH, Portlock TJ, Garnett JA, Cianciotto NP. 2018. Type II secretion-dependent aminopeptidase LapA and acyltransferase PlaC are redundant for nutrient acquisition during *Legionella pneumophila* intracellular infection of amoebas. *mBio* 9:e00528-18.

92. Wilton M, Halverson TWR, Charron-Mazenod L, Parkins MD, Lewenza S. 2018. Secreted phosphatase and deoxyribonuclease are required by *Pseudomonas aeruginosa* to defend against neutrophil extracellular traps. *Infect Immun* 86:e00403-18.

93. Chapon V, Czjzek M, El Hassouni M, Py B, Juy M, Barras F. 2001. Type II protein secretion in gram-negative pathogenic bacteria: the study of the structure/secretion relationships of the cellulase Cel5 (formerly EGZ) from *Erwinia chrysanthemi*. *J Mol Biol* 310:1055–1066.

94. Silva AJ, Pham K, Benitez JA. 2003. Haemagglutinin/protease expression and mucin gel penetration in El Tor biotype *Vibrio cholerae*. *Microbiology* 149:1883–1891.

95. Luo Q, Kumar P, Vickers TJ, Sheikh A, Lewis WG, Rasko DA, Sistrunk J, Fleckenstein JM. 2014. Enterotoxigenic *Escherichia coli* secretes a highly conserved mucin-degrading metalloprotease to effectively engage intestinal epithelial cells. *Infect Immun* 82:509–521.

96. Hews CL, Tran SL, Wegmann U, Brett B, Walsham ADS, Kavanaugh D, Ward NJ, Juge N, Schüller S. 2017. The StcE metalloprotease of enterohaemorrhagic *Escherichia coli*

reduces the inner mucus layer and promotes adherence to human colonic epithelium *ex vivo*. *Cell Microbiol* 19:e12717.

97. **Kooi C, Sokol PA.** 2009. *Burkholderia cenocepacia* zinc metalloproteases influence resistance to antimicrobial peptides. *Microbiology* 155:2818–2825.

98. **McCoy-Simandle K, Stewart CR, Dao J, DebRoy S, Rossier O, Bryce PJ, Cianciotto NP.** 2011. *Legionella pneumophila* type II secretion dampens the cytokine response of infected macrophages and epithelia. *Infect Immun* 79:1984–1997.

99. **Mallama CA, McCoy-Simandle K, Cianciotto NP.** 2017. The type II secretion system of *Legionella pneumophila* dampens the MyD88 and Toll-like receptor 2 signaling pathway in infected human macrophages. *Infect Immun* 85:e00897-16.

100. **Lathem WW, Grys TE, Witowski SE, Torres AG, Kaper JB, Tarr PI, Welch RA.** 2002. StcE, a metalloprotease secreted by *Escherichia coli* O157:H7, specifically cleaves C1 esterase inhibitor. *Mol Microbiol* 45:277–288.

101. **Szabady RL, Lokuta MA, Walters KB, Huttenlocher A, Welch RA.** 2009. Modulation of neutrophil function by a secreted mucinase of *Escherichia coli* O157:H7. *PLoS Pathog* 5:e1000320.

102. **Tilley D, Law R, Warren S, Samis JA, Kumar A.** 2014. CpaA a novel protease from *Acinetobacter baumannii* clinical isolates deregulates blood coagulation. *FEMS Microbiol Lett* 356:53–61.

103. **Waack U, Warnock M, Yee A, Huttinger Z, Smith S, Kumar A, Deroux A, Ginsburg D, Mobley HLT, Lawrence DA, Sandkvist M.** 2018. CpaA is a glycan-specific adamalysin-like protease secreted by *Acinetobacter baumannii* that inactivates coagulation factor XII. *mBio* 9:e01606-18.

104. **Overhage J, Lewenza S, Marr AK, Hancock RE.** 2007. Identification of genes involved in swarming motility using a *Pseudomonas aeruginosa* PAO1 mini-Tn5-*lux* mutant library. *J Bacteriol* 189:2164–2169.

105. **Duncan C, Prashar A, So J, Tang P, Low DE, Terebiznik M, Guyard C.** 2011. Lcl of *Legionella pneumophila* is an immunogenic GAG binding adhesin that promotes interactions with lung epithelial cells and plays a crucial role in biofilm formation. *Infect Immun* 79:2168–2181.

106. **Johnson TL, Fong JC, Rule C, Rogers A, Yildiz FH, Sandkvist M.** 2014. The type II secretion system delivers matrix proteins for biofilm formation by *Vibrio cholerae*. *J Bacteriol* 196:4245–4252.

107. **Fong JNC, Yildiz FH.** 2015. Biofilm matrix proteins. *Microbiol Spectr* 3:MB-0004-2014.

108. **Fong JC, Rogers A, Michael AK, Parsley NC, Cornell WC, Lin YC, Singh PK, Hartmann R, Drescher K, Vinogradov E, Dietrich LE, Partch CL, Yildiz FH.** 2017. Structural dynamics of RbmA governs plasticity of *Vibrio cholerae* biofilms. *eLife* 6:e26163.

109. **Ennouri H, d'Abzac P, Hakil F, Branchu P, Naïtali M, Lomenech AM, Oueslati R, Desbrières J, Sivadon P, Grimaud R.** 2017. The extracellular matrix of the oleolytic biofilms of *Marinobacter hydrocarbonoclasticus* comprises cytoplasmic proteins and T2SS effectors that promote growth on hydrocarbons and lipids. *Environ Microbiol* 19:159–173.

110. **Nouwen N, Ranson N, Saibil H, Wolpensinger B, Engel A, Ghazi A, Pugsley AP.** 1999. Secretin PulD: association with pilot PulS, structure, and ion-conducting channel formation. *Proc Natl Acad Sci U S A* 96:8173–8177.

111. **Nouwen N, Stahlberg H, Pugsley AP, Engel A.** 2000. Domain structure of secretin PulD revealed by limited proteolysis and electron microscopy. *EMBO J* 19:2229–2236.

112. **Chami M, Guilvout I, Gregorini M, Rémigy HW, Müller SA, Valerio M, Engel A, Pugsley AP, Bayan N.** 2005. Structural insights into the secretin PulD and its trypsin-resistant core. *J Biol Chem* 280:37732–37741.

113. **Tosi T, Estrozi LF, Job V, Guilvout I, Pugsley AP, Schoehn G, Dessen A.** 2014. Structural similarity of secretins from type II and type III secretion systems. *Structure* 22:1348–1355.

114. **Dunstan RA, Heinz E, Wijeyewickrema LC, Pike RN, Purcell AW, Evans TJ, Praszkier J, Robins-Browne RM, Strugnell RA, Korotkov KV, Lithgow T.** 2013. Assembly of the type II secretion system such as found in *Vibrio cholerae* depends on the novel pilotin AspS. *PLoS Pathog* 9:e1003117.

115. **Hu J, Worrall LJ, Hong C, Vuckovic M, Atkinson CE, Caveney N, Yu Z, Strynadka NCJ.** 2018. Cryo-EM analysis of the T3S injectisome reveals the structure of the needle and open secretin. *Nat Commun* 9:3840.

116. **Gu S, Rehman S, Wang X, Shevchik VE, Pickersgill RW.** 2012. Structural and functional insights into the pilotin-secretin complex of the type II secretion system. *PLoS Pathog* 8:e1002531.

117. **Strozen TG, Li G, Howard SP.** 2012. YghG (GspSβ) is a novel pilot protein required for localization of the GspSβ type II secretion system secretin of enterotoxigenic *Escherichia coli*. *Infect Immun* 80:2608–2622.

118. Collin S, Guilvout I, Nickerson NN, Pugsley AP. 2011. Sorting of an integral outer membrane protein via the lipoprotein-specific Lol pathway and a dedicated lipoprotein pilotin. *Mol Microbiol* **80**:655–665.

119. Collin S, Guilvout I, Chami M, Pugsley AP. 2007. YaeT-independent multimerization and outer membrane association of secretin PulD. *Mol Microbiol* **64**:1350–1357.

120. Dunstan RA, Hay ID, Wilksch JJ, Schittenhelm RB, Purcell AW, Clark J, Costin A, Ramm G, Strugnell RA, Lithgow T. 2015. Assembly of the secretion pores GspD, Wza and CsgG into bacterial outer membranes does not require the Omp85 proteins BamA or TamA. *Mol Microbiol* **97**:616–629.

121. Guilvout I, Brier S, Chami M, Hourdel V, Francetic O, Pugsley AP, Chamot-Rooke J, Huysmans GH. 2017. Prepore stability controls productive folding of the BAM-independent multimeric outer membrane secretin PulD. *J Biol Chem* **292**:328–338.

122. Korotkov KV, Gray MD, Kreger A, Turley S, Sandkvist M, Hol WGJ. 2009. Calcium is essential for the major pseudopilin in the type 2 secretion system. *J Biol Chem* **284**:25466–25470.

123. Cisneros DA, Bond PJ, Pugsley AP, Campos M, Francetic O. 2012. Minor pseudopilin self-assembly primes type II secretion pseudopilus elongation. *EMBO J* **31**:1041–1053.

124. Reindl S, Ghosh A, Williams GJ, Lassak K, Neiner T, Henche AL, Albers SV, Tainer JA. 2013. Insights into FlaI functions in archaeal motor assembly and motility from structures, conformations, and genetics. *Mol Cell* **49**:1069–1082.

125. Mancl JM, Black WP, Robinson H, Yang Z, Schubot FD. 2016. Crystal structure of a type IV pilus assembly ATPase: insights into the molecular mechanism of PilB from *Thermus thermophilus*. *Structure* **24**:1886–1897.

126. McCallum M, Tammam S, Khan A, Burrows LL, Howell PL. 2017. The molecular mechanism of the type IVa pilus motors. *Nat Commun* **8**:15091.

127. Gray MD, Bagdasarian M, Hol WGJ, Sandkvist M. 2011. In vivo cross-linking of EpsG to EpsL suggests a role for EpsL as an ATPase-pseudopilin coupling protein in the type II secretion system of *Vibrio cholerae*. *Mol Microbiol* **79**:786–798.

128. Nivaskumar M, Santos-Moreno J, Malosse C, Nadeau N, Chamot-Rooke J, Tran Van Nhieu G, Francetic O. 2016. Pseudopilin residue E5 is essential for recruitment by the type 2 secretion system assembly platform. *Mol Microbiol* **101**:924–941.

129. Shevchik VE, Robert-Baudouy J, Condemine G. 1997. Specific interaction between OutD, an *Erwinia chrysanthemi* outer membrane protein of the general secretory pathway, and secreted proteins. *EMBO J* **16**:3007–3016.

130. Bouley J, Condemine G, Shevchik VE. 2001. The PDZ domain of OutC and the N-terminal region of OutD determine the secretion specificity of the type II out pathway of *Erwinia chrysanthemi*. *J Mol Biol* **308**:205–219.

131. Douzi B, Ball G, Cambillau C, Tegoni M, Voulhoux R. 2011. Deciphering the Xcp *Pseudomonas aeruginosa* type II secretion machinery through multiple interactions with substrates. *J Biol Chem* **286**:40792–40801.

132. Reichow SL, Korotkov KV, Gonen M, Sun J, Delarosa JR, Hol WGJ, Gonen T. 2011. The binding of cholera toxin to the periplasmic vestibule of the type II secretion channel. *Channels (Austin)* **5**:215–218.

133. Pineau C, Guschinskaya N, Robert X, Gouet P, Ballut L, Shevchik VE. 2014. Substrate recognition by the bacterial type II secretion system: more than a simple interaction. *Mol Microbiol* **94**:126–140.

134. Michel-Souzy S, Douzi B, Cadoret F, Raynaud C, Quinton L, Ball G, Voulhoux R. 2018. Direct interactions between the secreted effector and the T2SS components GspL and GspM reveal a new effector-sensing step during type 2 secretion. *J Biol Chem* **293**:19441–19450.

135. Nunn D. 1999. Bacterial type II protein export and pilus biogenesis: more than just homologies? *Trends Cell Biol* **9**:402–408.

136. Forest KT. 2008. The type II secretion arrowhead: the structure of GspI-GspJ-GspK. *Nat Struct Mol Biol* **15**:428–430.

137. Nivaskumar M, Bouvier G, Campos M, Nadeau N, Yu X, Egelman EH, Nilges M, Francetic O. 2014. Distinct docking and stabilization steps of the pseudopilus conformational transition path suggest rotational assembly of type IV pilus-like fibers. *Structure* **22**:685–696.

138. Nivaskumar M, Francetic O. 2014. Type II secretion system: a magic beanstalk or a protein escalator. *Biochim Biophys Acta* **1843**:1568–1577.

139. O'Neal CJ, Amaya EI, Jobling MG, Holmes RK, Hol WGJ. 2004. Crystal structures of an intrinsically active cholera toxin mutant yield insight into the toxin activation mechanism. *Biochemistry* **43**:3772–3782.

140. Wedekind JE, Trame CB, Dorywalska M, Koehl P, Raschke TM, McKee M, FitzGerald D, Collier RJ, McKay DB. 2001. Refined crystallographic structure of *Pseudomonas*

aeruginosa exotoxin A and its implications for the molecular mechanism of toxicity. *J Mol Biol* **314:**823–837.

141. **Yoder MD, Jurnak F.** 1995. Protein motifs. 3. The parallel beta helix and other coiled folds. *FASEB J* **9:**335–342.

142. **Yu AC, Worrall LJ, Strynadka NC.** 2012. Structural insight into the bacterial mucinase StcE essential to adhesion and immune evasion during enterohemorrhagic *E. coli* infection. *Structure* **20:**707–717.

143. **Giglio KM, Fong JC, Yildiz FH, Sondermann H.** 2013. Structural basis for biofilm formation via the *Vibrio cholerae* matrix protein RbmA. *J Bacteriol* **195:**3277–3286.

144. **Maestre-Reyna M, Wu WJ, Wang AH.** 2013. Structural insights into RbmA, a biofilm scaffolding protein of *V. cholerae. PLoS One* **8:**e82458.

145. **Rule CS, Patrick M, Camberg JL, Maricic N, Hol WGJ, Sandkvist M.** 2016. Zinc coordination is essential for the function and activity of the type II secretion ATPase EpsE. *Microbiologyopen* **5:**870–882.

146. **Fulara A, Vandenberghe I, Read RJ, Devreese B, Savvides SN.** 2018. Structure and oligomerization of the periplasmic domain of GspL from the type II secretion system of *Pseudomonas aeruginosa. Sci Rep* **8:**16760.

147. **Zhang Y, Faucher F, Zhang W, Wang S, Neville N, Poole K, Zheng J, Jia Z.** 2018. Structure-guided disruption of the pseudopilus tip complex inhibits the type II secretion in *Pseudomonas aeruginosa. PLoS Pathog* **14:** e1007343.

148. **Korotkov KV, Delarosa JR, Hol WGJ.** 2013. A dodecameric ring-like structure of the N0 domain of the type II secretin from enterotoxigenic *Escherichia coli. J Struct Biol* **183:**354–362.

149. **Wretlind B, Pavlovskis OR.** 1984. Genetic mapping and characterization of *Pseudomonas aeruginosa* mutants defective in the formation of extracellular proteins. *J Bacteriol* **158:**801–808.

150. **Bo JN, Howard SP.** 1991. Mutagenesis and isolation of *Aeromonas hydrophila* genes which are required for extracellular secretion. *J Bacteriol* **173:**1241–1249.

151. **Paranjpye RN, Lara JC, Pepe JC, Pepe CM, Strom MS.** 1998. The type IV leader peptidase/N-methyltransferase of *Vibrio vulnificus* controls factors required for adherence to HEp-2 cells and virulence in iron-overloaded mice. *Infect Immun* **66:**5659–5668.

152. **Sikora AE, Zielke RA, Lawrence DA, Andrews PC, Sandkvist M.** 2011. Proteomic analysis of the *Vibrio cholerae* type II secretome reveals new proteins, including three

related serine proteases. *J Biol Chem* **286:** 16555–16566.

153. **Szabady RL, Yanta JH, Halladin DK, Schofield MJ, Welch RA.** 2011. TagA is a secreted protease of *Vibrio cholerae* that specifically cleaves mucin glycoproteins. *Microbiology* **157:**516–525.

154. **Golovkine G, Faudry E, Bouillot S, Voulhoux R, Attrée I, Huber P.** 2014. VE-cadherin cleavage by LasB protease from *Pseudomonas aeruginosa* facilitates type III secretion system toxicity in endothelial cells. *PLoS Pathog* **10:** e1003939.

155. **DuMont AL, Cianciotto NP.** 2017. *Stenotrophomonas maltophilia* serine protease StmPr1 induces matrilysis, anoikis, and protease-activated receptor 2 activation in human lung epithelial cells. *Infect Immun* **85:**e00544-17.

156. **Truchan HK, Christman HD, White RC, Rutledge NS, Cianciotto NP.** 2017. Type II secretion substrates of *Legionella pneumophila* translocate out of the pathogen-occupied vacuole via a semipermeable membrane. *mBio* **8:** e00870-17.

157. **Jha G, Rajeshwari R, Sonti RV.** 2005. Bacterial type two secretion system secreted proteins: double-edged swords for plant pathogens. *Mol Plant Microbe Interact* **18:**891–898.

158. **Nascimento R, Gouran H, Chakraborty S, Gillespie HW, Almeida-Souza HO, Tu A, Rao BJ, Feldstein PA, Bruening G, Goulart LR, Dandekar AM.** 2016. The type II secreted lipase/esterase LesA is a key virulence factor required for *Xylella fastidiosa* pathogenesis in grapevines. *Sci Rep* **6:**18598.

159. **Overbye LJ, Sandkvist M, Bagdasarian M.** 1993. Genes required for extracellular secretion of enterotoxin are clustered in *Vibrio cholerae. Gene* **132:**101–106.

160. **Francetic O, Belin D, Badaut C, Pugsley AP.** 2000. Expression of the endogenous type II secretion pathway in *Escherichia coli* leads to chitinase secretion. *EMBO J* **19:**6697–6703.

161. **Aragon V, Kurtz S, Cianciotto NP.** 2001. *Legionella pneumophila* major acid phosphatase and its role in intracellular infection. *Infect Immun* **69:**177–185.

162. **Ball G, Durand E, Lazdunski A, Filloux A.** 2002. A novel type II secretion system in *Pseudomonas aeruginosa. Mol Microbiol* **43:**475–485.

163. **Putker F, Tommassen-van Boxtel R, Stork M, Rodríguez-Herva JJ, Koster M, Tommassen J.** 2013. The type II secretion system (Xcp) of *Pseudomonas putida* is active and involved in the secretion of phosphatases. *Environ Microbiol* **15:**2658–2671.

164. **Rossier O, Dao J, Cianciotto NP.** 2009. A type II secreted RNase of *Legionella pneumophila*

facilitates optimal intracellular infection of *Hartmannella vermiformis. Microbiology* **155:** 882–890.

165. **Mulcahy H, Charron-Mazenod L, Lewenza S.** 2010. *Pseudomonas aeruginosa* produces an extracellular deoxyribonuclease that is required for utilization of DNA as a nutrient source. *Environ Microbiol* **12:**1621–1629.

166. **DiChristina TJ, Moore CM, Haller CA.** 2002. Dissimilatory Fe(III) and Mn(IV) reduction by *Shewanella putrefaciens* requires *ferE*, a homolog of the *pulE* (*gspE*) type II protein secretion gene. *J Bacteriol* **184:**142–151.

167. **Shi L, Deng S, Marshall MJ, Wang Z, Kennedy DW, Dohnalkova AC, Mottaz HM, Hill EA, Gorby YA, Beliaev AS, Richardson DJ, Zachara JM, Fredrickson JK.** 2008. Direct involvement of type II secretion system in extracellular translocation of *Shewanella oneidensis* outer membrane cytochromes MtrC and OmcA. *J Bacteriol* **190:**5512–5516.

168. **Kirn TJ, Jude BA, Taylor RK.** 2005. A colonization factor links *Vibrio cholerae* environmental survival and human infection. *Nature* **438:**863–866.

169. **Cadoret F, Ball G, Douzi B, Voulhoux R.** 2014. Txc, a new type II secretion system of *Pseudomonas aeruginosa* strain PA7, is regulated by the TtsS/TtsR two-component system and directs specific secretion of the CbpE chitin-binding protein. *J Bacteriol* **196:**2376–2386.

The Injectisome, a Complex Nanomachine for Protein Injection into Mammalian Cells

20

MARIA LARA-TEJERO[1] and JORGE E. GALÁN[1]

INTRODUCTION

Type III protein secretion systems (T3SSs) are multiprotein nanomachines present in many Gram-negative bacteria with a close relationship with a eukaryotic host. The primary function of these machines is the delivery of bacterially encoded effector proteins into target eukaryotic cells (1–4), to modulate a myriad of cell biological processes for the benefit of the bacteria that encode them (5, 6). T3SSs are widespread in nature, playing a central role in the pathogenic and symbiotic interactions between many bacteria and their hosts. Among the bacteria that encode T3SSs are many important human and plant pathogens. As the field has progressed, so has the amount of information available, precluding a comprehensive review of the literature. Therefore, here we focus on the structural and architectural aspects of the type III system. To reflect current knowledge, and to help the reader better understand the structural organization of this machine, we refer throughout to the complete type III secretion machine as the injectisome, and we describe in detail the different substructures that integrate it (i.e., the needle complex [NC], the export apparatus, and the sorting platform). Readers are referred to other reviews for more specific aspects of the structure and function of these secretion machines (1–4).

[1]Department of Microbial Pathogenesis, Yale University School of Medicine, New Haven, CT 06536

Protein Secretion in Bacteria

Edited by Maria Sandkvist, Eric Cascales, and Peter J. Christie

© 2019 American Society for Microbiology, Washington, DC

doi:10.1128/ecosalplus.ESP-0039-2018

THE INJECTISOME

The main structural element of the type III secretion system is the injectiome, a complex multiprotein structure composed of extracellular, envelope-associated, and cytoplasmic elements or substructures. To facilitate their description, each of these elements is discussed separately. However, it should be emphasized that the type III secretion machine is and operates as a single functional structural unit and that its separation into different subcomponents, while useful, is somewhat arbitrary. Also, we note that the increasing knowledge on the structural organization of this system has rendered some previous descriptions of this system more confusing. Furthermore, the plethora of gene names has made comparison of systems across species somewhat challenging; thus, when referring to specific components of the injectisome, we utilize a previously proposed universal nomenclature. Table 1 lists the specific names of proteins of the most studied T3SSs.

The NC

The NC is a major core element of the T3SS injectisome, and its discovery in 1998 constituted a major breakthrough in the understanding of type III secretion machines (7). Until then, T3SSs were simply a collection of genes required for protein secretion, without a clear framework to explain how they may constitute a protein secretion machine. Since the first visualization and isolation from *Salmonella enterica* serovar Typhimurium (7), it has been visualized in other bacteria, showing a conserved architecture (8–10). It consists of a multiring base substructure embedded in the bacterial envelope and a needle-like extension protruding several nanometers from the bacterial surface (Fig. 1). The needle is linked to the base through a substructure known as the inner rod, which docks into a socket-like structure within the NC composed of the export apparatus, which is thought to form a conduit to facilitate the passage of effector proteins through the bacterial inner membrane. At the distal side of the needle lies the needle tip complex, which senses the target host cell.

The base

The NC base is a multiring structure that spans the inner and outer membranes of the bacterial envelope. Despite its architectural complexity, the base of the NC is composed of a relatively small number of proteins. The inner rings (IR1 and IR2), located in the bacterial inner membrane and the cytoplasm, are composed of two proteins (SctJ and SctD form the IR1, while SctD alone forms the IR2), whereas the outer ring (OR) and the neck of the base are composed of a single protein (SctC), a member of the secretin family of outer membrane proteins. Single-particle cryo-electron microscopy (cryo-EM) of isolated complexes has revealed a detailed view of the NC of the *S.* Typhimurium pathogenicity island 1 (SPI-1)-encoded T3SS (11–13). This structure revealed that 15 SctC molecules form a double-walled stranded β-barrel complex in the outer membrane, which connects directly to 24 molecules of the IR components, amounting to a symmetry mismatch (12, 13). This mismatch confers flexibility on the neck, which may be important during NC assembly. The IRs are composed of two concentric rings: a larger peripheral ring formed by SctD encircling a smaller internal ring formed by SctJ. The smaller ring is shielded on the sides by the SctD ring and on top by the SctC neck. SctD and SctJ share a similar although inverted topology, with a periplasmic and a cytoplasmic domain separated by a single transmembrane segment. The N-terminal domain of SctJ is lipidated and located in the periplasm, and the C-terminal cytoplasmic domain is very short or absent in some homologues which also lack the transmembrane domain (14). In contrast, SctD has a longer N-terminal cytoplasmic domain that forms IR2 and links the NC base to the sorting platform (see below). The crystal structures of the soluble domains of some

TABLE 1 Principal components of most studied T3SSs

Universal nomenclature	Function	Yersinia	Salmonella SPI-1	Salmonella SPI-2	E. coli (EPEC/EHEC)	Shigella	Chlamydia	P. aeruginosa	P. syringae	Rhizobium	Flagellum
Needle complex (flagellar basal body)											
SctC	Secretin (OM ring)	YscC	InvG	SsaC	EscC	MxiD	CdsC	PscC	HrcC	RhcC1-RhcC2	NA[a]
	Secretin pilotin	YscW	InvH			MxiM		ExsB	HrpT		NA
SctD	IM ring	YscD	PrgH	SsaD	EscD	MxiG	CdsD	PscD	HrpQ	Y4yQ	FliG
SctJ	IM ring	YscJ	PrgK	SsaJ	EscJ	MxiJ	CdsJ	PscJ	HrcJ	NolT	FliF
SctF	Needle subunit	YscF	PrgI	SsaG	EscF	MxiH	CdsF	PscF	HrpA	NopA-NopB	FlgE (hook protein)
SctI	Inner rod subunit	YscI	PrgJ	SsaI	EscI	MxiI		PscI	HrpB	NolU	NA
Export apparatus (inner membrane proteins)											
SctU	Autoprotease/substrate switching	YscU	SpaS	SsaU	EscU	Spa40	CdsU	PscU	HrcU	RhcU	FlhB
SctV	Inner membrane channel	YscV	InvA	SsaV	EscV	MxiA	CdsV	PcrD	HrcV	Y4yR	FlhA
SctR	Inner membrane channel	YscR	SpaP	SsaR	EscR	Spa24	CdsR	PscR	HrcR	RhcR	FliP
SctS	Inner membrane channel	YscS	SpaQ	SsaS	EscS	Spa9	CdsD	PscS	HrcS	RhcS	FliQ
SctT	Inner membrane channel	YscT	SpaR	SsaT	EscT	Spa29	CdsT	PscT	HrcT	RhcT	FliR
Sorting platform (flagellar C-ring)											
SctQ	Core scaffold	YscQ	SpaO	SsaQ	SepQ	Spa33	CdsQ	PscQ	HrcQ	RhcQ	FliM-FliN
SctN	ATPase	YscN	InvC	SsaN	EscN	Spa47	CdsN	PscN	HrcN	RhcN	FliI
SctL	Core scaffold	YscL	OrgB	SsaK	EscL	MxiN	CdsL	PscL	HrpE	NolV	FliH
SctO	SctN/SctV linker	YscO	InvI	SsaO	EscO	Spa13	CdsO	PscO	HrpO	Y4yJ	FliJ
SctK	Core scaffold	YscK	OrgA	?	EscK	MxiK		PscK	HrpD		
Regulatory proteins											
SctP	Needle assembly regulator	YscP	InvJ	SsaP	EscP	Spa32	CdsP	PscP	HrpP		FliK
SctW	Regulator of translocase secretion	YopN-TyeA	InvE	SsaL	SepL	MxiC	CdsN	PopN	HrpJ		
	Initiation of needle assembly	?	OrgC	?	?	MxiL	?	?	?	?	
Translocators and tip complex protein											
SctB	Translocon	YopD	SipC	SsaE	EspB	IpaC	CopD	PopD	HrpK	NopX	
SctE	Translocon	YopB	SipB	SsaC	EspD	IpaB	CopB	PopB			
SctA	Tip protein	LcrV	SipD	SsaB	EspA	IpaD	CT584	PcrV			

[a]NA, not applicable.

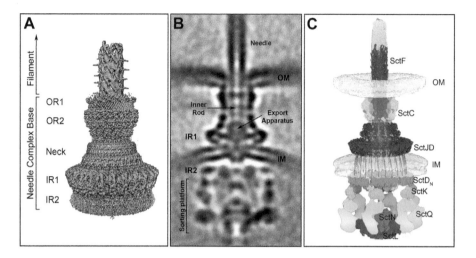

FIGURE 1 *Salmonella* Typhimurium SPI-1-encoded type III secretion system. (A) Surface view of the 3D reconstruction of the single-particle cryo-EM map of the needle complex (NC) substructure with the atomic structures of the different NC components docked. OR1, outer ring 1; OR2, outer ring 2; IR1, inner ring 1; IR2, inner ring 2. (B) Central section of an overall cryo-ET structure of the complete injectisome *in situ*. Of note is the location of IR2 in the cytosolic side of the bacterial envelope. IM, inner membrane; OM, outer membrane. (C) Molecular model of the organization of the injectisome *in situ*, with available atomic structures fitted into the model. Figure adapted from reference 96, with permission.

base components have been solved (i.e., the periplasmic domains of the outer membrane secretin EscC [SctC] [15], PrgH [SctD] [16], and EscJ [SctJ] [17]). In all cases, the atomic structures showed a modular architecture of three topologically similar α/β domains. A comparison of the structures revealed a strong similarity between them despite the lack of detectable sequence identity. The fact that these three proteins arrange in a ring led to the hypothesis that this modular fold may represent a ring-building motif (15). However, this domain is also present in proteins that do not organize in rings and has been shown to be dispensable for ring formation (18). Therefore, the relationship between the presence of this domain and the ability to organize in rings remains unclear. The structure of the NC *in situ* determined by cryo-electron tomography (cryo-ET) has shown that although the OR, neck, and IRs of the *in situ* and isolated structures are virtually identical, there are some unique features (19) (Fig. 1). The OR of the NC is inserted into the inner

leaflet of the outer membrane, resulting in an "inward pinch." Furthermore, although IR1, which is predicted to be located in the periplasmic space, completely overlaps in the isolated and in *in situ* structures, IR2, which is located in the cytoplasm, does not overlap. In the *in situ* structure this ring is pushed further away from IR1 to accommodate the inner membrane, which separates the two rings. Upon assembly of the sorting platform, IR2 undergoes a significant conformation change to adopt a "6 patch" organization to accommodate the 6 pods of the sorting platform (see below) (19).

The needle

The NC base has a several nanometer-long needle extension that confers on the NC its "syringe-like" appearance. The needle substructure is composed of a single, small protein subunit, SctF, polymerized in a helical fashion (20–23). In its native arrangement the length of the needle ranges between 30 and 70 nm, and its width ranges from 10 to 13 nm.

The atomic structure of the needle protein shows an α-helical hairpin arrangement with two α-helices of similar size separated by a short loop most often containing two proline residues separated by 2 amino acids (the PXXP motif) (24–26). More recent studies involving cryo-EM (27) and solid-state nuclear magnetic resonance (NMR) (23) of recombinant and native needle polymers indicate that the needle has a right-handed helical organization consisting of ~5.7 subunits per turn and a helical pitch of ~24 Å. The entire assembly is traversed by a channel ~25 Å in diameter. Although initially there were two incompatible atomic models for the needle polymer, several studies have now conclusively shown that the subunit orientation within the needle polymer places the extended N terminus of the protomer facing the outside of the polymer and the C terminus of the subunit facing the lumen. Residues that face the lumen of the needle are highly conserved and mostly polar, and analysis of their electrostatic potential reveals alternating positively and negatively charged regions. The implications of this observation are unclear, but it is possible that such an organization could play a role in the mechanisms of secretion through the needle channel. Within the filament, the individual proteins are stabilized by multiple inter- and intrasubunit contacts, resulting in a rather rigid structure. Attention has been placed on a small kink (residues Val-20 to Asn-22 in the case of the *S.* Typhimurium SPI-1 needle protein) that interrupts the N-terminal α-helix and that is not observed in the crystal structure of the soluble protomer. The implications of this observation have not been determined, but it is conceivable that it could play a role in signal transduction during activation of the injectisome upon contact with target cells.

The inner rod

The inner rod is a substructure that links the needle to the NC base by docking to the export apparatus (11). Although by analogy with the flagellar system this substructure was originally referred to as a rod, cross-linking as well as stoichiometry studies suggest that this structure is more likely akin to a "washer" rather than a rod since it is predicted to have ~6 subunits (28, 29). Like the needle substructure, it is built from a single small protein subunit, SctI. The atomic structure of the inner rod of *S.* Typhimurium has been solved by circular dichroism and NMR spectroscopy (30). In its soluble monomeric form this protein is largely unfolded, lacking tertiary structure. However, computational methods have suggested that the inner rod subunit shares a structure similar to that of the needle protein, an α-helical hairpin shape flanked by flexible regions. *In silico* modeling has also determined that the domains critical for filament assembly are well conserved between the needle and inner rod protein. However, the two subunits differ significantly at their N termini, and based on the needle filament structure, these differences would not allow the inner rod to elongate beyond one turn of the helix (~6 subunits) (28). This substructure has been implicated in substrate switching and needle length control (31, 32).

The needle tip complex and needle extension

The T3SS is inactive prior to contact with the eukaryotic host. In this inactive state, the needle filament is capped by a single protein that organizes in a tip complex (33, 34), or it is extended by another filament longer than the needle itself (35). The tip complex and the needle extension play a role in sensing the environment, preventing the premature unproductive secretion of effectors.

Based on their structure and biochemical properties, the proteins that make up the tip complex can be divided in two groups: the SipD/IpaD group (from *Salmonella* and *Shigella*, respectively) and the LcrV/PcrV group (from *Yersinia* and *Pseudomonas*). SipD/IpaD-like tip proteins are organized in an N-terminal α-helical hairpin, a long central coiled-coil domain, and a C-terminal region containing a mixture of α/β domains (36, 37). The central coiled-coil domain is characteristic of all tip proteins, although in LcrV/PcrV it is flanked by globular domains on the N and C termini that give them a dumbbell appearance (38). The central coiled-coil domain is important for the interaction of the tip protein with the needle filament, and the needle protomer is expected to bind at multiple sites on this domain. The N-terminal α-helical hairpin folds independently and has been shown to act as a self-chaperone preventing the untimely oligomerization of the SipD/IpaD subunits in the bacterial cytoplasm. This self-chaperoning domain is absent in the LcrV/PcrV family of tip proteins, in which a small cytoplasmic protein (LcrG/PcrG) functions as a chaperone instead (39).

Several crystal structures of different tip proteins have been solved, either alone (37, 38, 40), in combination with the needle protein (PrgI-SipD) (41), or in complex with bile salts (42, 43). What is needed, however, is an atomic-resolution view of the tip protein complex assembled at the needle tip. Presently there are low-resolution three-dimensional (3D) reconstructions from electron micrographs of the *Yersinia* LcrV tip (44) and the *Shigella* IpaD tip (33). The *Yersinia* LcrV tip complex showed a well-defined structure characterized for the presence of a base (formed by the N-terminal globular domain), a neck (formed by the coiled-coil region), and a head (comprising the C-terminal globular domain). This precise organization could be accounted for by a pentameric LcrV ring. The low-resolution structure of the IpaD tip complex is also compatible with a pentameric organization of the tip protein. This complex, however, does not show the characteristic morphology of the LcrV tip complex. It has been proposed that the *Shigella* tip complex may exhibit two different compositions, a homopentameric IpaD tip complex and a heteropentameric complex consisting of four IpaD molecules along with one IpaB molecule (45). Quantification of the relative abun-

dances of the two different complexes was not feasible; therefore, it is possible that the IpaD-IpaB complexes may represent a low proportion of injectisomes that have been activated prematurely, while the pentameric IpaD structure represents the resting tip complex.

A significant modification in the tip complex occurs in several bacterial species, including pathogenic *Escherichia coli* strains (enteropathogenic [EPEC] and enterohemorrhagic [EHEC]). These injectisomes are characterized by the presence of a filamentous extension to the needle substructure, which in *E. coli* is formed by a single protein, EspA (35). Similar to the mechanism by which flagellin assembles into the flagellum, the EspA filament assembles by coiled-coil interactions between EspA subunits. However, electron micrographs of negative stained EspA filaments showed that they are distinct from flagella (46). The 3D reconstruction of the EspA filaments shows that they consist of a helical tube ~120 Å wide, containing a hollow central channel of ~25 Å in diameter with a continuous channel through which effector proteins are translocated. The EspA filament shows helical symmetry, having 28 subunits in 5 turns for a 1-start helix (5.6 subunits/turn). A later study using cryo-EM of frozen hydrated filaments has shown that the EspA filament displays heterogeneity in the structure (47) due to a fixed rotation between subunits but a variable axial rise between adjacent subunits. How this variability in the structure relates to the function of the filaments has not yet been addressed.

The export apparatus

All T3SSs contain five conserved inner membrane proteins that are essential for their function, SctV, SctR, SctS, SctT, and SctU (48–51). Cryo-EM studies have correlated the presence of a defined density inside the inner membrane rings with the presence of the inner membrane proteins, indicating that at least a subset of these inner membrane proteins are located inside the NC structure, creating a channel through which the secreted proteins can traverse the inner membrane (52). More recently, a cryo-EM structure of an *in vitro*-assembled complex of 3 export apparatus components, SctR, SctS, and SctT, from the homologous flagellar export apparatus was solved, providing major insight into the organization of this substructure (53). The complex adopts a helical configuration and is organized with a stoichiometry of 5:4:1 for FliP (SctR), FliQ (SctS), and FliR (SctT), respectively. Remarkably, despite the presence of several predicted transmembrane domains, none of these proteins adopts a canonical membrane protein configuration. Rather, the complex is arranged in a helical configuration in which a FliP/FliQ pair and an additional FliP molecule combine with one copy of FliR. FliR is structurally equivalent to the FliP/FliQ pair, so the structure consists of six copies of a FliR-like element forming a single helical turn. The helical arrangement of these export apparatus components provides an optimal platform onto which the inner rod/needle filament can be effectively assembled. Two additional components of the export apparatus, SctV and SctU, whose locations within the NC have not been precisely determined, are likely to play a more specialized role in type III secretion. SctV has a large cytoplasmic domain that crystallizes as a circular nonamer (18) and by cryo-ET can be seen as toroidal shape density immediately below the cytoplasmic side of IR2 of the NC (19). It is possible that SctV may play a role in preparing the substrates for translocation through the export apparatus. It has also been suggested that the SctV family of proteins may work as a proton channel to energize the secretion process, although this activity has not been formally demonstrated (54). Another member of these inner membrane proteins, SctU, plays a role in substrate switching of the machine specificity from early substrates (i.e., NC components and regulators) to translocators and effectors. SctU, also known as the "switch protein," is a protease that undergoes autocatalytic cleavage (55–62). This autocleaving event was proposed to be the signal that triggers substrate switching from early substrates (needle and inner rod proteins and other accessory proteins) to middle and late substrates (translocators and effectors). However, more recent experiments have demonstrated that the cleavage event *per se* does not provide a signal for substrate switching and that the cleavage may simply be required to provide SctU with the appropriate conformation for its secretion function (63).

The Sorting Platform

There are several conserved cytosolic proteins, SctQ, SctK, SctL, and SctN, that are required for type III secretion. They arrange in a complex that operates as a sorting platform to organize the secretion process through the T3SS, establishing a hierarchy in the order in which substrates are secreted (64). The proteins that constitute the sorting platform are highly conserved among all T3SSs and have been shown to interact with one another (64–66). Recently, the structural organization of these proteins *in situ* was revealed by cryo-ET (19, 67) (Fig. 1). The sorting platform exhibits a cage-like architecture, enclosed by 6 pod-like structures that emerge from the NC and converge into a 6-spoke wheel-like structure at its cytoplasmic side. This structure serves as scaffold to place the associated ATPase SctN and SctO in close apposition to the export apparatus. These studies have also revealed that SctQ accounts for most of the protein density associated with the pods and is linked to the wheel-like capping structure formed by SctL on one side and SctK, which links the sorting platform to the NC, on the other. The presence of the 6-pod structure stands in contrast with the appearance of a related structure found in the flagellar apparatus known as the C-ring, which forms a closed ring stably linked to the flagellar basal body (68). These structural differences may reflect the fundamentally different roles they play. There is evidence indicating that the sorting platform exhibits a dynamic behavior with cycles of assembly and disassembly, although the specific role of this behavior in the function of T3SS and the mechanisms by which this behavior may be controlled are not currently known (69–71). It is possible that an alternatively translated product of the open reading frame encoding the core sorting platform component SctQ may be involved in this process (72–75). However, more research will be required to clarify these poorly understood aspects of type III secretion. The specific mechanisms by which the sorting platform may engage substrates are also poorly understood, although it is expected that the associated ATPase, SctN, with the help of SctO may play a central role in the process of bringing the substrates to the ring formed by the cytoplasmic domain of SctV on the cytosolic side of the NC base.

ASSEMBLY OF THE INJECTISOME

The assembly of the injectisome occurs in a stepwise manner (Fig. 2). First, the export apparatus components SctR, SctS, and SctT are inserted into the membrane (52) and subsequently form a pseudohexameric assembly (53). Later, two additional membrane proteins, SctU snd SctV, are added to the export apparatus. The assembled export apparatus then serves as a template for the assembly of the IRs of the NC (52). Nucleation of the IRs around the export apparatus may also result in the "extraction" of the export apparatus components from the plasma membrane (53).

The ORs and neck of the base are made of a single protein, SctC, belonging to the secretin family of outer membrane bacterial proteins (76). With the assistance of a "pilotin," SctC assembles into a pore in the outer membrane (77–80). Once the ORs and the IRs in association with export apparatus are independently assembled, they come together to form a complete NC base substructure. The cytoplasmic accessory proteins that form the sorting platform are recruited to the NC base, which starts to function as a T3SS dedicated to the secretion of early

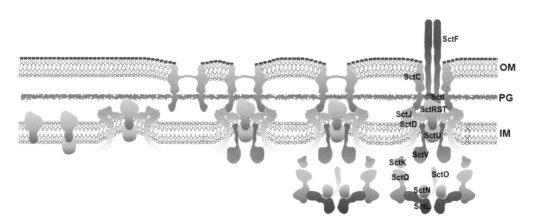

FIGURE 2 Model of the stepwise assembly of the injectisome. SctRST form a stable complex in the inner membrane, to which SctU is recruited. This complex nucleates the assembly of the IRs integrated by SctJ and SctD, which results in the extraction or "pulling" of the inner membrane components from the bacterial plasma membrane. At the same time, the secretin is independently assembled into the OR and the two structures come together to form the NC base substructure to which SctV is subsequently recruited. Once the NC base is formed, the cytoplasmic sorting platform is recruited to the cytoplasmic side of the NC base and the system starts to function as a type III secretion machine dedicated to the delivery of early substrates, such as the inner rod (SctI) and needle (SctF) subunits, to complete the assembly of the entire injectisome.

substrates (22, 81). These early substrates include proteins that will create the needle, SctF, and the inner rod, SctI, as well as accessory regulatory proteins SctP and the *S.* Typhimurium protein OrgC, which regulate the assembly process (82). Completion of the needle triggers substrate switching from early substrates (NC components and regulators of NC assembly) to translocators and effectors, and the injectisome is now ready to be activated by the contact with the host cell. An alternative outside-in assembly model has been proposed for *Yersinia* spp., although current evidence is not consistent with such a model (83).

ACTIVATION OF THE TYPE III SECRETION MACHINE

A distinctive feature of T3SSs is that they require an activating signal to secrete and translocate effectors. While the nature of the activating signal is poorly understood, in most cases it derives from the bacterial contact with target cells (84–86), although other agonists have also been described (87, 88). The activation step is presumably necessary to ensure that the effectors are not unproductively secreted into the extracellular environment prior to host cell contact. How cell contact triggers the activation of the T3SS is not known, but the tip complex is predicted to be involved in the process. In support of this hypothesis, it is possible to activate secretion *in vitro* by the addition of compounds, such as bile salts or Congo red, that bind the tip complex and presumably induce conformational changes similar to those that may occur upon cell contact (42, 43, 89). These conformational changes are thought to be transduced to the secretion machine through the needle and inner rod structures. Consistent with this hypothesis, mutations in the needle and inner rod proteins have been identified that result in constitutive or altered secretion phenotypes (28, 90–92). Activation of the secretion machine leads to secretion

of the translocators (SctB and SctE), which are deployed on the eukaryotic plasma membrane to form the translocation pore or translocon through which effector proteins directly reach the cytosol of the host cell (Fig. 3). Recent cryo-ET studies have provided insight into the organization of the translocon (93). These studies have revealed a well-defined "bend" on the target cell membrane in areas where the needle substructure makes contact with the host cell interface, reflecting the intimate association that is known to be required for optimal T3SS-mediated effector translocation (94). Notably, these studies showed the presence of a distinct density within the region of the target host cell membrane in close apposition to the needle tip of the T3SS injectisome, which was correlated to the presence of the translocon. Although the structure was not detailed enough to provide insight into the stoichiometry and/or molecular organization of the translocon, it showed a structure ~13.5 nm in diameter and 8 nm in thickness, which was smaller than the structure (~60 nm in diameter) of the EPEC translocons assembled from purified components on red blood cells (95). The reason for the different dimensions is unclear, but they may reflect the differences between the *in vivo* and *in vitro* deployment of the translocases. For the translocation to be productive, the translocases must be engaged by the secretion machinery preceding the effectors, through mechanisms involving the sorting platform. Consistent with this notion, only the translocases can be detected in complex with the sorting platform prior to host cell contact, and it is only in the absence of the translocases that the effectors can be detected at this location (64).

CONCLUDING REMARKS

Type III secretion protein machines have sparked the interest of scientists for more than two decades. Although there have been

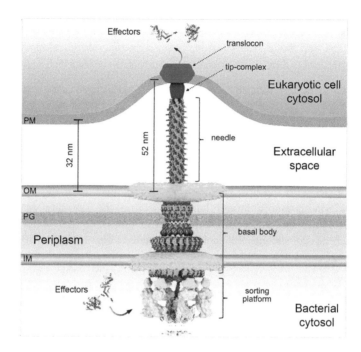

FIGURE 3 Model of the injectisome's interaction with a eukaryotic host cell. Activation of the injectisome leads to secretion of the translocators, which are deployed on the eukaryotic plasma membrane to form the translocon, which remains in contact with the needle to form a direct conduit between the bacterial and host cell cytosol that serves a passageway for the effector proteins. Figure adapted from reference 93, with permission.

remarkable advancements in the understanding of the structure and function of these machines, many knowledge gaps still remain. For example, atomic information on some essential elements of the injectisome is still missing, the precise mechanisms of substrate engagement by the secretion machine are poorly understood, and the mechanisms of eukaryotic cell sensing and signal transduction remain obscure. Addressing some of these fundamental issues most likely will require the development of novel experimental approaches to be able to study the function of these machines in live bacteria. The presence of these machines in many important bacterial pathogens has made them attractive targets for the development of next-generation antimicrobials that can be deployed to prevent and combat many important infectious diseases.

Citation

EcoSal Plus 2019; doi:10.1128/ecosalplus.ESP-0039-2018.

REFERENCES

1. Deng W, Marshall NC, Rowland JL, McCoy JM, Worrall LJ, Santos AS, Strynadka NCJ, Finlay BB. 2017. Assembly, structure, function and regulation of type III secretion systems. *Nat Rev Microbiol* **15:**323–337.
2. Galán JE, Lara-Tejero M, Marlovits TC, Wagner S. 2014. Bacterial type III secretion systems: specialized nanomachines for protein delivery into target cells. *Annu Rev Microbiol* **68:**415–438.
3. Notti RQ, Stebbins CE. 2016. The structure and function of type III secretion systems. *Microbiol Spectr* **4:**VMBF-0004-2015.
4. Wagner S, Grin I, Malmsheimer S, Singh N, Torres-Vargas CE, Westerhausen S. 2018. Bacterial type III secretion systems: a complex

device for the delivery of bacterial effector proteins into eukaryotic host cells. *FEMS Microbiol Lett* **365**:fny201.

5. **Hicks SW, Galán JE.** 2013. Exploitation of eukaryotic subcellular targeting mechanisms by bacterial effectors. *Nat Rev Microbiol* **11**:316–326.

6. **Pinaud L, Sansonetti PJ, Phalipon A.** 2018. Host cell targeting by enteropathogenic bacteria T3SS effectors. *Trends Microbiol* **26**:266–283.

7. **Kubori T, Matsushima Y, Nakamura D, Uralil J, Lara-Tejero M, Sukhan A, Galán JE, Aizawa SI.** 1998. Supramolecular structure of the *Salmonella typhimurium* type III protein secretion system. *Science* **280**:602–605.

8. **Blocker A, Jouihri N, Larquet E, Gounon P, Ebel F, Parsot C, Sansonetti P, Allaoui A.** 2001. Structure and composition of the *Shigella flexneri* "needle complex," a part of its type III secreton. *Mol Microbiol* **39**:652–663.

9. **Daniell SJ, Takahashi N, Wilson R, Friedberg D, Rosenshine I, Booy FP, Shaw RK, Knutton S, Frankel G, Aizawa S.** 2001. The filamentous type III secretion translocon of enteropathogenic *Escherichia coli*. *Cell Microbiol* **3**:865–871.

10. **Sekiya K, Ohishi M, Ogino T, Tamano K, Sasakawa C, Abe A.** 2001. Supermolecular structure of the enteropathogenic *Escherichia coli* type III secretion system and its direct interaction with the EspA-sheath-like structure. *Proc Natl Acad Sci U S A* **98**:11638–11643.

11. **Marlovits TC, Kubori T, Sukhan A, Thomas DR, Galán JE, Unger VM.** 2004. Structural insights into the assembly of the type III secretion needle complex. *Science* **306**:1040–1042.

12. **Schraidt O, Marlovits TC.** 2011. Three-dimensional model of *Salmonella*'s needle complex at subnanometer resolution. *Science* **331**:1192–1195.

13. **Worrall LJ, Hong C, Vuckovic M, Deng W, Bergeron JR, Majewski DD, Huang RK, Spreter T, Finlay BB, Yu Z, Strynadka NC.** 2016. Near-atomic-resolution cryo-EM analysis of the *Salmonella* T3S injectisome basal body. *Nature* **540**:597–601.

14. **Crepin VF, Prasannan S, Shaw RK, Wilson RK, Creasey E, Abe CM, Knutton S, Frankel G, Matthews S.** 2005. Structural and functional studies of the enteropathogenic *Escherichia coli* type III needle complex protein EscJ. *Mol Microbiol* **55**:1658–1670.

15. **Spreter T, Yip CK, Sanowar S, André I, Kimbrough TG, Vuckovic M, Pfuetzner RA, Deng W, Yu AC, Finlay BB, Baker D, Miller SI, Strynadka NC.** 2009. A conserved structural motif mediates formation of the periplasmic rings in the type III secretion system. *Nat Struct Mol Biol* **16**:468–476.

16. **Bergeron JR, Worrall LJ, Sgourakis NG, DiMaio F, Pfuetzner RA, Felise HB, Vuckovic M, Yu AC, Miller SI, Baker D, Strynadka NC.** 2013. A refined model of the prototypical *Salmonella* SPI-1 T3SS basal body reveals the molecular basis for its assembly. *PLoS Pathog* **9**:e1003307.

17. **Yip CK, Kimbrough TG, Felise HB, Vuckovic M, Thomas NA, Pfuetzner RA, Frey EA, Finlay BB, Miller SI, Strynadka NC.** 2005. Structural characterization of the molecular platform for type III secretion system assembly. *Nature* **435**:702–707.

18. **Abrusci P, Vergara-Irigaray M, Johnson S, Beeby MD, Hendrixson DR, Roversi P, Friede ME, Deane JE, Jensen GJ, Tang CM, Lea SM.** 2013. Architecture of the major component of the type III secretion system export apparatus. *Nat Struct Mol Biol* **20**:99–104.

19. **Hu B, Lara-Tejero M, Kong Q, Galan JE, Liu J.** 2017. In situ molecular architecture of the *Salmonella* type III secretion machine. *Cell* **168**:1065–1074.e10.

20. **Cordes FS, Komoriya K, Larquet E, Yang S, Egelman EH, Blocker A, Lea SM.** 2003. Helical structure of the needle of the type III secretion system of *Shigella flexneri*. *J Biol Chem* **278**:17103–17107.

21. **Galkin VE, Schmied WH, Schraidt O, Marlovits TC, Egelman EH.** 2010. The structure of the *Salmonella typhimurium* type III secretion system needle shows divergence from the flagellar system. *J Mol Biol* **396**:1392–1397.

22. **Kubori T, Sukhan A, Aizawa SI, Galán JE.** 2000. Molecular characterization and assembly of the needle complex of the *Salmonella typhimurium* type III protein secretion system. *Proc Natl Acad Sci U S A* **97**:10225–10230.

23. **Loquet A, Sgourakis NG, Gupta R, Giller K, Riedel D, Goosmann C, Griesinger C, Kolbe M, Baker D, Becker S, Lange A.** 2012. Atomic model of the type III secretion system needle. *Nature* **486**:276–279.

24. **Deane JE, Roversi P, Cordes FS, Johnson S, Kenjale R, Daniell S, Booy F, Picking WD, Picking WL, Blocker AJ, Lea SM.** 2006. Molecular model of a type III secretion system needle: implications for host-cell sensing. *Proc Natl Acad Sci U S A* **103**:12529–12533.

25. **Poyraz O, Schmidt H, Seidel K, Delissen F, Ader C, Tenenboim H, Goosmann C, Laube B, Thünemann AF, Zychlinsky A, Baldus M, Lange A, Griesinger C, Kolbe M.** 2010. Protein refolding is required for assembly of the type three secretion needle. *Nat Struct Mol Biol* **17**:788–792.

26. **Zhang L, Wang Y, Picking WL, Picking WD, De Guzman RN.** 2006. Solution structure of monomeric BsaL, the type III secretion needle protein of *Burkholderia pseudomallei. J Mol Biol* 359:322–330.

27. **Hu J, Worrall LJ, Hong C, Vuckovic M, Atkinson CE, Caveney N, Yu Z, Strynadka NCJ.** 2018. Cryo-EM analysis of the T3S injectisome reveals the structure of the needle and open secretin. *Nat Commun* 9:3840.

28. **Lefebre MD, Galán JE.** 2014. The inner rod protein controls substrate switching and needle length in a *Salmonella* type III secretion system. *Proc Natl Acad Sci U S A* 111:817–822.

29. **Zilkenat S, Franz-Wachtel M, Stierhof YD, Galán JE, Macek B, Wagner S.** 2016. Determination of the stoichiometry of the complete bacterial type III secretion needle complex using a combined quantitative proteomic approach. *Mol Cell Proteomics* 15:1598–1609.

30. **Zhong D, Lefebre M, Kaur K, McDowell MA, Gdowski C, Jo S, Wang Y, Benedict SH, Lea SM, Galan JE, De Guzman RN.** 2012. The *Salmonella* type III secretion system inner rod protein PrgJ is partially folded. *J Biol Chem* 287:25303–25311.

31. **Marlovits TC, Kubori T, Lara-Tejero M, Thomas D, Unger VM, Galán JE.** 2006. Assembly of the inner rod determines needle length in the type III secretion injectisome. *Nature* 441:637–640.

32. **Wood SE, Jin J, Lloyd SA.** 2008. YscP and YscU switch the substrate specificity of the *Yersinia* type III secretion system by regulating export of the inner rod protein YscI. *J Bacteriol* 190:4252–4262.

33. **Epler CR, Dickenson NE, Bullitt E, Picking WL.** 2012. Ultrastructural analysis of IpaD at the tip of the nascent MxiH type III secretion apparatus of *Shigella flexneri. J Mol Biol* 420:29–39.

34. **Mueller CA, Broz P, Müller SA, Ringler P, Erne-Brand F, Sorg I, Kuhn M, Engel A, Cornelis GR.** 2005. The V-antigen of *Yersinia* forms a distinct structure at the tip of injectisome needles. *Science* 310:674–676.

35. **Knutton S, Rosenshine I, Pallen MJ, Nisan I, Neves BC, Bain C, Wolff C, Dougan G, Frankel G.** 1998. A novel EspA-associated surface organelle of enteropathogenic *Escherichia coli* involved in protein translocation into epithelial cells. *EMBO J* 17:2166–2176.

36. **Erskine PT, Knight MJ, Ruaux A, Mikolajek H, Wong Fat Sang N, Withers J, Gill R, Wood SP, Wood M, Fox GC, Cooper JB.** 2006. High resolution structure of BipD: an invasion protein associated with the type III secretion system of *Burkholderia pseudomallei. J Mol Biol* 363:125–136.

37. **Johnson S, Roversi P, Espina M, Olive A, Deane JE, Birket S, Field T, Picking WD, Blocker AJ, Galyov EE, Picking WL, Lea SM.** 2007. Self-chaperoning of the type III secretion system needle tip proteins IpaD and BipD. *J Biol Chem* 282:4035–4044.

38. **Derewenda U, Mateja A, Devedjiev Y, Routzahn KM, Evdokimov AG, Derewenda ZS, Waugh DS.** 2004. The structure of *Yersinia pestis* V-antigen, an essential virulence factor and mediator of immunity against plague. *Structure* 12:301–306.

39. **Chaudhury S, de Azevedo Souza C, Plano GV, De Guzman RN.** 2015. The LcrG tip chaperone protein of the *Yersinia pestis* type III secretion system is partially folded. *J Mol Biol* 427:3096–3109.

40. **Chaudhury S, Battaile KP, Lovell S, Plano GV, De Guzman RN.** 2013. Structure of the *Yersinia pestis* tip protein LcrV refined to 1.65 Å resolution. *Acta Crystallogr Sect F Struct Biol Cryst Commun* 69:477–481.

41. **Lunelli M, Hurwitz R, Lambers J, Kolbe M.** 2011. Crystal structure of PrgI-SipD: insight into a secretion competent state of the type three secretion system needle tip and its interaction with host ligands. *PLoS Pathog* 7:e1002163.

42. **Wang Y, Nordhues BA, Zhong D, De Guzman RN.** 2010. NMR characterization of the interaction of the *Salmonella* type III secretion system protein SipD and bile salts. *Biochemistry* 49:4220–4226.

43. **Chatterjee S, Zhong D, Nordhues BA, Battaile KP, Lovell S, De Guzman RN.** 2011. The crystal structures of the *Salmonella* type III secretion system tip protein SipD in complex with deoxycholate and chenodeoxycholate. *Protein Sci* 20:75–86.

44. **Broz P, Mueller CA, Müller SA, Philippsen A, Sorg I, Engel A, Cornelis GR.** 2007. Function and molecular architecture of the *Yersinia* injectisome tip complex. *Mol Microbiol* 65:1311–1320.

45. **Cheung M, Shen DK, Makino F, Kato T, Roehrich AD, Martinez-Argudo I, Walker ML, Murillo I, Liu X, Pain M, Brown J, Frazer G, Mantell J, Mina P, Todd T, Sessions RB, Namba K, Blocker AJ.** 2015. Three-dimensional electron microscopy reconstruction and cysteine-mediated crosslinking provide a model of the type III secretion system needle tip complex. *Mol Microbiol* 95:31–50.

46. **Daniell SJ, Kocsis E, Morris E, Knutton S, Booy FP, Frankel G.** 2003. 3D structure of EspA filaments from enteropathogenic *Escherichia coli. Mol Microbiol* 49:301–308.

47. **Wang YA, Yu X, Yip C, Strynadka NC, Egelman EH.** 2006. Structural polymorphism

in bacterial EspA filaments revealed by cryo-EM and an improved approach to helical reconstruction. *Structure* **14:**1189–1196.

48. **Allaoui A, Woestyn S, Sluiters C, Cornelis GR.** 1994. YscU, a *Yersinia enterocolitica* inner membrane protein involved in Yop secretion. *J Bacteriol* **176:**4534–4542.

49. **Galán JE, Ginocchio C, Costeas P.** 1992. Molecular and functional characterization of the *Salmonella* invasion gene *invA*: homology of InvA to members of a new protein family. *J Bacteriol* **174:**4338–4349.

50. **Ginocchio CC, Galán JE.** 1995. Functional conservation among members of the *Salmonella typhimurium* InvA family of proteins. *Infect Immun* **63:**729–732.

51. **Groisman EA, Ochman H.** 1993. Cognate gene clusters govern invasion of host epithelial cells by *Salmonella typhimurium* and *Shigella flexneri*. *EMBO J* **12:**3779–3787.

52. **Wagner S, Königsmaier L, Lara-Tejero M, Lefebre M, Marlovits TC, Galán JE.** 2010. Organization and coordinated assembly of the type III secretion export apparatus. *Proc Natl Acad Sci U S A* **107:**17745–17750.

53. **Kuhlen L, Abrusci P, Johnson S, Gault J, Deme J, Caesar J, Dietsche T, Mebrhatu MT, Ganief T, Macek B, Wagner S, Robinson CV, Lea SM.** 2018. Structure of the core of the type III secretion system export apparatus. *Nat Struct Mol Biol* **25:**583–590.

54. **Lee PC, Rietsch A.** 2015. Fueling type III secretion. *Trends Microbiol* **23:**296–300.

55. **Edqvist PJ, Olsson J, Lavander M, Sundberg L, Forsberg A, Wolf-Watz H, Lloyd SA.** 2003. YscP and YscU regulate substrate specificity of the *Yersinia* type III secretion system. *J Bacteriol* **185:**2259–2266.

56. **Lavander M, Sundberg L, Edqvist PJ, Lloyd SA, Wolf-Watz H, Forsberg A.** 2002. Proteolytic cleavage of the FlhB homologue YscU of *Yersinia pseudotuberculosis* is essential for bacterial survival but not for type III secretion. *J Bacteriol* **184:**4500–4509.

57. **Ferris HU, Furukawa Y, Minamino T, Kroetz MB, Kihara M, Namba K, Macnab RM.** 2005. FlhB regulates ordered export of flagellar components via autocleavage mechanism. *J Biol Chem* **280:**41236–41242.

58. **Zarivach R, Deng W, Vuckovic M, Felise HB, Nguyen HV, Miller SI, Finlay BB, Strynadka NC.** 2008. Structural analysis of the essential self-cleaving type III secretion proteins EscU and SpaS. *Nature* **453:**124–127.

59. **Deane JE, Graham SC, Mitchell EP, Flot D, Johnson S, Lea SM.** 2008. Crystal structure of Spa40, the specificity switch for the *Shigella*

flexneri type III secretion system. *Mol Microbiol* **69:**267–276.

60. **Wiesand U, Sorg I, Amstutz M, Wagner S, van den Heuvel J, Lührs T, Cornelis GR, Heinz DW.** 2009. Structure of the type III secretion recognition protein YscU from *Yersinia enterocolitica*. *J Mol Biol* **385:**854–866.

61. **Lountos GT, Austin BP, Nallamsetty S, Waugh DS.** 2009. Atomic resolution structure of the cytoplasmic domain of *Yersinia pestis* YscU, a regulatory switch involved in type III secretion. *Protein Sci* **18:**467–474.

62. **Björnfot AC, Lavander M, Forsberg A, Wolf-Watz H.** 2009. Autoproteolysis of YscU of *Yersinia pseudotuberculosis* is important for regulation of expression and secretion of Yop proteins. *J Bacteriol* **191:**4259–4267.

63. **Monjarás Feria JV, Lefebre MD, Stierhof YD, Galán JE, Wagner S.** 2015. Role of autocleavage in the function of a type III secretion specificity switch protein in *Salmonella enterica* serovar Typhimurium. *mBio* **6:**e01459-15.

64. **Lara-Tejero M, Kato J, Wagner S, Liu X, Galán JE.** 2011. A sorting platform determines the order of protein secretion in bacterial type III systems. *Science* **331:**1188–1191.

65. **Jackson MW, Plano GV.** 2000. Interactions between type III secretion apparatus components from *Yersinia pestis* detected using the yeast two-hybrid system. *FEMS Microbiol Lett* **186:**85–90.

66. **Spaeth KE, Chen YS, Valdivia RH.** 2009. The *Chlamydia* type III secretion system C-ring engages a chaperone-effector protein complex. *PLoS Pathog* **5:**e1000579.

67. **Hu B, Morado DR, Margolin W, Rohde JR, Arizmendi O, Picking WL, Picking WD, Liu J.** 2015. Visualization of the type III secretion sorting platform of *Shigella flexneri*. *Proc Natl Acad Sci U S A* **112:**1047–1052.

68. **Thomas D, Morgan DG, DeRosier DJ.** 2001. Structures of bacterial flagellar motors from two FliF-FliG gene fusion mutants. *J Bacteriol* **183:**6404–6412.

69. **Diepold A, Kudryashev M, Delalez NJ, Berry RM, Armitage JP.** 2015. Composition, formation, and regulation of the cytosolic C-ring, a dynamic component of the type III secretion injectisome. *PLoS Biol* **13:**e1002039.

70. **Diepold A, Sezgin E, Huseyin M, Mortimer T, Eggeling C, Armitage JP.** 2017. A dynamic and adaptive network of cytosolic interactions governs protein export by the T3SS injectisome. *Nat Commun* **8:**15940.

71. **Zhang Y, Lara-Tejero M, Bewersdorf J, Galán JE.** 2017. Visualization and character-

ization of individual type III protein secretion machines in live bacteria. *Proc Natl Acad Sci U S A* **114**:6098–6103.

72. **Bzymek KP, Hamaoka BY, Ghosh P.** 2012. Two translation products of *Yersinia yscQ* assemble to form a complex essential to type III secretion. *Biochemistry* **51**:1669–1677.

73. **McDowell MA, Marcoux J, McVicker G, Johnson S, Fong YH, Stevens R, Bowman LA, Degiacomi MT, Yan J, Wise A, Friede ME, Benesch JL, Deane JE, Tang CM, Robinson CV, Lea SM.** 2016. Characterisation of *Shigella* Spa33 and *Thermotoga* FliM/N reveals a new model for C-ring assembly in T3SS. *Mol Microbiol* **99**:749–766.

74. **Yu XJ, Liu M, Matthews S, Holden DW.** 2011. Tandem translation generates a chaperone for the *Salmonella* type III secretion system protein SsaQ. *J Biol Chem* **286**:36098–36107.

75. **Notti RQ, Bhattacharya S, Lilic M, Stebbins CE.** 2015. A common assembly module in injectisome and flagellar type III secretion sorting platforms. *Nat Commun* **6**:7125.

76. **Majewski DD, Worrall LJ, Strynadka NC.** 2018. Secretins revealed: structural insights into the giant gated outer membrane portals of bacteria. *Curr Opin Struct Biol* **51**:61–72.

77. **Daefler S, Russel M.** 1998. The *Salmonella typhimurium* InvH protein is an outer membrane lipoprotein required for the proper localization of InvG. *Mol Microbiol* **28**:1367–1380.

78. **Crago AM, Koronakis V.** 1998. *Salmonella* InvG forms a ring-like multimer that requires the InvH lipoprotein for outer membrane localization. *Mol Microbiol* **30**:47–56.

79. **Burghout P, Beckers F, de Wit E, van Boxtel R, Cornelis GR, Tommassen J, Koster M.** 2004. Role of the pilot protein YscW in the biogenesis of the YscC secretin in *Yersinia enterocolitica*. *J Bacteriol* **186**:5366–5375.

80. **Okon M, Moraes TF, Lario PI, Creagh AL, Haynes CA, Strynadka NC, McIntosh LP.** 2008. Structural characterization of the type-III pilot-secretin complex from *Shigella flexneri*. *Structure* **16**:1544–1554.

81. **Magdalena J, Hachani A, Chamekh M, Jouihri N, Gounon P, Blocker A, Allaoui A.** 2002. Spa32 regulates a switch in substrate specificity of the type III secreton of *Shigella flexneri* from needle components to Ipa proteins. *J Bacteriol* **184**:3433–3441.

82. **Kato J, Dey S, Soto JE, Butan C, Wilkinson MC, De Guzman RN, Galan JE.** 2018. A protein secreted by the *Salmonella* type III secretion system controls needle filament assembly. *eLife* **7**:e35886.

83. **Diepold A, Amstutz M, Abel S, Sorg I, Jenal U, Cornelis GR.** 2010. Deciphering the assembly of the *Yersinia* type III secretion injectisome. *EMBO J* **29**:1928–1940.

84. **Ménard R, Sansonetti P, Parsot C.** 1994. The secretion of the *Shigella flexneri* Ipa invasins is activated by epithelial cells and controlled by IpaB and IpaD. *EMBO J* **13**:5293–5302.

85. **Zierler MK, Galán JE.** 1995. Contact with cultured epithelial cells stimulates secretion of *Salmonella typhimurium* invasion protein InvJ. *Infect Immun* **63**:4024–4028.

86. **Mounier J, Bahrani FK, Sansonetti PJ.** 1997. Secretion of *Shigella flexneri* Ipa invasins on contact with epithelial cells and subsequent entry of the bacterium into cells are growth stage dependent. *Infect Immun* **65**:774–782.

87. **Bahrani FK, Sansonetti PJ, Parsot C.** 1997. Secretion of Ipa proteins by *Shigella flexneri*: inducer molecules and kinetics of activation. *Infect Immun* **65**:4005–4010.

88. **Olive AJ, Kenjale R, Espina M, Moore DS, Picking WL, Picking WD.** 2007. Bile salts stimulate recruitment of IpaB to the *Shigella flexneri* surface, where it colocalizes with IpaD at the tip of the type III secretion needle. *Infect Immun* **75**:2626–2629.

89. **Dickenson NE, Zhang L, Epler CR, Adam PR, Picking WL, Picking WD.** 2011. Conformational changes in IpaD from *Shigella flexneri* upon binding bile salts provide insight into the second step of type III secretion. *Biochemistry* **50**:172–180.

90. **Kenjale R, Wilson J, Zenk SF, Saurya S, Picking WL, Picking WD, Blocker A.** 2005. The needle component of the type III secreton of *Shigella* regulates the activity of the secretion apparatus. *J Biol Chem* **280**:42929–42937.

91. **Cherradi Y, Schiavolin L, Moussa S, Meghraoui A, Meksem A, Biskri L, Azarkan M, Allaoui A, Botteaux A.** 2013. Interplay between predicted inner-rod and gatekeeper in controlling substrate specificity of the type III secretion system. *Mol Microbiol* **87**:1183–1199.

92. **Veenendaal AK, Hodgkinson JL, Schwarzer L, Stabat D, Zenk SF, Blocker AJ.** 2007. The type III secretion system needle tip complex mediates host cell sensing and translocon insertion. *Mol Microbiol* **63**:1719–1730.

93. **Park D, Lara-Tejero M, Waxham MN, Li W, Hu B, Galán JE, Liu J.** 2018. Visualization of the type III secretion mediated *Salmonella*-host cell interface using cryo-electron tomography. *eLife* **7**:39514.

94. **Lara-Tejero M, Galán JE.** 2009. *Salmonella enterica* serovar Typhimurium pathogenicity

island 1-encoded type III secretion system translocases mediate intimate attachment to nonphagocytic cells. *Infect Immun* **77:**2635–2642.

95. **Ide T, Laarmann S, Greune L, Schillers H, Oberleithner H, Schmidt MA.** 2001. Charac-

terization of translocation pores inserted into plasma membranes by type III-secreted Esp proteins of enteropathogenic *Escherichia coli*. *Cell Microbiol* **3:**669–679.

96. **Galán JE, Waksman G.** 2018. Protein-injection machines in bacteria. *Cell* **172:**1306–1318.

Promises and Challenges of the Type Three Secretion System Injectisome as an Antivirulence Target

21

ALYSSA C. FASCIANO,[1] LAMYAA SHABAN,[2] and JOAN MECSAS[3]

INTRODUCTION

Antibiotic resistance is a great and growing threat to public health, motivating scientists to find innovative strategies to cure infections (1–3). An alternative approach to classical antibiotics is to target virulence factors (4): bacterial factors required for infection or damage but not for growth outside the host (2, 5, 6). An antivirulence factor should render the bacteria non-pathogenic by neutralizing a critical virulence element, thereby allowing clearance of the pathogen by the host immune system (5–8).

The type 3 secretion system injectisome (T3SSi) is expressed in a broad spectrum of Gram-negative bacteria and is usually crucial for virulence (4, 9). This needle-and-syringe-like apparatus functions as a conduit for the delivery of effector proteins from the bacterial cytoplasm into host cells (Fig. 1A). These T3SSi systems share homology with 8 essential core components of flagellar T3SS and contain an additional 20 to 30 proteins involved in expression,

[1]Program in Immunology, Sackler School of Graduate Biomedical Sciences, Tufts University School of Medicine, Boston, MA 02111
[2]Program in Molecular Microbiology, Sackler School of Graduate Biomedical Sciences, Tufts University School of Medicine, Boston, MA 02111
[3]Department of Molecular Biology and Microbiology, Tufts University School of Medicine, Boston, MA 02111

Protein Secretion in Bacteria
Edited by Maria Sandkvist, Eric Cascales, and Peter J. Christie
© 2019 American Society for Microbiology, Washington, DC
doi:10.1128/ecosalplus.ESP-0032-2018

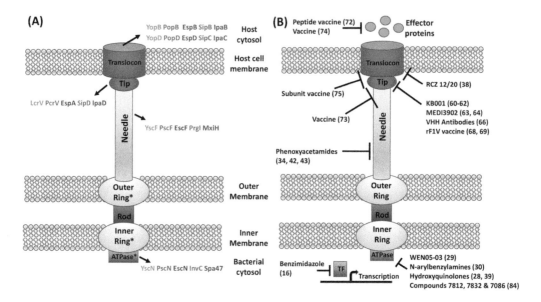

FIGURE 1 Structure of T3SSi.(A) Asterisks indicate regions with conserved components between T3SSi and flagella. Orange, *Yersinia*; blue, *Pseudomonas*; purple, EPEC/EHEC; green, *Salmonella*; red, *Shigella*. (B) Potential targets of compounds based on inhibition of T3SSi function, biochemical or binding studies, genetic resistance, or animal studies.

secretion, and translocation of effector proteins (9–11). Therapeutic strategies against the T3SSi have been pursued, including interfering with transcriptional regulation, chaperone-effector interaction, assembly of various structures (outer ring, needle, or tip complex), or effector translocation or function (4, 5, 12–18).

Targeting the T3SSi as an effective means of curtailing infection has been rationalized in several ways. Since the injectisome is absent in many resident microbiota, one proposed advantage is that more of the microbiome would be preserved during treatment. Furthermore, the likelihood of developing resistance in resident microbiota that can be transferred by horizontal gene transfer to pathogenic bacteria is minimal. However, due to the homology between some components of the T3SSi and flagella, some inhibitors also affect flagella (13, 19, 20), an observation that may mitigate this advantage. Another potential benefit is that since these antivirulence agents should minimally affect bacterial

growth, they may exert low selective pressure in the environment, and therefore, drug resistance may develop infrequently. To our knowledge this has not been experimentally tested in an animal model of infection. On the other hand, disadvantages to be considered include that anti-T3SSi agents may not impede bacterial growth in infected immunocompromised individuals and that some infections require bactericidal agents. Nonetheless, discovering and studying reagents that inhibit the T3SSi remains attractive both for the potential therapeutic benefits and for their use as important tools to elucidate the structure-function relationships of this complex machinery.

This review focuses on advances in T3SSi-targeted therapies in the past 4 years (Tables 1 and 2), including small molecules, antibodies, and vaccines, whose molecular targets are known (Fig. 1B). Excellent in-depth reviews covering progress of the field until 2014–2015 and structure of molecules include references 2, 21, and 22. Some previously well-studied compounds are also summarized in Table 1.

SMALL MOLECULES

Many studies use high-throughput screens (HTSs) to identify small-molecule inhibitors of T3SSi via phenotypic readouts of T3SSi functions, including inhibition of T3SSi expression in bacteria (13, 15, 23–25), secretion of effectors into the extracellular supernatant (14, 17, 25–27), or translocation of effector proteins into host cells (14, 18). A benefit of such approaches is that identified molecules are effective in the context of the bacterium. However, complications include the fact that the inhibitors may target more than one protein, may target a host protein, or may alter T3SSi function by generally affecting bacterial cell physiology rather than a specific component of the machinery. Consequently, identification of the specific targets of many small-molecule inhibitors has lagged and structure-activity relationship (SAR) studies are complicated if the molecule targets several proteins.

Recently, several exciting advances have been made in both target identification and identification of lead compounds with sufficiently low 50% inhibitory concentrations for *in vivo* studies. More classical pharmacological approaches that identify compounds that bind to a protein or inhibit its biochemical activity have been fruitfully employed (16, 28–30). Increasingly, the structures of T3SS components are being exploited to elucidate the design of potential inhibitors to these proteins (31–34).

Salicylidene Acylhydrazides

Salicylidene acylhydrazides (SAHs) are the first identified and most widely studied class of synthetic small molecules that target the T3SSi across many bacterial species (13, 14). Several studies suggest that some of these molecules have multiple targets or act indi-

rectly on the T3SSi by impacting bacterial physiology (19, 25, 35–37). Of the derivatives generated, many show promising results. Modifications to improve stability and selectivity of SAH ME0055 resulted in two new synthesized compounds, RCZ12 and RCZ20, that inhibit secretion of EHEC T3SS translocon protein, EspD, as effectively as ME0055 (Fig. 1B). Unlike the parent compound, RCZ12 and RCZ20 have no effect on bacterial growth, suggesting that they are more specific (38). Affinity chromatography experiments revealed coiled-coil domain 1 of EspD as the inhibitors' key domain-binding site (38). These compounds show dual functionality by also downregulating transcription of the locus of enterocyte effacement that encodes the T3SS (38). Recent mechanistic analysis of another SAH, INP0341, showed that it prevents T3SS expression in *Pseudomonas aeruginosa* clinical isolates without affecting growth (39).

A very recent study employed a multiple-assay approach to elucidate the mechanism of action of a group of previously identified T3SS inhibitors (40). Compound SAH INP0007 disrupts YscD punctum formation, suggesting interference with needle assembly, and significantly decreases flagellar motility. Whether inhibition occurs by direct binding to a common core component between the T3SSi and flagella or by interference with other processes that render bacteria less able to build both systems is still unknown (40). Compound 4 (C4), a haloid-containing sulfonamidobenzamide (SAB), which was originally identified along with SAHs as inhibitors of the T3SS (13), is now postulated to have an indirect effect on T3SS transcription by inhibiting the secretion process (40).

Compounds Targeting the T3SS ATPase

Using the known structure of the enteropathogenic *Escherichia coli* (EPEC) EscN ATPase, a computational HTS identified compounds predicted to block the protein's

TABLE 1 Possible targets and function of small-molecule inhibitors of the T3SS[a]

Compound(s)	Organism(s)	Target	Inhibits bacterial growth?	Toxic to cells?	In vivo studies?	Phenotype/readout	Reference(s)
SAH (C1, C2), SAB C4	*Yersinia pseudotuberculosis*		No	NT	No	Inhibit T3SS transcription and Yop secretion; C2 and C4 inhibit flagellar motility	13
SAH (C1–C23), C1-INP0007	*Y. pseudotuberculosis*		No	No	No	Inhibit secretion and translocation	14
SAH C1-INP0007, SAH C11-INP0403	*Salmonella enterica*		No	No	Yes	Inhibit secretion and blocks invasion; first study to validate SAH *in vivo* using bovine intestinal ligated loops	27
SAH C11-INP0403 (ME0053)	*S. enterica*	Suggested indirect effect: iron chelation	No	No	No	Inhibits T3SS transcription and secretion; upregulation of iron acquisition	25
SAH INP0341, SAH INP0400	*Chlamydia trachomatis*	Suggested indirect effect: iron chelation	No	No	Yes	Inhibit T3SS transcription; upregulation of iron acquisition; protect mice against vaginal infection when administered topically	35, 80, 81
SAH INP0341	*C. trachomatis*		No	No	No	Mutations isolated in HemG suggesting indirect effect on T3SS	37
SAH INP0400, SAH INP0402 (C15)	*Shigella flexneri*	Suggested to inhibit T3SS basal needle assembly	No	No	No	Inhibit secretion and blocks invasion; assembly of fewer and shorter needles	17
SAH ME0052 (C8, INP0010), SAH ME0053 (C11, INP0403), SAH ME0054 (C10, INP0401), SAH ME0055 (C17, INP0031)	EHEC	Suggested to inhibit T3SS regulators	No	No	No	Inhibit secretion	15
SAH ME0052 (C8, INP0010), SAH ME0055 (C17, INP0031)	*Y. pseudotuberculosis*, *Escherichia coli*		No	No	No	Inhibit secretion; pulldown assays showed that WrbA, FolX, and Tpx bind to SAH, suggesting indirect effect on T3SS	36
SAH INP0404, SAH INP0405	*S. enterica*		No	No	No	Mutations isolated in FlhA gene suggest targeting of T3SS basal body	19

Compound	Organism	Target				Description	Ref
SAH INP0341	Pseudomonas aeruginosa		No	No	No	Inhibits T3SS transcription and ExoS secretion	39
SAH RCZ212 and RCZ20	EHEC	EspD (needle pore protein)	No	No	No	Inhibit EspD secretion; assembly of fewer and shorter needles	38
SAB C4	Y. pseudotuberculosis		No	No	No	Inhibits secretion	40
SAH INP0007	Y. pseudotuberculosis		No	No	No	Affects YscD punctum formation	40
SAH INP0010	Y. pseudotuberculosis		No	Yes	No	Affects YscD punctum formation	40
Salicylideneanilide C3	Y. pseudotuberculosis		No	NT	No	Inhibits secretion and transcription	13
Salicylideneanilide	EPEC		No	No	No	Inhibits T3SS transcription and EspB secretion	26
Benzimidazole	Y. pseudotuberculosis	LcrF (T3SS master regulator)	No	No	Yes	Reduces cytotoxicity in infected cells; protective in a murine model	16
C15, C19, C22, C24, and C38	Y. pseudotuberculosis, P. aeruginosa		No	No	No	Inhibit effector translocation	18
C20	Y. pseudotuberculosis, P. aeruginosa	Suggested to interfere with adherence	No	No	No	Inhibits effector translocation	18
Compound D	Y. pseudotuberculosis, Yersinia pestis, P. aeruginosa	Suggested to target YopD (translocon)	NT	Yes	NT	Inhibits effector secretion	82
Thiazolidinones	S. enterica, P. aeruginosa, Yersinia enterocolitica, Pseudomonas syringae	Inhibits T2SS, suggesting common target with T3SS such as secretin	No	No	Yes; tobacco plants	Inhibit transcription and secretion; reduce needle complex formation; reduce hypersensitivity response in plant leaves	83
Phenoxyacetamides	P. aeruginosa	Suggested to target PscF (needle protein)	No	No	No	Isolation of pscF mutants resistant to phenoxyacetamide inhibitors	34, 42, 43
Phenoxyacetamides	P. aeruginosa	NT	NT	NT	Yes	Reduce abscess size in mouse model of P. aeruginosa abscess formation	44
Piericidins	Y. pseudotuberculosis		No	No	No	Inhibit T3SS-dependent NF-κB activation	45
Piericidin A1	Y. pseudotuberculosis	Suggested to target YscF (needle protein)	NT	NT	No	Reduces number of needles present	46
Library of compounds	Salmonella spp.	SipD (tip protein), SipB (translocon protein)	NT	NT	No	Surface plasmon resonance screen to find compounds that bind to SipD and SipB	48
Library of compounds	Shigella spp.	IpaD (tip protein)	NT	NT	No	Surface plasmon resonance screen to find compounds that bind to IpaD	49

(Continued on next page)

TABLE 1 Possible targets and function of small-molecule inhibitors of the T3SS[a] (Continued)

Compound(s)	Organism(s)	Target	Inhibits bacterial growth?	Toxic to cells?	In vivo studies?	Phenotype/readout	Reference(s)
Malic diamide	Y. pseudotuberculosis		No	No	No	Inhibits secretion of YopB and YopD	40
Flavonoids	S. enterica	Covalent labeling of SPI-1 substrates	No	NT	No	Inhibit bacterial invasion of host cells	47
Compounds 7812, 7832, and 7086	Y. pestis	YscN (T3SS ATPase)	No, except for 7086	No	No	Inhibit secretion	84
WEN05-03	EPEC	EscN (T3SS ATPase)	No	No	No	Inhibits ATP hydrolysis; reduces toxicity to infected HeLa cells	29
N-Arylbenzylamines	C. trachomatis	Suggested to target SctN (T3SS ATPase)	No	No	No	Reduce secretion and chlamydial inclusions in host cells	30
HQsINP1750, INP1767, and INP1855	C. trachomatis, Y. pseudotuberculosis		No	No	No	Inhibit cytotoxicity	41
HQ INP1855	P. aeruginosa	Suggested to target T3SS ATPase	No	No	Yes	Reduces cytotoxicity on host cells; reduces bacterial burden and lung pathology in infected mice; reduces activity of homologous T3SS ATPase YscN	28
HQ INP1750	P. aeruginosa, Y. pseudotuberculosis	Suggested to target T3SS ATPase	No	No	No	Inhibits secretion and flagellar motility; reduces activity of Yersinia T3SS ATPase YscN	39
Licoflavonol	S. enterica		No	NT	No	Reduces expression of chaperone sicA and invF (transcriptional regulator for SPI-1 effector proteins)	50
Epigallocatechin gallate	EPEC/EHEC, S. enterica, Y. pseudotuberculosis		No	NT	No	Reduces adherence of EHEC/EPEC; reduces Salmonella invasion into host cells; reduces Yersinia-induced cell death	52
Epigallocatechin gallate	S. enterica		No	NT	No	Reduces Salmonella invasion into host cells	51
Psidium guajava leaf extract	EPEC/EHEC, S. enterica, Y. pseudotuberculosis		No	NT	No	Reduces adherence of EHEC/EPEC; reduces Salmonella invasion into host cells; reduces Yersinia-induced cell death	53

Sanguinarine chloride	S. enterica	No	Yes at higher concns	No	Inhibits bacterial invasion of host cells	54
Thymol	S. enterica	Slightly at higher concns	Slightly at higher concns	Yes	Inhibits bacterial invasion of host cells; protects mice against infection	85
Obovatol	S. enterica	No	NT	No	Reduces hemolysis of sheep red blood cells	55
7-Hydroxycoumarin (umbelliferone)	Ralstonia solanacearum	Yes (86)	NT	Yes; tobacco plants	Reduces expression of T3SS effector genes; reduces disease progression on tobacco plants	87
SAHs	R. solanacearum	Minimal	NT	Yes; tomato plants	Inhibit translocation; reduce bacterial growth on tomato plants	56
SAHs	Erwinia amylovora	No	NT	Yes; apple plants	Reduce expression of T3SS genes; reduce disease symptoms on apple plants	57
Phenols	Xanthomonas oryzae	No	NT	Yes; rice plants	Reduce expression of hrpG and hrpX (regulators of hrp genes which regulate T3SS effector expression); reduce disease symptoms on rice plants	58
Thiazolidin-2-cyanamide derivatives	X. oryzae	No	NT	Yes; rice plants	Reduce expression of hrpG and hrpX (regulators of hrp genes which regulate T3SS effector expression); reduce disease symptoms on rice plants	59

aNT, not tested; EHEC, enterohemorrhagic *Escherichia coli*; EPEC, enteropathogenic *Escherichia coli*; T3SS, type III secretion system.

TABLE 2 Antibodies, vaccines, and peptomers against T3SS components[a]

Class	Organism(s)	Target(s)	Phenotype/readout	Therapeutic potential	References
Antibody (KB001)	Pseudomonas aeruginosa	PcrV (tip)	Protects host cells against T3SS-mediated toxicity and protects mice against acute pulmonary infection (reviewed in reference 88)	Did not meet efficacy endpoints in phase II clinical trials	60–62
Bispecific antibody (MEDI3902)	P. aeruginosa	PcrV (tip), Psl (exopolysaccharide)	In vitro cytotoxicity protection and in vivo protection of acute pneumonia model in mice	Currently in phase II clinical trials	63, 64
Single-VH domain antibodies	Shigella flexneri	IpaD (tip)	Reduces hemolysis of sheep red blood cells		66
rF1V vaccine	Yersinia pestis	LcrV (tip), F1 protein	Enhances survival of cynomolgus macaques infected with lethal aerosol challenge of Y. pestis	Orphan drug designation by FDA	69
Rabbit polyclonal antisera	STEC	STEC$_{O103}$ T3SS proteins	Block adherence of STEC to host cells; immunized mice not protected against fecal shedding		71
Peptide vaccine	Salmonella enterica	SseI (effector)	Protects mice against acute infection		72
Vaccine	S. enterica	PrgI (needle), SipD (tip)	Protects mice against infection		73
Vaccine	S. enterica	SseB (effector), flagellin	Protects mice against infection		74
Subunit vaccine	S. enterica	S1 (fusion of SipD and SipB [tip and translocon]), S2 (fusion of SseB and SseC [tip and translocon])	Protects mice against lethal challenge		75
Polypeptide	S. enterica, S. flexneri	SipB (translocon), IpaB (translocon)	Inhibits bacterial invasion into host cells	Polypeptide too large for therapeutic potential	76
Peptides	EPEC	EspA (tip)	Inhibit EspA polymerization, thereby preventing A/E lesions		77
Peptides	EHEC, Citrobacter rodentium	EspA (tip)	Protect mice against colon damage after C. rodentium challenge		78
Peptomers (phepropeptin D derivatives)	Yersinia pseudotuberculosis, P. aeruginosa		Inhibit secretion of T3SS proteins; inhibit Yersinia YopM effector translocation and reduce cell rounding		79

[a]STEC, Shiga toxin-producing Escherichia coli; A/E, attaching/effacing.

active site (29). One lead compound (WEN05-03) competitively inhibits hydrolysis of ATP by EscN and reduces toxicity to infected HeLa cells (29). Another study using molecular docking and virtual screening identified a series of *N*-arylbenzylamines predicted to target the SctN T3SS ATPase of *Chlamydia trachomatis* (30). Two of these compounds block translocation of the T3SS effector, IncA, into cultured cells and reduce chlamydial survival in these cells (30). Hydroxyquinoline (HQ) derivatives were first described as inhibitors of T3SSi gene expression in *Yersinia pseudotuberculosis* and *C. trachomatis* (41). HQ INP1855 inhibits YscN ATPase activity *in vitro* as well as impairing flagellar motility, providing evidence that it might target conserved ATPases found in the T3SS and flagella (28). In addition, HQ INP1855 reduces *P. aeruginosa* T3SS-mediated cytotoxicity in cultured cells, blocks secretion of ExoS effector protein, and enhances survival and reduces bacterial burden and lung pathology of mice infected intranasally with *P. aeruginosa* (28). HQ INP1750 acts similarly to HQ INP1855 and inhibits both ExoS secretion and flagellar motility (39). However, a direct interaction between these HQ derivatives and T3SS ATPases remains to be shown.

Compounds Targeting Needles or Needle Assembly

Phenoxyacetamide (PXA) was first discovered as an inhibitor of the T3SSi in *P. aeruginosa*, and SAR analysis demonstrated strict stereoselectivity, suggesting an interaction with a specific target or site (42). Isolation of several mutants of PscF resistant to PXA inhibitors provides genetic evidence that PXAs target the needle protein (34, 43). Modeling of PXA inhibitors supports the idea that these molecules intercalate within the needle and interact simultaneously with several assembled PscF subunits; however, biochemical and structural studies are needed to demonstrate a direct interaction. Importantly, injection of PXA (MBX2359) into abscesses

formed by *P. aeruginosa* significantly reduces abscess size, providing evidence that these inhibitors are efficacious in infection models in mammals (44).

Piericidins, a class of compounds derived from *Actinomycetales*, inhibit translocation of YopM into cultured cells (45). A follow-up study showed that *Yersinia* treated with piericidin A1 has fewer needles, suggesting that piericidin A1 inhibits a step prior to or during needle assembly (46). The related Psc T3SS of *P. aeruginosa* and the Ysa T3SS of *Y. enterocolitica* are not inhibited, indicating its specificity but potentially limiting its usefulness without additional SAR analysis (46).

Compounds Targeting Translocon and/or Effector Secretion and Activity

Using click chemistry, the flavonoids baicalein and quercetin were found to covalently modify *Salmonella enterica* serovar Typhimurium translocases and effectors, resulting in changes to stability or activity (47). The N-terminal chaperone-binding domain is proposed to be the modified site (47). These flavonoids inhibit invasion of *S*. Typhimurium into cultured cells but have no effect on effector secretion or needle assembly (47). Screening libraries for compounds that bind to *Salmonella* SipD (48) or *Shigella* IpaD tip proteins (49) identified a new class of small molecules based on the indole scaffold as potential inhibitors of the T3SSi. Malic diamide (42), a compound structurally related to PXA, significantly inhibits the secretion of YopB and YopD proteins required for translocation, without disrupting needle YscF punctum formation, indicating that it targets the translocon (40).

In the past few years, several natural compounds have been identified, typically in screens for secretion (50–53) or translocation into target cells (54) or by inhibiting the effects on T3SSi-mediated functions on targeted host cells (55). Potentially promising compounds are listed in Table 1, but to our knowledge, the specificity against T3SSi or protein targets has not been investigated in depth.

Anti-T3SS Compounds Tested against Plant Pathogens

Plants are also susceptible to infection by bacteria harboring T3SSs, and there have been several recent exciting findings. Natural and synthetic compounds were screened for the ability to reduce expression of the *Ralstonia solanacearum* T3SS pilus gene *hrpY* (56). The most potent inhibitors were SAHs, which inhibit secretion of T3SS effector AvrA and limit bacterial growth on tomato plants (56). SAHs also reduce the expression of T3SS genes of *Erwinia amylovora* and reduce disease symptoms on inoculated crab apple pistils (57). Phenolic compounds repress the expression of T3SS transcriptional regulators *hrpG* and *hrpX* of *Xanthomonas oryzae* and reduce disease symptoms on rice leaves (58). Thiazolidine-2-cyanamide compounds also reduce relative expression of *X. oryzae hrpG* and *hrpX* and disease symptoms on rice (59).

ANTIBODIES, VACCINES, AND PEPTIDES

Recent advances in targeting T3SSi using antibodies, vaccines, and polypeptides are summarized below and in Table 2.

Antibodies

A monoclonal antibody, KB001, that binds to the *P. aeruginosa* T3SS tip protein, PcrV, initially showed promise in the treatment of patients with airway-associated *P. aeruginosa* infection or colonization but failed in phase II clinical trials for not meeting efficacy endpoints (60–62). By contrast, a bispecific antibody, MEDI3902, against *P. aeruginosa* PcrV and the Psl exopolysaccharide is effective against a wide range of clinical isolates and is currently in phase II clinical trials for prevention of ventilator nosocomial pneumonia (63, 64).

Single-domain antibodies that consist of the N-terminal variable region of an immunoglobulin heavy chain (VHH) but not the light chain can be isolated from camelid species (65). A panel of VHH single-domain antibodies was raised against the *Shigella flexneri* IpaD tip protein (66). Four such antibodies that bound IpaD significantly inhibit hemolysis of sheep red blood cells, a measure of T3SS translocon functionality (66). Structural binding analysis revealed that these inhibitory VHHs mostly bound to the distal domain of IpaD, suggesting the importance of this region in T3SS function (66).

Vaccines

Work towards a plague vaccine has led to testing a recombinant vaccine consisting of the *Yersinia pestis* F1 protein and the T3SS tip protein LcrV, reviewed in reference 67. The FDA has granted orphan drug status for the development of this rF1V vaccine as a prophylactic for high-risk individuals (68, 69). Efforts to lessen Shiga toxin-producing *Escherichia coli* (STEC) disease burden in cattle to reduce transmission to humans are ongoing. Cohorts of cattle immunized against serotype O157 have reduced shedding of O157 but not of other STEC serotypes due to serotype specificity (70). To develop vaccines against a different prevalent serotype, antisera to five T3SS proteins, EspA, EspB, EspF, NleA, and Tir, of STEC serotype O103 were studied. These antisera block STEC adherence to HEp-2 cells (71). In efficacy studies, mice developed strong serum IgG titers against four of these five proteins but still shed O103 after oral administration, indicating that the bacteria could still be transmitted (71).

Recent attempts to develop T3SS-targeted vaccines against *Salmonella enterica* show some success in mouse studies. A peptide vaccine that elicits a CD4 T cell response against T3SS effector protein SseI protects mice against acute infection, a tantalizing result given that only a single peptide elicits protection (72). Mice were immunized by different routes with *Salmonella* T3SS proteins SipD and PrgI in combination or alone; oral immunization with SipD provides the highest level of protection against lethal challenge (73). Increased protection is observed when flagellin is added to a vaccine against *Salmonella* T3SS protein SseB (74). A subunit vaccine against *Salmonella* consisting of two components, S1 (a genetic fusion of SPI-1 translocon proteins SipB and SipD) and S2 (a genetic fusion of SPI-2 proteins SseB and SseC), elicits strong IgG titers to all four proteins in mice (75). These mice are significantly protected against challenge with *S.* Typhimurium and *S. enterica* serovar Enteritidis and experience reduced cecal inflammation (75). These results warrant studies on long-term protection.

Peptides

Anti-T3SS peptides (Table 2) have been identified against *Salmonella* (76), EPEC (77), enterohemorrhagic *E. coli* (EHEC) (78), and more recently, against *Yersinia* (79). Derivatives of the natural compound phepropeptin D that contained various peptoid substitutions on the cyclic peptide backbone significantly inhibit NF-kB signaling, secretion of the effector protein YopE, and translocation of YopM into HeLa cells by *Yersinia* (79). The peptomers do not affect *Yersinia* growth or flagellar motility, indicating their potential specificity to the T3SSi. Several derivatives also inhibit secretion of the *P. aeruginosa* effector protein ExoU, suggesting that they might target a conserved component of these two injectisome systems (79).

CONCLUSION AND PERSPECTIVE

Discovery of and research into inhibitors of the T3SSi is a highly active area of research, with many candidates from different classes that are effective in blocking the function of T3SS. Although antibodies and vaccines are further along in the pipeline, many small-molecule inhibitors show promise. Some molecules have a narrower spectrum of activity, while others have broader spectra, including those that target components conserved between the T3SSi and flagella. Both have benefits and disadvantages. For instance, an effective but narrow-spectrum molecule against the T3SSi of the multidrug-resistant *P. aeruginosa* could save many lives each year. By contrast, a narrow-spectrum molecule effective towards *Y. pestis* would not save many lives annually unless a major outbreak occurred. Yet importantly, study of such a molecule could help elucidate structure-function relations of the T3SSi and be used as a platform to develop molecules highly effective against homologous components in other T3SSis. Resistance mutants, biochemical assays, structural modeling, and rational designs are helping to identify targets and generate more potent inhibitors. Validation of their efficacy in animal systems is ongoing. Both basic science and clinical translational research from academic and pharmaceutical groups is crucial to the advancement of these molecules to combat the rising threat of antibiotic resistance.

ACKNOWLEDGMENTS

We thank Anne McCabe for useful discussions and critical reading of the manuscript.

A.C.F. was supported in part by NIH grant T32 AI007077, L.S. was supported in part by NIH grant AI007422, and J.M. was supported by NIH grants R01 AI113166, STTR R41 AI22433, and NIH U19 AI131126. J.M. has an ongoing NIH-funded collaboration with Paratek Inc. (grant STTR R41 AI22433).

CITATION

EcoSal Plus 2019; doi:10.1128/ecosalplus. ESP-0032-2018.

REFERENCES

1. **Bassetti M, Merelli M, Temperoni C, Astilean A.** 2013. New antibiotics for bad bugs: where are we? *Ann Clin Microbiol Antimicrob* **12**:22.

2. **McShan AC, De Guzman RN.** 2015. The bacterial type III secretion system as a target for developing new antibiotics. *Chem Biol Drug Des* **85**:30–42.

3. **Czaplewski L, Bax R, Clokie M, Dawson M, Fairhead H, Fischetti VA, Foster S, Gilmore BF, Hancock RE, Harper D, Henderson IR, Hilpert K, Jones BV, Kadioglu A, Knowles D, Ólafsdóttir S, Payne D, Projan S, Shaunak S, Silverman J, Thomas CM, Trust TJ, Warn P, Rex JH.** 2016. Alternatives to antibiotics-a pipeline portfolio review. *Lancet Infect Dis* **16**:239–251.

4. **Keyser P, Elofsson M, Rosell S, Wolf-Watz H.** 2008. Virulence blockers as alternatives to antibiotics: type III secretion inhibitors against Gram-negative bacteria. *J Intern Med* **264**:17–29.

5. **Baron C.** 2010. Antivirulence drugs to target bacterial secretion systems. *Curr Opin Microbiol* **13**:100–105.

6. **Allen RC, Popat R, Diggle SP, Brown SP.** 2014. Targeting virulence: can we make evolution-proof drugs? *Nat Rev Microbiol* **12**:300–308.

7. **Clatworthy AE, Pierson E, Hung DT.** 2007. Targeting virulence: a new paradigm for antimicrobial therapy. *Nat Chem Biol* **3**:541–548.

8. **Maura D, Ballok AE, Rahme LG.** 2016. Considerations and caveats in anti-virulence drug development. *Curr Opin Microbiol* **33**:41–46.

9. **Cornelis GR.** 2006. The type III secretion injectisome. *Nat Rev Microbiol* **4**:811–825.

10. **Galán JE, Wolf-Watz H.** 2006. Protein delivery into eukaryotic cells by type III secretion machines. *Nature* **444**:567–573.

11. **Galán JE, Lara-Tejero M, Marlovits TC, Wagner S.** 2014. Bacterial type III secretion systems: specialized nanomachines for protein delivery into target cells. *Annu Rev Microbiol* **68**:415–438.

12. **Titball RW, Howells AM, Oyston PC, Williamson ED.** 1997. Expression of the *Yersinia pestis* capsular antigen (F1 antigen) on the surface of an *aroA* mutant of *Salmonella typhimurium* induces high levels of protection against plague. *Infect Immun* **65**:1926–1930.

13. **Kauppi AM, Nordfelth R, Uvell H, Wolf-Watz H, Elofsson M.** 2003. Targeting bacterial virulence: inhibitors of type III secretion in *Yersinia. Chem Biol* **10**:241–249.

14. **Nordfelth R, Kauppi AM, Norberg HA, Wolf-Watz H, Elofsson M.** 2005. Small-molecule inhibitors specifically targeting type III secretion. *Infect Immun* **73**:3104–3114.

15. **Tree JJ, Wang D, McInally C, Mahajan A, Layton A, Houghton I, Elofsson M, Stevens MP, Gally DL, Roe AJ.** 2009. Characterization of the effects of salicylidene acylhydrazide compounds on type III secretion in *Escherichia coli* O157:H7. *Infect Immun* **77**:4209–4220.

16. **Garrity-Ryan LK, Kim OK, Balada-Llasat JM, Bartlett VJ, Verma AK, Fisher ML, Castillo C, Songsungthong W, Tanaka SK, Levy SB, Mecsas J, Alekshun MN.** 2010. Small molecule inhibitors of LcrF, a *Yersinia pseudotuberculosis* transcription factor, attenuate virulence and limit infection in a murine pneumonia model. *Infect Immun* **78**:4683–4690.

17. **Veenendaal AK, Sundin C, Blocker AJ.** 2009. Small-molecule type III secretion system inhibitors block assembly of the *Shigella* type III secreton. *J Bacteriol* **191**:563–570.

18. **Harmon DE, Davis AJ, Castillo C, Mecsas J.** 2010. Identification and characterization of small-molecule inhibitors of Yop translocation in *Yersinia pseudotuberculosis. Antimicrob Agents Chemother* **54**:3241–3254.

19. **Martinez-Argudo I, Veenendaal AK, Liu X, Roehrich AD, Ronessen MC, Franzoni G, van Rietschoten KN, Morimoto YV, Saijo-Hamano Y, Avison MB, Studholme DJ, Namba K, Minamino T, Blocker AJ.** 2013. Isolation of *Salmonella* mutants resistant to the inhibitory effect of salicylidene acylhydrazides on flagella-mediated motility. *PLoS One* **8**:e52179.

20. **Negrea A, Bjur E, Ygberg SE, Elofsson M, Wolf-Watz H, Rhen M.** 2007. Salicylidene acylhydrazides that affect type III protein secretion in *Salmonella enterica* serovar Typhimurium. *Antimicrob Agents Chemother* **51**:2867–2876.

21. **Duncan MC, Linington RG, Auerbuch V.** 2012. Chemical inhibitors of the type three secretion system: disarming bacterial pathogens. *Antimicrob Agents Chemother* **56**:5433–5441.

22. **Charro N, Mota LJ.** 2015. Approaches targeting the type III secretion system to treat or prevent bacterial infections. *Expert Opin Drug Discov* **10**:373–387.

23. **Wolf K, Betts HJ, Chellas-Géry B, Hower S, Linton CN, Fields KA.** 2006. Treatment of *Chlamydia trachomatis* with a small molecule inhibitor of the *Yersinia* type III secretion system disrupts progression of the chlamydial developmental cycle. *Mol Microbiol* **61**:1543–1555.

24. Muschiol S, Bailey L, Gylfe A, Sundin C, Hultenby K, Bergström S, Elofsson M, Wolf-Watz H, Normark S, Henriques-Normark B. 2006. A small-molecule inhibitor of type III secretion inhibits different stages of the infectious cycle of *Chlamydia trachomatis*. *Proc Natl Acad Sci U S A* **103**:14566–14571.

25. Layton AN, Hudson DL, Thompson A, Hinton JC, Stevens JM, Galyov EE, Stevens MP. 2010. Salicylidene acylhydrazide-mediated inhibition of type III secretion system-1 in *Salmonella enterica* serovar Typhimurium is associated with iron restriction and can be reversed by free iron. *FEMS Microbiol Lett* **302**:114–122.

26. Gauthier A, Robertson ML, Lowden M, Ibarra JA, Puente JL, Finlay BB. 2005. Transcriptional inhibitor of virulence factors in enteropathogenic *Escherichia coli*. *Antimicrob Agents Chemother* **49**:4101–4109.

27. Hudson DL, Layton AN, Field TR, Bowen AJ, Wolf-Watz H, Elofsson M, Stevens MP, Galyov EE. 2007. Inhibition of type III secretion in *Salmonella enterica* serovar Typhimurium by small-molecule inhibitors. *Antimicrob Agents Chemother* **51**:2631–2635.

28. Anantharajah A, Faure E, Buyck JM, Sundin C, Lindmark T, Mecsas J, Yahr TL, Tulkens PM, Mingeot-Leclercq MP, Guery B, Van Bambeke F. 2016. Inhibition of the injectisome and flagellar type III secretion systems by INP1855 impairs *Pseudomonas aeruginosa* pathogenicity and inflammasome activation. *J Infect Dis* **214**:1105–1116.

29. Bzdzion L, Krezel H, Wrzeszcz K, Grzegorek I, Nowinska K, Chodaczek G, Swietnicki W. 2017. Design of small molecule inhibitors of type III secretion system ATPase EscN from enteropathogenic *Escherichia coli*. *Acta Biochim Pol* **64**:49–63.

30. Grishin AV, Luyksaar SI, Kapotina LN, Kirsanov DD, Zayakin ES, Karyagina AS, Zigangirova NA. 2018. Identification of chlamydial T3SS inhibitors through virtual screening against T3SS ATPase. *Chem Biol Drug Des* **91**:717–727.

31. Galán JE, Collmer A. 1999. Type III secretion machines: bacterial devices for protein delivery into host cells. *Science* **284**:1322–1328.

32. Dean P. 2011. Functional domains and motifs of bacterial type III effector proteins and their roles in infection. *FEMS Microbiol Rev* **35**:1100–1125.

33. Abrusci P, McDowell MA, Lea SM, Johnson S. 2014. Building a secreting nanomachine: a structural overview of the T3SS. *Curr Opin Struct Biol* **25**:111–117.

34. Bowlin NO, Williams JD, Knoten CA, Torhan MC, Tashjian TF, Li B, Aiello D, Mecsas J, Hauser AR, Peet NP, Bowlin TL, Moir DT. 2014. Mutations in the *Pseudomonas aeruginosa* needle protein gene pscF confer resistance to phenoxyacetamide inhibitors of the type III secretion system. *Antimicrob Agents Chemother* **58**:2211–2220.

35. Slepenkin A, Enquist PA, Hägglund U, de la Maza LM, Elofsson M, Peterson EM. 2007. Reversal of the antichlamydial activity of putative type III secretion inhibitors by iron. *Infect Immun* **75**:3478–3489.

36. Wang D, Zetterström CE, Gabrielsen M, Beckham KS, Tree JJ, Macdonald SE, Byron O, Mitchell TJ, Gally DL, Herzyk P, Mahajan A, Uvell H, Burchmore R, Smith BO, Elofsson M, Roe AJ. 2011. Identification of bacterial target proteins for the salicylidene acylhydrazide class of virulence-blocking compounds. *J Biol Chem* **286**:29922–29931.

37. Engström P, Nguyen BD, Normark J, Nilsson I, Bastidas RJ, Gylfe A, Elofsson M, Fields KA, Valdivia RH, Wolf-Watz H, Bergström S. 2013. Mutations in *hemG* mediate resistance to salicylidene acylhydrazides, demonstrating a novel link between protoporphyrinogen oxidase (HemG) and *Chlamydia trachomatis* infectivity. *J Bacteriol* **195**:4221–4230.

38. Zambelloni R, Connolly JPR, Huerta Uribe A, Burgess K, Marquez R, Roe AJ. 2017. Novel compounds targeting the enterohemorrhagic *Escherichia coli* type three secretion system reveal insights into mechanisms of secretion inhibition. *Mol Microbiol* **105**:606–619.

39. Anantharajah A, Buyck JM, Sundin C, Tulkens PM, Mingeot-Leclercq MP, Van Bambeke F. 2017. Salicylidene acylhydrazides and hydroxyquinolines act as inhibitors of type three secretion systems in *Pseudomonas aeruginosa* by distinct mechanisms. *Antimicrob Agents Chemother* **61**:e02566-16.

40. Morgan JM, Lam HN, Delgado J, Luu J, Mohammadi S, Isberg RR, Wang H, Auerbuch V. 2018. An experimental pipeline for initial characterization of bacterial type III secretion system inhibitor mode of action using enteropathogenic *Yersinia*. *Front Cell Infect Microbiol* **8**:404.

41. Enquist PA, Gylfe A, Hägglund U, Lindström P, Norberg-Scherman H, Sundin C, Elofsson M. 2012. Derivatives of 8-hydroxyquinoline—antibacterial agents that target intra- and extracellular Gram-negative pathogens. *Bioorg Med Chem Lett* **22**:3550–3553.

42. Aiello D, Williams JD, Majgier-Baranowska H, Patel I, Peet NP, Huang J, Lory S, Bowlin

TL, Moir DT. 2010. Discovery and characterization of inhibitors of *Pseudomonas aeruginosa* type III secretion. *Antimicrob Agents Chemother* **54**:1988–1999.

43. **Williams JD, Torhan MC, Neelagiri VR, Brown C, Bowlin NO, Di M, McCarthy CT, Aiello D, Peet NP, Bowlin TL, Moir DT.** 2015. Synthesis and structure-activity relationships of novel phenoxyacetamide inhibitors of the *Pseudomonas aeruginosa* type III secretion system (T3SS). *Bioorg Med Chem* **23**:1027–1043.

44. **Berube BJ, Murphy KR, Torhan MC, Bowlin NO, Williams JD, Bowlin TL, Moir DT, Hauser AR.** 2017. Impact of type III secretion effectors and of phenoxyacetamide inhibitors of type III secretion on abscess formation in a mouse model of *Pseudomonas aeruginosa* infection. *Antimicrob Agents Chemother* **61**:e01207.

45. **Duncan MC, Wong WR, Dupzyk AJ, Bray WM, Linington RG, Auerbuch V.** 2014. An NF-κB-based high-throughput screen identifies piericidins as inhibitors of the *Yersinia pseudotuberculosis* type III secretion system. *Antimicrob Agents Chemother* **58**:1118–1126.

46. **Morgan JM, Duncan MC, Johnson KS, Diepold A, Lam H, Dupzyk AJ, Martin LR, Wong WR, Armitage JP, Linington RG, Auerbuch V.** 2017. Piericidin A1 blocks *Yersinia* Ysc type III secretion system needle assembly. *mSphere* **2**:e00030-17.

47. **Tsou LK, Lara-Tejero M, RoseFigura J, Zhang ZJ, Wang YC, Yount JS, Lefebre M, Dossa PD, Kato J, Guan F, Lam W, Cheng YC, Galán JE, Hang HC.** 2016. Antibacterial flavonoids from medicinal plants covalently inactivate type III protein secretion substrates. *J Am Chem Soc* **138**:2209–2218.

48. **McShan AC, Anbanandam A, Patnaik S, De Guzman RN.** 2016. Characterization of the binding of hydroxyindole, indoleacetic acid, and morpholinoaniline to the *Salmonella* type III secretion system proteins SipD and SipB. *ChemMedChem* **11**:963–971.

49. **Dey S, Anbanandam A, Mumford BE, De Guzman RN.** 2017. Characterization of small-molecule scaffolds that bind to the *Shigella* type III secretion system protein IpaD. *ChemMedChem* **12**:1534–1541.

50. **Guo Z, Li X, Li J, Yang X, Zhou Y, Lu C, Shen Y.** 2016. Licoflavonol is an inhibitor of the type three secretion system of *Salmonella enterica* serovar Typhimurium. *Biochem Biophys Res Commun* **477**:998–1004.

51. **Tsou LK, Yount JS, Hang HC.** 2017. Epigallocatechin-3-gallate inhibits bacterial virulence and invasion of host cells. *Bioorg Med Chem* **25**:2883–2887.

52. **Nakasone N, Higa N, Toma C, Ogura Y, Suzuki T, Yamashiro T.** 2017. Epigallocatechin gallate inhibits the type III secretion system of Gram-negative enteropathogenic bacteria under model conditions. *FEMS Microbiol Lett* **364**:fnx111.

53. **Nakasone N, Ogura Y, Higa N, Toma C, Koizumi Y, Takaesu G, Suzuki T, Yamashiro T.** 2018. Effects of *Psidium guajava* leaf extract on secretion systems of Gram-negative enteropathogenic bacteria. *Microbiol Immunol* **62**:444–453.

54. **Zhang Y, Liu Y, Wang T, Deng X, Chu X.** 2018. Natural compound sanguinarine chloride targets the type III secretion system of *Salmonella enterica* serovar Typhimurium. *Biochem Biophys Rep* **14**:149–154.

55. **Choi WS, Lee TH, Son SJ, Kim TG, Kwon BM, Son HU, Kim SU, Lee SH.** 2017. Inhibitory effect of obovatol from *Magnolia obovata* on the *Salmonella* type III secretion system. *J Antibiot (Tokyo)* **70**:1065–1069.

56. **Puigvert M, Solé M, López-Garcia B, Coll NS, Beattie KD, Davis RA, Elofsson M, Valls M.** 2018. Type III secretion inhibitors for the management of bacterial plant diseases. *Mol Plant Pathol.*

57. **Yang F, Korban SS, Pusey PL, Elofsson M, Sundin GW, Zhao Y.** 2014. Small-molecule inhibitors suppress the expression of both type III secretion and amylovoran biosynthesis genes in *Erwinia amylovora*. *Mol Plant Pathol* **15**:44–57.

58. **Fan S, Tian F, Li J, Hutchins W, Chen H, Yang F, Yuan X, Cui Z, Yang CH, He C.** 2017. Identification of phenolic compounds that suppress the virulence of *Xanthomonas oryzae* on rice via the type III secretion system. *Mol Plant Pathol* **18**:555–568.

59. **Xiang X, Tao H, Jiang S, Zhang LH, Cui ZN.** 2018. Synthesis and bioactivity of thiazolidin-2-cyanamide derivatives against type III secretion system of *Xanthomonas oryzae* on rice. *Pestic Biochem Physiol* **149**:89–97.

60. **François B, Luyt CE, Dugard A, Wolff M, Diehl JL, Jaber S, Forel JM, Garot D, Kipnis E, Mebazaa A, Misset B, Andremont A, Ploy MC, Jacobs A, Yarranton G, Pearce T, Fagon JY, Chastre J.** 2012. Safety and pharmacokinetics of an anti-PcrV PEGylated monoclonal antibody fragment in mechanically ventilated patients colonized with *Pseudomonas aeruginosa*: a randomized, double-blind, placebo-controlled trial. *Crit Care Med* **40**:2320–2326.

61. **Milla CE, Chmiel JF, Accurso FJ, VanDevanter DR, Konstan MW, Yarranton**

G, Geller DE, Group KBS, KB001 Study Group. 2014. Anti-PcrV antibody in cystic fibrosis: a novel approach targeting *Pseudomonas aeruginosa* airway infection. *Pediatr Pulmonol* **49**:650–658.

62. **Jain R, Beckett VV, Konstan MW, Accurso FJ, Burns JL, Mayer-Hamblett N, Milla C, VanDevanter DR, Chmiel JF, Group KAS, KB001-A Study Group.** 2018. KB001-A, a novel anti-inflammatory, found to be safe and well-tolerated in cystic fibrosis patients infected with *Pseudomonas aeruginosa. J Cyst Fibros* **17**:484–491.

63. **DiGiandomenico A, Keller AE, Gao C, Rainey GJ, Warrener P, Camara MM, Bonnell J, Fleming R, Bezabeh B, Dimasi N, Sellman BR, Hilliard J, Guenther CM, Datta V, Zhao W, Gao C, Yu XQ, Suzich JA, Stover CK.** 2014. A multifunctional bispecific antibody protects against *Pseudomonas aeruginosa. Sci Transl Med* **6**:262ra155.

64. **Tabor DE, Oganesyan V, Keller AE, Yu L, McLaughlin RE, Song E, Warrener P, Rosenthal K, Esser M, Qi Y, Ruzin A, Stover CK, DiGiandomenico A.** 2018. *Pseudomonas aeruginosa* PcrV and Psl, the molecular targets of bispecific antibody MEDI3902, are conserved among diverse global clinical isolates. *J Infect Dis* **218**:1983–1994.

65. **Harmsen MM, De Haard HJ.** 2007. Properties, production, and applications of camelid single-domain antibody fragments. *Appl Microbiol Biotechnol* **77**:13–22.

66. **Barta ML, Shearer JP, Arizmendi O, Tremblay JM, Mehzabeen N, Zheng Q, Battaile KP, Lovell S, Tzipori S, Picking WD, Shoemaker CB, Picking WL.** 2017. Single-domain antibodies pinpoint potential targets within *Shigella* invasion plasmid antigen D of the needle tip complex for inhibition of type III secretion. *J Biol Chem* **292**:16677–16687.

67. **Verma SK, Tuteja U.** 2016. Plague vaccine development: current research and future trends. *Front Immunol* **7**:602.

68. **Price JL, Manetz TS, Shearer JD, House RV.** 2013. Preclinical safety assessment of a recombinant plague vaccine (rF1V). *Int J Toxicol* **32**:327–335.

69. **Fellows P, Price J, Martin S, Metcalfe K, Krile R, Barnewall R, Hart MK, Lockman H.** 2015. Characterization of a cynomolgus macaque model of pneumonic plague for evaluation of vaccine efficacy. *Clin Vaccine Immunol* **22**:1070–1078.

70. **Smith DR.** 2014. Vaccination of cattle against *Escherichia coli* O157:H7. *Microbiol Spectr* **2**: EHEC-0006-2013.

71. **Desin TS, Townsend HG, Potter AA.** 2015. Antibodies directed against Shiga-toxin producing *Escherichia coli* serotype O103 type III secreted proteins block adherence of heterologous STEC serotypes to HEp-2 cells. *PLoS One* **10**:e0139803.

72. **Kurtz JR, Petersen HE, Frederick DR, Morici LA, McLachlan JB.** 2014. Vaccination with a single CD4 T cell peptide epitope from a *Salmonella* type III-secreted effector protein provides protection against lethal infection. *Infect Immun* **82**:2424–2433.

73. **Jneid B, Moreau K, Plaisance M, Rouaix A, Dano J, Simon S.** 2016. Role of T3SS-1 SipD protein in protecting mice against non-typhoidal *Salmonella* Typhimurium. *PLoS Negl Trop Dis* **10**:e0005207.

74. **Lee SJ, Benoun J, Sheridan BS, Fogassy Z, Pham O, Pham QM, Puddington L, McSorley SJ.** 2017. Dual immunization with SseB/flagellin provides enhanced protection against *Salmonella* infection mediated by circulating memory cells. *J Immunol* **199**:1353–1361.

75. **Martinez-Becerra FJ, Kumar P, Vishwakarma V, Kim JH, Arizmendi O, Middaugh CR, Picking WD, Picking WL.** 2018. Characterization and protective efficacy of type III secretion proteins as a broadly protective subunit vaccine against *Salmonella enterica* serotypes. *Infect Immun* **86**:e00473-17.

76. **Hayward RD, Hume PJ, McGhie EJ, Koronakis V.** 2002. A *Salmonella* SipB-derived polypeptide blocks the 'trigger' mechanism of bacterial entry into eukaryotic cells. *Mol Microbiol* **45**:1715–1727.

77. **Larzábal M, Mercado EC, Vilte DA, Salazar-González H, Cataldi A, Navarro-Garcia F.** 2010. Designed coiled-coil peptides inhibit the type three secretion system of enteropathogenic *Escherichia coli. PLoS One* **5**:e9046.

78. **Larzábal M, Zotta E, Ibarra C, Rabinovitz BC, Vilte DA, Mercado EC, Cataldi Á.** 2013. Effect of coiled-coil peptides on the function of the type III secretion system-dependent activity of enterohemorrhagic *Escherichia coli* O157:H7 and *Citrobacter rodentium. Int J Med Microbiol* **303**:9–15.

79. **Lam H, Schwochert J, Lao Y, Lau T, Lloyd C, Luu J, Kooner O, Morgan J, Lokey S, Auerbuch V.** 2017. Synthetic cyclic peptomers as type III secretion system inhibitors. *Antimicrob Agents Chemother* **61**:e00060-17.

80. **Slepenkin A, Chu H, Elofsson M, Keyser P, Peterson EM.** 2011. Protection of mice from a *Chlamydia trachomatis* vaginal infection using a salicylidene acylhydrazide, a potential microbicide. *J Infect Dis* **204**:1313–1320.

81. Pedersen C, Slepenkin A, Andersson SB, Fagerberg JH, Bergström CA, Peterson EM. 2014. Formulation of the microbicide INP0341 for in vivo protection against a vaginal challenge by *Chlamydia trachomatis*. *PLoS One* 9:e110918.

82. Jessen DL, Bradley DS, Nilles ML. 2014. A type III secretion system inhibitor targets YopD while revealing differential regulation of secretion in calcium-blind mutants of *Yersinia pestis*. *Antimicrob Agents Chemother* 58:839–850.

83. Felise HB, Nguyen HV, Pfuetzner RA, Barry KC, Jackson SR, Blanc MP, Bronstein PA, Kline T, Miller SI. 2008. An inhibitor of gram-negative bacterial virulence protein secretion. *Cell Host Microbe* 4:325–336.

84. Swietnicki W, Carmany D, Retford M, Guelta M, Dorsey R, Bozue J, Lee MS, Olson MA. 2011. Identification of small-molecule inhibitors of *Yersinia pestis* type III secretion system YscN ATPase. *PLoS One* 6:e19716.

85. Zhang Y, Liu Y, Qiu J, Luo ZQ, Deng X. 2018. The herbal compound thymol protects mice from lethal infection by *Salmonella* Typhimurium. *Front Microbiol* 9:1022.

86. Yang L, Ding W, Xu Y, Wu D, Li S, Chen J, Guo B. 2016. New insights into the antibacterial activity of hydroxycoumarins against *Ralstonia solanacearum*. *Molecules* 21:468.

87. Yang L, Li S, Qin X, Jiang G, Chen J, Li B, Yao X, Liang P, Zhang Y, Ding W. 2017. Exposure to umbelliferone reduces *Ralstonia solanacearum* biofilm formation, transcription of type III secretion system regulators and effectors and virulence on tobacco. *Front Microbiol* 8:1234.

88. Sawa T, Ito E, Nguyen VH, Haight M. 2014. Anti-PcrV antibody strategies against virulent *Pseudomonas aeruginosa*. *Hum Vaccin Immunother* 10:2843–2852.

Biological and Structural Diversity of Type IV Secretion Systems

22

YANG GRACE LI,[1] BO HU,[1] and PETER J. CHRISTIE[1]

INTRODUCTION

The bacterial type IV secretion systems (T4SSs) are a large, versatile family of macromolecular translocation systems functioning in Gram-negative (G⁻) and Gram-positive (G⁺) bacteria (1). These systems mediate the transfer of DNA or monomeric or multimeric protein substrates to a large range of prokaryotic and eukaryotic cell types (Fig. 1A). Conjugation systems, the earliest described subfamily of T4SSs (2), transfer mobile genetic elements (MGEs) between bacteria. They pose an enormous medical problem because MGEs often harbor cargoes of antibiotic resistance genes and fitness traits that endow pathogens with antibiotic resistance and other growth advantages under selective pressures (3–5). Effector translocators, a more recently described T4SS subfamily (6, 7), are deployed by pathogenic bacteria to deliver effector proteins to eukaryotic cells during the course of infection (8–11). The conjugation and effector translocator systems, as well as newly discovered interbacterial killing systems, transmit their cargos through direct donor-target cell contact (12–14). A few other T4SSs designated uptake or release systems acquire DNA substrates from the milieu or release DNA or protein substrates into the milieu (Fig. 1A) (1, 6).

[1]Department of Microbiology and Molecular Genetics, McGovern Medical School, Houston, TX 77030
Protein Secretion in Bacteria
Edited by Maria Sandkvist, Eric Cascales, and Peter J. Christie
© 2019 American Society for Microbiology, Washington, DC
doi:10.1128/microbiolspec.PSIB-0012-2018

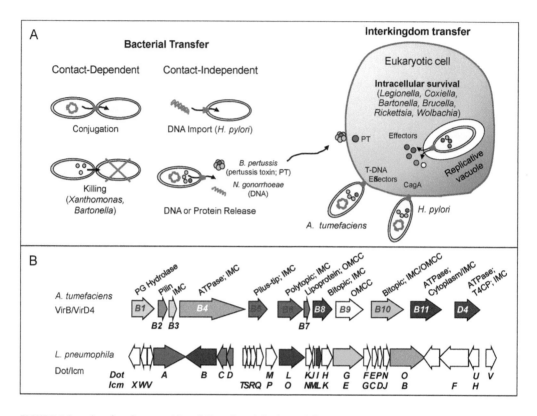

FIGURE 1 Functional and compositional diversity of the bacterial type IV secretion systems (T4SSs). (A) (Left) Contact-dependent conjugation systems and recently described killing systems deliver DNA or protein substrates directly to bacterial target cells. Contact-independent systems mediate DNA import, DNA export, or export of the multimeric pertussis toxin. (Right) Various pathogenic bacteria and symbionts have evolved T4SSs to deliver effector proteins or DNA-protein complexes into eukaryotic host cells to subvert host physiological processes. (B) Gene arrangements and architectures of the *A. tumefaciens* VirB/VirD4 and *L. pneumophila* Dot/Icm secretion systems, with color-coding of the genes encoding homologous subunits; unshaded genes are unique to the Dot/Icm system. The VirB/VirD4 subunit enzymatic functions and associations with inner membrane complex (IMC), outer membrane core complex (OMCC), or pilus are listed. PG Hydrolase, peptidoglycan hydrolase; T4CP, type IV coupling protein.

The T4SSs are defined by the presence of a minimum set of conserved or "signature" subunits (8). The *Agrobacterium tumefaciens* VirB/VirD4 T4SS, whose Vir subunit nomenclature is widely adopted in this field when referring to the conserved subunits of T4SSs, is assembled from VirB1 through VirB11 and VirD4 (Fig. 1B) (15). In G⁻ species, T4SSs are composed of these Vir-like subunits, although many systems have appropriated other subunits or domains from unknown ancestries presumably for specialized functions (16–18). In G⁺ species, six VirB/D4-like subunits (VirB1,

VirB3, VirB4, VirB6, VirB8, and VirD4) are required to build "minimized" systems spanning the single cytoplasmic membrane and cell wall (8, 19). Vir subunits can be grouped as (i) two or three conserved ATPases (VirB4, VirB11, and VirD4) that coordinate the recruitment and processing of substrates, catalyze structural changes in the T4SS channel necessary for substrate passage, and in some cases regulate pilus biogenesis (12, 20–23), (ii) integral inner membrane (IM) subunits (VirB3, VirB6, and VirB8) that presumptively form an IM channel, (iii) a transglycosylase (VirB1) that contributes to (G⁻ species) or is required for (G⁺ species) assembly of the channel across the murein layer (8, 24, 25), and (iv) outer membrane (OM)-associated subunits (VirB7, VirB9, and VirB10) that form a structural scaffold for the portion of the channel spanning the periplasm and OM of G⁻ species (26, 27). Here, we summarize results of recent mechanistic and high-resolution microscopy studies that are providing exciting new insights into the biogenesis and structural arrangement of T4SSs and how they have evolved such extreme biological diversity.

SUBSTRATE RECOGNITION: SUBSTRATE SIGNALS AND ADAPTORS/CHAPERONES

T4SSs recruit specific repertoires of substrates through recognition of translocation signals (TSs) and accessory factors bound to substrates. For conjugative DNA transfer, the recruitment and delivery of DNA substrates through cognate T4SSs can be summarized briefly as follows. First, an accessory factor binds the origin-of-transfer (*oriT*) sequence carried by an MGE. Accessory factors generally fall into the ribbon-helix-helix family of DNA binding proteins, as exemplified by the TraM protein encoded by F plasmid and TrwA encoded by plasmid R388 (hereafter, the origin of the named T4SS or protein will appear in subscript, e.g., $TraM_F$ and $TrwA_{R388}$) (28–31). The accessory factor, through a combination of DNA bending and direct protein-protein interactions, recruits a protein termed the relaxase to *oriT* to form the catalytically active relaxosome (32, 33). The relaxase cleaves the DNA strand destined for transfer DNA (T-strand), and as a consequence of nicking, the relaxase remains covalently bound to the 5′ end of the T-strand. The accessory factor and relaxase together promote docking of the relaxosome with a cognate VirD4 substrate receptor. In the F plasmid transfer system, the $TraM_F$ accessory factor binds a short motif at the C terminus of VirD4-like TraD, whereas the $TraI_F$ relaxase

carries internal TSs that form only when TraI adopts its tertiary structure (34). Other relaxases may additionally or alternatively carry C-terminal TSs, which are typically composed of clusters of positively charged or hydrophobic residues (35–40). These TSs mediate relaxase interactions with the VirD4 receptors, although the structural bases for these interactions are not yet defined. Other specialized accessory factors are members of the ParA superfamily of partitioning proteins; these proteins also physically couple the relaxosome with the VirD4 receptors through establishment of multiple protein-protein contacts (41–43). Once the relaxosome docks with the VirD4 substrate receptor, by mechanisms that are not yet known, the relaxase is unfolded (23) and the accessory factor(s) is released, and the relaxase pilots the covalently bound T-strand through the T4SS to recipient cells. In recipient cells, the relaxase catalyzes recircularization of the T-strand through a reversal of the strand-breaking reaction, followed by second-strand synthesis and replication of the transferred element.

Among the effector translocators, some systems translocate only one or a few effector proteins, whereas others deliver several hundred into eukaryotic target cells, where they function in a myriad of ways to subvert host cell physiologies (Fig. 1A) (1, 6, 9). As with the conjugation systems, effector translocators recognize their substrate repertoires through a combination of internal and C-terminal TSs carried by the effectors and binding of adaptors or chaperones associated with the effectors. In addition to their role in physically coupling the effector with the VirD4 receptor, adaptors and chaperones block effector aggregation or prevent nonproductive protein interactions in the bacterium prior to effector translocation (44–49). Until recently, effector translocators were thought to function exclusively to deliver protein effectors to eukaryotic cells, where they disrupt host cell physiological processes that aid in infection. Recently, however, members of this subfamily were shown to translocate toxin components of toxin-antitoxin modules to kill other bacteria in the vicinity (Fig. 1A). *Xanthomonas* spp., for example, deploy a VirB/VirD4-like T4SS to deliver toxins whose bacteriolytic activities can degrade peptidoglycan in target cells lacking the corresponding antitoxin (13). In *Bartonella* spp., the VbhT toxin is similarly transmitted via a VirB-like T4SS to target bacteria (14). Interestingly, VbhT carries a C-terminal TS identical to that previously determined to be involved in T4SS trafficking of interkingdom effectors during the course of *Bartonella* infections. These findings establish an evolutionary link between toxins transmitted interbacterially for niche establishment and effectors delivered to eukaryotic cells for pathogenic ends (14).

The VirD4 Receptor

VirD4-like substrate receptors couple DNA or protein substrates to the T4SS; consequently, these subunits are also called type IV coupling proteins (T4CPs) (Fig. 2A and B) (50, 51). T4CPs are phylogenetically related to the SpoIIIE and FtsK ATPases functioning in DNA transport across membranes during sporulation and cell division, respectively (52, 53). These ATPases are typically configured as homohexamers with an N-terminal transmembrane domain and a nucleotide binding/hydrolysis domain (NBD), giving rise to an overall F1-ATPase architecture (53, 54). T4CPs also carry a sequence variable all-alpha domain (AAD) situated at the base of the NBD, and many carry a second, variable C-terminal domain (CTD) (51, 54). The AAD functions in substrate binding and specificity (55, 56), whereas the CTDs can also confer substrate specificity (28) or spatiotemporal control over substrate selection and translocation through the T4SS (57, 58). Two structural studies have shed important light on interactions between T4CP CTDs and cognate substrates. First, in the F plasmid transfer system, the accessory factor TraM$_F$ was shown to physically couple the F relaxosome to the TraD receptor through simultaneous

binding of F's *oriT* sequence and a 13-residue sequence at the end of TraD's CTD (28, 59). Second, in the more complex *Legionella pneumophila* Dot/Icm system, which is capable of delivering over 300 effectors to eukaryotic target cells during infection, the VirD4-like DotL receptor has a long (~200-residue) CTD that binds DotM, DotN, the IcmSW adaptor complex, and LvgA (Fig. 2B) (48, 49, 58, 60, 61). These bound factors collectively stabilize DotL and mediate recruitment of distinct subsets of effectors to the Dot/Icm T4SS. Modeling of the elongated CTD-adaptor subassembly onto DotL's NBD hexameric sphere gives rise to a bipartitate bell-shape structure that presumably sits at the base of the translocation channel to recruit and feed substrates into the channel (Fig. 2B) (58). Interestingly, however, to date neither the DotL-adaptor complex nor other VirD4 hexamers have been visualized in association with cognate T4SS machines (see below) (18). These and other biochemical findings (62–64) have supported a model that T4CPs associate only transiently with cognate channels as a function of substrate binding.

Structural Advances: Subunits and Subassemblies

Structures now exist for intact or soluble domains of the conserved ATPases (VirD4, VirB4, and VirB11) and several channel/pilus subunits (VirB5, VirB8, VirB9, and VirB10), obtained by X-ray crystallography, nuclear magnetic resonance, or negative-stain electron microscopy (nsEM) (54, 65–72). Larger subassemblies termed outer membrane core complexes (OMCCs) have also been visualized by nsEM from systems phylogenetically related to the VirB/VirD4 T4SS (27, 73) as well as more distantly related systems, e.g., the *L. pneumophila* Dot/Icm and *H. pylori* Cag T4SSs (74, 75). The OMCCs of the VirB/VirD4-like systems are ~1.1-MDa complexes composed of 14 copies of the VirB7-, VirB9-, and VirB10-like subunits. These complexes are arranged as large barrels of ~185 Å in width and height (27). The outer layer of the OMCC from the pKM101-encoded T4SS (Tra$_{pKM101}$) was solved by X-ray crystallography, revealing a network of intra- and intersubunit contacts and a distal cap composed of 14 copies of a helix-loop-helix domain of VirB10 termed the antenna projection (26). The cap is postulated to span the OM, and its central pore of ~32 Å is postulated to comprise the OM channel through which substrates pass to the cell exterior. Interestingly, chimeric T4SSs composed of the IM-spanning portion of the Tra$_{pKM101}$ T4SS joined to heterologous OMCCs from other VirB/VirD4-like T4SSs are capable of translocating DNA substrates between bacteria, confirming that the observed structural conservation of OMCCs from these systems extends to the level of function (73). Very recently, a structure of the entire OMCC of a *Xanthomonas citri* T4SS was solved at 3.3 Å by cryo-electron microscopy (cryo-EM); it shows in unprecedented atomic detail an extensive VirB7-VirB9-VirB10 interaction network and also identifies flexible linkers and weak contacts that are postulated to account for intrinsic flexibility of the OMCC necessary for signal-activated channel gating (20, 76, 77).

The VirB$_{3-10}$ structure

A much larger substructure elaborated by the Trw$_{R388}$ T4SS was solved by nsEM at a resolution of 20 Å (78). This structure is composed of homologs of the VirB3-VirB10 subunits and was designated the VirB$_{3-10}$ or T4SS$_{3-10}$ complex (Fig. 2C and E). The ~3.5-MDa structure consists of the OMCC joined by a thin stalk to an even larger IM complex, designated the IMC, of 25.5 nm in diameter and 10.5 nm in thickness. The entire structure, with a length of ~34 nm, spans the cell envelope such that the OMCC's cap forms the OM pore and the upper portion of the IMC spans the IM. The IMC platform is composed of 12 copies each of VirB3, VirB5, and VirB8 and 24 copies of VirB6. This platform connects to two barrel-like structures of 10.5 nm in width and 13.5 nm

in height that correspond to two side-by-side hexamers of VirB4 extending into the cytoplasm (Fig. 2C and E). The VirB$_{3-10}$ structure lacks the conjugative pilus elaborated by conjugation machines in G$^-$ species (see below), as well as the VirB2, VirB11, and VirD4 homologs required for substrate transfer. Interestingly, however, in a recent update a VirB$_{3-10}$ structure was solved that additionally has one or two dimers of VirD4 situated between the VirB4 hexameric barrels (79). These dimers might correspond to an assembly intermediate of the T4CP that engages with the channel in the absence of bound substrate. It is also intriguing to consider that an early X-ray structure of a soluble domain of the TrwB$_{R388}$ coupling protein (54), which has guided our thinking for many years regarding the hexameric structure of T4CPs, might not reflect the oligomeric and active states of the VirD4 ATPases assembled *in vivo* (79).

The overall VirB$_{3-10}$/VirD4 dimer structure lacks a detectable channel. However, results of a chromatin immunoprecipitation-based cross-linking assay termed transfer DNA immunoprecipitation using the model *A. tumefaciens* VirB/VirD4 system allowed for provisional assignments of channel composition. As DNA substrates are translocated through the VirB/VirD4 channel, they can be formaldehyde cross-linked sequentially with the VirD4 and VirB11 ATPases, then with the VirB6 and VirB8 IMC subunits, and finally with the VirB9 and VirB2 pilin subunits in the periplasm (Fig. 2A) (20, 80–83). In subsequent studies, evidence also was presented for DNA substrate close contacts with VirB4-like subunits (71, 84). Thus, seven VirB/VirD4 subunits depicted in Fig. 2A are envisioned to comprise the translocation channel, while the remaining VirB components contribute indirectly to channel assembly or function.

The *L. pneumophila* Dot/Icm Structure

The *L. pneumophila* Dot/Icm T4SS is assembled from VirB-like subunits as well as

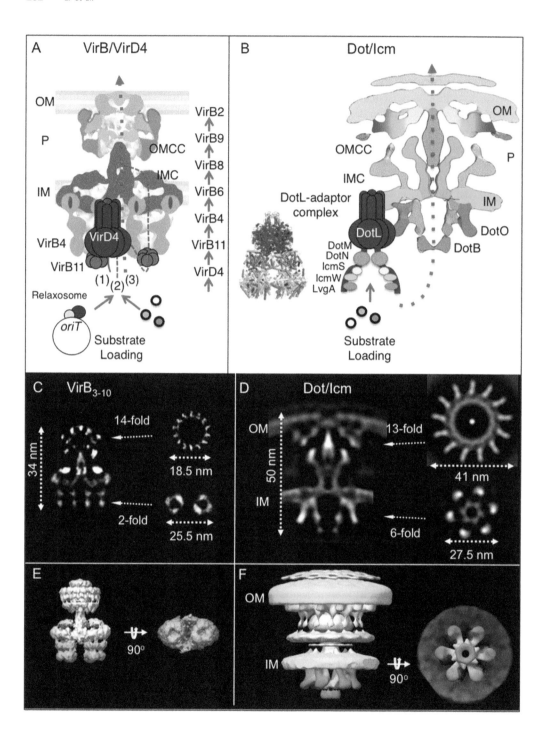

approximately 20 additional subunits (Fig. 1B) (85, 86). Not surprisingly, therefore, the Dot/Icm structure recently visualized by *in situ* cryo-electron tomography (cryo-ET) is much larger than the VirB$_{3-10}$ substructure (Fig. 2D and F) (18, 87). The OMCC is 42 nm wide and 31 nm high and presents as a wheel-like structure with 13-fold symmetry, as opposed to the 14-fold symmetries of the VirB/VirD4 OMCCs (18). The entire wheel is embedded in the inner leaflet of the OM, and a central pore of 6 nm projects across the outer leaflet of the OM. The wheel extends into the periplasm, where it is connected to a cylinder that extends to the IM, establishing contact with the IMC. Most strikingly, refinement of the IMC showed that it adopts a 6-fold symmetry and forms two concentric rings of 16 nm and 27.5 nm at the cytoplasmic entrance to the translocation channel (Fig. 2D) (18). In side view, 6 inverted V structures extend into the cytoplasm, such that the inner arms of the V's form the inner ring and the outer arms form the outer ring. These V structures are composed of VirB4-like DotO, and thus, the cytoplasmic complex consists of a hexamer of 6 DotO dimers (Fig. 2F). Furthermore, VirB11-like DotB was shown to dynamically associate at the base of the DotO inner ring by a mechanism dependent on ATP hydrolysis. The cytoplasmic complex is therefore composed of a central DotO hexamer onto which the DotB hexamer binds, presumably when the machine is activated for substrate transfer. This symmetric IMC architecture differs strikingly from the asymmetric IMC of the VirB$_{3-10}$ structure marked by side-by-side VirB4 barrels (Fig. 2E and F). Gratifyingly, the Dot/Icm structure identifies for the first time a continuous T4SS channel extending from the cytoplasmic entrance (marked by the DotB lumen) to the cell surface (marked by the OMCC pore) (Fig. 2B, D, and F) (18). It is also interesting to note that the Dot/Icm T4SS

FIGURE 2 **Architectures of the phylogenetically distant VirB/VirD4 and Dot/Icm T4SSs. (A) A schematic of the VirB$_{3-10}$ structure elaborated by the Trw$_{R388}$ T4SS and solved by single-particle nsEM. A hexamer of the VirD4 receptor is fitted between the two hexameric barrels of the VirB4 ATPase. The VirD4 receptor recruits MGEs, such as conjugative plasmids, through recognition of components of the relaxosome (relaxase and accessory factors) assembled at the origin-of-transfer (*oriT*) sequence. VirD4 recruits protein substrates (colored dots) through direct or adaptor-mediated contacts. Substrates engage with the VirD4 receptor and are then delivered sequentially through a translocation channel composed of the VirB proteins listed at the right, as deduced from the transfer DNA immunoprecipitation assay (78). The route of transfer across the IM is not known; substrates might be conveyed through the VirD4 hexamer (route 1, solid line), the VirB4 hexamer (route 2, small dashed line), or a channel composed of the VirB6 and VirB8 subunits (route 3, dotted line). Substrates then pass through the periplasm and across the OM via an OMCC channel. (B) A schematic of the *L. pneumophila* Dot/Icm T4SS solved by *in situ* cryo-ET (16). The centrally stacked hexamers of the VirB4-like DotO and DotB form the cytoplasmic entrance to a channel that spans the entire cell envelope. The bell-shaped DotL-adaptor complex is comprised of the hexameric nucleotide binding domain (purple) and C-terminal domain bound with DotN (brown), IcmS (yellow), IcmW (aqua), and LvgA (green); reprinted with permission by Kwak et al. (58). The DotL-adaptor receptor complex was not part of the visualized Dot/Icm T4SS (16) and is provisionally positioned adjacent to the Dot/Icm T4SS. Upon loading of substrates, the DotL-adaptor complex is postulated to present effectors to the DotB/DotO energy center for delivery through the central channel. (C to F) Comparison of the R388-encoded VirB$_{3-10}$ substructure and the *L. pneumophila* Dot/Icm T4SS. (C) A central section through the longitudinal plane of the VirB$_{3-10}$ single-particle reconstruction with cross sections of the OMCC and IMC at the positions indicated. (D) A central section through longitudinal plane of a global average structure of *L. pneumophila* Dot/Icm T4SS with cross sections at the positions indicated. (E) A three-dimensional (3D) surface rendering of the VirB$_{3-10}$ substructure shown in side and bottom views. The side-by-side hexameric barrels of the VirB4 ATPase are colored pink. (F) A 3D surface rendering of the Dot/Icm T4SS shown in side and bottom views. The bacterial membranes are in green and the DotO and DotB hexameric ATPases comprising the entrance to the translocation channel are in shades of pink and purple, respectively.**

assembles at the cell poles and that polar delivery of effectors into the eukaryotic host cell evidently is required for successful *L. pneumophila* infection (88).

T4SS-Associated Mechanisms for Target Cell Attachment

Conjugation systems of G⁻ species elaborate conjugative pili that extend for as long as 20 μm from the donor cell (12). Flexible pili elaborated by F plasmids extend or retract to draw potential recipients into direct contact (89), whereas more brittle pili produced by other conjugative plasmids are either sloughed or broken from the cell, where they accumulate and induce cellular aggregation (12, 90). To date, only one effector translocator system, the *H. pylori* Cag T4SS, also has been shown to elaborate pili in the presence of host epithelial cells (91–93). Interestingly, this Cag T4SS additionally elaborates large sheathed structures, or "membrane tubes" (94), that were recently visualized by *in situ* cryo-ET. Features of these tube structures led the authors to suggest they might arise by the extension of a pilus from an IM platform. As the pilus protrudes across the OM, the distorted membrane surrounds the pilus, forming a sheath or tube that projects from the cell surface (95).

Surprisingly, the role of the pilus in substrate transfer is still not firmly established. On the one hand, there is some evidence in the F plasmid transfer system for DNA transfer between distant cells attached together by the F pilus, suggesting that the F pilus can function as a translocation channel (96). The structure of the F pilus was recently solved by cryo-EM, and strikingly, the lumen is lined with IM phospholipid (PL). This discovery has important implications regarding the mechanism of F pilus assembly and retraction, but the presence of PL also imparts an overall weak negative charge to the inner lumen of possible importance for conveyance of the DNA substrate through the pilus (97). On the other hand, several observations argue against

a role for the pilus as a conduit for substrates. First, in the *A. tumefaciens* VirB/VirD4 T4SS and related T4SSs, "uncoupling" mutations have been isolated that selectively block pilus production without impeding substrate transfer, strongly indicating that extended pili are not required for DNA transfer (73, 83, 98). Second, conjugation systems functioning in G⁺ species do not elaborate pili yet can transfer DNA between cells at very high frequencies (8, 99, 100). Third, recently it was shown that the *E. coli* pKM101 conjugation system employs cell surface adhesins as an alternative to conjugative pili to mediate formation of donor-target cell contacts (101). This finding is of special interest given the paucity of evidence for pilus production by effector translocators other than the *H. pylori* Cag T4SS. It is enticing to propose that most members of the effector translocator subfamily have dispensed with energetically costly pilus production in favor of appropriating chromosomally encoded surface adhesins to establish productive bacterial-eukaryotic cell membrane contacts.

CONCLUDING REMARKS

Mechanistic and structural studies are rapidly shaping a deeper understanding of how the T4SS superfamily evolved such functional diversity. Most notably, recent advances have been made in structural definition of T4SSs that are phylogenetically distantly related to the "canonical" VirB/VirD4-like T4SSs, e.g., *L. pneumophila* Dot/Icm. Despite these advances, many fundamental questions remain: (i) How do substrates dock with the T4SS, and how are they processed for transfer? (ii) What is the route of translocation, and what are the signaling requirements for channel activation? (iii) What mechanisms mediate productive donor-target cell contacts, and what is the architecture of the mating junction? As basic studies continue to investigate these questions, we also note with excitement the emergence of translational studies aimed at designing T4SS machine inhibitors or re-purposing T4SSs as therapeutic delivery systems (102–104). The integration of basic and translational approaches ensures a bright and vibrant future for the fascinating T4SS nanomachines as well as the scientists devoted to their study.

ACKNOWLEDGMENTS

We thank members of the Christie and Hu laboratories for helpful discussions.

Work in the Christie laboratory is supported by NIH grants R01GM48476 and 1R21AI137918. Work in the Hu laboratory is supported by grant AU-1953-20180324 from the Welch Foundation. Work in the Christie and Hu laboratories is supported by R21AI142378 and in part by NIH grant DK056338, which supports the Texas Medical Center Digestive Diseases Center.

CITATION

Li YG, Hu B, Christie PJ. 2019. Biological and structural diversity of type IV secretion systems. Microbiol Spectrum 7(2):PSIB-0012-2018.

REFERENCES

1. **Grohmann E, Christie PJ, Waksman G, Backert S.** 2018. Type IV secretion in Gram-negative and Gram-positive bacteria. *Mol Microbiol* **107**:455–471.
2. **Lederberg J, Tatum EL.** 1946. Novel genotypes in mixed cultures of biochemical mutants of bacteria. *Cold Spring Harb Symp Quant Biol* **11**:113–114.
3. **Bellanger X, Payot S, Leblond-Bourget N, Guédon G.** 2014. Conjugative and mobilizable genomic islands in bacteria: evolution and diversity. *FEMS Microbiol Rev* **38**:720–760.
4. **von Wintersdorff CJ, Penders J, van Niekerk JM, Mills ND, Majumder S, van Alphen LB, Savelkoul PH, Wolffs PF.** 2016. Dissemination of antimicrobial resistance in microbial ecosystems through horizontal gene transfer. *Front Microbiol* **7**:173.
5. **Koraimann G.** 2018. Spread and persistence of virulence and antibiotic resistance genes: a ride on the F plasmid conjugation module. *EcoSal Plus* **8**:ESP-0003-2018.
6. **Cascales E, Christie PJ.** 2003. The versatile bacterial type IV secretion systems. *Nat Rev Microbiol* **1**:137–149.
7. **Winans SC, Burns DL, Christie PJ.** 1996. Adaptation of a conjugal transfer system for the export of pathogenic macromolecules. *Trends Microbiol* **4**:64–68.
8. **Bhatty M, Laverde Gomez JA, Christie PJ.** 2013. The expanding bacterial type IV secretion lexicon. *Res Microbiol* **164**:620–639.
9. **Gonzalez-Rivera C, Bhatty M, Christie PJ.** 2016. Mechanism and function of type IV secretion during infection of the human host. *Microbiol Spectr* **4**:VMBF-0024-2015.
10. **Qiu J, Luo ZQ.** 2017. *Legionella* and *Coxiella* effectors: strength in diversity and activity. *Nat Rev Microbiol* **15**:591–605.
11. **Dehio C, Tsolis RM.** 2017. Type IV effector secretion and subversion of host functions by *Bartonella* and *Brucella* species. *Curr Top Microbiol Immunol* **413**:269–295.
12. **Arutyunov D, Frost LS.** 2013. F conjugation: back to the beginning. *Plasmid* **70**:18–32.
13. **Souza DP, Oka GU, Alvarez-Martinez CE, Bisson-Filho AW, Dunger G, Hobeika L, Cavalcante NS, Alegria MC, Barbosa LR, Salinas RK, Guzzo CR, Farah CS.** 2015. Bacterial killing via a type IV secretion system. *Nat Commun* **6**:6453.
14. **Harms A, Liesch M, Körner J, Québatte M, Engel P, Dehio C.** 2017. A bacterial toxin-antitoxin module is the origin of inter-bacterial and inter-kingdom effectors of *Bartonella*. *PLoS Genet* **13**:e1007077.
15. **Berger BR, Christie PJ.** 1994. Genetic complementation analysis of the *Agrobacterium tumefaciens virB* operon: *virB2* through *virB11* are essential virulence genes. *J Bacteriol* **176**:3646–3660.
16. **Christie PJ.** 2016. The mosaic type IV secretion systems. *EcoSal Plus* **7**:ESP-0020-2015.
17. **Christie PJ, Gomez Valero L, Buchrieser C.** 2017. Biological diversity and evolution of type IV secretion systems. *Curr Top Microbiol Immunol* **413**:1–30.
18. **Chetrit D, Hu B, Christie PJ, Roy CR, Liu J.** 2018. A unique cytoplasmic ATPase complex defines the *Legionella pneumophila* type IV secretion channel. *Nat Microbiol* **3**:678–686.
19. **Grohmann E, Keller W, Muth G.** 2017. Mechanisms of conjugative transfer and type IV secretion-mediated effector transport in Gram-positive bacteria. *Curr Top Microbiol Immunol* **413**:115–141.

20. Cascales E, Atmakuri K, Sarkar MK, Christie PJ. 2013. DNA substrate-induced activation of the *Agrobacterium* VirB/VirD4 type IV secretion system. *J Bacteriol* 195:2691–2704.

21. Ripoll-Rozada J, Zunzunegui S, de la Cruz F, Arechaga I, Cabezón E. 2013. Functional interactions of VirB11 traffic ATPases with VirB4 and VirD4 molecular motors in type IV secretion systems. *J Bacteriol* 195:4195–4201.

22. Chandran Darbari V, Waksman G. 2015. Structural biology of bacterial type IV secretion systems. *Annu Rev Biochem* 84:603–629.

23. Trokter M, Waksman G. 2018. Translocation through the conjugative type 4 secretion system requires unfolding of its protein substrate. *J Bacteriol* 200:e00615-17.

24. Arends K, Celik EK, Probst I, Goessweiner-Mohr N, Fercher C, Grumet L, Soellue C, Abajy MY, Sakinc T, Broszat M, Schiwon K, Koraimann G, Keller W, Grohmann E. 2013. TraG encoded by the pIP501 type IV secretion system is a two-domain peptidoglycan-degrading enzyme essential for conjugative transfer. *J Bacteriol* 195:4436–4444.

25. Laverde Gomez JA, Bhatty M, Christie PJ. 2014. PrgK, a multidomain peptidoglycan hydrolase, is essential for conjugative transfer of the pheromone-responsive plasmid pCF10. *J Bacteriol* 196:527–539.

26. Chandran V, Fronzes R, Duquerroy S, Cronin N, Navaza J, Waksman G. 2009. Structure of the outer membrane complex of a type IV secretion system. *Nature* 462:1011–1015.

27. Fronzes R, Schäfer E, Wang L, Saibil HR, Orlova EV, Waksman G. 2009. Structure of a type IV secretion system core complex. *Science* 323:266–268.

28. Lu J, Wong JJ, Edwards RA, Manchak J, Frost LS, Glover JN. 2008. Structural basis of specific TraD-TraM recognition during F plasmid-mediated bacterial conjugation. *Mol Microbiol* 70:89–99.

29. Moncalián G, Grandoso G, Llosa M, de la Cruz F. 1997. oriT-processing and regulatory roles of TrwA protein in plasmid R388 conjugation. *J Mol Biol* 270:188–200.

30. Moncalián G, de la Cruz F. 2004. DNA binding properties of protein TrwA, a possible structural variant of the Arc repressor superfamily. *Biochim Biophys Acta* 1701:15–23.

31. Tato I, Matilla I, Arechaga I, Zunzunegui S, de la Cruz F, Cabezon E. 2007. The ATPase activity of the DNA transporter TrwB is modulated by protein TrwA: implications for a common assembly mechanism of DNA translocating motors. *J Biol Chem* 282:25569–25576.

32. de la Cruz F, Frost LS, Meyer RJ, Zechner EL. 2010. Conjugative DNA metabolism in Gram-negative bacteria. *FEMS Microbiol Rev* 34:18–40.

33. Zechner EL, Lang S, Schildbach JF. 2012. Assembly and mechanisms of bacterial type IV secretion machines. *Philos Trans R Soc Lond B Biol Sci* 367:1073–1087.

34. Redzej A, Ilangovan A, Lang S, Gruber CJ, Topf M, Zangger K, Zechner EL, Waksman G. 2013. Structure of a translocation signal domain mediating conjugative transfer by type IV secretion systems. *Mol Microbiol* 89:324–333.

35. Vergunst AC, van Lier MC, den Dulk-Ras A, Stüve TA, Ouwehand A, Hooykaas PJ. 2005. Positive charge is an important feature of the C-terminal transport signal of the VirB/D4-translocated proteins of *Agrobacterium*. *Proc Natl Acad Sci U S A* 102:832–837.

36. van Kregten M, Lindhout BI, Hooykaas PJ, van der Zaal BJ. 2009. *Agrobacterium*-mediated T-DNA transfer and integration by minimal VirD2 consisting of the relaxase domain and a type IV secretion system translocation signal. *Mol Plant Microbe Interact* 22:1356–1365.

37. Alperi A, Larrea D, Fernández-González E, Dehio C, Zechner EL, Llosa M. 2013. A translocation motif in relaxase TrwC specifically affects recruitment by its conjugative type IV secretion system. *J Bacteriol* 195:4999–5006.

38. Lang S, Kirchberger PC, Gruber CJ, Redzej A, Raffl S, Zellnig G, Zangger K, Zechner EL. 2011. An activation domain of plasmid R1 TraI protein delineates stages of gene transfer initiation. *Mol Microbiol* 82:1071–1085.

39. Lang S, Zechner EL. 2012. General requirements for protein secretion by the F-like conjugation system R1. *Plasmid* 67:128–138.

40. Meyer R. 2015. Mapping type IV secretion signals on the primase encoded by the broad-host-range plasmid R1162 (RSF1010). *J Bacteriol* 197:3245–3254.

41. Atmakuri K, Cascales E, Burton OT, Banta LM, Christie PJ. 2007. *Agrobacterium* ParA/MinD-like VirC1 spatially coordinates early conjugative DNA transfer reactions. *EMBO J* 26:2540–2551.

42. Hamilton HL, Domínguez NM, Schwartz KJ, Hackett KT, Dillard JP. 2005. *Neisseria gonorrhoeae* secretes chromosomal DNA via a novel type IV secretion system. *Mol Microbiol* 55:1704–1721.

43. Gruber CJ, Lang S, Rajendra VK, Nuk M, Raffl S, Schildbach JF, Zechner EL. 2016. Conjugative DNA transfer is enhanced by plas-

mid R1 partitioning proteins. *Front Mol Biosci* **3:**32.

44. **Vergunst AC, Schrammeijer B, den Dulk-Ras A, de Vlaam CM, Regensburg-Tuïnk TJ, Hooykaas PJ.** 2000. VirB/D4-dependent protein translocation from *Agrobacterium* into plant cells. *Science* **290:**979–982.

45. **Zhao Z, Sagulenko E, Ding Z, Christie PJ.** 2001. Activities of *virE1* and the VirE1 secretion chaperone in export of the multifunctional VirE2 effector via an *Agrobacterium* type IV secretion pathway. *J Bacteriol* **183:**3855–3865.

46. **Atmakuri K, Ding Z, Christie PJ.** 2003. VirE2, a type IV secretion substrate, interacts with the VirD4 transfer protein at cell poles of *Agrobacterium tumefaciens*. *Mol Microbiol* **49:**1699–1713.

47. **Dym O, Albeck S, Unger T, Jacobovitch J, Branzburg A, Michael Y, Frenkiel-Krispin D, Wolf SG, Elbaum M.** 2008. Crystal structure of the *Agrobacterium* virulence complex VirE1-VirE2 reveals a flexible protein that can accommodate different partners. *Proc Natl Acad Sci U S A* **105:**11170–11175.

48. **Vincent CD, Friedman JR, Jeong KC, Sutherland MC, Vogel JP.** 2012. Identification of the DotL coupling protein subcomplex of the *Legionella* Dot/Icm type IV secretion system. *Mol Microbiol* **85:**378–391.

49. **Sutherland MC, Nguyen TL, Tseng V, Vogel JP.** 2012. The *Legionella* IcmSW complex directly interacts with DotL to mediate translocation of adaptor-dependent substrates. *PLoS Pathog* **8:**e1002910.

50. **Cabezón E, Sastre JI, de la Cruz F.** 1997. Genetic evidence of a coupling role for the TraG protein family in bacterial conjugation. *Mol Gen Genet* **254:**400–406.

51. **Alvarez-Martinez CE, Christie PJ.** 2009. Biological diversity of prokaryotic type IV secretion systems. *Microbiol Mol Biol Rev* **73:**775–808.

52. **Burton B, Dubnau D.** 2010. Membrane-associated DNA transport machines. *Cold Spring Harb Perspect Biol* **2:**a000406.

53. **Gomis-Rüth FX, Solà M, de la Cruz F, Coll M.** 2004. Coupling factors in macromolecular type-IV secretion machineries. *Curr Pharm Des* **10:**1551–1565.

54. **Gomis-Rüth FX, Moncalián G, Pérez-Luque R, González A, Cabezón E, de la Cruz F, Coll M.** 2001. The bacterial conjugation protein TrwB resembles ring helicases and F1-ATPase. *Nature* **409:**637–641.

55. **de Paz HD, Larrea D, Zunzunegui S, Dehio C, de la Cruz F, Llosa M.** 2010. Functional dissection of the conjugative coupling protein TrwB. *J Bacteriol* **192:**2655–2669.

56. **Whitaker N, Chen Y, Jakubowski SJ, Sarkar MK, Li F, Christie PJ.** 2015. The all-alpha domains of coupling proteins from the *Agrobacterium tumefaciens* VirB/VirD4 and *Enterococcus faecalis* pCF10-encoded type IV secretion systems confer specificity to binding of cognate DNA substrates. *J Bacteriol* **197:**2335–2349.

57. **Whitaker N, Berry TM, Rosenthal N, Gordon JE, Gonzalez-Rivera C, Sheehan KB, Truchan HK, VieBrock L, Newton IL, Carlyon JA, Christie PJ.** 2016. Chimeric coupling proteins mediate transfer of heterologous type IV effectors through the *Escherichia coli* pKM101-encoded conjugation machine. *J Bacteriol* **198:**2701–2718.

58. **Kwak MJ, Kim JD, Kim H, Kim C, Bowman JW, Kim S, Joo K, Lee J, Jin KS, Kim YG, Lee NK, Jung JU, Oh BH.** 2017. Architecture of the type IV coupling protein complex of *Legionella pneumophila*. *Nat Microbiol* **2:**17114.

59. **Wong JJ, Lu J, Glover JN.** 2012. Relaxosome function and conjugation regulation in F-like plasmids—a structural biology perspective. *Mol Microbiol* **85:**602–617.

60. **Meir A, Chetrit D, Liu L, Roy CR, Waksman G.** 2018. *Legionella* DotM structure reveals a role in effector recruiting to the type 4B secretion system. *Nat Commun* **9:**507.

61. **Xu J, Xu D, Wan M, Yin L, Wang X, Wu L, Liu Y, Liu X, Zhou Y, Zhu Y.** 2017. Structural insights into the roles of the IcmS-IcmW complex in the type IVb secretion system of *Legionella pneumophila*. *Proc Natl Acad Sci U S A* **114:**13543–13548.

62. **Schröder G, Krause S, Zechner EL, Traxler B, Yeo HJ, Lurz R, Waksman G, Lanka E.** 2002. TraG-like proteins of DNA transfer systems and of the *Helicobacter pylori* type IV secretion system: inner membrane gate for exported substrates? *J Bacteriol* **184:**2767–2779.

63. **Schröder G, Lanka E.** 2003. TraG-like proteins of type IV secretion systems: functional dissection of the multiple activities of TraG (RP4) and TrwB (R388). *J Bacteriol* **185:**4371–4381.

64. **Larrea D, de Paz HD, Arechaga I, de la Cruz F, Llosa M.** 2013. Structural independence of conjugative coupling protein TrwB from its type IV secretion machinery. *Plasmid* **70:**146–153.

65. **Yeo HJ, Savvides SN, Herr AB, Lanka E, Waksman G.** 2000. Crystal structure of the hexameric traffic ATPase of the *Helicobacter pylori* type IV secretion system. *Mol Cell* **6:**1461–1472.

66. **Yeo H-J, Yuan Q, Beck MR, Baron C, Waksman G.** 2003. Structural and functional

characterization of the VirB5 protein from the type IV secretion system encoded by the conjugative plasmid pKM101. *Proc Natl Acad Sci U S A* **100**:15947–15952.

67. **Terradot L, Bayliss R, Oomen C, Leonard GA, Baron C, Waksman G.** 2005. Structures of two core subunits of the bacterial type IV secretion system, VirB8 from *Brucella suis* and ComB10 from *Helicobacter pylori*. *Proc Natl Acad Sci U S A* **102**:4596–4601.

68. **Hare S, Bayliss R, Baron C, Waksman G.** 2006. A large domain swap in the VirB11 ATPase of *Brucella suis* leaves the hexameric assembly intact. *J Mol Biol* **360**:56–66.

69. **Bayliss R, Harris R, Coutte L, Monier A, Fronzes R, Christie PJ, Driscoll PC, Waksman G.** 2007. NMR structure of a complex between the VirB9/VirB7 interaction domains of the pKM101 type IV secretion system. *Proc Natl Acad Sci U S A* **104**:1673–1678.

70. **Walldén K, Williams R, Yan J, Lian PW, Wang L, Thalassinos K, Orlova EV, Waksman G.** 2012. Structure of the VirB4 ATPase, alone and bound to the core complex of a type IV secretion system. *Proc Natl Acad Sci U S A* **109**:11348–11353.

71. **Peña A, Matilla I, Martín-Benito J, Valpuesta JM, Carrascosa JL, de la Cruz F, Cabezón E, Arechaga I.** 2012. The hexameric structure of a conjugative VirB4 protein ATPase provides new insights for a functional and phylogenetic relationship with DNA translocases. *J Biol Chem* **287**:39925–39932.

72. **Prevost MS, Waksman G.** 2018. X-ray crystal structures of the type IVb secretion system DotB ATPases. *Protein Sci* **27**:1464–1475.

73. **Gordon JE, Costa TRD, Patel RS, Gonzalez-Rivera C, Sarkar MK, Orlova EV, Waksman G, Christie PJ.** 2017. Use of chimeric type IV secretion systems to define contributions of outer membrane subassemblies for contact-dependent translocation. *Mol Microbiol* **105**:273–293.

74. **Kubori T, Koike M, Bui XT, Higaki S, Aizawa S, Nagai H.** 2014. Native structure of a type IV secretion system core complex essential for *Legionella* pathogenesis. *Proc Natl Acad Sci U S A* **111**:11804–11809.

75. **Frick-Cheng AE, Pyburn TM, Voss BJ, McDonald WH, Ohi MD, Cover TL.** 2016. Molecular and structural analysis of the *Helicobacter pyloricag* type IV secretion system core complex. *mBio* **7**:e02001-15.

76. **Sgro GG, Costa TR, Cenens W, Souza DP, Cassago A, Coutinho de Oliveira L, Salinas RK, Portugal RV, Farah CS, Waksman G.** 2018. CryoEM structure of the core complex of a bacterial killing type IV secretion system. *Nat Microbiol* **3**:1429–1440.

77. **Cascales E, Christie PJ.** 2004. *Agrobacterium* VirB10, an ATP energy sensor required for type IV secretion. *Proc Natl Acad Sci U S A* **101**:17228–17233.

78. **Low HH, Gubellini F, Rivera-Calzada A, Braun N, Connery S, Dujeancourt A, Lu F, Redzej A, Fronzes R, Orlova EV, Waksman G.** 2014. Structure of a type IV secretion system. *Nature* **508**:550–553.

79. **Redzej A, Ukleja M, Connery S, Trokter M, Felisberto-Rodrigues C, Cryar A, Thalassinos K, Hayward RD, Orlova EV, Waksman G.** 2017. Structure of a VirD4 coupling protein bound to a VirB type IV secretion machinery. *EMBO J* **36**:3080–3095.

80. **Cascales E, Christie PJ.** 2004. Definition of a bacterial type IV secretion pathway for a DNA substrate. *Science* **304**:1170–1173.

81. **Atmakuri K, Cascales E, Christie PJ.** 2004. Energetic components VirD4, VirB11 and VirB4 mediate early DNA transfer reactions required for bacterial type IV secretion. *Mol Microbiol* **54**:1199–1211.

82. **Jakubowski SJ, Krishnamoorthy V, Cascales E, Christie PJ.** 2004. *Agrobacterium tumefaciens* VirB6 domains direct the ordered export of a DNA substrate through a type IV secretion system. *J Mol Biol* **341**:961–977.

83. **Jakubowski SJ, Cascales E, Krishnamoorthy V, Christie PJ.** 2005. *Agrobacterium tumefaciens* VirB9, an outer-membrane-associated component of a type IV secretion system, regulates substrate selection and T-pilus biogenesis. *J Bacteriol* **187**:3486–3495.

84. **Li F, Alvarez-Martinez C, Chen Y, Choi KJ, Yeo HJ, Christie PJ.** 2012. *Enterococcus faecalis* PrgJ, a VirB4-like ATPase, mediates pCF10 conjugative transfer through substrate binding. *J Bacteriol* **194**:4041–4051.

85. **Christie PJ, Vogel JP.** 2000. Bacterial type IV secretion: conjugation systems adapted to deliver effector molecules to host cells. *Trends Microbiol* **8**:354–360.

86. **Kubori T, Nagai H.** 2016. The type IVB secretion system: an enigmatic chimera. *Curr Opin Microbiol* **29**:22–29.

87. **Ghosal D, Chang YW, Jeong KC, Vogel JP, Jensen GJ.** 2017. *In situ* structure of the *Legionella* Dot/Icm type IV secretion system by electron cryotomography. *EMBO Rep* **18**:726–732.

88. **Jeong KC, Ghosal D, Chang YW, Jensen GJ, Vogel JP.** 2017. Polar delivery of *Legionella* type IV secretion system substrates is essential for virulence. *Proc Natl Acad Sci U S A* **114**:8077–8082.

89. **Clarke M, Maddera L, Harris RL, Silverman PM.** 2008. F-pili dynamics by live-cell imaging. *Proc Natl Acad Sci U S A* **105:**17978–17981.

90. **Bradley DE.** 1980. Morphological and serological relationships of conjugative pili. *Plasmid* **4:**155–169.

91. **Kwok T, Zabler D, Urman S, Rohde M, Hartig R, Wessler S, Misselwitz R, Berger J, Sewald N, König W, Backert S.** 2007. *Helicobacter* exploits integrin for type IV secretion and kinase activation. *Nature* **449:**862–866.

92. **Shaffer CL, Gaddy JA, Loh JT, Johnson EM, Hill S, Hennig EE, McClain MS, McDonald WH, Cover TL.** 2011. *Helicobacter pylori* exploits a unique repertoire of type IV secretion system components for pilus assembly at the bacteria-host cell interface. *PLoS Pathog* **7:** e1002237.

93. **Johnson EM, Gaddy JA, Voss BJ, Hennig EE, Cover TL.** 2014. Genes required for assembly of pili associated with the *Helicobacter pylori* cag type IV secretion system. *Infect Immun* **82:**3457–3470.

94. **Rohde M, Püls J, Buhrdorf R, Fischer W, Haas R.** 2003. A novel sheathed surface organelle of the *Helicobacter pylori* cag type IV secretion system. *Mol Microbiol* **49:**219–234.

95. **Chang YW, Shaffer CL, Rettberg LA, Ghosal D, Jensen GJ.** 2018. *In vivo* structures of the *Helicobacter pylori* cag type IV secretion system. *Cell Rep* **23:**673–681.

96. **Babic A, Lindner AB, Vulic M, Stewart EJ, Radman M.** 2008. Direct visualization of horizontal gene transfer. *Science* **319:**1533–1536.

97. **Costa TRD, Ilangovan A, Ukleja M, Redzej A, Santini JM, Smith TK, Egelman EH, Waksman G.** 2016. Structure of the bacterial sex F pilus reveals an assembly of a stoichiometric protein-phospholipid complex. *Cell* **166:**1436–1444.e10.

98. **Sagulenko E, Sagulenko V, Chen J, Christie PJ.** 2001. Role of *Agrobacterium* VirB11 ATPase in T-pilus assembly and substrate selection. *J Bacteriol* **183:**5813–5825.

99. **Bhatty M, Cruz MR, Frank KL, Gomez JA, Andrade F, Garsin DA, Dunny GM, Kaplan HB, Christie PJ.** 2015. *Enterococcus faecalis* pCF10-encoded surface proteins PrgA, PrgB (aggregation substance) and PrgC contribute to plasmid transfer, biofilm formation and virulence. *Mol Microbiol* **95:**660–677.

100. **Schmitt A, Jiang K, Camacho MI, Jonna VR, Hofer A, Westerlund F, Christie PJ, Berntsson RP.** 2018. PrgB promotes aggregation, biofilm formation, and conjugation through DNA binding and compaction. *Mol Microbiol* **109:**291–305.

101. **González-Rivera C, Khara P, Awad D, Patel R, Li YG, Bogisch M, Christie PJ.** 2019. Two pKM101-encoded proteins, the pilus-tip protein TraC and Pep, assemble on the *Escherichia coli* cell surface as adhesins required for efficient conjugative DNA transfer. *Mol Microbiol* **111:**96–117.

102. **Guzmán-Herrador DL, Steiner S, Alperi A, González-Prieto C, Roy CR, Llosa M.** 2017. DNA delivery and genomic integration into mammalian target cells through type IV A and B secretion systems of human pathogens. *Front Microbiol* **8:**1503.

103. **Smith MA, Coinçon M, Paschos A, Jolicoeur B, Lavallée P, Sygusch J, Baron C.** 2012. Identification of the binding site of *Brucella* VirB8 interaction inhibitors. *Chem Biol* **19:**1041–1048.

104. **Shaffer CL, Good JA, Kumar S, Krishnan KS, Gaddy JA, Loh JT, Chappell J, Almqvist F, Cover TL, Hadjifrangiskou M.** 2016. Peptidomimetic small molecules disrupt type IV secretion system activity in diverse bacterial pathogens. *mBio* **7:**e00221-16.

Hostile Takeover: Hijacking of Endoplasmic Reticulum Function by T4SS and T3SS Effectors Creates a Niche for Intracellular Pathogens

23

APRIL Y. TSAI,[1] BEVIN C. ENGLISH,[1] and RENÉE M. TSOLIS[1]

INTRODUCTION

Multiple intracellular pathogens utilize the type III secretion system (T3SS) and type IV secretion system (T4SS) to target functions of the host cell's endoplasmic reticulum (ER). While pathogens such as *Legionella pneumophila* and *Brucella abortus* have long been known to replicate in association with the ER (1, 2), the connection of vacuoles containing other intracellular pathogens, such as *Coxiella burnetii* (3, 4), *Anaplasma* spp. (5, 6), and *Chlamydia trachomatis* and its relatives (7, 8), with the ER has been recognized relatively recently. However, manipulation of ER function is not limited to pathogens that replicate within a vacuole, as cytosolic pathogens such as *Orientia tsutsugamushi* (9, 10) and *Rickettsia rickettsii* (11) also target ER-based functions via secreted effectors to promote their intracellular growth.

Recent progress in large-scale analyses of secreted proteins and in genetic analysis of previously intractable intracellular bacteria such as *C. trachomatis*, *C. burnetii*, and *Ricksttsia* spp. has led to an explosion in identification of new T3SS and T4SS effectors, and for some of these effectors, exciting recent advances have revealed how their interactions with host components contribute to the intracellular replication cycle of these organisms. This review

[1]Department of Medical Microbiology & Immunology, University of California, Davis, Davis, CA 95616
Protein Secretion in Bacteria
Edited by Maria Sandkvist, Eric Cascales, and Peter J. Christie
© 2019 American Society for Microbiology, Washington, DC
doi:10.1128/microbiolspec.PSIB-0027-2019

focuses on recent progress in understanding how interactions with the ER mediated by secreted effectors (primarily of T4SS and T3SS) promote infection by intracellular bacteria.

THE ER: A BIOSYNTHESIS AND SIGNALING HUB OF THE CELL

The ER performs multiple functions that are critical to cellular homeostasis. Approximately one-third of the mammalian cell's proteome is targeted to the ER, and accordingly, its best-characterized role is that of the "factory" for correct folding of proteins that ultimately function in the plasma membrane, the extracellular space, or secretory compartments such as the ER itself, the Golgi, secretory vesicles, and lysosomes. Within the ER lumen, protein folding is assisted by ER-resident chaperones, such as the Hsp70 chaperone BiP, which binds hydrophobic protein regions, thereby preventing their aggregation (reviewed in reference 12). The majority of secretory proteins are further modified by addition of glycans to asparagine residues, referred to as N-linked glycosylation. This modification increases the solubility and stability of hydrophobic proteins and promotes their cellular targeting and function (reviewed in reference 13). As protein folding proceeds, resident ER proteins and chaperones also perform quality control to ensure that misfolded or aggregated proteins do not accumulate, as they can disrupt ER function. If a protein is terminally misfolded and cannot be refolded to a functional conformation, it is targeted to the ER-associated degradation (ERAD) pathway, wherein the misfolded protein is extracted from the ER membrane to the cytosol while being tagged with polyubiquitin chains, resulting in proteosomal degradation (reviewed in reference 14).

In addition to its role in protein folding, the ER is site of lipid biosynthesis and central regulator of lipid levels throughout the cell (reviewed in reference 15). The ER produces the main phospholipids composing cellular membranes, as well as less abundant membrane components. Enzymes that synthesize cholesterol are also located in the ER. After their synthesis, these lipids are distributed from the ER to their sites of function in the cell via the secretory pathway or via membrane contact sites with other organelles (see below). Further, under conditions of excess nutrition, ER-localized enzymes synthesize triacylglycerides for energy storage within lipid droplets in the cell. Together, these ER-based functions are critical for maintaining cellular lipid homeostasis. As vacuolar pathogens replicate, their vacuole needs to expand, and thus an association with the ER could provide membrane lipids needed to enlarge the intracellular niche. Lipids produced by the ER might also provide biosynthetic material to intracellular pathogens for generation of membrane lipids or for energy (16).

Within the structure of the ER, specialized membrane domains are organized to carry out specific functions. Specific subdomains of the ER give rise to peroxisomes, organelles that sequester enzymes for β-oxidation of very-long-chain fatty acids as well as for metabolism of cholesterol, bile acids, and polyamines (17, 18). Another set of specialized ER domains are the membrane contact sites (MCS) that form between ER and other organelles in the cell, including mitochondria, the Golgi apparatus, the plasma membrane, endosomes, and peroxisomes (reviewed in reference 19). These are sites where organelles are tethered to each other via interactions between proteins in apposing membranes. The MCS between ER and mitochondria, for example, are extensive and play essential roles in mitochondrial division (20) and calcium signaling between the ER and mitochondria (21, 22). The ER proteins VAPA and VAPB tether multiple organelles in the cell to the ER via MCS, including the Golgi, endosomes, and the plasma membrane (19). Of particular interest for thinking about how pathogens could associate with the ER after uptake, it is now appreciated that endosomes associate with the ER, and these contacts become more extensive as endosomes mature; in fact, endosomes remain tightly associated with the ER throughout their trafficking (23), suggesting a potential point of contact between pathogen-containing endosomes and the ER that might be exploited by pathogens.

In response to a stimulus such as amino acid starvation, yet another specialized ER domain known as the omegasome forms, providing one of the pathways to initiate autophagy, a process in which cellular components are recycled to provide nutrition to the cell (reviewed in reference 24). The omegasome contains the protein DFCP1 and is enriched in phosphatidylinositol-3-phosphate, which is thought to increase the membrane curvature to initiate formation of the phagophore, the double membrane that is characteristic of autophagosomes (25).

BACTERIAL STRATEGIES FOR CO-OPTING ER FUNCTION

Recent work has identified how the ER functions outlined above can be subverted by intracellular bacterial pathogens to generate a replicative niche, gain nutrients for growth, or spread from cell to cell. While *Brucella abortus* was recently shown to replicate with the ER lumen (26), other pathogens, including *Legionella pneumophila*, *Chamydia* spp., *Simkania negevensis*, *Anaplasma* spp., and *C. burnetii*, reside in a vacuole that during some part of their replicative cycle is tethered to the ER via membrane contact sites between the pathogen-containing vacuole and the ER (2, 4, 6–8). Yet another group of pathogens, exemplified by *R. rickettsii* and *Orientia tsutsugamushi*, reside in the host cell's cytosol and secrete effectors that target ER functions (11, 27). The following section reviews recent advances in our understanding of how pathogen effectors interact with the ER.

Subversion of Vesicular Trafficking Between the ER and Golgi Apparatus

Brucella abortus, a zoonotic pathogen causing abortion in ruminants and febrile infections in humans, utilizes its T4SS to replicate intracellularly in multiple cell types, with the macrophage being the best studied (reviewed in reference 28). After uptake by macrophages, *B. abortus* is able to avoid degradation in lysosomes (reviewed in reference 29) and replicates within the ER (26). To establish this replicative niche, *B. abortus* utilizes its T4SS to interact with ER exit sites, where ER-to-Golgi transport is initiated, in a manner that is dependent on the small GTPase Sar1 (30), though the effectors mediating the association with ER exit sites have not yet been identified (Fig. 1). *B. abortus* also requires Golgi-to-ER transport for the maintenance of its replicative niche, as the small GTPase Rab2 contributes to intracellular replication of *B. abortus* (31). To date, approximately 15 T4SS effectors have been shown to be translocated into infected host cells (28). Recently, the T4SS effector BspB (Table 1) was found to alter secretory trafficking from both ER to Golgi and from the Golgi to the ERGIC (ER-to-Golgi intermediate compartment) to the ER by interacting with the conserved oligomeric Golgi (COG) complex (24). This interaction between BspB and the COG complex diverts Golgi-derived vesicles to *Brucella*'s replicative compartment, thereby promoting its intracellular replication—possibly by providing membrane for expansion of the bacterial niche (Fig. 1). In addition to BspB, several *B. abortus* effectors, including BspA, BspD, BspK, and VceC, accumulate in the ER after ectopic expression (32–35), suggesting that they may perturb the early secretory pathway, but how these effectors function remains to be determined.

Legionella pneumophila, which naturally infects amoebae but causes opportunistic respiratory infections in humans, also uses its T4SS (called Dot/Icm) to target trafficking

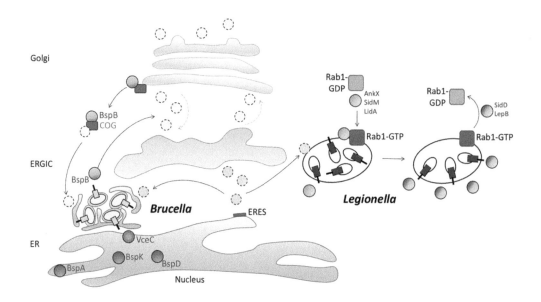

FIGURE 1 Hijacking of vesicular traffic between ER and Golgi by T4SS effectors of *B. abortus* (left) and *L. pneumophila* (right). It has been proposed that like *B. abortus*, *L. pneumophila* also intercepts Golgi-ER traffic (90). Abbreviations: ERGIC, ER-to-Golgi intermediate compartment; ERES, ER exit site; COG, conserved oligomeric Golgi complex.

between the Golgi and the ER (reviewed in reference 36). Of the over 300 Dot/Icm effectors identified to date, a subset targets the function of the early secretory pathway. RalF, the first *L. pneumophila* T4SS effector to be identified, acts as a guanine nucleotide exchange factor (GEF) for ARF1, a small GTPase that regulates secretory membrane transport, primarily between the Golgi and ER (37). Several effectors target Rab1, the small, membrane-associated GTPase that regulates ER-to-Golgi vesicular transport. Rab1 cycles between GDP-bound (inactive) and GTP-bound (active) forms, with the assistance of multiple cellular factors (Fig. 1 and Table 1). GEFs activate Rab1 by converting GDP-Rab1 into GTP-Rab1. GTP-Rab1 then interacts with its target proteins in the membrane transport pathway to promote tethering and fusion of membrane vesicles (reviewed in reference 38). To inactivate GTP-Rab1, GTPase-activating proteins (GAPs) stimulate the GTPase activity of Rab1 to convert it to

inactive GDP-Rab1. The interaction of GDP-Rab1 with membranes is regulated by GDP dissociation inhibitor (Rab-GDI), which extracts it from membranes, and by a GDI displacement factor (GDF), which targets Rab1-GDP to membranes to restart the Rab cycle. Rab1 is recruited to the *Legionella*-containing vacuole (LCV) via the activity of the Dot/Icm effector SidM (also known as DrrA [39]). Biochemical analysis of SidM has revealed multiple activities for modulating Rab1 activity (reviewed in references 36 and 40). A C-terminal phosphatidylinositol-4-phosphate (PI4P) binding domain of SidM mediates its association with the LCV after its secretion by the T4SS (41). The central domain of SidM has GDF/GEF activity for Rab1, which displaces Rab-GDI from Rab1-GDP and mediates GDP exchange for GTP, leading to its association with the membrane of the LCV. After recruiting Rab1 to the LCV, SidM uses its N-terminal domain, which contains adenylyltransferase activity, to covalently modify Rab1 by AMPylation (ade-nylylation) of a tyrosine residue. This modifi-cation of Rab1 prevents its interaction with GAPs, and as a result, Rab1 remains in its GTP-bound form and becomes constitutively active (42). Modification and recruitment of Rab1 by SidM are assisted by a second Dot/Icm effector, LidA (39, 43), which interacts with GTP-bound Rab1 (and with other Rab-GTP complexes as well [44, 45]). This manipulation of Rab1 activity enables *L. pneumophila* to recruit ER-derived vesicles, thereby remod-eling its phagosome into a compartment supporting its replication (46).

Over the time of cellular infection by *L. pneumophila*, additional effectors are trans-located that act in an antagonistic manner to SidM and LidA on Rab1 activity. SidD acts to de-AMPylate Rab1 (47), which restores its GTPase activity. Subsequently, LepB pro-motes hydrolysis of GTP by Rab1 (48), thereby enabling its extraction from the LCV by host Rab-GDI proteins. The time course of effec-tor secretion and Rab1 modulation by *L. pneumophila* suggests that early during infec-tion, recruitment and activation of Rab1 are beneficial for replication. However, prolonged activation of Rab1 may elicit cellular responses that are detrimental to intracellular infection, since ectopic expression of SidM/DrrA is cytotoxic to cells (42).

An additional covalent modification of Rab1, phosphocholination, is mediated by AnkX (49). Interestingly, while AMPylation targets a conserved tyrosine residue in Rab1, AnkX targets the adjacent serine residue. AnkX contains both ankyrin repeat domains and a FIC (filamentation induced by cyclic AMP) domain, which utilizes CDP-choline as a substrate for phosphocholination of Rab1 (50). It appears that Rab1 can only be either AMPylated or phosphocholinated at once, as only one or the other modification was iden-tified per Rab1 molecule (50). Phosphocholi-nation of Rab1 appears to promote its activity in a manner similar to AMPylation. Similarly to AMPylation, AnkX-mediated phospho-cholination can be reversed by a second

effector, Lem3 (51, 52). It was recently found that an endogenous host protein, transform-ing growth factor β-activated kinase (TAK1), regulates Rab1 by phosphorylation at the same site as modified by AnkX and SidM, suggesting that these T4SS effectors mimic the host's own regulatory mechanism to co-opt Rab1 function (53).

One puzzling observation is that despite the multiple effectors that modulate Rab1 ac-tivity during *Legionella* infection, Rab1 itself appears to be dispensable for intracellular rep-lication. A possible explanation for this find-ing is that a subset of effectors, such as SidM, AnkX, and LidA, appear to target multiple GTPases (44, 45, 50) and that these additional activities may act in parallel with perturba-tion of Rab1 function to promote the intra-cellular life cycle of *L. pneumophila*.

Tethering of Pathogen-Containing Vacuoles to the ER

Several intracellular pathogens, including *L. pneumophila*, *C. burnetii* (3, 4), *Anaplasma* spp. (5, 6), and *Chlamydia* spp., replicate in vacuoles that are closely associated to the ER but do not appear to fuse with it. This lifestyle is shared by *Simkania negevensis*, an organism related phylogenetically to *Chla-mydia* and that naturally infects amoebae and, similar to *L. pneumophila*, causes opportunis-tic respiratory tract infection (7, 8). While the effectors mediating association with *C. bur-netii*, *Anaplasma* spp., and *S. negevensis* with the ER remain to be identified, recent work has identified T3SS and T4SS effectors that promote association of vacuoles containing *C. trachomatis* and *L. pneumophila* with the ER (Fig. 2).

Chlamydia spp. are obligately intracellular pathogens that cause genital tract and ocular (*C. trachomatis*) or respiratory tract (*C. pneu-moniae* and *C. psittaci*) infections. Both path-ogens replicate within a vacuole termed an inclusion that has membrane contact sites with the ER. In *C. trachomatis*, the T3SS substrate IncD localizes to the inclusion membrane and

TABLE 1 Secreted pathogen effectors that localize to the ER or modulate its function

Pathogen (secretion system)	Effector	Activity	Reference(s)
Brucella abortus (VirB T4ASS)	BspB	Impairs ER-to-Golgi secretory trafficking; interacts with the COG complex in the Golgi and redirects membrane vesicles from the Golgi to Brucella vacuole	91
	VceC	Localizes to ER (ectopic expression); induces ER stress; interacts with BiP/GRP70	32–34
	BtpA (TcpB/Btp1)	Induces ER stress; binds microtubules; inhibits TLR signaling	75–78
	BspA	Unknown; localizes to ER on ectopic expression	35
	BspD	Unknown; localizes to ER on ectopic expression	35
	BspK	Unknown; localizes to ER on ectopic expression	35
Legionella pneumophila (Dot/Icm T4BSS)	Lgt1	Inhibits the IRE1 pathway of the UPR; inhibits translation elongation by glucosylation of eukaryotic elongation factor 1A	82–84
	Lgt2	Inhibits the IRE1 pathway of the UPR; inhibits translation elongation by glucosylation of eukaryotic elongation factor 1A	82, 84, 92
	SidE	Localizes to the cytoplasmic face of the LCV; regulates ER tubule rearrangement and recruitment of ER markers to the LCV via modulating ubiquitination	64, 93–95
	SidC	Ubiquitin ligase and PI4P binding activity; promotes the association of LCVs with the ER by recruiting ER vesicles	66, 68, 96
	SdeA, -B, -C	Promote ER reorganization by progressive ADP-ribosylation of ubiquitin and transfer of phosphoribosyl moiety to Rtn4	64, 94, 95, 97
	Ceg9	Tethers the LCV to the ER via association with Rtn4	61
	SidM/DrrA	Recruits Rab1 to LCV; acts as a GEF to recruit vacuoles to the LCV and as a GDF for Rab1; AMPylates Rab1	39, 41, 42, 47, 50, 98–100
	LidA	Interacts with GTP-Rab1 to maintain it in the active conformation	43, 45, 101
	SidD	Catalyzes AMP release from Rab1	47
	AnkX	Transfers phosphocholine to Rab1	50
	RalF	Acts as a GEF to activate ARF	37
	Lem3	Reverses activity of AnkX by removing phosphocholine from Rab1	52
	SetA	Glycosylates Rab1	102, 103
	LepB	Inactivates Rab1 via RabGAP activity; manipulates phosphoinositide composition of the Legionella-containing vacuole via phosphatidylinositide 4-kinase activity	48, 104, 105
	Lpg1137	Cleaves syntaxin 17 at the mitochondrion-associated ER membrane and blocks autophagy	106
	RavZ	Delipidates Atg8 (LC3-II) at the phagophore to inhibit autophagosome formation	70
	LpSpl	Sphingosine-1-phosphate lyase disrupts host sphingolipid biosynthesis and inhibits autophagy during infection	71

(Continued on next page)

TABLE 1 Secreted pathogen effectors that localize to the ER or modulate its function *(Continued)*

Pathogen (secretion system)	Effector	Activity	Reference(s)
Chlamydia trachomatis (T3SS)	CT229	*C. pneumoniae* homolog Cpn0585 recruits Rab1 from the ER to the inclusion membrane (effector that recruits Rab1 to *C. trachomatis* inclusion has yet to be identified)	107
	IncD	Mediates contact with the ER at MCS via binding to ceramide transfer protein CERT	54, 108
	IncV	Tethers the *C. trachomatis* inclusion to the ER via interactions with VAPs	55
	MrcA	Interacts with the Ca^{2+} channel inositol-1,4,5-trisphosphate receptor, type 3, to promote release of bacteria from infected cells	109
Coxiella burnetii (Dot/Icm T4BSS)	ElpA	Localizes to the ER on ectopic expression and blocks secretory traffic	59
Anaplasma spp. (VirB T4ASS)	Ats-1	Nucleates autophagosomes by interacting with Beclin and recruitment of DFCP1 and ATG14L to the pathogen-containing vacuole	5, 72
Orientia tsutsugamushi (T1SS)	Ank4	Interacts with eukaryotic chaperone Bat3 to transiently impede ER-associated protein degradation	9
	Ank9	Destabilizes the ER and Golgi by binding COPB2 and induces ATF4-dependent UPR	27
Rickettsia rickettsii (Rvh T4ASS)	RARP-2	Forms membranous structures in association with the ER; contributes to lysis of infected host cells.	11

mediates contact with the ER at membrane contact sites that also contain VAPA/B, the lipid transfer protein CERT, and the ER calcium sensor STIM1 (7, 54). A second T3SS effector, IncV, interacts with VAPA/B at the membrane contact sites between the inclusion and the ER (55). A *C. trachomatis incV* mutant exhibited decreased association of its inclusion with the ER but no overall intracellular growth defect, suggesting both the importance of IncV in ER tethering and the involvement of additional effectors in this process. Association of the chlamydial inclusion with the ER, especially with CERT, may promote acquisition of lipids to promote replication of *Chlamydia* either for nutrition or for expansion of the inclusion membrane during bacterial replication.

C. burnetii is a zoonotic pathogen that causes Q fever, which can manifest with both acute and chronic pathologies (56). It utilizes a T4SS related to the *Legionella* Dot/Icm apparatus to promote its intracellular repli-

cation (57, 58). Over 100 *C. burnetii* Dot/Icm substrates have been identified to date, but only a few have been characterized functionally (56). The vacuole containing *C. burnetii*, termed the parasitophorous vacuole, is decorated with calnexin and is tethered to the ER via membrane contacts that contain the host sterol-binding protein ORP1L, a protein that interacts with VAPA/B at ER MCS (4). While the effectors mediating this tethering remain to be identified, multiple T4SS effectors, including Cbu0372, Cbu1576, and ElpA, localize on ectopic expression to the ER (59, 60), and Cbu0635 interferes with the secretory pathway (58), suggesting that these and/or other T4SS effectors may play a role in interactions with the ER.

Recent evidence suggests tethering of vacuoles containing *Legionella* to the ER during the early stage of infection (Fig. 2). The Dot/Icm effector Ceg9 interacts with the ER protein reticulon 4 (Rtn4) shortly after uptake of bacteria, suggesting that recruit-

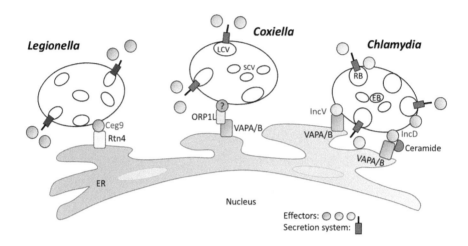

FIGURE 2 Role of pathogen effectors in tethering of pathogen-containing vacuoles to the ER. Abbreviations: Rtn4, reticulon 4; LCV, *Coxiella* large-cell variant; SCV, *Coxiella* small-cell variant; RB, *Chlamydia* reticulate body; EB, *Chlamydia* elementary body.

ment of the ER helps to develop the replicative niche for *L. pneumophila* (61). In the host cell, Rtn4 helps generate the tubular morphology that is characteristic of the peripheral ER (62) and participates in the formation of plasma membrane-ER MCS that function in cellular Ca^{2+} homeostasis (63). Like Rab1, Rtn4 is targeted by multiple Dot/Icm effectors, including the SdeA to -C proteins, which modulate Rtn4 function via ubiquitination (64). Intriguingly, the Sde proteins perform a sequential set of reactions to transfer ubiquitin to Rtn4:ADP-ribosyl transfer to ubiquitin, followed by a nucleotidase/phosphohydrolase reaction that removes AMP and transfers phosphoribosylated ubiquitin to Rtn4 (64). The early targeting of Rtn4 after *L. pneumophila* entry to the cell and the localization of Rtn4 to plasma membrane-ER MCS raise the possibility that *L. pneumophila* could co-opt these MCS early during cellular infection to associate with the ER. Another Dot/Icm substrate involved in tethering the LCV to the ER is SidC, which is anchored to the

cytosolic face of the LCV via binding of PI4P (65–67). Recruitment of ER proteins to the LCV by SidC requires a ubiquitin ligase activity in its N terminus (68). Taken together, these findings suggest that pathogens use multiple strategies to tether their replicative vacuoles to the ER.

Subversion of Autophagy Initiation at the ER

The *L. pneumophila* T4SS effector protein RavZ is secreted from the LCV and targets to omegasomes via its ability to interact with the lipid phosphatidylinositol-3-phosphate, which is enriched at these sites (69). There, the cysteine protease activity of RavZ irreversibly deconjugates lipids from ATG8 proteins (LC3-II) in the early-stage autophagosomal structures. As a result, the biogenesis of autophagosomes at the ER is inhibited (70). Since a *ravZ* mutant does not have a replication defect (70), it is unknown whether this activity promotes intracellular replication of *L. pneumophila*; however, RavZ may act in concert with other effectors, such as the inhibitor of sphingolipid biosynthesis LpSpl (71), to modulate autophagy. In contrast to *L. pneumophila*, *Anaplasma phagocytophilum*, which replicates within neutrophils, activates autophagy via a T4SS effector, Ats-1, to promote its replication (5). Ats-1 has an N-terminal domain that nucleates autophagosomes by interacting with Beclin 1, a protein crucial to initiation of autophagy, to recruit the ER-localized autophagy initiation proteins ATG14L and DFCP1 to the *A. phagocytophilum* inclusion (5). This subversion of autophagy initiation recruits autophagosomes to the inclusion, effectively delivering nutrients for intracellular replication of *A. phagocytophilum* (72).

Effector Modulation of the ER UPR

The unfolded-protein response (UPR) is a response to perturbation of ER function (broadly termed ER stress) that is initially cytoprotective and promotes return to homeostasis but can lead to apoptosis in the case of unresolved stress. The cellular response to ER stress is transmitted via three membrane sensors, IRE1α (inositol-requiring enzyme 1), ATF6 (activating transcription factor 6), and PERK (protein kinase RNA-like ER kinase). This response is linked to innate immunity via signaling through cytosolic pathways (reviewed in references 73 and 74). Secreted effectors have been identified that both induce and inhibit the UPR during intracellular infection. The *B. abortus* T4SS effector VceC localizes on ectopic expression to the ER, and during infection it activates IRE1α, initiating a proinflammatory arm of the UPR (32, 34) that activates NF-κB. This response could be beneficial to *B. abortus* in the bovine placenta, as placental inflammation in this context triggers abortion, driving transmission of the pathogen in its natural reservoir (32, 34). A second T4SS effector, BtpA (also called TcpB or Btp1) (75), triggers all three branches of the UPR during *Brucella*

melitensis infection (76), but rather than localizing to the ER, BtpA binds microtubules (77, 78). It is not known how microtubule stabilization by BtpA links to UPR induction, but potential mechanisms could include altering interaction of integral ER membrane proteins with microtubules or effects on microtubule-dependent vesicular transport in the secretory pathway (79).

Like *B. abortus*, *O. tsutsugamushi*, the obligate intracellular agent of scrub typhus, activates the UPR (9, 27). Two T1SS-secreted effectors have been implicated (Table 1): Ank4, an ankyrin repeat protein, interacts at the cytosolic face of the ER with Bat3, a host cytosolic chaperone involved in ERAD (80), to inhibit UPR-induced ERAD during the early (nonreplicative) phase of *O. tsutsugamushi* infection. Later, Ank4 expression is downregulated, which releases repression of ERAD, making amino acids available for intracellular replication of *O. tsutsugamushi*, which provides an important source of nutrition, since this bacterium is auxotrophic for several amino acids (81). Ectopically expressed Ank9 binds the Golgi protein COPB2, involved in Golgi-to-ER vesicular trafficking, and Ank9 also traffics from Golgi to the ER, where it disrupts organelle morphology and induces the UPR. Ectopic Ank9 expression phenocopies disruption of the Golgi and ER, as well as inhibition of protein secretion, observed in cells infected with *O. tsutsugamushi* (27).

L. pneumophila activates the UPR at the transcriptional level (82) but suppresses the downstream translation of UPR target transcripts by translocating five T4SS effectors (Lgt1 to -3, SidI, and SidL) that inhibit translation elongation. The *L. pneumophila* effectors function by glycosylation of a conserved serine residue in host elongation factor 1A (83). Translation inhibition effectively reduces the basal load of protein entering the ER for protein folding, which is a physiologic activator of IRE1α (82, 84); thus, the outcome of this interaction is to inhibit the IRE1α pathway. The ability to block IRE1α signaling is shared by *S. negevensis*, which encodes both the T3SS and T4SS in its genome; however, the effectors that mediate this activity remain to be discovered (8). Blockade of the IRE1α pathway may be beneficial in the context of bacterial infection either to reduce innate immune signaling downstream of this pathway or to block induction of apoptosis in response to uncontrolled ER stress. Interestingly, blockade of the UPR is a strategy shared by viral pathogens, which, via their subversion of the ER for production of virions, trigger ER stress (85).

CONCLUSIONS AND PERSPECTIVE

The biosynthetic capacity of the ER and its extensive network of contacts with other cellular organelles make it a logical target for exploitation by T3SSs and T4SSs of different intracellular pathogens. Pathogens such as *L. pneumophila* and *C. burnetii* have dedicated a substantial part of their genome coding capacities to secreted effectors that modulate their host cells, which highlights the importance of these interactions to their biology (56, 86, 87). Recent progress in understanding the biology of T3SS and T4SS effectors has revealed novel mechanisms utilized by intracellular bacteria to co-opt multiple functions of the ER, including protein secretion, lipid biosynthesis, membrane tethering, and autophagy initiation, to promote their replication. While our understanding of how individual effectors modulate ER function is growing, for the majority of effectors, the molecular mechanisms of action remain unknown.

One of the challenges to identifying effector functions and understanding their roles in the context of infection has been the redundancy of effector function. However, elegant approaches have been employed in *C. trachomatis* and *Mycobacterium tuberculosis* to generate interaction networks (88, 89) for secreted effectors that, together with newly developed methodologies for genetic manipu-

lation of the obligate intracellular pathogens, will facilitate functional and mechanistic studies of effector proteins and uncover new strategies by which they manipulate ER biology.

CITATION

Tsai AY, English BC, Tsolis RM. 2019. Hostile takeover: hijacking of endoplasmic reticulum function by T4SS and T3SS effectors creates a niche for intracellular pathogens. Microbiol Spectrum 7(3):PSIB-0027-2019.

REFERENCES

1. **Anderson TD, Cheville NF.** 1986. Ultrastructural morphometric analysis of *Brucella abortus*-infected trophoblasts in experimental placentitis. Bacterial replication occurs in rough endoplasmic reticulum. *Am J Pathol* **124**:226–237.
2. **Swanson MS, Isberg RR.** 1995. Association of *Legionella pneumophila* with the macrophage endoplasmic reticulum. *Infect Immun* **63**:3609–3620.
3. **Campoy EM, Zoppino FC, Colombo MI.** 2011. The early secretory pathway contributes to the growth of the *Coxiella*-replicative niche. *Infect Immun* **79**:402–413.
4. **Justis AV, Hansen B, Beare PA, King KB, Heinzen RA, Gilk SD.** 2017. Interactions between the *Coxiella burnetii* parasitophorous vacuole and the endoplasmic reticulum involve the host protein ORP1L. *Cell Microbiol* **19**:e12637.
5. **Niu H, Xiong Q, Yamamoto A, Hayashi-Nishino M, Rikihisa Y.** 2012. Autophagosomes induced by a bacterial Beclin 1 binding protein facilitate obligatory intracellular infection. *Proc Natl Acad Sci U S A* **109**:20800–20807.
6. **Truchan HK, Cockburn CL, Hebert KS, Magunda F, Noh SM, Carlyon JA.** 2016. The pathogen-occupied vacuoles of *Anaplasma phagocytophilum* and *Anaplasma marginale* interact with the endoplasmic reticulum. *Front Cell Infect Microbiol* **6**:22.
7. **Derré I.** 2015. Chlamydiae interaction with the endoplasmic reticulum: contact, function and consequences. *Cell Microbiol* **17**:959–966.
8. **Mehlitz A, Karunakaran K, Herweg JA, Krohne G, van de Linde S, Rieck E, Sauer M, Rudel T.** 2014. The chlamydial organism *Simkania negevensis* forms ER vacuole contact

sites and inhibits ER-stress. *Cell Microbiol* **16:**1224–1243.

9. **Rodino KG, VieBrock L, Evans SM, Ge H, Richards AL, Carlyon JA.** 2017. *Orientia tsutsugamushi* modulates endoplasmic reticulum-associated degradation to benefit its growth. *Infect Immun* **86:**e00596-17.

10. **VieBrock L, Evans SM, Beyer AR, Larson CL, Beare PA, Ge H, Singh S, Rodino KG, Heinzen RA, Richards AL, Carlyon JA.** 2015. *Orientia tsutsugamushi* ankyrin repeat-containing protein family members are type 1 secretion system substrates that traffic to the host cell endoplasmic reticulum. *Front Cell Infect Microbiol* **4:**186.

11. **Lehman SS, Noriea NF, Aistleitner K, Clark TR, Dooley CA, Nair V, Kaur SJ, Rahman MS, Gillespie JJ, Azad AF, Hackstadt T.** 2018. The rickettsial ankyrin repeat protein 2 is a type IV secreted effector that associates with the endoplasmic reticulum. *mBio* **9:**e00975-18.

12. **McCaffrey K, Braakman I.** 2016. Protein quality control at the endoplasmic reticulum. *Essays Biochem* **60:**227–235.

13. **Helenius A, Aebi M.** 2004. Roles of N-linked glycans in the endoplasmic reticulum. *Annu Rev Biochem* **73:**1019–1049.

14. **Preston GM, Brodsky JL.** 2017. The evolving role of ubiquitin modification in endoplasmic reticulum-associated degradation. *Biochem J* **474:**445–469.

15. **Jacquemyn J, Cascalho A, Goodchild RE.** 2017. The ins and outs of endoplasmic reticulum-controlled lipid biosynthesis. *EMBO Rep* **18:**1905–1921.

16. **Toledo A, Benach JL.** 2015. Hijacking and use of host lipids by intracellular pathogens. *Microbiol Spectr* **3:**VMBF-0001-2014.

17. **Kim PK, Mullen RT, Schumann U, Lippincott-Schwartz J.** 2006. The origin and maintenance of mammalian peroxisomes involves a de novo PEX16-dependent pathway from the ER. *J Cell Biol* **173:**521–532.

18. **Joshi AS, Zhang H, Prinz WA.** 2017. Organelle biogenesis in the endoplasmic reticulum. *Nat Cell Biol* **19:**876–882.

19. **Wu H, Carvalho P, Voeltz GK.** 2018. Here, there, and everywhere: the importance of ER membrane contact sites. *Science* **361:**eaan5835.

20. **Friedman JR, Lackner LL, West M, DiBenedetto JR, Nunnari J, Voeltz GK.** 2011. ER tubules mark sites of mitochondrial division. *Science* **334:**358–362.

21. **Cárdenas C, Miller RA, Smith I, Bui T, Molgó J, Müller M, Vais H, Cheung KH, Yang J, Parker I, Thompson CB, Birnbaum MJ, Hallows KR, Foskett JK.** 2010. Essential regulation of cell bioenergetics by constitutive InsP3 receptor Ca2+ transfer to mitochondria. *Cell* **142:**270–283.

22. **Rizzuto R, Pinton P, Carrington W, Fay FS, Fogarty KE, Lifshitz LM, Tuft RA, Pozzan T.** 1998. Close contacts with the endoplasmic reticulum as determinants of mitochondrial Ca2+ responses. *Science* **280:**1763–1766.

23. **Friedman JR, Dibenedetto JR, West M, Rowland AA, Voeltz GK.** 2013. Endoplasmic reticulum-endosome contact increases as endosomes traffic and mature. *Mol Biol Cell* **24:**1030–1040.

24. **Miller C, Celli J.** 2016. Avoidance and subversion of eukaryotic homeostatic autophagy mechanisms by bacterial pathogens. *J Mol Biol* **428:**3387–3398.

25. **Axe EL, Walker SA, Manifava M, Chandra P, Roderick HL, Habermann A, Griffiths G, Ktistakis NT.** 2008. Autophagosome formation from membrane compartments enriched in phosphatidylinositol 3-phosphate and dynamically connected to the endoplasmic reticulum. *J Cell Biol* **182:**685–701.

26. **Sedzicki J, Tschon T, Low SH, Willemart K, Goldie KN, Letesson JJ, Stahlberg H, Dehio C.** 2018. 3D correlative electron microscopy reveals continuity of *Brucella*-containing vacuoles with the endoplasmic reticulum. *J Cell Sci* **131:**jcs210799.

27. **Beyer AR, Rodino KG, VieBrock L, Green RS, Tegels BK, Oliver LD Jr, Marconi RT, Carlyon JA.** 2017. *Orientia tsutsugamushi* Ank9 is a multifunctional effector that utilizes a novel GRIP-like Golgi localization domain for Golgi-to-endoplasmic reticulum trafficking and interacts with host COPB2. *Cell Microbiol* **19:**e12727.

28. **Dehio C, Tsolis RM.** 2017. Type IV effector secretion and subversion of host functions by *Bartonella* and *Brucella* species. *Curr Top Microbiol Immunol* **413:**269–295.

29. **Celli J.** 2015. The changing nature of the *Brucella*-containing vacuole. *Cell Microbiol* **17:**951–958.

30. **Celli J, Salcedo SP, Gorvel JP.** 2005. Brucella coopts the small GTPase Sar1 for intracellular replication. *Proc Natl Acad Sci U S A* **102:**1673–1678.

31. **Fugier E, Salcedo SP, de Chastellier C, Pophillat M, Muller A, Arce-Gorvel V, Fourquet P, Gorvel JP.** 2009. The glyceraldehyde-3-phosphate dehydrogenase and the small GTPase Rab 2 are crucial for *Brucella* replication. *PLoS Pathog* **5:**e1000487.

32. **de Jong MF, Starr T, Winter MG, den Hartigh AB, Child R, Knodler LA, van Dijl**

JM, Celli J, Tsolis RM. 2013. Sensing of bacterial type IV secretion via the unfolded protein response. *mBio* 4:e00418-12.

33. de Jong MF, Sun YH, den Hartigh AB, van Dijl JM, Tsolis RM. 2008. Identification of VceA and VceC, two members of the VjbR regulon that are translocated into macrophages by the *Brucella* type IV secretion system. *Mol Microbiol* 70:1378–1396.

34. Keestra-Gounder AM, Byndloss MX, Seyffert N, Young BM, Chávez-Arroyo A, Tsai AY, Cevallos SA, Winter MG, Pham OH, Tiffany CR, de Jong MF, Kerrinnes T, Ravindran R, Luciw PA, McSorley SJ, Bäumler AJ, Tsolis RM. 2016. NOD1 and NOD2 signalling links ER stress with inflammation. *Nature* 532:394–397.

35. Myeni S, Child R, Ng TW, Kupko JJ III, Wehrly TD, Porcella SF, Knodler LA, Celli J. 2013. Brucella modulates secretory trafficking via multiple type IV secretion effector proteins. *PLoS Pathog* 9:e1003556.

36. Sherwood RK, Roy CR. 2016. Autophagy evasion and endoplasmic reticulum subversion: the yin and yang of *Legionella* intracellular infection. *Annu Rev Microbiol* 70:413–433.

37. Nagai H, Kagan JC, Zhu X, Kahn RA, Roy CR. 2002. A bacterial guanine nucleotide exchange factor activates ARF on *Legionella* phagosomes. *Science* 295:679–682.

38. Stenmark H. 2009. Rab GTPases as coordinators of vesicle traffic. *Nat Rev Mol Cell Biol* 10:513–525.

39. Machner MP, Isberg RR. 2006. Targeting of host Rab GTPase function by the intravacuolar pathogen *Legionella pneumophila*. *Dev Cell* 11:47–56.

40. Neunuebel MR, Machner MP. 2012. The taming of a Rab GTPase by *Legionella pneumophila*. *Small GTPases* 3:28–33.

41. Brombacher E, Urwyler S, Ragaz C, Weber SS, Kami K, Overduin M, Hilbi H. 2009. Rab1 guanine nucleotide exchange factor SidM is a major phosphatidylinositol 4-phosphate-binding effector protein of *Legionella pneumophila*. *J Biol Chem* 284:4846–4856.

42. Müller MP, Peters H, Blümer J, Blankenfeldt W, Goody RS, Itzen A. 2010. The *Legionella* effector protein DrrA AMPylates the membrane traffic regulator Rab1b. *Science* 329:946–949.

43. Derré I, Isberg RR. 2005. LidA, a translocated substrate of the *Legionella pneumophila* type IV secretion system, interferes with the early secretory pathway. *Infect Immun* 73:4370–4380.

44. So EC, Schroeder GN, Carson D, Mattheis C, Mousnier A, Broncel M, Tate EW, Frankel G.

2016. The Rab-binding profiles of bacterial virulence factors during infection. *J Biol Chem* 291:5832–5843.

45. Schoebel S, Cichy AL, Goody RS, Itzen A. 2011. Protein LidA from *Legionella* is a Rab GTPase supereffector. *Proc Natl Acad Sci U S A* 108:17945–17950.

46. Arasaki K, Toomre DK, Roy CR. 2012. The *Legionella pneumophila* effector DrrA is sufficient to stimulate SNARE-dependent membrane fusion. *Cell Host Microbe* 11:46–57.

47. Neunuebel MR, Chen Y, Gaspar AH, Backlund PS Jr, Yergey A, Machner MP. 2011. De-AMPylation of the small GTPase Rab1 by the pathogen *Legionella pneumophila*. *Science* 333:453–456.

48. Ingmundson A, Delprato A, Lambright DG, Roy CR. 2007. *Legionella pneumophila* proteins that regulate Rab1 membrane cycling. *Nature* 450:365–369.

49. Pan X, Lührmann A, Satoh A, Laskowski-Arce MA, Roy CR. 2008. Ankyrin repeat proteins comprise a diverse family of bacterial type IV effectors. *Science* 320:1651–1654.

50. Mukherjee S, Liu X, Arasaki K, McDonough J, Galán JE, Roy CR. 2011. Modulation of Rab GTPase function by a protein phosphocholine transferase. *Nature* 477:103–106.

51. Goody PR, Heller K, Oesterlin LK, Müller MP, Itzen A, Goody RS. 2012. Reversible phosphocholination of Rab proteins by *Legionella pneumophila* effector proteins. *EMBO J* 31:1774–1784.

52. Tan Y, Arnold RJ, Luo ZQ. 2011. *Legionella pneumophila* regulates the small GTPase Rab1 activity by reversible phosphorylcholination. *Proc Natl Acad Sci U S A* 108:21212–21217.

53. Levin RS, Hertz NT, Burlingame AL, Shokat KM, Mukherjee S. 2016. Innate immunity kinase TAK1 phosphorylates Rab1 on a hotspot for posttranslational modifications by host and pathogen. *Proc Natl Acad Sci U S A* 113:E4776–E4783.

54. Derré I, Swiss R, Agaisse H. 2011. The lipid transfer protein CERT interacts with the *Chlamydia* inclusion protein IncD and participates to ER-*Chlamydia* inclusion membrane contact sites. *PLoS Pathog* 7:e1002092.

55. Stanhope R, Flora E, Bayne C, Derré I. 2017. IncV, a FFAT motif-containing *Chlamydia* protein, tethers the endoplasmic reticulum to the pathogen-containing vacuole. *Proc Natl Acad Sci U S A* 114:12039–12044.

56. van Schaik EJ, Chen C, Mertens K, Weber MM, Samuel JE. 2013. Molecular pathogenesis of the obligate intracellular bacterium *Coxiella burnetii*. *Nat Rev Microbiol* 11:561–573.

57. **Beare PA, Gilk SD, Larson CL, Hill J, Stead CM, Omsland A, Cockrell DC, Howe D, Voth DE, Heinzen RA.** 2011. Dot/Icm type IVB secretion system requirements for *Coxiella burnetii* growth in human macrophages. *mBio* 2:e00175-11.

58. **Carey KL, Newton HJ, Lührmann A, Roy CR.** 2011. The *Coxiella burnetii* Dot/Icm system delivers a unique repertoire of type IV effectors into host cells and is required for intracellular replication. *PLoS Pathog* 7:e1002056.

59. **Graham JG, Winchell CG, Sharma UM, Voth DE.** 2015. Identification of ElpA, a *Coxiella burnetii* pathotype-specific Dot/Icm type IV secretion system substrate. *Infect Immun* 83: 1190–1198.

60. **Weber MM, Chen C, Rowin K, Mertens K, Galvan G, Zhi H, Dealing CM, Roman VA, Banga S, Tan Y, Luo ZQ, Samuel JE.** 2013. Identification of *Coxiella burnetii* type IV secretion substrates required for intracellular replication and *Coxiella*-containing vacuole formation. *J Bacteriol* 195:3914–3924.

61. **Haenssler E, Ramabhadran V, Murphy CS, Heidtman MI, Isberg RR.** 2015. Endoplasmic reticulum tubule protein reticulon 4 associates with the *Legionella pneumophila* vacuole and with translocated substrate Ceg9. *Infect Immun* 83:3479–3489.

62. **Voeltz GK, Prinz WA, Shibata Y, Rist JM, Rapoport TA.** 2006. A class of membrane proteins shaping the tubular endoplasmic reticulum. *Cell* 124:573–586.

63. **Jozsef L, Tashiro K, Kuo A, Park EJ, Skoura A, Albinsson S, Rivera-Molina F, Harrison KD, Iwakiri Y, Toomre D, Sessa WC.** 2014. Reticulon 4 is necessary for endoplasmic reticulum tubulation, STIM1-Orai1 coupling, and store-operated calcium entry. *J Biol Chem* 289:9380–9395.

64. **Kotewicz KM, Ramabhadran V, Sjoblom N, Vogel JP, Haenssler E, Zhang M, Behringer J, Scheck RA, Isberg RR.** 2017. A single *Legionella* effector catalyzes a multistep ubiquitination pathway to rearrange tubular endoplasmic reticulum for replication. *Cell Host Microbe* 21:169–181.

65. **Luo ZQ, Isberg RR.** 2004. Multiple substrates of the *Legionella pneumophila* Dot/Icm system identified by interbacterial protein transfer. *Proc Natl Acad Sci U S A* 101:841–846.

66. **Ragaz C, Pietsch H, Urwyler S, Tiaden A, Weber SS, Hilbi H.** 2008. The *Legionella pneumophila* phosphatidylinositol-4 phosphate-binding type IV substrate SidC recruits endoplasmic reticulum vesicles to a replication-permissive vacuole. *Cell Microbiol* 10:2416–2433.

67. **Weber SS, Ragaz C, Reus K, Nyfeler Y, Hilbi H.** 2006. *Legionella pneumophila* exploits PI(4) P to anchor secreted effector proteins to the replicative vacuole. *PLoS Pathog* 2:e46.

68. **Hsu F, Luo X, Qiu J, Teng YB, Jin J, Smolka MB, Luo ZQ, Mao Y.** 2014. The *Legionella* effector SidC defines a unique family of ubiquitin ligases important for bacterial phagosomal remodeling. *Proc Natl Acad Sci U S A* 111:10538–10543.

69. **Horenkamp FA, Kauffman KJ, Kohler LJ, Sherwood RK, Krueger KP, Shteyn V, Roy CR, Melia TJ, Reinisch KM.** 2015. The *Legionella* anti-autophagy effector RavZ targets the autophagosome via PI3P- and curvature-sensing motifs. *Dev Cell* 34:569–576.

70. **Choy A, Dancourt J, Mugo B, O'Connor TJ, Isberg RR, Melia TJ, Roy CR.** 2012. The *Legionella* effector RavZ inhibits host autophagy through irreversible Atg8 deconjugation. *Science* 338:1072–1076.

71. **Rolando M, Escoll P, Nora T, Botti J, Boitez V, Bedia C, Daniels C, Abraham G, Stogios PJ, Skarina T, Christophe C, Dervins-Ravault D, Cazalet C, Hilbi H, Rupasinghe TW, Tull D, McConville MJ, Ong SY, Hartland EL, Codogno P, Levade T, Naderer T, Savchenko A, Buchrieser C.** 2016. *Legionella pneumophila* S1P-lyase targets host sphingolipid metabolism and restrains autophagy. *Proc Natl Acad Sci U S A* 113:1901–1906.

72. **Rikihisa Y.** 2017. Role and function of the type IV secretion system in *Anaplasma* and *Ehrlichia* species. *Curr Top Microbiol Immunol* 413:297–321.

73. **Byndloss MX, Keestra-Gounder AM, Bäumler AJ, Tsolis RM.** 2016. NOD1 and NOD2: new functions linking endoplasmic reticulum stress and inflammation. *DNA Cell Biol* 35:311–313.

74. **Celli J, Tsolis RM.** 2015. Bacteria, the endoplasmic reticulum and the unfolded protein response: friends or foes? *Nat Rev Microbiol* 13:71–82.

75. **Salcedo SP, Marchesini MI, Lelouard H, Fugier E, Jolly G, Balor S, Muller A, Lapaque N, Demaria O, Alexopoulou L, Comerci DJ, Ugalde RA, Pierre P, Gorvel JP.** 2008. *Brucella* control of dendritic cell maturation is dependent on the TIR-containing protein Btp1. *PLoS Pathog* 4:e21.

76. **Smith JA, Khan M, Magnani DD, Harms JS, Durward M, Radhakrishnan GK, Liu YP, Splitter GA.** 2013. *Brucella* induces an unfolded protein response via TcpB that supports intracellular replication in macrophages. *PLoS Pathog* 9:e1003785.

77. **Felix C, Kaplan Türköz B, Ranaldi S, Koelblen T, Terradot L, O'Callaghan D,**

Vergunst AC. 2014. The *Brucella* TIR domain containing proteins BtpA and BtpB have a structural WxxxE motif important for protection against microtubule depolymerisation. *Cell Commun Signal* **12**:53.

78. **Radhakrishnan GK, Harms JS, Splitter GA.** 2011. Modulation of microtubule dynamics by a TIR domain protein from the intracellular pathogen *Brucella melitensis*. *Biochem J* **439**: 79–83.

79. **Gurel PS, Hatch AL, Higgs HN.** 2014. Connecting the cytoskeleton to the endoplasmic reticulum and Golgi. *Curr Biol* **24**:R660–R672.

80. **Claessen JH, Ploegh HL.** 2011. BAT3 guides misfolded glycoproteins out of the endoplasmic reticulum. *PLoS One* **6**:e28542.

81. **Min CK, Yang JS, Kim S, Choi MS, Kim IS, Cho NH.** 2008. Genome-based construction of the metabolic pathways of *Orientia tsutsugamushi* and comparative analysis within the *Rickettsiales* order. *Comp Funct Genomics* **2008**:623145.

82. **Treacy-Abarca S, Mukherjee S.** 2015. *Legionella* suppresses the host unfolded protein response via multiple mechanisms. *Nat Commun* **6**:7887.

83. **Belyi Y, Niggeweg R, Opitz B, Vogelsgesang M, Hippenstiel S, Wilm M, Aktories K.** 2006. *Legionella pneumophila* glucosyltransferase inhibits host elongation factor 1A. *Proc Natl Acad Sci U S A* **103**:16953–16958.

84. **Hempstead AD, Isberg RR.** 2015. Inhibition of host cell translation elongation by *Legionella pneumophila* blocks the host cell unfolded protein response. *Proc Natl Acad Sci U S A* **112**: E6790–E6797.

85. **Smith JA.** 2014. A new paradigm: innate immune sensing of viruses via the unfolded protein response. *Front Microbiol* **5**:222.

86. **Gomez-Valero L, Rusniok C, Carson D, Mondino S, Pérez-Cobas AE, Rolando M, Pasricha S, Reuter S, Demirtas J, Crumbach J, Descorps-Declere S, Hartland EL, Jarraud S, Dougan G, Schroeder GN, Frankel G, Buchrieser C.** 2019. More than 18,000 effectors in the *Legionella* genus genome provide multiple, independent combinations for replication in human cells. *Proc Natl Acad Sci U S A* **116**:2265–2273.

87. **Ensminger AW.** 2016. *Legionella pneumophila*, armed to the hilt: justifying the largest arsenal of effectors in the bacterial world. *Curr Opin Microbiol* **29**:74–80.

88. **Mirrashidi KM, Elwell CA, Verschueren E, Johnson JR, Frando A, Von Dollen J, Rosenberg O, Gulbahce N, Jang G, Johnson T, Jäger S, Gopalakrishnan AM, Sherry J,** **Dunn JD, Olive A, Penn B, Shales M, Cox JS, Starnbach MN, Derre I, Valdivia R, Krogan NJ, Engel J.** 2015. Global mapping of the Inc-human interactome reveals that retromer restricts *Chlamydia* infection. *Cell Host Microbe* **18**:109–121.

89. **Penn BH, Netter Z, Johnson JR, Von Dollen J, Jang GM, Johnson T, Ohol YM, Maher C, Bell SL, Geiger K, Golovkine G, Du X, Choi A, Parry T, Mohapatra BC, Storck MD, Band H, Chen C, Jager S, Shales M, Portnoy DA, Hernandez R, Coscoy L, Cox JS, Krogan NJ.** 2018. An Mtb-human protein-protein interaction map identifies a switch between host antiviral and antibacterial responses. *Mol Cell* **71**:637–648.e635.

90. **Bärlocher K, Welin A, Hilbi H.** 2017. Formation of the *Legionella* replicative compartment at the crossroads of retrograde trafficking. *Front Cell Infect Microbiol* **7**:482.

91. **Miller CN, Smith EP, Cundiff JA, Knodler LA, Bailey Blackburn J, Lupashin V, Celli J.** 2017. A *Brucella* type IV effector targets the COG tethering complex to remodel host secretory traffic and promote intracellular replication. *Cell Host Microbe* **22**:317–329.e317.

92. **Belyi Y, Tabakova I, Stahl M, Aktories K.** 2008. Lgt: a family of cytotoxic glucosyltransferases produced by *Legionella pneumophila*. *J Bacteriol* **190**:3026–3035.

93. **Jeong KC, Sexton JA, Vogel JP.** 2015. Spatiotemporal regulation of a *Legionella pneumophila* T4SS substrate by the metaeffector SidJ. *PLoS Pathog* **11**:e1004695.

94. **Qiu J, Sheedlo MJ, Yu K, Tan Y, Nakayasu ES, Das C, Liu X, Luo ZQ.** 2016. Ubiquitination independent of E1 and E2 enzymes by bacterial effectors. *Nature* **533**:120–124.

95. **Sheedlo MJ, Qiu J, Tan Y, Paul LN, Luo ZQ, Das C.** 2015. Structural basis of substrate recognition by a bacterial deubiquitinase important for dynamics of phagosome ubiquitination. *Proc Natl Acad Sci U S A* **112**:15090–15095.

96. **Weber S, Wagner M, Hilbi H.** 2014. Live-cell imaging of phosphoinositide dynamics and membrane architecture during *Legionella* infection. *mBio* **5**:e00839-13.

97. **Bardill JP, Miller JL, Vogel JP.** 2005. IcmS-dependent translocation of SdeA into macrophages by the *Legionella pneumophila* type IV secretion system. *Mol Microbiol* **56**:90–103.

98. **Hubber A, Arasaki K, Nakatsu F, Hardiman C, Lambright D, De Camilli P, Nagai H, Roy CR.** 2014. The machinery at endoplasmic reticulum-plasma membrane contact sites contributes to spatial regulation of multiple *Legionella* effector proteins. *PLoS Pathog* **10**:e1004222.

99. **Machner MP, Isberg RR.** 2007. A bifunctional bacterial protein links GDI displacement to Rab1 activation. *Science* **318:**974–977.

100. **Murata T, Delprato A, Ingmundson A, Toomre DK, Lambright DG, Roy CR.** 2006. The *Legionella pneumophila* effector protein DrrA is a Rab1 guanine nucleotide-exchange factor. *Nat Cell Biol* **8:**971–977.

101. **Neunuebel MR, Mohammadi S, Jarnik M, Machner MP.** 2012. *Legionella pneumophila* LidA affects nucleotide binding and activity of the host GTPase Rab1. *J Bacteriol* **194:**1389–1400.

102. **Wang Z, McCloskey A, Cheng S, Wu M, Xue C, Yu Z, Fu J, Liu Y, Luo ZQ, Liu X.** 2018. Regulation of the small GTPase Rab1 function by a bacterial glucosyltransferase. *Cell Discov* **4:**53.

103. **Heidtman M, Chen EJ, Moy MY, Isberg RR.** 2009. Large-scale identification of *Legionella pneumophila* Dot/Icm substrates that modulate host cell vesicle trafficking pathways. *Cell Microbiol* **11:**230–248.

104. **Chen J, Reyes M, Clarke M, Shuman HA.** 2007. Host cell-dependent secretion and translocation of the LepA and LepB effectors of *Legionella pneumophila*. *Cell Microbiol* **9:**1660–1671.

105. **Dong N, Niu M, Hu L, Yao Q, Zhou R, Shao F.** 2016. Modulation of membrane phosphoinositide dynamics by the phosphatidylinositide 4-kinase activity of the *Legionella* LepB effector. *Nat Microbiol* **2:**16236.

106. **Arasaki K, Mikami Y, Shames SR, Inoue H, Wakana Y, Tagaya M.** 2017. *Legionella* effector Lpg1137 shuts down ER-mitochondria communication through cleavage of syntaxin 17. *Nat Commun* **8:**15406.

107. **Cortes C, Rzomp KA, Tvinnereim A, Scidmore MA, Wizel B.** 2007. *Chlamydia pneumoniae* inclusion membrane protein Cpn0585 interacts with multiple Rab GTPases. *Infect Immun* **75:**5586–5596.

108. **Agaisse H, Derré I.** 2014. Expression of the effector protein IncD in *Chlamydia trachomatis* mediates recruitment of the lipid transfer protein CERT and the endoplasmic reticulum-resident protein VAPB to the inclusion membrane. *Infect Immun* **82:**2037–2047.

109. **Nguyen PH, Lutter EI, Hackstadt T.** 2018. *Chlamydia trachomatis* inclusion membrane protein MrcA interacts with the inositol 1,4,5-trisphosphate receptor type 3 (ITPR3) to regulate extrusion formation. *PLoS Pathog* **14:**e1006911.

Type V Secretion in Gram-Negative Bacteria

24

HARRIS D. BERNSTEIN[1]

INTRODUCTION

Type V, or "autotransporter," secretion is an umbrella term that is often used to refer to a group of distinct but conceptually related protein export pathways that are widely distributed in Gram-negative bacteria. Autotransporters are generally single polypeptides that contain a signal peptide that promotes translocation across the inner membrane (IM) via the Sec pathway, an extracellular ("passenger") domain, and a domain that anchors the protein to the outer membrane (OM). Passenger domains have a wide variety of functions, but they often promote virulence (1). In the archetypical, or "classical" (type Va), autotransporter pathway, which was discovered in 1987, the passenger domain is located at the N terminus of the protein adjacent to the signal peptide (2). Although passenger domains range in size from ~20 to 300 kDa and are highly diverse in sequence (3), X-ray crystallographic and *in silico* studies predict that they usually fold into a repetitive structure known as a β helix (4–8) (Fig. 1). The membrane anchor domains are ~30 kDa and are also highly diverse in sequence but contain short conserved sequence motifs (3, 9). Like most membrane-spanning segments associated with OM proteins (OMPs), these domains fold into a

[1]Genetics and Biochemistry Branch, National Institute of Diabetes and Digestive and Kidney Diseases, National Institutes of Health, Bethesda, MD 20892

Protein Secretion in Bacteria
Edited by Maria Sandkvist, Eric Cascales, and Peter J. Christie
© 2019 American Society for Microbiology, Washington, DC
doi:10.1128/ecosalplus.ESP-0031-2018

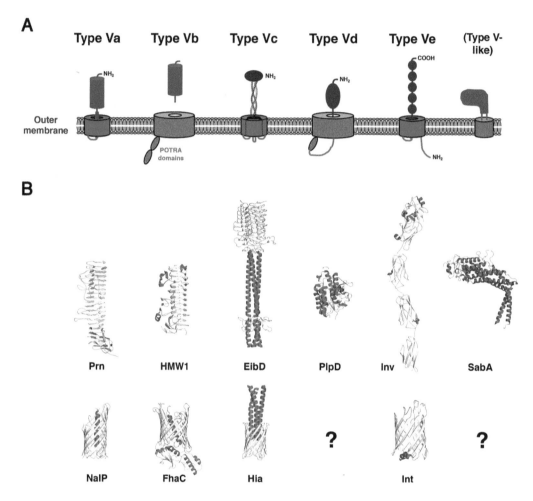

FIGURE 1 Illustration of type V secretion pathways. (A) Proteins in type V (and type V-like) secretion pathways consist of a 12-stranded (red), 16-stranded (green), or predicted 8-stranded (pink) β barrel domain and an extracellular ("passenger") domain that typically folds into a β-helical (blue), mixed coiled-coil/β roll/β prism (purple) or globular (brown) structure. The 16-stranded β barrel domains are members of the Omp85 superfamily and contain periplasmic POTRA domains. In most cases the β barrel and passenger domains are covalently linked, but in the type Vb pathway the β barrel domain and the extracellular component ("exoprotein") are separate polypeptides. In the type Vc pathway both domains are formed through the assembly of three identical subunits. The passenger domain is located at the N terminus of the protein in the type Va, Vb, Vc, and Vd pathways, but it is found at the C terminus in the type Ve pathway. In the type V-like pathway the extracellular domain is located in a loop that connects the first two β strands of the β barrel domain. (B) Crystal structures of representative polypeptides from each pathway are shown. α-helical segments are colored red and β strands are colored yellow. The structures include the pertactin (Prn) passenger domain (4) (PDB code 1DAB), a fragment of the HMW1 exoprotein (98) (PDB code 2ODL), a fragment of the EibD passenger domain (24) (PDB code 2XQH), the phospholipase D (PlpD) passenger domain (34) (PDB code 5FYA), the invasin (Inv) passenger domain (28) (1CWV), the SabA extracellular domain (36) (PDB code 4O5J), and the NalP, FhaC, Hia, and intimin (Int) β barrel domains (PDB codes 1UYO, 4QKY, 2GR7, and 4E1S) (10, 18, 29, 99). The helix inside the FhaC β barrel was generated from a neighboring asymmetric unit in the crystal lattice. No structures of β barrel domains of type Vd or type V-like proteins have been reported. Modified from *Molecular Microbiology* (100), with permission.

closed, amphipathic β sheet or "β barrel" structure. The C-terminal domains that have been crystallized to date all form nearly superimposable 12-stranded β barrels (10–15). The two domains are connected by a short α-helical "linker" that is embedded inside the β barrel domain (10, 12, 13, 16). Many passenger domains are released from the cell surface by a proteolytic cleavage following their secretion (17).

Several other pathways have been described that appear to be variations on the same theme (Fig. 1). Trimeric autotransporters (type Vc pathway) are comprised of three identical subunits that each contain an N-terminal passenger domain that can exceed 4,000 residues in length and an ~80-residue C-terminal segment that contributes four β strands to a single 12-stranded β barrel. Although the structure of the β barrel domains is very similar to those of classical autotransporters (18, 19), the three passenger domains assemble into a long coiled-coil "stalk" that emerges from the β barrel domain. The stalk is interspersed with and/or terminated by globular β-roll or β-prism "head" domains that function as adhesins (20–27). In the intimin/invasin (type Ve) pathway, the order of the domains is reversed. These "inverted autotransporters" contain a 12-stranded β barrel domain at (or near) the N terminus and a passenger domain comprised of multiple immunoglobulin (Ig)-like repeats at the C terminus (28–30). Although the structure of the β barrel domain resembles that of classical and trimeric autotransporters, the linker does not form an α helix (29). In the type Vb, or two-partner secretion (TPS), pathway, a single "exoprotein" is secreted by a coordinately expressed OM transporter. While exoproteins have the same β-helical architecture as the passenger domains of classical autotransporters, the transporters are members of the Omp85 superfamily, a group of proteins that have 16-stranded β barrel domains and 1 to 7 periplasmic POTRA (polypeptide transport-associated) domains that are believed to mediate protein-protein interactions (31, 32). The TPS pathway is the only type V pathway in which a β barrel protein secretes a non-covalently linked polypeptide (for details, *see* Chapter 25) (101). The type Vd pathway is related to the type Vb pathway in that the C-terminal domains are similar to TpsB proteins, but the covalently linked passenger domains are patatin-like lipases that are released into the environment (33–35). Finally, a family of *Helicobacter pylori* proteins (at least some of which are adhesins) has been described in which an extracellular α-helical domain of up to ~1,000 amino acids is situated between β strands 1 and 2 of a putative 8-stranded β barrel (36–40). These proteins have been proposed to represent a "type V-like" pathway based on their modular organization (40), but they do not have a clear phylogenetic relationship to other autotransporters and their structure is unique.

AUTOTRANSPORTER ASSEMBLY AND THE MECHANISM OF PASSENGER DOMAIN SECRETION

Although the first classical autotransporter was discovered over 30 years ago (2), the mechanism(s) by which passenger domains are translocated across the OM through the type V pathways is still not well understood. It was originally proposed that passenger domains are secreted through a channel formed solely by the covalently linked β barrel domain (hence the name "autotransporter") (2). Indeed, it is easy to imagine how translocation in the type Va pathway, which proceeds in a C- to N-terminal direction (41, 42), might involve the insertion of a C-terminal hairpin into the β barrel pore followed by the progressive secretion of more distal segments. The resolution of the hairpin following the completion of translocation would explain why the two domains are connected by an intrabarrel linker. It should be noted, however, that the self-transport model was proposed before significant insights into the biogenesis of bacterial OMPs had emerged. Our view of autotransporter secretion has evolved considerably in recent years and has been strongly influenced both by new experimental data (that focus primarily on the type Va pathway) and by the identification and characterization of the machinery that catalyzes OMP assembly.

Based on all of the available evidence, it now appears that the β barrel domain does play a role in translocation but that the process by which passenger domains are transported across the OM is more complex than originally envisioned. On a fundamental level, the finding that translocation is abolished by the replacement of the C terminus of an autotransporter with the β barrel of another OMP suggests that the native β barrel domain does not simply target the passenger domain to an unlinked transporter (43). Furthermore, the finding that mutations that slow the folding and/or membrane integration of the β barrel domain concomitantly delay the initiation of passenger domain translocation also suggests that the β barrel domain promotes the transport reaction (9, 44). The idea that autotransporters are completely autonomous secretion systems, however, was first challenged by two contradictory lines of evidence. Crystal structures revealed that the β barrel pore of classical autotransporters is only ~10 Å in diameter and therefore only wide enough to accommodate a single α helix or a hairpin in an extended conformation (10–15). Molecular dynamic simulations also indicated that β barrel domains are relatively rigid and are unlikely to expand significantly without an input of energy (45, 46). Paradoxically, considerable evidence has emerged that polypeptides that have local tertiary structures can be secreted by the type Va pathway. A subset of native type Va and type Ve passenger domains undergo disulfide bonding in the periplasm, and at least some ~10- to 20-kDa heterologous polypeptides that fold in the periplasm are secreted effectively when they are fused to passenger domains (29, 47–51). An analysis of the secretion of peptides that vary in length and structural complexity also suggests that the translocation channel is ~17 to 20 Å wide (52). Furthermore, evidence that the linker is already embedded inside the β barrel in an α-helical conformation during translocation strongly suggests that the active transport channel contains at least an α helix and an extended polypeptide (53, 54). Finally, several studies have indicated that the β barrel domain reaches its native state only after the passenger domain is completely secreted (55–57). Taken together, the results imply that during translocation the β barrel domain is in an open or distorted conformation that would be incompatible with stable integration into a lipid bilayer.

A plausible alternative to the self-transport hypothesis that accounts for the secretion of folded polypeptides arose from an analysis of stalled translocation intermediates. One study exploited the fortuitous discovery that the insertion of a peptide linker near the middle of the passenger domain of a classical autotransporter (the *Escherichia coli* O157:H7 EspP protein) did not affect the initiation of translocation but transiently stalled translocation when the inserted peptide was in the vicinity of the transport channel (41). Site-specific photo-cross-linking experiments showed that passenger domain residues located near the site of stalling are in close proximity to BamA, a member of the Omp85 superfamily. BamA is an essential component of the barrel assembly machinery (Bam) complex, a hetero-oligomer that catalyzes the membrane insertion of essentially all β barrel proteins, including autotransporters (58–61). In a second study, chemical cross-linking experiments showed that a related autotransporter (Hbp) was close to the Bam complex when the secretion of the passenger domain was stalled by a different method (55). Interestingly, the crystal structure of BamA together with molecular dynamics simulations strongly suggests that the BamA β barrel can open laterally (62). Although the function of the BamA lateral gate is still unclear, the results raise the intriguing possibility that passenger domains are secreted through a hybrid channel composed of open forms of both the linked β barrel domain and the BamA β barrel. Such a channel would presumably be wide enough to accommodate the transport of polypeptides that have local tertiary structures.

The analysis of these and other assembly intermediates has led to a detailed model for the biogenesis of classical autotransporters. The finding that the EspP linker becomes protected from proteolysis and chemical modification (53) prior to the initiation of passenger domain translocation suggests that the β barrel domain begins to fold in the periplasm (Fig. 2, step I). Consistent with this idea, a recent study indicated that the tri-

meric β barrel of a type Vc autotransporter begins to assemble in the periplasm (63). The observation that the linker is required for the membrane integration of the EspP β barrel domain *in vivo* (53) and accelerates assembly in an *in vitro* assay (64) suggests that it nucleates early folding events. Photo-cross-linking experiments (44, 56) have shown that at this stage the EspP β barrel domain interacts with the periplasmic chaperone Skp, a jellyfish-like homotrimer that binds to both small and large β barrels in a 1:1 or 2:1 ratio (65, 66). Subsequently, the β barrel domain is targeted to the Bam complex (Fig. 2, step II). Cross-links between specific residues of the EspP β barrel domain and two lipoprotein subunits

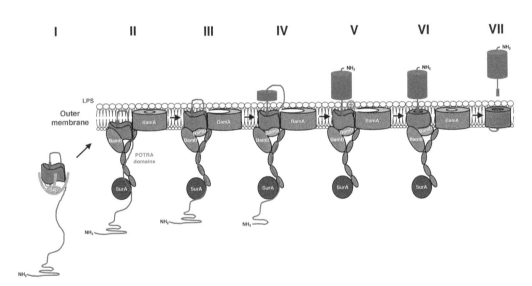

FIGURE 2 Model for the assembly of a classical autotransporter. Available evidence suggests that the β barrel domain (red) begins to fold in the periplasm (step I) and incorporates the C terminus of the passenger domain (blue) in a hairpin conformation. At this stage the β barrel domain interacts with the molecular chaperone Skp. The partially folded β barrel domain is then targeted to the OM, where it binds to BamA, BamB, and BamD in a stereospecific fashion (step II). The surface exposure of the passenger domain and the initiation of translocation require an additional assembly step in which the β barrel domain moves into the membrane (step III). Both autotransporter and BamA β barrels are in an open conformation at this stage. Translocation involves the progressive movement of passenger domain segments from the chaperone SurA to the POTRA domains of BamA to the transport channel and is driven at least in part by vectorial folding (step IV). Following the completion of translocation the hairpin is resolved (step V), and an unusual lipid-facing basic or large polar residue found in at least a subset of autotransporters facilitates the completion of β barrel domain assembly (step VI). The β barrel domain is then released from the Bam complex, and, in some cases, the two domains are separated by an intrabarrel cleavage or an extrabarrel cleavage mediated by a *trans*-acting protease (step VII). In *E. coli* the Bam complex contains five subunits, but BamC and BamE have been omitted for clarity. Modified from *Molecular Microbiology* (57, 100), with permission.

of the Bam complex, BamB and BamD, can be detected at this step (44). Interestingly, a map generated by projecting the molecular interactions implied by the photo-cross-linking experiments onto the crystal structure of the Bam holocomplex supports the idea that the β barrel domain is already folded into a cylinder-like structure (67). The initiation of passenger domain translocation requires an additional assembly step that appears to correspond to the movement (but not full integration) of the β barrel domain into the OM (44, 68) (Fig. 2, step III). As suggested above, the passenger domain might be transported through a hybrid channel that contains the BamA β barrel in an open conformation. Available evidence indicates that translocation involves a stepwise transfer of passenger domain segments from the chaperone SurA, which binds to the first POTRA domain (69), to membrane-proximal POTRA domains and then to the transport channel (44) (Fig. 2, step IV). In the TPS pathway, exoproteins use a similar path to traverse the cognate transporter (70). Following the completion of translocation (Fig. 2, step V), a surface-exposed basic or large polar residue stimulates a final step in the folding of at least some classical autotransporters that may correspond to the closing of the β barrel domain (57) (Fig. 2, step VI). Ultimately, the β barrel domain is released from the Bam complex and the passenger domain is cleaved (Fig. 2, step VII).

ENERGETICS OF PASSENGER DOMAIN SECRETION

Because the periplasm is devoid of ATP and there is no electrochemical gradient across the OM, the source of energy for passenger domain translocation has remained unclear. It is possible that in some cases an interaction between the passenger domain or specialized components of the OM transport machinery and an energized IM protein drives the translocation reaction. The observation that the Bam complex and SurA are sufficient to promote passenger domain translocation into proteoliposomes, however, suggests that autotransporter assembly does not strictly require an exogenous energy source (71).

To explain the energetics of autotransporter secretion, it was proposed years ago that small segments of the passenger domain passively diffuse through the transport channel and then fold in the extracellular space (72). Folding would trap the passenger domain on the cell surface and thereby provide the driving force for translocation. This hypothesis is especially attractive given that most passenger domains are composed of modular β helices that might fold in a stepwise fashion. A subset of passenger domains contain so-called passenger-associated transport repeats that might also contribute to progressive folding (73). Indeed, even the passenger domain of a classical autotransporter that has a globular structure has been predicted to fold sequentially based on the arrangement of its secondary-structure elements (12). During the last decade the "vectorial folding" hypothesis has been supported by several observations. Studies that have analyzed the refolding or unfolding of passenger domains in vitro or the effect of mutations on passenger domain secretion in vivo have demonstrated that the folding of a conserved ~20- to 25-kDa "stable core" segment located at the C termini of many passenger domains plays a key role in driving translocation (6, 54, 68, 74–76). The results of kinetic simulations also suggest that passenger domain secretion is driven by the free energy of folding in the extracellular milieu (77). Furthermore, an analysis of insertions and deletions in the intimin passenger domain suggested that secretion by the type Ve pathway is driven by sequential folding of the Ig-like domains (78). A recent study provided intriguing evidence that the folding of classical autotransporter passenger domains on the cell surface is not spontaneous but is nucleated by the fifth extracellular loop of the β barrel domain (79).

Despite the evidence that supports the vectorial folding model, several observations have strongly suggested that autotransporter secretion is not driven solely by passenger domain folding. It has been shown, for example, that multiple point mutations introduced into the middle of the EspP passenger domain destabilize the protein but only moderately impair translocation (80). In addition, an intrinsically disordered polypeptide fused to the C terminus of EspP was secreted as rapidly and efficiently as the native passenger domain (51). The disordered polypeptide is unusually acidic, and the neutralization of multiple acidic amino acids was shown to stall translocation. Taken together with the finding that many native passenger domains are acidic, this observation suggests that charge interactions and/or the Donnan potential across the OM (81) might help to drive translocation. Furthermore, it seems likely that the secretion of ~10- to 20-kDa folded polypeptides that has been reported would also require the input of energy from an alternative source. In this regard, it is noteworthy that structural studies on the OM transporters associated with the chaperone/usher and type VIII secretion pathways suggest that they drive translocation by defining a low-energy pathway or using an entropy-based diffusion mechanism (82, 83).

ACCESSORY FACTORS IN AUTOTRANSPORTER ASSEMBLY

Like all OMPs, autotransporters must remain in an assembly-competent conformation in the periplasm. Consistent with this expectation, several periplasmic chaperones that play a broad role in OMP biogenesis, including DegP, FkpA, Skp, and SurA, have been shown to interact with autotransporters *in vivo* and/or *in vitro* (41, 56, 84, 85). Presumably because periplasmic chaperones have redundant or at least partially overlapping functions (86), however, the requirement for specific chaperones in autotransporter assembly appears to be protein-, organism-, and condition-dependent (41, 63, 84, 87, 88).

It is still unclear whether factors other than periplasmic chaperones and the Bam complex play a general role in autotransporter assembly. The finding that both EspP and the *E. coli* autotransporter Ag43 can be assembled into proteoliposomes that contain only the Bam complex in purified-protein- or spheroplast-based assays (71, 89) certainly suggests that no other factors are absolutely essential. There is evidence, however, that a member of the Omp85 superfamily (TamA) and an IM-anchored protein (TamB) that interacts with TamA facilitate the assembly of a subset of autotransporters, including the *Citrobacter* p1121 protein and Ag43 (90, 91). Interestingly, TamA/B has also been implicated in the assembly of inverse autotransporters (92). Although different models for the function of TamA/B have been proposed (67, 93), the observation that the Tam system facilitates the biogenesis of the fimbrial usher protein FimD (94) suggests that it is not an autotransporter-specific assembly factor. In addition, the efficient assembly of a subset of trimeric autotransporters requires the activity of a trimeric IM lipoprotein that is encoded in the same operon (95).

CONCLUDING REMARKS

Studies on type V secretion suggest that autotransporter β barrel domains are not autonomous transporters but that the assembly of the β barrel domain and the secretion of the passenger domain are catalyzed by the Bam complex in a concerted reaction. Although it is possible that BamA promotes passenger domain translocation indirectly by keeping the autotransporter β barrel domain in an open conformation, the fact that some members of the Omp85 superfamily catalyze secretion reactions suggests that BamA might play a direct role in translocation. In any case, to obtain a better understanding of autotransporter biogenesis, it will likely be necessary to elucidate the function of the Bam complex, which so far has remained elusive (96). Of course the degree to which the assembly of type Vc-Ve and "autotransporter-like" proteins resembles the assembly of the better-characterized classical autotransporters also remains to be investigated. Indeed, it should be interesting to determine why features of the Omp85 superfamily are found only in the type Vd pathway and if the folding of predominantly α-helical passenger domains drives their secretion. Given that the loops of most OMPs are relatively short (<75 residues) and that at least in some cases only small insertions are tolerated (97), the presence of the large loop structures in "autotransporter-like" proteins is intriguing. Based on the discovery of these proteins, the structural similarity of the β barrel domains found in multiple type V pathways, and the unique structures of passenger domains, it is tempting to speculate that the Bam complex evolved to facilitate the efficient export of a range of specialized polypeptides that are paired with specific types of β barrels.

ACKNOWLEDGMENTS

I thank Matt Doyle for helping to generate Fig. 1.

Work conducted in my laboratory is supported by the NIDDK Intramural Research Program (Z01-DK052037).

CITATION

EcoSal Plus 2019; doi:10.1128/ecosalplus.ESP-0031-2018.

REFERENCES

1. **Henderson IR, Nataro JP.** 2001. Virulence functions of autotransporter proteins. *Infect Immun* **69:**1231–1243.
2. **Pohlner J, Halter R, Beyreuther K, Meyer TF.** 1987. Gene structure and extracellular secretion of *Neisseria gonorrhoeae* IgA protease. *Nature* **325:**458–462.
3. **Celik N, Webb CT, Leyton DL, Holt KE, Heinz E, Gorrell R, Kwok T, Naderer T, Strugnell RA, Speed TP, Teasdale RD, Likić VA, Lithgow T.** 2012. A bioinformatic strategy for the detection, classification and analysis of bacterial autotransporters. *PLoS One* **7:**e43245.
4. **Emsley P, Charles IG, Fairweather NF, Isaacs NW.** 1996. Structure of *Bordetella pertussis* virulence factor P.69 pertactin. *Nature* **381:**90–92.
5. **Otto BR, Sijbrandi R, Luirink J, Oudega B, Heddle JG, Mizutani K, Park SY, Tame JR.** 2005. Crystal structure of hemoglobin protease, a heme binding autotransporter protein from pathogenic *Escherichia coli*. *J Biol Chem* **280:**17339–17345.
6. **Junker M, Schuster CC, McDonnell AV, Sorg KA, Finn MC, Berger B, Clark PL.** 2006. Pertactin beta-helix folding mechanism suggests common themes for the secretion and folding of autotransporter proteins. *Proc Natl Acad Sci U S A* **103:**4918–4923.
7. **Gangwer KA, Mushrush DJ, Stauff DL, Spiller B, McClain MS, Cover TL, Lacy DB.** 2007. Crystal structure of the *Helicobacter pylori* vacuolating toxin p55 domain. *Proc Natl Acad Sci U S A* **104:**16293–16298.
8. **Heras B, Totsika M, Peters KM, Paxman JJ, Gee CL, Jarrott RJ, Perugini MA, Whitten AE, Schembri MA.** 2014. The antigen 43 structure reveals a molecular Velcro-like mechanism of autotransporter-mediated bacterial clumping. *Proc Natl Acad Sci U S A* **111:**457–462.
9. **Leyton DL, Johnson MD, Thapa R, Huysmans GH, Dunstan RA, Celik N, Shen HH, Loo D, Belousoff MJ, Purcell AW, Henderson IR, Beddoe T, Rossjohn J, Martin LL, Strugnell RA, Lithgow T.** 2014. A mortise-tenon joint in the transmembrane domain modulates autotransporter assembly into bacterial outer membranes. *Nat Commun* **5:**4239.
10. **Oomen CJ, van Ulsen P, van Gelder P, Feijen M, Tommassen J, Gros P.** 2004. Structure of the translocator domain of a bacterial autotransporter. *EMBO J* **23:**1257–1266.
11. **Barnard TJ, Dautin N, Lukacik P, Bernstein HD, Buchanan SK.** 2007. Autotransporter structure reveals intra-barrel cleavage followed by conformational changes. *Nat Struct Mol Biol* **14:**1214–1220.
12. **van den Berg B.** 2010. Crystal structure of a full-length autotransporter. *J Mol Biol* **396:**627–633.
13. **Tajima N, Kawai F, Park SY, Tame JR.** 2010. A novel intein-like autoproteolytic mechanism in autotransporter proteins. *J Mol Biol* **402:**645–656.
14. **Zhai Y, Zhang K, Huo Y, Zhu Y, Zhou Q, Lu J, Black I, Pang X, Roszak AW, Zhang X, Isaacs NW, Sun F.** 2011. Autotransporter passenger domain secretion requires a hydro-

phobic cavity at the extracellular entrance of the β-domain pore. *Biochem J* **435:**577–587.

15. **Gawarzewski I, DiMaio F, Winterer E, Tschapek B, Smits SHJ, Jose J, Schmitt L.** 2014. Crystal structure of the transport unit of the autotransporter adhesin involved in diffuse adherence from *Escherichia coli. J Struct Biol* **187:**20–29.

16. **Barnard TJ, Gumbart J, Peterson JH, Noinaj N, Easley NC, Dautin N, Kuszak AJ, Tajkhorshid E, Bernstein HD, Buchanan SK.** 2012. Molecular basis for the activation of a catalytic asparagine residue in a self-cleaving bacterial autotransporter. *J Mol Biol* **415:**128–142.

17. **Dautin N, Bernstein HD.** 2007. Protein secretion in gram-negative bacteria via the autotransporter pathway. *Annu Rev Microbiol* **61:**89–112.

18. **Meng G, Surana NK, St Geme JW III, Waksman G.** 2006. Structure of the outer membrane translocator domain of the *Haemophilus influenzae* Hia trimeric autotransporter. *EMBO J* **25:**2297–2304.

19. **Shahid SA, Bardiaux B, Franks WT, Krabben L, Habeck M, van Rossum BJ, Linke D.** 2012. Membrane-protein structure determination by solid-state NMR spectroscopy of microcrystals. *Nat Methods* **9:**1212–1217.

20. **Nummelin H, Merckel MC, Leo JC, Lankinen H, Skurnik M, Goldman A.** 2004. The *Yersinia* adhesin YadA collagen-binding domain structure is a novel left-handed parallel β-roll. *EMBO J* **23:**701–711.

21. **Szczesny P, Linke D, Ursinus A, Bär K, Schwarz H, Riess TM, Kempf VA, Lupas AN, Martin J, Zeth K.** 2008. Structure of the head of the *Bartonella* adhesin BadA. *PLoS Pathog* **4:**e1000119.

22. **Edwards TE, Phan I, Abendroth J, Dieterich SH, Masoudi A, Guo W, Hewitt SN, Kelley A, Leibly D, Brittnacher MJ, Staker BL, Miller SI, Van Voorhis WC, Myler PJ, Stewart LJ.** 2010. Structure of a *Burkholderia pseudomallei* trimeric autotransporter adhesin head. *PLoS One* **5:**e12803.

23. **Agnew C, Borodina E, Zaccai NR, Conners R, Burton NM, Vicary JA, Cole DK, Antognozzi M, Virji M, Brady RL.** 2011. Correlation of in situ mechanosensitive responses of the *Moraxella catarrhalis* adhesin UspA1 with fibronectin and receptor CEACAM1 binding. *Proc Natl Acad Sci U S A* **108:**15174–15178.

24. **Leo JC, Lyskowski A, Hattula K, Hartmann MD, Schwarz H, Butcher SJ, Linke D, Lupas AN, Goldman A.** 2011. The structure of *E. coli* IgG-binding protein D suggests a general model for bending and binding in trimeric autotransporter adhesins. *Structure* **19:**1021–1030.

25. **Hartmann MD, Grin I, Dunin-Horkawicz S, Deiss S, Linke D, Lupas AN, Hernandez Alvarez B.** 2012. Complete fiber structures of complex trimeric autotransporter adhesins conserved in enterobacteria. *Proc Natl Acad Sci U S A* **109:**20907–20912.

26. **Malito E, Biancucci M, Faleri A, Ferlenghi I, Scarselli M, Maruggi G, Lo Surdo P, Veggi D, Liguori A, Santini L, Bertoldi I, Petracca R, Marchi S, Romagnoli G, Cartocci E, Vercellino I, Savino S, Spraggon G, Norais N, Pizza M, Rappuoli R, Masignani V, Bottomley MJ.** 2014. Structure of the meningococcal vaccine antigen NadA and epitope mapping of a bactericidal antibody. *Proc Natl Acad Sci U S A* **111:**17128–17133.

27. **Koiwai K, Hartmann MD, Linke D, Lupas AN, Hori K.** 2016. Structural basis for toughness and flexibility in the C-terminal passenger domain of an *Acinetobacter* trimeric autotransporter adhesin. *J Biol Chem* **291:**3705–3724.

28. **Hamburger ZA, Brown MS, Isberg RR, Bjorkman PJ.** 1999. Crystal structure of invasin: a bacterial integrin-binding protein. *Science* **286:**291–295.

29. **Fairman JW, Dautin N, Wojtowicz D, Liu W, Noinaj N, Barnard TJ, Udho E, Przytycka TM, Cherezov V, Buchanan SK.** 2012. Crystal structures of the outer membrane domain of intimin and invasin from enterohemorrhagic *E. coli* and enteropathogenic *Y. pseudotuberculosis. Structure* **20:**1233–1243.

30. **Leo JC, Oberhettinger P, Schütz M, Linke D.** 2015. The inverse autotransporter family: intimin, invasin and related proteins. *Int J Med Microbiol* **305:**276–282.

31. **Gentle IE, Burri L, Lithgow T.** 2005. Molecular architecture and function of the Omp85 family of proteins. *Mol Microbiol* **58:**1216–1225.

32. **Arnold T, Zeth K, Linke D.** 2010. Omp85 from the thermophilic cyanobacterium *Thermosynechococcus elongatus* differs from proteobacterial Omp85 in structure and domain composition. *J Biol Chem* **285:**18003–18015.

33. **Salacha R, Kovačić F, Brochier-Armanet C, Wilhelm S, Tommassen J, Filloux A, Voulhoux R, Bleves S.** 2010. The *Pseudomonas aeruginosa* patatin-like protein PlpD is the archetype of a novel type V secretion system. *Environ Microbiol* **12:**1498–1512.

34. **da Mata Madeira PV, Zouhir S, Basso P, Neves D, Laubier A, Salacha R, Bleves S, Faudry E, Contreras-Martel C, Dessen A.**

2016. Structural basis of lipid targeting and destruction by the type V secretion system of *Pseudomonas aeruginosa*. *J Mol Biol* **428**(9 Part A):1790–1803.

35. **Casasanta MA, Yoo CC, Smith HB, Duncan AJ, Cochrane K, Varano AC, Allen-Vercoe E, Slade DJ.** 2017. A chemical and biological toolbox for type Vd secretion: characterization of the phospholipase A1 autotransporter FplA from *Fusobacterium nucleatum*. *J Biol Chem* **292**:20240–20254.

36. **Pang SS, Nguyen ST, Perry AJ, Day CJ, Panjikar S, Tiralongo J, Whisstock JC, Kwok T.** 2014. The three-dimensional structure of the extracellular adhesion domain of the sialic acid-binding adhesin SabA from *Helicobacter pylori. J Biol Chem* **289**:6332–6340.

37. **Hage N, Howard T, Phillips C, Brassington C, Overman R, Debreczeni J, Gellert P, Stolnik S, Winkler GS, Falcone FH.** 2015. Structural basis of Lewis(b) antigen binding by the *Helicobacter pylori* adhesin BabA. *Sci Adv* **1**:e1500315.

38. **Javaheri A, Kruse T, Moonens K, Mejías-Luque R, Debraekeleer A, Asche CI, Tegtmeyer N, Kalali B, Bach NC, Sieber SA, Hill DJ, Königer V, Hauck CR, Moskalenko R, Haas R, Busch DH, Klaile E, Slevogt H, Schmidt A, Backert S, Remaut H, Singer BB, Gerhard M.** 2016. *Helicobacter pylori* adhesin HopQ engages in a virulence-enhancing interaction with human CEACAMs. *Nat Microbiol* **2**:16189.

39. **Moonens K, Gideonsson P, Subedi S, Bugaytsova J, Romaõ E, Mendez M, Nordén J, Fallah M, Rakhimova L, Shevtsova A, Lahmann M, Castaldo G, Brännström K, Coppens F, Lo AW, Ny T, Solnick JV, Vandenbussche G, Oscarson S, Hammarström L, Arnqvist A, Berg DE, Muyldermans S, Borén T, Remaut H.** 2016. Structural insights into polymorphic ABO glycan binding by *Helicobacter pylori. Cell Host Microbe* **19**:55–66.

40. **Coppens F, Castaldo G, Debraekeleer A, Subedi S, Moonens K, Lo A, Remaut H.** 2018. Hop-family *Helicobacter* outer membrane adhesins form a novel class of type 5-like secretion proteins with an interrupted β-barrel domain. *Mol Microbiol* **110**:33–46.

41. **Ieva R, Bernstein HD.** 2009. Interaction of an autotransporter passenger domain with BamA during its translocation across the bacterial outer membrane. *Proc Natl Acad Sci U S A* **106**:19120–19125.

42. **Junker M, Besingi RN, Clark PL.** 2009. Vectorial transport and folding of an autotransporter virulence protein during outer membrane secretion. *Mol Microbiol* **71**:1323–1332.

43. **Saurí A, Oreshkova N, Soprova Z, Jong WS, Sani M, Peters PJ, Luirink J, van Ulsen P.** 2011. Autotransporter β-domains have a specific function in protein secretion beyond outer-membrane targeting. *J Mol Biol* **412**:553–567.

44. **Pavlova O, Peterson JH, Ieva R, Bernstein HD.** 2013. Mechanistic link between β barrel assembly and the initiation of autotransporter secretion. *Proc Natl Acad Sci U S A* **110**:E938–E947.

45. **Khalid S, Sansom MS.** 2006. Molecular dynamics simulations of a bacterial autotransporter: NalP from *Neisseria meningitidis. Mol Membr Biol* **23**:499–508.

46. **Tian P, Bernstein HD.** 2010. Molecular basis for the structural stability of an enclosed β-barrel loop. *J Mol Biol* **402**:475–489.

47. **Veiga E, de Lorenzo V, Fernández LA.** 2004. Structural tolerance of bacterial autotransporters for folded passenger protein domains. *Mol Microbiol* **52**:1069–1080.

48. **Skillman KM, Barnard TJ, Peterson JH, Ghirlando R, Bernstein HD.** 2005. Efficient secretion of a folded protein domain by a monomeric bacterial autotransporter. *Mol Microbiol* **58**:945–958.

49. **Swanson KA, Taylor LD, Frank SD, Sturdevant GL, Fischer ER, Carlson JH, Whitmire WM, Caldwell HD.** 2009. *Chlamydia trachomatis* polymorphic membrane protein D is an oligomeric autotransporter with a higher-order structure. *Infect Immun* **77**:508–516.

50. **Leyton DL, Sevastsyanovich YR, Browning DF, Rossiter AE, Wells TJ, Fitzpatrick RE, Overduin M, Cunningham AF, Henderson IR.** 2011. Size and conformation limits to secretion of disulfide-bonded loops in autotransporter proteins. *J Biol Chem* **286**:42283–42291.

51. **Kang'ethe W, Bernstein HD.** 2013. Charge-dependent secretion of an intrinsically disordered protein via the autotransporter pathway. *Proc Natl Acad Sci U S A* **110**:E4246–E4255.

52. **Saurí A, Ten Hagen-Jongman CM, van Ulsen P, Luirink J.** 2012. Estimating the size of the active translocation pore of an autotransporter. *J Mol Biol* **416**:335–345.

53. **Ieva R, Skillman KM, Bernstein HD.** 2008. Incorporation of a polypeptide segment into the β-domain pore during the assembly of a bacterial autotransporter. *Mol Microbiol* **67**:188–201.

54. **Peterson JH, Tian P, Ieva R, Dautin N, Bernstein HD.** 2010. Secretion of a bacterial virulence factor is driven by the folding of a C-terminal segment. *Proc Natl Acad Sci U S A* **107**:17739–17744.

55. **Saurí A, Soprova Z, Wickström D, de Gier JW, Van der Schors RC, Smit AB, Jong WS, Luirink J.** 2009. The Bam (Omp85) complex is

involved in secretion of the autotransporter haemoglobin protease. *Microbiology* **155**:3982–3991.

56. **Ieva R, Tian P, Peterson JH, Bernstein HD.** 2011. Sequential and spatially restricted interactions of assembly factors with an autotransporter β domain. *Proc Natl Acad Sci U S A* **108**: E383–E391.

57. **Peterson JH, Hussain S, Bernstein HD.** 2018. Identification of a novel post-insertion step in the assembly of a bacterial outer membrane protein. *Mol Microbiol* **110**:143–159.

58. **Voulhoux R, Bos MP, Geurtsen J, Mols M, Tommassen J.** 2003. Role of a highly conserved bacterial protein in outer membrane protein assembly. *Science* **299**:262–265.

59. **Wu T, Malinverni J, Ruiz N, Kim S, Silhavy TJ, Kahne D.** 2005. Identification of a multicomponent complex required for outer membrane biogenesis in *Escherichia coli*. *Cell* **121**:235–245.

60. **Jain S, Goldberg MB.** 2007. Requirement for YaeT in the outer membrane assembly of autotransporter proteins. *J Bacteriol* **189**:5393–5398.

61. **Hagan CL, Kim S, Kahne D.** 2010. Reconstitution of outer membrane protein assembly from purified components. *Science* **328**:890–892.

62. **Noinaj N, Kuszak AJ, Gumbart JC, Lukacik P, Chang H, Easley NC, Lithgow T, Buchanan SK.** 2013. Structural insight into the biogenesis of β-barrel membrane proteins. *Nature* **501**:385–390.

63. **Sikdar R, Peterson JH, Anderson DE, Bernstein HD.** 2017. Folding of a bacterial integral outer membrane protein is initiated in the periplasm. *Nat Commun* **8**:1309.

64. **Hussain S, Bernstein HD.** 2018. The Bam complex catalyzes efficient insertion of bacterial outer membrane proteins into membrane vesicles of variable lipid composition. *J Biol Chem* **293**:2959–2973.

65. **Walton TA, Sandoval CM, Fowler CA, Pardi A, Sousa MC.** 2009. The cavity-chaperone Skp protects its substrate from aggregation but allows independent folding of substrate domains. *Proc Natl Acad Sci U S A* **106**:1772–1777.

66. **Schiffrin B, Calabrese AN, Devine PWA, Harris SA, Ashcroft AE, Brockwell DJ, Radford SE.** 2016. Skp is a multivalent chaperone of outer-membrane proteins. *Nat Struct Mol Biol* **23**:786–793.

67. **Albenne C, Ieva R.** 2017. Job contenders: roles of the β-barrel assembly machinery and the translocation and assembly module in autotransporter secretion. *Mol Microbiol* **106**:505–517.

68. **Soprova Z, Sauri A, van Ulsen P, Tame JR, den Blaauwen T, Jong WS, Luirink J.** 2010. A conserved aromatic residue in the autochaperone domain of the autotransporter Hbp is critical for initiation of outer membrane translocation. *J Biol Chem* **285**:38224–38233.

69. **Bennion D, Charlson ES, Coon E, Misra R.** 2010. Dissection of β-barrel outer membrane protein assembly pathways through characterizing BamA POTRA 1 mutants of *Escherichia coli*. *Mol Microbiol* **77**:1153–1171.

70. **Baud C, Guérin J, Petit E, Lesne E, Dupré E, Locht C, Jacob-Dubuisson F.** 2014. Translocation path of a substrate protein through its Omp85 transporter. *Nat Commun* **5**:5271.

71. **Roman-Hernandez G, Peterson JH, Bernstein HD.** 2014. Reconstitution of bacterial autotransporter assembly using purified components. *eLife* **3**:e04234.

72. **Klauser T, Pohlner J, Meyer TF.** 1992. Selective extracellular release of cholera toxin B subunit by *Escherichia coli*: dissection of *Neisseria* Iga β-mediated outer membrane transport. *EMBO J* **11**:2327–2335.

73. **Doyle MT, Tran EN, Morona R.** 2015. The passenger-associated transport repeat promotes virulence factor secretion efficiency and delineates a distinct autotransporter subtype. *Mol Microbiol* **97**:315–329.

74. **Velarde JJ, Nataro JP.** 2004. Hydrophobic residues of the autotransporter EspP linker domain are important for outer membrane translocation of its passenger. *J Biol Chem* **279**:31495–31504.

75. **Renn JP, Clark PL.** 2008. A conserved stable core structure in the passenger domain β-helix of autotransporter virulence proteins. *Biopolymers* **89**:420–427.

76. **Baclayon M, Ulsen P, Mouhib H, Shabestari MH, Verzijden T, Abeln S, Roos WH, Wuite GJ.** 2016. Mechanical unfolding of an autotransporter passenger protein reveals the secretion starting point and processive transport intermediates. *ACS Nano* **10**:5710–5719.

77. **Besingi RN, Chaney JL, Clark PL.** 2013. An alternative outer membrane secretion mechanism for an autotransporter protein lacking a C-terminal stable core. *Mol Microbiol* **90**:1028–1045.

78. **Leo JC, Oberhettinger P, Yoshimoto S, Udatha DB, Morth JP, Schütz M, Hori K, Linke D.** 2016. Secretion of the intimin passenger domain is driven by protein folding. *J Biol Chem* **291**:20096–20112.

79. **Yuan X, Johnson MD, Zhang J, Lo AW, Schembri MA, Wijeyewickrema LC, Pike RN, Huysmans GHM, Henderson IR, Leyton DL.** 2018. Molecular basis for the folding of β-helical autotransporter passenger domains. *Nat Commun* **9**:1395.

80. Kang'ethe W, Bernstein HD. 2013. Stepwise folding of an autotransporter passenger domain is not essential for its secretion. *J Biol Chem* **288**:35028–35038.

81. Stock JB, Rauch B, Roseman S. 1977. Periplasmic space in *Salmonella typhimurium* and *Escherichia coli*. *J Biol Chem* **252**:7850–7861.

82. Geibel S, Procko E, Hultgren SJ, Baker D, Waksman G. 2013. Structural and energetic basis of folded-protein transport by the FimD usher. *Nature* **496**:243–246.

83. Goyal P, Krasteva PV, Van Gerven N, Gubellini F, Van den Broeck I, Troupiotis-Tsaïlaki A, Jonckheere W, Péhau-Arnaudet G, Pinkner JS, Chapman MR, Hultgren SJ, Howorka S, Fronzes R, Remaut H. 2014. Structural and mechanistic insights into the bacterial amyloid secretion channel CsgG. *Nature* **516**:250–253.

84. Ruiz-Perez F, Henderson IR, Leyton DL, Rossiter AE, Zhang Y, Nataro JP. 2009. Roles of periplasmic chaperone proteins in the biogenesis of serine protease autotransporters of *Enterobacteriaceae*. *J Bacteriol* **191**:6571–6583.

85. Ruiz-Perez F, Henderson IR, Nataro JP. 2010. Interaction of FkpA, a peptidyl-prolyl cis/trans isomerase with EspP autotransporter protein. *Gut Microbes* **1**:339–344.

86. Rizzitello AE, Harper JR, Silhavy TJ. 2001. Genetic evidence for parallel pathways of chaperone activity in the periplasm of *Escherichia coli*. *J Bacteriol* **183**:6794–6800.

87. Purdy GE, Fisher CR, Payne SM. 2007. IcsA surface presentation in *Shigella flexneri* requires the periplasmic chaperones DegP, Skp, and SurA. *J Bacteriol* **189**:5566–5573.

88. Peterson JH, Plummer AM, Fleming KG, Bernstein HD. 2017. Selective pressure for rapid membrane integration constrains the sequence of bacterial outer membrane proteins. *Mol Microbiol* **106**:777–792.

89. Norell D, Heuck A, Tran-Thi TA, Götzke H, Jacob-Dubuisson F, Clausen T, Daley DO, Braun V, Müller M, Fan E. 2014. Versatile *in vitro* system to study translocation and functional integration of bacterial outer membrane proteins. *Nat Commun* **5**:5396.

90. Selkrig J, Mosbahi K, Webb CT, Belousoff MJ, Perry AJ, Wells TJ, Morris F, Leyton DL, Totsika M, Phan MD, Celik N, Kelly M, Oates C, Hartland EL, Robins-Browne RM, Ramarathinam SH, Purcell AW, Schembri MA, Strugnell RA, Henderson IR, Walker D, Lithgow T. 2012. Discovery of an archetypal protein transport system in bacterial outer membranes. *Nat Struct Mol Biol* **19**:506–510, S1.

91. Shen HH, Leyton DL, Shiota T, Belousoff MJ, Noinaj N, Lu J, Holt SA, Tan K, Selkrig J, Webb CT, Buchanan SK, Martin LL, Lithgow T. 2014. Reconstitution of a nanomachine driving the assembly of proteins into bacterial outer membranes. *Nat Commun* **5**:5078.

92. Heinz E, Stubenrauch CJ, Grinter R, Croft NP, Purcell AW, Strugnell RA, Dougan G, Lithgow T. 2016. Conserved features in the structure, mechanism, and biogenesis of the inverse autotransporter protein family. *Genome Biol Evol* **8**:1690–1705.

93. Bamert RS, Lundquist K, Hwang H, Webb CT, Shiota T, Stubenrauch CJ, Belousoff MJ, Goode RJA, Schittenhelm RB, Zimmerman R, Jung M, Gumbart JC, Lithgow T. 2017. Structural basis for substrate selection by the translocation and assembly module of the β-barrel assembly machinery. *Mol Microbiol* **106**:142–156.

94. Stubenrauch C, Belousoff MJ, Hay ID, Shen HH, Lillington J, Tuck KL, Peters KM, Phan MD, Lo AW, Schembri MA, Strugnell RA, Waksman G, Lithgow T. 2016. Effective assembly of fimbriae in *Escherichia coli* depends on the translocation assembly module nanomachine. *Nat Microbiol* **1**:16064.

95. Grin I, Hartmann MD, Sauer G, Hernandez Alvarez B, Schütz M, Wagner S, Madlung J, Macek B, Felipe-Lopez A, Hensel M, Lupas A, Linke D. 2014. A trimeric lipoprotein assists in trimeric autotransporter biogenesis in enterobacteria. *J Biol Chem* **289**:7388–7398.

96. Noinaj N, Gumbart JC, Buchanan SK. 2017. The β-barrel assembly machinery in motion. *Nat Rev Microbiol* **15**:197–204.

97. Janssen R, Tommassen J. 1994. PhoE protein as a carrier for foreign epitopes. *Int Rev Immunol* **11**:113–121.

98. Yeo HJ, Yokoyama T, Walkiewicz K, Kim Y, Grass S, Geme JW III. 2007. The structure of the *Haemophilus influenzae* HMW1 pro-piece reveals a structural domain essential for bacterial two-partner secretion. *J Biol Chem* **282**:31076–31084.

99. Clantin B, Delattre AS, Rucktooa P, Saint N, Méli AC, Locht C, Jacob-Dubuisson F, Villeret V. 2007. Structure of the membrane protein FhaC: a member of the Omp85-TpsB transporter superfamily. *Science* **317**:957–961.

100. Bernstein HD. 2015. Looks can be deceiving: recent insights into the mechanism of protein secretion by the autotransporter pathway. *Mol Microbiol* **97**:205–215.

101. Nash ZM, Cotter PA. 2019. *Bordetella* filamentous hemagglutinin, a model for the two partner secretion pathway. *Microbiol Spectr* **7**: PSIB-0024-2019.

Bordetella Filamentous Hemagglutinin, a Model for the Two-Partner Secretion Pathway

25

ZACHARY M. NASH[1] and PEGGY A. COTTER[1]

INTRODUCTION

Bacteria use surface molecules to interact with inanimate objects during biofilm development, other bacteria during sociomicrobiological community activities, and host organisms during mutualistic, commensal, and parasitic symbioses. Among the mechanisms for delivering proteins to the surface of Gram-negative bacteria are type V secretion systems (T5SS) (1–3). T5SS comprise a passenger domain and an associated β-barrel transporter domain that, once integrated into the outer membrane via the Bam assembly complex, is sufficient for export of the passenger from the periplasm to the cell surface. Based on domain architecture, T5SS are categorized into five classes, with type Vb or two-partner secretion (TPS) pathway systems being distinct because the passenger domain (referred to generically as a TpsA protein) is synthesized independently from the transporter domain (the TpsB protein). This arrangement requires a mechanism for passenger-transporter recognition in the periplasm and may allow reuse of the transporter for export of multiple copies of the same, or closely related, passenger proteins.

TpsB proteins are members of the Omp85 superfamily, which includes the β-barrel outer membrane insertases BamA in bacteria, Tob55/Sam50 in mitochondria, and Toc75 in chloroplasts (4, 5). The C-terminal ~350 amino

[1]Department of Microbiology and Immunology, University of North Carolina—Chapel Hill, Chapel Hill, NC 27599
Protein Secretion in Bacteria
Edited by Maria Sandkvist, Eric Cascales, and Peter J. Christie
© 2019 American Society for Microbiology, Washington, DC
doi:10.1128/microbiolspec.PSIB-0024-2018

acids (aa) of Omp85 family proteins form a β-barrel pore composed of 16 β-strands linked by loops and turns (Fig. 1A). At or near the N termini are periplasmically located polypeptide transport-associated (POTRA) domains (two in TpsB proteins), each about 80 aa in length with a βααββ structure, that mediate recognition of translocation substrates (6). TpsA proteins are large, often longer than 3,000 aa. They contain highly conserved N-terminal ~250-aa "TPS domains" that fold into right-handed β-helices, with each turn of the helix formed from three β-strands in a triangular arrangement (7–10) (Fig. 1B). The remaining portions of TpsA proteins are not highly conserved but are predicted to contain substantial β-helical regions, usually within the N-terminal half of the protein, as well as regions of undetermined structure and distinct globular domains located near the C terminus.

TPS systems are widespread among Gram-negative bacteria, and TpsA proteins perform a broad range of activities that includes adherence, cytolysis, iron acquisition, and interbacterial signaling and intoxication (11). The best-characterized TPS system is filamentous hemagglutinin (FHA) and its transporter FhaC of *Bordetella pertussis*, the causal agent of human whooping cough, and *Bordetella bronchiseptica*, which infects the respiratory tracts of a broad range of mammals (12). FHA is a key *Bordetella* virulence factor. It is essential for establishing infection in the lower respiratory tract by mediating adherence to respiratory epithelia, and it plays an important role in suppressing the initial inflammatory response to infection, promoting bacterial persistence (13–15). In this chapter, we discuss what is known about FHA secretion as a prototypical TPS pathway protein. Although differences surely exist, many aspects of FHA secretion are likely to be conserved among TPS family members.

INITIAL STEPS IN THE SECRETION PROCESS

FHA is first synthesized as an ~370-kDa pre-proprotein (3,591 aa in *B. pertussis* Tohama I and 3,710 aa in *B. bronchiseptica* RB50) called

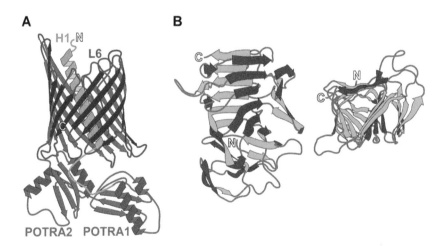

FIGURE 1 Structures of FhaC and the TPS domain of FhaB. (A) Helix 1 (H1; orange) and loop 6 (L6; fuchsia) are located within the pore of the 16-stranded β barrel (blue) of FhaC when the transporter is in the "closed" state. The POTRA domains (POTRA1 and POTRA2; red) remain periplasmic for selective recognition of the FhaB TPS domain. **(B)** The TPS domain of FhaB adopts a triangular β-helical structure, shown from the side of the helix (left) and top down in a C-terminal to N-terminal direction (right). Termini are indicated by outlined letters.

FhaB. Its 71-aa signal sequence, which is removed during Sec-mediated translocation across the cytoplasmic membrane, contains a 25-aa N-terminal extended signal peptide region (ESPR) (16). ESPRs are highly conserved and present on a large number of T5SS proteins (17, 18), and they have been implicated in regulating cotranslational and posttranslational Sec-mediated translocation (19–23). For FhaB, the ESPR appears to slow Sec-mediated posttranslational translocation across the cytoplasmic membrane (24), although how it contributes to the overall secretion mechanism remains unknown.

Once in the periplasm, FhaB must find FhaC. Early studies showed that polypeptides corresponding to the N-terminal ~300 aa of FhaB are efficiently secreted into culture supernatants of *B. pertussis* in an FhaC-dependent manner and can be secreted by *Escherichia coli* when coproduced with FhaC (7, 25, 26). These results indicated that the TPS domain contains all of the information necessary for recognition and translocation (of at least the TPS domain) by FhaC and led to further use of such models to investigate initial steps in the TPS pathway. Using *E. coli* liposomes containing FhaC, Fan et al. showed that FhaC is, reciprocally, the only outer membrane protein required for translocation of the N-terminal 370 aa of FhaB and provided evidence that folding of the FhaB N terminus occurs after translocation (27). Using a variety of protein-protein interaction approaches, Jacob-Dubuisson and colleagues showed that the POTRA domains of FhaC bind unfolded, but not folded, β-strands within the FhaB TPS domain, possibly via β augmentation (28, 29). Moreover, mutational analyses revealed that amino acids at the tip of loop six of FhaC, which is located within the channel in the crystal structure (Fig. 1A), are critical for secretion, perhaps by recognizing the presence of a secretion substrate (30). These studies identified regions within FhaC and the FhaB TPS domain that are important for recognition

and secretion, but they could not address the dynamics of TpsA translocation. To investigate the movement of FhaB across the outer membrane, the Jacob-Dubuisson group designed an approach to trap and characterize secretion intermediates (31). This study was the first to show translocation of a (truncated) substrate through the channel of an Omp85 family β-barrel (31). It revealed interactions between specific amino acids within FhaB and the POTRA domains, the interior of the channel, and the four-stranded (B5 to B8) extracellular β-sheet of FhaC. The data allowed Baud et al. (31) to propose a model (Fig. 2) in which FhaC alternates between "open" (helix 1 at the N terminus of FhaC in the periplasm) and "closed" (helix 1 in the channel) conformations in the absence of its substrate. Binding of extended amphipathic segments within the TPS domain of FhaB to the POTRA domains stabilizes the open conformation and allows the N terminus of FhaB to thread through the channel. Specific antiparallel β-strands near the N terminus of the TPS domain then bind to the extracellular β-sheet of FhaC (containing strands B5 to B8), while adjacent β-strands are translocated through the pore and begin folding into a β-helix on the cell surface, the stability of which prevents backsliding through the channel. Continued translocation of FhaB β-strands then extends the β-helix into the extracellular milieu.

MECHANISTIC INSIGHT DERIVED FROM TOPOLOGICAL ANALYSIS OF FhaB/FHA

Experiments using the TPS domain provided substantial insight into the TPS translocation mechanism. Indeed, the ratcheted diffusion model proposed by Baud et al. (31) helps explain how proteins can be moved across the hydrophobic outer membrane in the absence of an obvious chemical energy source (there is no ATP in the periplasm)—a major outstanding question for all T5SS. However, translocation of full-length TpsA proteins, the

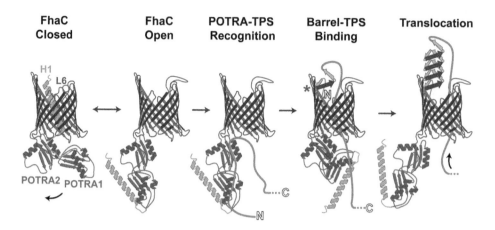

FIGURE 2 Initial steps in secretion of FhaB. FhaC alternates between a closed state, in which H1 (orange) and L6 (fuchsia) plug the channel, and an open state, in which H1 and L6 localize to the periplasm and the extracellular space, respectively. The POTRA domains of FhaC (red) bind the unfolded FhaB TPS domain (green line), stabilizing FhaC in the open state. The N terminus of FhaB then binds the interior of the FhaC barrel at β-strands B5 to B8 (blue asterisk) and forms into a β-helix as the protein is translocated, preventing backsliding through the channel.

C-terminal halves of which are not predicted to fold into β-helices, is likely to be more complicated. Moreover, for FhaB (and possibly other TpsA proteins), the C-terminal ~1,200 aa (called the prodomain) are removed at some point during the secretion process to form the ~250-kDa FHA protein, which is both cell associated and released into the extracellular milieu (32, 33). The prodomain does not appear to function as an independent protein, as it is degraded rapidly and not detectable as a separate polypeptide in whole-cell lysates or culture supernatants (26, 34). It was therefore proposed to function as some sort of chaperone, perhaps keeping the N terminus of FhaB in an unfolded state within the periplasm (26), with FHA being the "mature" molecule that is functional during infection. X-ray crystallography, electron microscopy, and molecular modeling analyses indicate that the N-terminal ~2,000 aa of FHA form a rigid β-helical rod, while the C-terminal ~500 aa (the mature C-terminal domain [MCD]) form a globular domain that appears to be bilobed in the released form of the protein (35–37).

In search of a protease responsible for cleavage of FhaB to create FHA, Coutte

et al. identified SphB1, a subtilisin-like serine protease and classical autotransporter (type Va secretion system) protein (38). Deletion of *sphB1* in *B. pertussis* results in the production of a slightly longer FHA molecule (called FHA*) that is released much less efficiently into culture supernatants than FHA is (38). Coutte et al. proposed a model in which FhaB emerges from FhaC on the cell surface in an N- to C-terminal direction until the C terminus of the MCD is exposed and cleaved by SphB1, resulting in formation and release of FHA into the extracellular milieu (Fig. 3A). Although attractive, this model did not account for the fact that a large amount of FHA is retained on the cell surface and could not explain the biophysics of moving the globular ~500-aa MCD through FhaC. While seeking to understand the role of released FHA during infection, our lab found that, as in *B. pertussis*, Δ*sphB1* mutants of *B. bronchiseptica* produce a slightly larger FHA protein (FHA′) that is poorly released from the cell surface (34). Moreover and unexpectedly, fractionation, limited proteolysis, immunoblotting, and fluorescence microscopy approaches showed that FHA is anchored to the cell by its N terminus, with

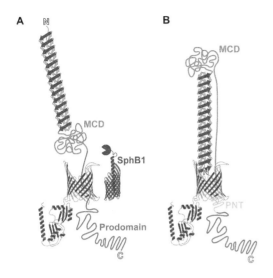

A

N

MCD

SphB1

Prodomain

C

B

MCD

N

PNT

C

FIGURE 3 **Two models for FhaB secretion: distal N-terminus versus hairpin. In the model proposed by Coutte et al. (38) (A), the N terminus of FhaB is pushed away from the membrane as more of the polypeptide translocates through FhaC. The protease SphB1 cleaves between the mature C-terminal domain (MCD) and the periplasmic prodomain, causing release of FHA. In the alternative "hairpin" model proposed by Mazar and Cotter (34) (B), the N terminus of FhaB remains bound to FhaC during secretion, and the MCD is located at the distal end of the β-helix. A portion of the MCD spans the helix length, as it is tethered to the periplasmic prodomain. The prodomain N terminus (PNT) prevents translocation of the prodomain through FhaC.**

the MCD located distally from the cell surface (34). Consistent with this orientation, antibodies against the MCD, but not antibodies recognizing the N terminus, block FHA-dependent adherence to epithelial cells and macrophages (34). Additional studies showed that the prodomain remains in the periplasm due to a region called the prodomain N terminus (PNT), and cysteine accessibility experiments showed that the prodomain is required for the MCD to achieve its normal conformation (39). These data, together with those of Baud et al. (31), support a model in which FhaB is translocated through FhaC as a "hairpin," with its N terminus anchored to the extracellular β-sheet (containing strands B5 to B8) of FhaC

and its β-helical shaft elongating by the addition of β-strands to the distal end (Fig. 3B). The stability of the β-helix, which ultimately extends about 40 nm, allows the globular MCD to be "pulled" to the surface without backsliding until movement is stopped by the PNT reaching the periplasmic side of FhaC, through which it cannot pass. The MCD then samples limited conformations before folding into its final form, followed at some point by cleavage by SphB1, degradation of the prodomain, and release of FHA into the extracellular milieu (39).

ROLE OF THE FhaB PRODOMAIN

Although the prodomain remains in the periplasm, it appears not to be required for keeping the N terminus of FhaB in a secretion-competent state, because secretion and release of FHA occur in strains producing truncated FhaB proteins that lack the prodomain (34). However, given its large size, it seems unlikely that prodomain function is limited to tethering the C terminus of the MCD to the membrane, which might require only the ~145-aa PNT region. FhaB contains two distinct domains at its C terminus: the extreme C terminus (ECT), composed of the C-terminal ~100 aa that are invariant among all predicted FhaB proteins, and the highly conserved proline-rich region (PRR), which is immediately N terminal to the ECT. Deletion of the sequence encoding the ECT results in increased conversion of FhaB to FHA, while deletion of the sequence encoding the PRR results in no discernible phenotype in bacteria grown in culture in the laboratory (13). However, both ΔECT and ΔPRR mutants are cleared much faster than wild-type bacteria from the lungs of intranasally inoculated mice, despite being indistinguishable from wild-type bacteria in their ability to adhere to the respiratory epithelium (13). Because the prodomain does not appear to be an independently functional polypeptide, these data suggest that full-length FhaB is not simply an intermediate molecule in the maturation pathway of FHA but plays an important role in the ability of the bacteria to resist inflammation-mediated clearance in the lower respiratory tract (13).

The fact that deletion of *sphB1* results in altered (rather than lack of) cleavage of FhaB (34, 38), together with the observation that the increased degradation of the prodomain that occurs in ΔECT mutants also occurs in ΔECT Δ*sphB1* double mutants (13), suggests the existence of a periplasmic protease that degrades the prodomain in a manner that is somehow negatively regulated by the ECT. *B. bronchiseptica* gene BB0300, annotated as *ctpA*, is predicted to encode a member of the Prc/Tsp/CtpA family of periplasmic proteases that target C termini. Deletion of *ctpA* abrogates increased conversion of FhaB to FHA in the ΔECT strain, as well as formation of FHA′ in a Δ*sphB1* mutant (Z. M. Nash and P. A. Cotter, in revision). Moreover, although a version of FHA that is even longer than FHA′ is produced in a Δ*ctpA* Δ*sphB1* double mutant (indicating the existence of yet another protease that can cleave the prodomain in the periplasm), no form of FHA is detected in culture supernatants, indicating that cleavage by either CtpA or SphB1 must occur for FHA (or FHA′) to be released. By inducing synthesis of FhaB synchronously, we found that cleavage of FhaB occurs in a stepwise manner, with CtpA-dependent degradation of the prodomain occurring first, followed by SphB1-mediated cleavage to form FHA (Nash and Cotter, submitted). We also found that the central portion of the prodomain (the region between the PNT and the PRR) is required for ECT-dependent control of CtpA-mediated degradation of the prodomain. Together, these data support a model (Fig. 4) in which full-length FhaB resides in the outer membrane, with the MCD located distally from the cell surface and the prodomain in the periplasm with the ECT preventing CtpA-mediated degradation. In response to a signal that is propagated from the MCD to the prodomain, an unknown protease (P3 in Fig. 4) removes

the ECT, exposing a C terminus that is susceptible to degradation by CtpA, which occurs in a processive manner, ultimately creating FHA′ (or FHA* in *B. pertussis*), the C terminus of which is located in the middle of the PNT. Removal of the C-terminal half of the PNT weakens its translocation-blocking activity, allowing the C terminus of the MCD to move through FhaC to the surface, where it is cleaved by SphB1. The small peptide that is created (extending from the SphB1 cleavage site to the end of FHA′ or FHA*) is then free to slide back into the periplasm, the N terminus of FHA disassociates from the β-sheet of FhaC (containing strands B5 to B8), and FHA is released from the cell surface.

A HYPOTHESIS: TPS AS A MECHANISM FOR REGULATED, SELECTIVE TOXIN DELIVERY

Questions raised by the model described above include the nature of the signal that is sensed by the MCD and the role of controlled degradation of the prodomain in FhaB/FHA function. The fact that ΔECT and ΔPRR mutants of *B. bronchiseptica* are indistinguishable from wild-type bacteria in their ability to adhere to epithelial cells *in vitro* and *in vivo* but are unable to persist in the lower respiratory tract suggests that full-length FhaB is required specifically for resisting clearance by phagocytic cells (13). Adenylate cyclase toxin (ACT) is a calmodulin-activated adenylyl cyclase that is produced by all classical bordetellae (*B. pertussis*, *B. bronchiseptica*, and *B. parapertussis*) that also contributes to defense against inflammation-mediated clearance (40, 41). ACT inhibits phagocytosis and oxidative burst in neutrophils (42), blocks complement-dependent phagocytosis in macrophages (43), and suppresses activation and chemotaxis in T lymphocytes (44). Although exported by a dedicated type 1 secretion system, ACT binds tightly to the MCD of FhaB/FHA on the cell surface (45), and there is evidence that it is newly secreted ACT, not that which has been

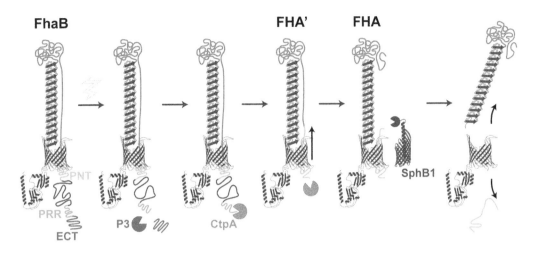

FIGURE 4 Model for stepwise processing and release of FhaB. Upon receipt of an unknown maturation signal (yellow bolt), an as-yet-unidentified protease (P3) removes the extreme C terminus (ECT) and exposes a substrate for the protease CtpA. CtpA processively degrades the prodomain through a portion of the PNT, forming FHA′ and shifting the polypeptide to expose the cleavage site of SphB1. FHA is formed from SphB1-dependent cleavage of FHA′, and it is retained at the membrane until the remaining portion of the prodomain exits FhaC (gray barrel).

released into culture supernatants, that is most efficient at intoxicating host cells (46). Moreover, both FHA and ACT have been shown to bind receptors, such as CD11b/CD18 and very late antigen 5 or leukocyte response integrin–integrin-associated protein, that are present on neutrophils, dendritic cells, and macrophages (47–50). We hypothesize that ACT forms a complex with FhaB on the cell surface and that binding of this complex to receptors on phagocytes causes a conformational change in FhaB that is propagated across the membrane, resulting in cleavage and degradation of the prodomain by the unknown protease and CtpA, followed by delivery of ACT to, and intoxication of, the host cell, hence promoting persistence of the bacteria in the face of a robust inflammatory response. In this way, FhaB may serve as an efficient ACT delivery system that prevents indiscriminant intoxication of unintended host cells, such as epithelial cells, and wasteful dispersal of ACT into the extracellular milieu. Once FHA is released from the cell surface, FhaC can be loaded with a new FhaB protein for another round of toxin delivery. The possibil-

ity that many, if not all, TPS systems function in this manner is supported by a recent report by Ruhe et al. (51), who proposed a somewhat similar mechanism for a TPS system that mediates interbacterial competition. In that system, the toxin that is delivered is the C terminus of the TpsA protein and the target cells are nonsibling bacteria, but the concept that receptor binding by a region in the middle of the TpsA protein triggers a conformational change and toxin delivery is the same. Future experiments will be required to determine the generality of this strategy, as well as the mechanistic variations among TPS systems.

AN OUTSTANDING QUESTION: DOES FHA, EITHER SURFACE ASSOCIATED OR RELEASED, PLAY A ROLE IN VIRULENCE?

Because of FHA's first-identified role, and demonstrated importance, as an adhesin, the efficient release of FHA into culture supernatants (and assumed release of FHA *in vivo*) has been enigmatic. Indeed, a desire to determine if FHA release is important during infection is what initially motivated our laboratory to investigate the FhaB/FHA secretion mechanism. Although Δ*sphB1* mutants are severely defective in mouse models (reference 52 and our unpublished data), their *in vivo* phenotype(s) cannot be wholly attributed to lack of FHA release because SphB1 cleaves multiple surface proteins in addition to FhaB (our unpublished data). The discovery that FhaB/FHA secretion, processing, and release are much more complicated than initially imagined also reveals the challenge of creating a mutant in which FhaB and surface-associated FHA remain functional while FHA release is prevented, which would be required to query the function of FHA release *in vivo*. Moreover, although the fact that ΔECT and ΔPRR mutants are defective specifically for resisting inflammation-mediated clearance supports a role for FhaB during infection, the fact that these mutants do not differ from wild-type bacteria in their ability to adhere to the respiratory epithelium *in vitro* and *in vivo* or to suppress the initial inflammatory response does not rule out the possibility that it is FhaB, and not FHA, that is important for these functions. Hence, whether "mature" FHA plays any role during the *Bordetella*-mammalian host interaction is now an open question. Continued investigation of the mechanism of FhaB/FHA secretion and processing, with accompanying infection experiments, will hopefully answer the outstanding questions regarding this important virulence factor, as well as provide a roadmap for investigating other members of the large TPS family.

CITATION

Nash ZM, Cotter PA. 2019. *Bordetella* filamentous hemagglutinin, a model for the two-partner secretion pathway. Microbiol Spectrum 7(2):PSIB-0024-2018.

REFERENCES

1. **Leo JC, Grin I, Linke D.** 2012. Type V secretion: mechanism(s) of autotransport through the bacterial outer membrane. *Philos Trans R Soc Lond B Biol Sci* **367**:1088–1101.

2. **Costa TRD, Felisberto-Rodrigues C, Meir A, Prevost MS, Redzej A, Trokter M, Waksman G.** 2015. Secretion systems in Gram-negative bacteria: structural and mechanistic insights. *Nat Rev Microbiol* **13**:343–359.

3. **Fan E, Chauhan N, Udatha DBRKG, Leo JC, Linke D.** 2016. Type V secretion systems in bacteria. *Microbiol Spectr* **4**:305–335.

4. **Ulrich T, Rapaport D.** 2015. Biogenesis of beta-barrel proteins in evolutionary context. *Int J Med Microbiol* **305**:259–264.

5. **Gentle I, Gabriel K, Beech P, Waller R, Lithgow T.** 2004. The Omp85 family of proteins is essential for outer membrane biogenesis in mitochondria and bacteria. *J Cell Biol* **164**:19–24.

6. **Simmerman RF, Dave AM, Bruce BD.** 2014. Structure and function of POTRA domains of Omp85/TPS superfamily. *Int Rev Cell Mol Biol* **308**:1–34.

7. **Clantin B, Hodak H, Willery E, Locht C, Jacob-Dubuisson F, Villeret V.** 2004. The crystal structure of filamentous hemagglutinin secretion domain and its implications for the two-partner secretion pathway. *Proc Natl Acad Sci U S A* **101**:6194–6199.

8. **Yeo HJ, Yokoyama T, Walkiewicz K, Kim Y, Grass S, Geme JW III.** 2007. The structure of the *Haemophilus influenzae* HMW1 pro-piece reveals a structural domain essential for bacterial two-partner secretion. *J Biol Chem* **282**:31076–31084.

9. **Weaver TM, Smith JA, Hocking JM, Bailey LJ, Wawrzyn GT, Howard DR, Sikkink LA, Ramirez-Alvarado M, Thompson JR.** 2009. Structural and functional studies of truncated hemolysin A from *Proteus mirabilis*. *J Biol Chem* **284**:22297–22309.

10. **Baelen S, Dewitte F, Clantin B, Villeret V.** 2013. Structure of the secretion domain of HxuA from *Haemophilus influenzae*. *Acta Crystallogr Sect F Struct Biol Cryst Commun* **69**:1322–1327.

11. **Guérin J, Bigot S, Schneider R, Buchanan SK, Jacob-Dubuisson F.** 2017. Two-partner secretion: combining efficiency and simplicity in the secretion of large proteins for bacteria-host and bacteria-bacteria interactions. *Front Cell Infect Microbiol* **7**:148.

12. **Melvin JA, Scheller EV, Miller JF, Cotter PA.** 2014. *Bordetella pertussis* pathogenesis: current and future challenges. *Nat Rev Microbiol* **12**:274–288.

13. **Melvin JA, Scheller EV, Noël CR, Cotter PA.** 2015. New insight into filamentous hemagglutinin secretion reveals a role for full-length FhaB in *Bordetella* virulence. *mBio* **6**:e01189-15.

14. **Inatsuka CS, Julio SM, Cotter PA.** 2005. *Bordetella* filamentous hemagglutinin plays a critical role in immunomodulation, suggesting a mechanism for host specificity. *Proc Natl Acad Sci U S A* **102:**18578–18583.

15. **Alonso S, Pethe K, Mielcarek N, Raze D, Locht C.** 2001. Role of ADP-ribosyltransferase activity of pertussis toxin in toxin-adhesin redundancy with filamentous hemagglutinin during *Bordetella pertussis* infection. *Infect Immun* **69:**6038–6043.

16. **Jacob-Dubuisson F, Buisine C, Mielcarek N, Clément E, Menozzi FD, Locht C.** 1996. Amino-terminal maturation of the *Bordetella pertussis* filamentous haemagglutinin. *Mol Microbiol* **19:**65–78.

17. **Henderson IR, Navarro-Garcia F, Nataro JP.** 1998. The great escape: structure and function of the autotransporter proteins. *Trends Microbiol* **6:**370–378.

18. **Desvaux M, Cooper LM, Filenko NA, Scott-Tucker A, Turner SM, Cole JA, Henderson IR.** 2006. The unusual extended signal peptide region of the type V secretion system is phylogenetically restricted. *FEMS Microbiol Lett* **264:**22–30.

19. **Sijbrandi R, Urbanus ML, ten Hagen-Jongman CM, Bernstein HD, Oudega B, Otto BR, Luirink J.** 2003. Signal recognition particle (SRP)-mediated targeting and Sec-dependent translocation of an extracellular *Escherichia coli* protein. *J Biol Chem* **278:**4654–4659.

20. **Peterson JH, Woolhead CA, Bernstein HD.** 2003. Basic amino acids in a distinct subset of signal peptides promote interaction with the signal recognition particle. *J Biol Chem* **278:**46155–46162.

21. **Desvaux M, Scott-Tucker A, Turner SM, Cooper LM, Huber D, Nataro JP, Henderson IR.** 2007. A conserved extended signal peptide region directs posttranslational protein translocation via a novel mechanism. *Microbiology* **153:**59–70.

22. **Szabady RL, Peterson JH, Skillman KM, Bernstein HD.** 2005. An unusual signal peptide facilitates late steps in the biogenesis of a bacterial autotransporter. *Proc Natl Acad Sci U S A* **102:**221–226.

23. **Peterson JH, Szabady RL, Bernstein HD.** 2006. An unusual signal peptide extension inhibits the binding of bacterial presecretory proteins to the signal recognition particle, trigger factor, and the SecYEG complex. *J Biol Chem* **281:**9038–9048.

24. **Chevalier N, Moser M, Koch H-G, Schimz K-L, Willery E, Locht C, Jacob-Dubuisson F, Müller M.** 2004. Membrane targeting of a bacterial virulence factor harbouring an extended signal peptide. *J Mol Microbiol Biotechnol* **8:**7–18.

25. **Jacob-Dubuisson F, Buisine C, Willery E, Renauld-Mongénie G, Locht C.** 1997. Lack of functional complementation between *Bordetella pertussis* filamentous hemagglutinin and *Proteus mirabilis* HpmA hemolysin secretion machineries. *J Bacteriol* **179:**775–783.

26. **Renauld-Mongénie G, Cornette J, Mielcarek N, Menozzi FD, Locht C.** 1996. Distinct roles of the N-terminal and C-terminal precursor domains in the biogenesis of the *Bordetella pertussis* filamentous hemagglutinin. *J Bacteriol* **178:**1053–1060.

27. **Fan E, Fiedler S, Jacob-Dubuisson F, Müller M.** 2012. Two-partner secretion of gram-negative bacteria: a single β-barrel protein enables transport across the outer membrane. *J Biol Chem* **287:**2591–2599.

28. **Hodak H, Clantin B, Willery E, Villeret V, Locht C, Jacob-Dubuisson F.** 2006. Secretion signal of the filamentous haemagglutinin, a model two-partner secretion substrate. *Mol Microbiol* **61:**368–382.

29. **Delattre A-S, Saint N, Clantin B, Willery E, Lippens G, Locht C, Villeret V, Jacob-Dubuisson F.** 2011. Substrate recognition by the POTRA domains of TpsB transporter FhaC. *Mol Microbiol* **81:**99–112.

30. **Clantin B, Delattre AS, Rucktooa P, Saint N, Méli AC, Locht C, Jacob-Dubuisson F, Villeret V.** 2007. Structure of the membrane protein FhaC: a member of the Omp85-TpsB transporter superfamily. *Science* **317:**957–961.

31. **Baud C, Guérin J, Petit E, Lesne E, Dupré E, Locht C, Jacob-Dubuisson F.** 2014. Translocation path of a substrate protein through its Omp85 transporter. *Nat Commun* **5:**5271.

32. **Aricò B, Nuti S, Scarlato V, Rappuoli R.** 1993. Adhesion of *Bordetella pertussis* to eukaryotic cells requires a time-dependent export and maturation of filamentous hemagglutinin. *Proc Natl Acad Sci U S A* **90:**9204–9208.

33. **Domenighini M, Relman D, Capiau C, Falkow S, Prugnola A, Scarlato V, Rappuoli R.** 1990. Genetic characterization of *Bordetella pertussis* filamentous haemagglutinin: a protein processed from an unusually large precursor. *Mol Microbiol* **4:**787–800.

34. **Mazar J, Cotter PA.** 2006. Topology and maturation of filamentous haemagglutinin suggest a new model for two-partner secretion. *Mol Microbiol* **62:**641–654.

35. **Kajava AV, Cheng N, Cleaver R, Kessel M, Simon MN, Willery E, Jacob-Dubuisson F, Locht C, Steven AC.** 2001. Beta-helix model for the filamentous haemagglutinin adhesin of

Bordetella pertussis and related bacterial secretory proteins. *Mol Microbiol* **42**:279–292.

36. **Makhov AM, Hannah JH, Brennan MJ, Trus BL, Kocsis E, Conway JF, Wingfield PT, Simon MN, Steven AC.** 1994. Filamentous hemagglutinin of *Bordetella pertussis.* A bacterial adhesin formed as a 50-nm monomeric rigid rod based on a 19-residue repeat motif rich in beta strands and turns. *J Mol Biol* **241**:110–124.

37. **Kajava AV, Steven AC.** 2006. The turn of the screw: variations of the abundant beta-solenoid motif in passenger domains of type V secretory proteins. *J Struct Biol* **155**:306–315.

38. **Coutte L, Antoine R, Drobecq H, Locht C, Jacob-Dubuisson F.** 2001. Subtilisin-like autotransporter serves as maturation protease in a bacterial secretion pathway. *EMBO J* **20**:5040–5048.

39. **Noël CR, Mazar J, Melvin JA, Sexton JA, Cotter PA.** 2012. The prodomain of the *Bordetella* two-partner secretion pathway protein FhaB remains intracellular yet affects the conformation of the mature C-terminal domain. *Mol Microbiol* **86**:988–1006.

40. **Fedele G, Schiavoni I, Adkins I, Klimova N, Sebo P.** 2017. Invasion of dendritic cells, macrophages and neutrophils by the *Bordetella* adenylate cyclase toxin: a subversive move to fool host immunity. *Toxins (Basel)* **9**:293.

41. **Novak J, Cerny O, Osickova A, Linhartova I, Masin J, Bumba L, Sebo P, Osicka R.** 2017. Structure-function relationships underlying the capacity of *Bordetella* adenylate cyclase toxin to disarm host phagocytes. *Toxins (Basel)* **9**:300.

42. **Confer DL, Eaton JW.** 1982. Phagocyte impotence caused by an invasive bacterial adenylate cyclase. *Science* **217**:948–950.

43. **Kamanova J, Kofronova O, Masin J, Genth H, Vojtova J, Linhartova I, Benada O, Just I, Sebo P.** 2008. Adenylate cyclase toxin subverts phagocyte function by RhoA inhibition and unproductive ruffling. *J Immunol* **181**:5587–5597.

44. **Paccani SR, Dal Molin F, Benagiano M, Ladant D, D'Elios MM, Montecucco C, Baldari CT.** 2008. Suppression of T-lymphocyte activation and chemotaxis by the adenylate cyclase toxin of *Bordetella pertussis. Infect Immun* **76**:2822–2832.

45. **Hoffman C, Eby J, Gray M, Heath Damron F, Melvin J, Cotter P, Hewlett E.** 2016. *Bordetella* adenylate cyclase toxin interacts with filamentous haemagglutinin to inhibit biofilm formation in vitro. *Mol Microbiol* **103**:214–228.

46. **Gray MC, Donato GM, Jones FR, Kim T, Hewlett EL.** 2004. Newly secreted adenylate cyclase toxin is responsible for intoxication of target cells by *Bordetella pertussis. Mol Microbiol* **53**:1709–1719.

47. **Guermonprez P, Khelef N, Blouin E, Rieu P, Ricciardi-Castagnoli P, Guiso N, Ladant D, Leclerc C.** 2001. The adenylate cyclase toxin of *Bordetella pertussis* binds to target cells via the alpha(M)beta(2) integrin (CD11b/CD18). *J Exp Med* **193**:1035–1044.

48. **Van Strijp JA, Russell DG, Tuomanen E, Brown EJ, Wright SD.** 1993. Ligand specificity of purified complement receptor type three (CD11b/CD18, alpha m beta 2, Mac-1). Indirect effects of an Arg-Gly-Asp (RGD) sequence. *J Immunol* **151**:3324–3336.

49. **Ishibashi Y, Yoshimura K, Nishikawa A, Claus S, Laudanna C, Relman DA.** 2002. Role of phosphatidylinositol 3-kinase in the binding of *Bordetella pertussis* to human monocytes. *Cell Microbiol* **4**:825–833.

50. **Ishibashi Y, Nishikawa A.** 2002. *Bordetella pertussis* infection of human respiratory epithelial cells up-regulates intercellular adhesion molecule-1 expression: role of filamentous hemagglutinin and pertussis toxin. *Microb Pathog* **33**:115–125.

51. **Ruhe ZC, Subramanian P, Song K, Nguyen JY, Stevens TA, Low DA, Jensen GJ, Hayes CS.** 2018. Programmed secretion arrest and receptor-triggered toxin export during antibacterial contact-dependent growth inhibition. *Cell* **175**:921–933.e14.

52. **Coutte L, Alonso S, Reveneau N, Willery E, Quatannens B, Locht C, Jacob-Dubuisson F.** 2003. Role of adhesin release for mucosal colonization by a bacterial pathogen. *J Exp Med* **197**:735–742.

Structure and Activity of the Type VI Secretion System

26

YASSINE CHERRAK,[1,*] NICOLAS FLAUGNATTI,[1,*,†] ERIC DURAND,[1]
LAURE JOURNET,[1] and ERIC CASCALES[1]

INTRODUCTION

The type VI secretion system (T6SS) is a multiprotein machine that belongs to the versatile family of contractile injection systems (CISs) (1–4). CISs deliver effectors into target cells using a spring-like mechanism (4–6). Briefly, CISs assemble a needle-like structure, loaded with effectors, wrapped into a sheath built in an extended, metastable conformation (Fig. 1). Contraction of the sheath propels the needle toward the competitor cell. Genomes of Gram-negative bacteria usually encode one or several T6SSs, with an overrepresentation in *Proteobacteria* and *Bacteroidetes* (8–10; for a review on the role of T6SS in gut-associated *Bacteroidales*, see the chapter by Coyne and Comstock [7]). The broad arsenal of effectors delivered by T6SSs includes antibacterial proteins such as peptidoglycan hydrolases, eukaryotic effectors that act on cell cytoskeleton, and toxins that can target all cell types, such as DNases, phospholipases, and NAD^+ hydrolases (11–14). Consequently, the T6SS plays a

[1]Laboratoire d'Ingénierie des Systèmes Macromoléculaires (LISM), Institut de Microbiologie de la Méditerranée (IMM), Aix-Marseille Université, CNRS, UMR 7255, 13402 Marseille Cedex 20, France
[*]These authors contributed equally.
[†]Current address: Laboratory of Molecular Microbiology, Global Health Institute, School of Life Sciences, Ecole Polytechnique Fédérale de Lausanne (EPFL), Lausanne, Switzerland.

Protein Secretion in Bacteria
Edited by Maria Sandkvist, Eric Cascales, and Peter J. Christie
© 2019 American Society for Microbiology, Washington, DC
doi:10.1128/microbiolspec.PSIB-0031-2019

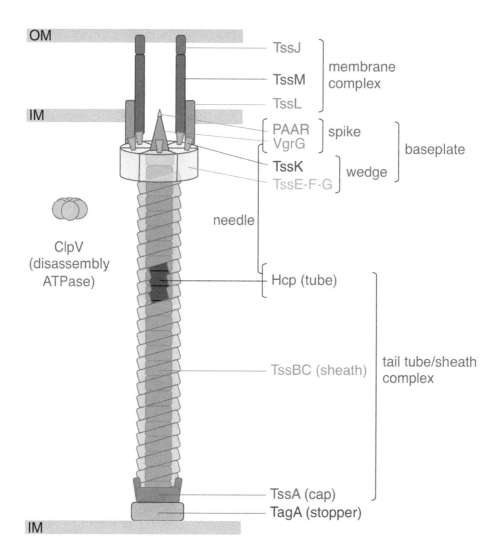

FIGURE 1 Schematic representation of the T6SS. The different subunits are labeled, as are the different subcomplexes. IM, inner membrane; OM, outer membrane.

critical role in reshaping bacterial communities and, directly or indirectly, in pathogenesis (15–19). Destroying bacterial competitors also provides exogenous DNA that can be acquired in naturally competent bacteria and that serves as a reservoir for antibiotic resistance gene spread (20). This chapter lists the major effector families and summarizes the current knowledge on the assembly and mode of action of the T6SS.

TYPE VI SECRETION SYSTEM EFFECTORS

Several T6SSs have been shown to target eukaryotic cells (21–23). By promoting or preventing cytoskeleton rearrangements through the action of specific effectors that target actin or tubulin, the T6SSs of *Vibrio cholerae*, *Aeromonas hydrophila*, and *Pseudomonas aeruginosa* disable phagocytic cells or stimulate internalization into nonphagocytic cells (21, 22, 24–26). Other T6SSs have been demonstrated to manipulate host cells, although the molecular determinants are not yet entirely understood (27–30). However, T6SS gene clusters are widespread in Gram-negative bacterial genomes and not restricted to pathogens (10). Most of them encode proteins with potent antibacterial activities, such as enzymes that cleave essential macromolecules such as DNA, phospholipids, or the peptidoglycan mesh or essential metabolites such as $NAD^+/NADP^+$ (31–36). Additional T6SS antibacterial effectors include ADP-ribosyltransferases that specifically target the Z ring and hence inhibit cell division (37). Antibacterial effectors are active in the periplasm or cytoplasm of the target cell and are coproduced with immunity proteins that remain in the producing cell and act as antitoxins to prevent autointoxication during dueling between sister cells (11–13). More recently, T6SS effectors that collect manganese or zinc in the environment to provide metals to the cell have been described (38–40). By deploying antibacterial effectors or scavenging metals, T6SSs play an important role in bacterial communities, and hence T6SS gene clusters are usually highly represented in species present in multispecies microbiota such as the human gut (7, 16–18, 41). In general, the regulatory mechanisms and signals underlying expression of T6SS genes, production of T6SS subunits, or post-translational activation of the secretion apparatus are tightly linked to environmental cues in the niche in which the T6SS is required to destroy competitors (42–45).

TYPE VI SECRETION MECHANISM OF ACTION

T6SSs use a contractile mechanism to inject effectors (Fig. 2). This mechanism is shared with all CISs: a sheath, assembled in an extended conformation, wraps a needle. Contraction of the sheath into a stable state propels the needle (1, 3–5). The needle is composed of an inner tube capped by the spike complex that pierces the membrane of the target cell (Fig. 1). The tail tube/sheath complex (TTC) is built on an assembly platform named the baseplate (BP) (Fig. 1). The TTC and BP are collectively called the tail, a structure that is conserved among all CISs. In addition to this theme common to all CISs, T6SSs have evolved (i) a membrane complex (MC), which docks the tail to the cell envelope and serves as a channel for the passage of the needle upon sheath contraction, and (ii) a specialized BP component to properly orient the needle toward the cell exterior, by recognizing and binding the MC (2–5, 46–48) (Fig. 1).

T6SS biogenesis starts with the assembly of the MC in the cell envelope and that of the BP in the cytoplasm (49–51) (Fig. 2). Once the BP is docked to the MC, the inner tube and sheath are coordinately assembled (49–52) (Fig. 2).

ARCHITECTURE OF THE TYPE VI SECRETION SYSTEM

The Membrane Complex

The vast majority of T6SS gene clusters of proteobacterial species encode three membrane proteins: TssJ, TssL, and TssM (8–10, 53) (Fig. 1). TssJ is an outer membrane-associated lipoprotein that protrudes in the periplasm (54). TssL and TssM are anchored in the inner membrane (55–57). The structures of several TssJ homologues have been reported: they all share a classical transthyretin fold with an additional loop, of variable length and composition, located between β-strands 1 and 2 (58–60). TssL bears a single C-terminal membrane-spanning segment (56) and a cytoplasmic domain that comprises two bundles of α-helices (61–63). TssM possesses three transmembrane helices followed by a large periplasmic region (55, 57). The periplasmic region of TssM comprises three domains, including the C-terminal domain that engages in interaction with the TssJ

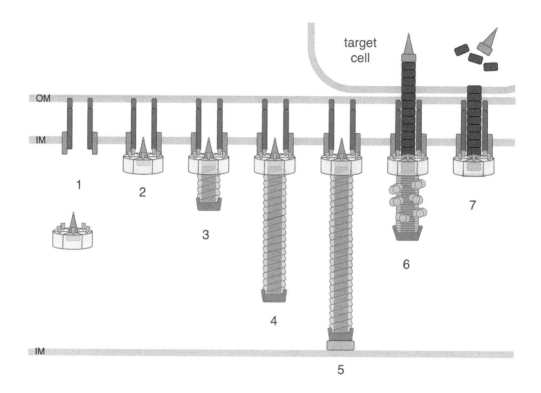

FIGURE 2 Assembly and mechanism of firing of the T6SS. T6SS biogenesis starts with the positioning and assembly of the membrane complex and the assembly of the BP (1). The recruitment and docking of the BP on the membrane complex (2) initiate the TssA-mediated polymerization of the tail tube/sheath tubular structure (3 to 5), which is stopped when hitting the opposite membrane by the TagA stopper (5). Sheath contraction propels the tube/spike needle into the target (6). The ClpV ATPase is recruited to the contracted sheath to recycle sheath subunits (6). Needle components, and effectors associated with them, are delivered inside the target (7).

extra loop (49, 58). TssL and TssM interact through their transmembrane segments (55, 64, 65). The cytoplasmic domains of TssL and TssM mediate contacts with the BP (50, 57, 64, 66, 67).

The electron microscopy structure of the fully assembled 1.7-MDa TssJLM MC from enteroaggregative *Escherichia coli* has been reported (49, 68, 69). The complex has a rocket-like structure: a large base, which contains the cytoplasmic and membrane domains of TssL and TssM, is followed by arches and pillars which correspond to the TssM periplasmic domains and TssJ (68). The TssJLM complex, which has 5-fold symmetry *in vivo* and after purification,

comprises 15 copies of TssJ and 10 copies of TssL and of TssM (49, 58). The MC delimits an internal lumen with a diameter insufficient for the passage of the tail tube. In addition, this lumen is partly occluded by a periplasmic constriction gate, suggesting that large conformational changes occur upon BP docking or sheath contraction (49, 58).

The MC can be accessorized by additional subunits, such as peptidoglycan-binding proteins (53, 70, 71). MC anchorage to the cell wall likely stabilizes the MC to resist the forces generated during sheath contraction (70). Finally, recent studies have shown that proper assembly of the MC requires the activity of peptidoglycan-degrading enzymes (72, 73).

Interestingly, while the tail complex is evolutionarily related to contractile injection machines, the evolution history of the MC is less clear. TssL and TssM present significant homologies with two accessory subunits associated with type IVb secretion systems, DotU and IcmF, respectively (8, 9). No homologue of TssJ is associated with DotU/IcmF complexes, suggesting that TssJ is from a different ancestry. Indeed, while essential when present, TssJ is lacking in some T6SSs, such as those of *Agrobacterium* and *Acinetobacter*. Further studies are required to understand whether other proteins can compensate for the absence of TssJ in these species. The fact that the MC has a distinct history compared to the tail is also exemplified by the observation that no TssJLM complex is present in *Bacteroidales* T6SSs (74, 75). However, putative uncharacterized membrane proteins are encoded within these T6SS gene clusters, suggesting that a different transenvelope complex has been domesticated to anchor the tail (74, 75).

The Tail

The baseplate

The BP (Fig. 1) is a large complex, 2.7 MDa, comprising >60 polypeptides of at least six different proteins (50). The role of the BP is to initiate the polymerization of the TTC. While it has not been formally shown yet, the T6SS BP is believed to trigger sheath contraction, as demonstrated in other CISs. A specific role of the T6SS BP is to anchor the TTC to the MC. The BP is composed of six wedge subcomplexes organized around the central hub, i.e., the N-terminal domain of the VgrG spike (76, 77) (Fig. 1). VgrG hence belongs to two tail subcomplexes: it constitutes the tip of the needle and the hub for the BP. The wedge complex is composed of four proteins: TssE, -F, -G, and -K. These four proteins assemble a structure of 1:2:1:6 stoichiometry, the TssG peptide being the central core (77–79). Two TssF subunits wrap TssG to form a triangular shape called the

trifurcation unit, whereas two extensions of TssG make contacts with two TssK trimers (77). TssE, -F, -G are, respectively, homologues of phage T4 gp25, gp6, and gp7 and phage Mu Mup46, Mup47, and Mup48 (50, 77, 79, 80), which also constitute the inner part of phage BPs (79–81). TssK has no homologue in *Myoviridae* but shares architectural homologies with receptor-binding proteins (RBP) of *Siphoviridae* phages (67). The structure of the N-terminal domain of TssK is superimposable with that of *Siphoviridae* RBP shoulder domains that are anchored into the BP (67). Indeed, the TssK N-terminal domain establishes extensive contacts with the TssF$_2$G complex (67). The TssK C-terminal domain has evolved to bind to the MC and specifically to the TssL and TssM cytoplasmic domains (57, 64, 66, 67). Similar to the MC, the BP can be accessorized by additional subunits, such as TssA1 in *P. aeruginosa* (82), that may stabilize the complex or provide additional functions.

The tail tube/sheath complex

The TTC comprises the needle and the contractile sheath (Fig. 1). It forms an ~1-µm-long tubular structure in the cytoplasm that is assembled in 30 to 50 s (52, 83).

The needle is composed of the inner tube topped by the spike complex. The inner tube is made of hexamers of the Hcp protein (84–86). These donut-shaped hexameric Hcp rings (87, 88) stack on each other in a head-to-tail orientation to form a hollow tube (86). Interestingly, despite very low sequence similarities between T6SS Hcps and tube proteins from other CISs, their structure is strictly conserved (5). Hcp tube polymerization starts at the BP, through direct recruitment of the first ring to the base of the VgrG hub/spike (89). The spike complex is composed of a trimer of the VgrG protein and, in most instances, of the PAAR repeat protein (85, 90). VgrG contains several conserved domains (24, 85). The N-terminal domain resembles the phage T4 gp27 protein and acts as a symmetry adaptor between the 6-fold symmetry of the inner tube and the 3-fold symmetry of the VgrG central and C-terminal domains, which share homologies with the phage T4 gp5 N-terminal and β-prism domains (89, 91, 92). The VgrG β-prism domain is a triangular β-helix that forms, together with the conical PAAR protein, the penetration device of the T6SS needle (90, 93). The VgrG trimer and the PAAR protein can be extended by additional domains that may act as effectors or as adaptors for effectors (24, 90).

The sheath polymerizes from the BP. It is proposed that, similarly to its gp25-like homologues in *Myoviridae*, the TssE BP subunit constitutes the sheath polymerization initiator (79, 91). In contrast to other CISs, the T6SS sheath is composed of two proteins, TssB and TssC (1, 52, 85, 94, 95), forming a stable dimer that is the repeat unit for sheath polymerization (96–98). Six TssBC dimers form a strand that wraps an Hcp hexameric ring. The TssBC dimer can be divided into three regions: domains 1 and 2 resemble CIS sheath proteins, and domain 3 is inserted into domain 2 (99, 100). Extensive contacts between TssBC dimers from the same strand and from the neighboring −1 and +1 strands stabilize the extended conformation of the sheath polymer (100, 101).

In the T6SS, assembly of the inner tube and that of the extended sheath are interdependent (86, 102). The TssA protein coordinates the polymerization of the TTC (103) (Fig. 2). TssA localizes at the distal extremity of the growing tail tube/sheath (103), at the location in which hexameric tube rings and TssBC strands are incorporated (104). TssA presents a 6-arm starfish-like structure with a central core (103). Protein-protein interaction studies have suggested that the central core of TssA may undergo large conformational changes to insert new Hcp hexamers, whereas the arms may facilitate sheath polymerization (103, 105). Tail tube/sheath polymerization proceeds in the cytoplasm and is stopped when

the distal end hits the membrane on the opposite membrane of the bacterial cell (104, 106). A recent study has identified TagA, a protein that interacts with TssA to stop the assembly of the tail and to maintain the sheath under the extended conformation (106) (Fig. 2). However, the TssA cap protein and the TagA stopper are not conserved in T6SS gene clusters, suggesting that different mechanisms control tail tube/sheath assembly and termination in different T6SS$^+$ species (105–107).

Contraction of the T6SS sheath, which occurs in less than 2–5 ms, is believed to start at the BP. The cryo-electron microscopy structure of the *Vibrio cholerae* T6SS sheath has been solved in the two states, extended and contracted, allowing a reconstitution of the molecular events leading to contraction (99, 100). Contraction consists of a reorganization of the TssBC strands and, notably, an outward rotation of the sheath subunits (100). By doing so, the sheath compacts on the BP and contacts with the inner tube are abolished, thus promoting its expulsion (5, 100, 101). The free energy released during contraction is estimated to >44,000 kcal·mol^{-1} for a 1-μm-long sheath (100).

After contraction, the sheath is disassembled by a dedicated AAA$^+$ ATPase, ClpV (94, 102) (Fig. 2). ClpV binds to an N-terminal helix of TssC that belongs to sheath domain 3 (108, 109), which is accessible only in the contracted conformation (98, 100). Although this is not clearly established, it is proposed that contracted sheath subunits are recycled rather than conveyed to degradation (102).

LOADING AND TRANSPORT OF EFFECTORS

As summarized above, a broad repertoire of antibacterial and antihost activities has been described already for T6SS effectors. In addition, the mode of loading and transfer of these effectors into target cells is also variable. The common theme is that

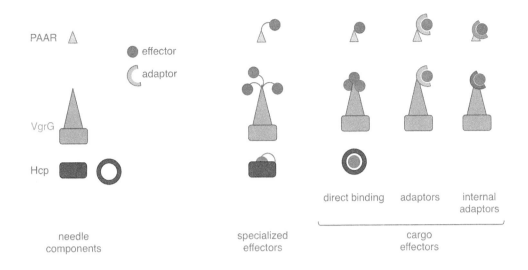

FIGURE 3 Schematic representation of the mechanisms of effector loading. Effectors are depicted as red circles. Specialized effectors are chimeric needle proteins with extensions encoding the effector. Cargo effectors are independent proteins that associate with needle components (Hcp, VgrG, and PAAR). Binding of cargo effectors to needle components could be direct or mediated by adaptor modules that are independent proteins (adaptors) or extensions of VgrG and PAAR (internal adaptors).

these effectors are associated with needle components, as the needle is the only portion of the T6SS to be propelled into the target cell (12, 13) (Fig. 3). Effectors can be additional domains fused to needle components such as Hcp, VgrG, or PAAR or independent proteins that directly or indirectly bind to Hcp, VgrG, or PAAR (12, 13). Recruitment of these independent cargo effectors to Hcp, VgrG, or PAAR can be mediated by adaptors, which are themselves domains of the needle components, or independent proteins (110) (Fig. 3).

Specialized Hcp, VgrG, and PAAR

When the effector module is on the same polypeptide as the needle component, the T6SS subunit is called "specialized" or "evolved." Although effectors fused to Hcp or PAAR have been described (36, 90, 111), the best-characterized examples are C-terminal extensions of specialized VgrGs such as *V. cholerae* VgrG1, which cross-links actin, and VgrG3, which has peptidoglycan

glycoside hydrolase activity; *A. hydrophila* VgrG1, which ADP-ribosylates actin; *P. aeruginosa* VgrG2b, which interacts with the tubule cap complex; and *Burkholderia pseudomallei* VgrG5, which induces host cell membrane fusion (21–26, 112, 113).

Cargo Effectors

Cargo effectors are independent proteins that need to recognize their Hcp, VgrG, or PAAR carrier for transfer. This recognition could be direct, such as the case of effectors that bind Hcp, or may require an additional adaptor module that binds VgrG or PAAR (12, 13, 110) (Fig. 3). Usually the effector genes are genetically linked to genes encoding their vehicle, their adaptors (if any), and, in the case of antibacterial toxins, their immunity proteins. These genetic elements could be found within T6SS gene clusters or as Hcp-VgrG islands scattered on the genome (9, 10).

When associated with Hcp, the effector is embedded in the lumen of the hexameric ring and is thus likely found inside the channel of the inner tube during T6SS assembly (16, 114). Consequently, it is protected and stabilized (114, 115). However, the available space in the Hcp ring lumen limits the size of the effector to be transported, which is estimated to be <25 kDa (114).

Adaptors can be isolated proteins or domains fused to the cargo or the vehicle (110). Adaptors from distinct families, such as DUF1795 (EagT6 and EagR), DUF2169, DUF4123 (Tap-1 and Tec), transthyretin, and recombination hot spot (Rhs), have been described and studied (35, 90, 116–125). When several copies of VgrG or PAAR proteins are encoded within the genome, these adaptor modules specify the carriers on which the effector should be mounted (112, 118, 122, 123, 126). In addition to loading the effector on the vehicle, some of these adaptors have been shown to act as chaperones to stabilize the effector or to wrap hydrophobic transmembrane segments to prevent effector aggregation (112, 124, 125).

CONCLUDING REMARKS

Although the T6SS is one of the most recently identified secretion apparatuses, we now have a detailed view on how the system is assembled, how it is structurally arranged, and how effectors are loaded and transported. The broad repertoire of effectors has only recently started to emerge, and it is likely that many effectors with interesting activities will be identified and characterized in the next years. The discovery of the T6SS 13 years ago and its role as an antibacterial weapon have altered our view of bacterial communities. It is now broadly admitted that not only do bacteria cohabitate peacefully but also complex interactions are established to maintain stable ecosystems, such as the human gut microbiota. Further fundamental and translational works are required to better understand how T6SS activation or inhibition may impact micro-bial communities and may perturb complex ecosystems.

ACKNOWLEDGMENTS

We thank the laboratory members for helpful discussions and support.

Our work on the T6SS is supported by the Centre National de la Recherche Scientifique (CNRS), the Aix-Marseille Université (AMU), and grants from the Agence Nationale de la Recherche (ANR-14-CE14-0006 and ANR-17-CE11-0039 to E.C., ANR-18-CE15-0013 to L.J., and ANR-18-CE11-0023 to E.D.), the Fondation pour la Recherche Médicale (DEQ20180339165), and the Fondation Bettencourt-Schueller. Y.C. is supported by the FRM (FRM-ECO20160736014).

CITATION

Cherrak Y, Flaugnatti N, Durand E, Journet L, Cascales E. 2019. Structure and activity of the type VI secretion system. Microbiol Spectrum 7(4):PSIB-0031-2019.

REFERENCES

1. **Bönemann G, Pietrosiuk A, Mogk A.** 2010. Tubules and donuts: a type VI secretion story. *Mol Microbiol* **76:**815–821.
2. **Zoued A, Brunet YR, Durand E, Aschtgen MS, Logger L, Douzi B, Journet L, Cambillau C, Cascales E.** 2014. Architecture and assembly of the type VI secretion system. *Biochim Biophys Acta* **1843:**1664–1673.
3. **Cascales E.** 2017. Microbiology: and *Amoebophilus* invented the machine gun! *Curr Biol* **27:**R1170–R1173.
4. **Taylor NMI, van Raaij MJ, Leiman PG.** 2018. Contractile injection systems of bacteriophages and related systems. *Mol Microbiol* **108:**6–15.
5. **Brackmann M, Nazarov S, Wang J, Basler M.** 2017. Using force to punch holes: mechanics of contractile nanomachines. *Trends Cell Biol* **27:**623–632.
6. **Cianfanelli FR, Monlezun L, Coulthurst SJ.** 2016. Aim, load, fire: the type VI secretion system, a bacterial nanoweapon. *Trends Microbiol* **24:**51–62.
7. **Coyne MJ, Comstock LE.** 2019. Type VI secretion systems and the gut microbiota. *Microbiol Spectr* **7:**PSIB-0009-2018.

8. **Bingle LE, Bailey CM, Pallen MJ.** 2008. Type VI secretion: a beginner's guide. *Curr Opin Microbiol* **11:**3–8.

9. **Cascales E.** 2008. The type VI secretion toolkit. *EMBO Rep* **9:**735–741.

10. **Boyer F, Fichant G, Berthod J, Vandenbrouck Y, Attree I.** 2009. Dissecting the bacterial type VI secretion system by a genome wide in silico analysis: what can be learned from available microbial genomic resources? *BMC Genomics* **10:**104.

11. **Russell AB, Peterson SB, Mougous JD.** 2014. Type VI secretion system effectors: poisons with a purpose. *Nat Rev Microbiol* **12:**137–148.

12. **Durand E, Cambillau C, Cascales E, Journet L.** 2014. VgrG, Tae, Tle, and beyond: the versatile arsenal of type VI secretion effectors. *Trends Microbiol* **22:**498–507.

13. **Alcoforado Diniz J, Liu YC, Coulthurst SJ.** 2015. Molecular weaponry: diverse effectors delivered by the type VI secretion system. *Cell Microbiol* **17:**1742–1751.

14. **Hachani A, Wood TE, Filloux A.** 2016. Type VI secretion and anti-host effectors. *Curr Opin Microbiol* **29:**81–93.

15. **Fu Y, Waldor MK, Mekalanos JJ.** 2013. Tn-Seq analysis of *Vibrio cholerae* intestinal colonization reveals a role for T6SS-mediated antibacterial activity in the host. *Cell Host Microbe* **14:**652–663.

16. **Sana TG, Flaugnatti N, Lugo KA, Lam LH, Jacobson A, Baylot V, Durand E, Journet L, Cascales E, Monack DM.** 2016. *Salmonella* Typhimurium utilizes a T6SS-mediated antibacterial weapon to establish in the host gut. *Proc Natl Acad Sci U S A* **113:**E5044–E5051.

17. **Sana TG, Lugo KA, Monack DM.** 2017. T6SS: the bacterial "fight club" in the host gut. *PLoS Pathog* **13:**e1006325.

18. **Chassaing B, Cascales E.** 2018. Antibacterial weapons: targeted destruction in the microbiota. *Trends Microbiol* **26:**329–338.

19. **García-Bayona L, Comstock LE.** 2018. Bacterial antagonism in host-associated microbial communities. *Science* **361:**eaat2456.

20. **Veening JW, Blokesch M.** 2017. Interbacterial predation as a strategy for DNA acquisition in naturally competent bacteria. *Nat Rev Microbiol* **15:**621–629.

21. **Pukatzki S, Ma AT, Sturtevant D, Krastins B, Sarracino D, Nelson WC, Heidelberg JF, Mekalanos JJ.** 2006. Identification of a conserved bacterial protein secretion system in *Vibrio cholerae* using the *Dictyostelium* host model system. *Proc Natl Acad Sci U S A* **103:**1528–1533.

22. **Sana TG, Baumann C, Merdes A, Soscia C, Rattei T, Hachani A, Jones C, Bennett KL,** Filloux A, Superti-Furga G, Voulhoux R, Bleves S. 2015. Internalization of *Pseudomonas aeruginosa* strain PAO1 into epithelial cells is promoted by interaction of a T6SS effector with the microtubule network. *mBio* **6:**e00712-15.

23. **Schwarz S, Singh P, Robertson JD, LeRoux M, Skerrett SJ, Goodlett DR, West TE, Mougous JD.** 2014. VgrG-5 is a *Burkholderia* type VI secretion system-exported protein required for multinucleated giant cell formation and virulence. *Infect Immun* **82:**1445–1452.

24. **Pukatzki S, Ma AT, Revel AT, Sturtevant D, Mekalanos JJ.** 2007. Type VI secretion system translocates a phage tail spike-like protein into target cells where it cross-links actin. *Proc Natl Acad Sci U S A* **104:**15508–15513.

25. **Suarez G, Sierra JC, Erova TE, Sha J, Horneman AJ, Chopra AK.** 2010. A type VI secretion system effector protein, VgrG1, from *Aeromonas hydrophila* that induces host cell toxicity by ADP ribosylation of actin. *J Bacteriol* **192:**155–168.

26. **Durand E, Derrez E, Audoly G, Spinelli S, Ortiz-Lombardía M, Raoult D, Cascales E, Cambillau C.** 2012. Crystal structure of the VgrG1 actin cross-linking domain of the *Vibrio cholerae* type VI secretion system. *J Biol Chem* **287:**38190–38199.

27. **Aubert DF, Xu H, Yang J, Shi X, Gao W, Li L, Bisaro F, Chen S, Valvano MA, Shao F.** 2016. A *Burkholderia* type VI effector deamidates Rho GTPases to activate the pyrin inflammasome and trigger inflammation. *Cell Host Microbe* **19:**664–674.

28. **Eshraghi A, Kim J, Walls AC, Ledvina HE, Miller CN, Ramsey KM, Whitney JC, Radey MC, Peterson SB, Ruhland BR, Tran BQ, Goo YA, Goodlett DR, Dove SL, Celli J, Veesler D, Mougous JD.** 2016. Secreted effectors encoded within and outside of the *Francisella* pathogenicity island promote intramacrophage growth. *Cell Host Microbe* **20:**573–583.

29. **Ledvina HE, Kelly KA, Eshraghi A, Plemel RL, Peterson SB, Lee B, Steele S, Adler M, Kawula TH, Merz AJ, Skerrett SJ, Celli J, Mougous JD.** 2018. A phosphatidylinositol 3-kinase effector alters phagosomal maturation to promote intracellular growth of *Francisella. Cell Host Microbe* **24:**285–295.e8.

30. **Lennings J, West TE, Schwarz S.** 2019. The *Burkholderia* type VI secretion system 5: composition, regulation and role in virulence. *Front Microbiol* **9:**3339.

31. **Hood RD, Singh P, Hsu F, Güvener T, Carl MA, Trinidad RR, Silverman JM, Ohlson BB, Hicks KG, Plemel RL, Li M, Schwarz S, Wang WY, Merz AJ, Goodlett DR, Mougous JD.**

2010. A type VI secretion system of *Pseudomonas aeruginosa* targets a toxin to bacteria. *Cell Host Microbe* 7:25–37.

32. **Russell AB, Hood RD, Bui NK, LeRoux M, Vollmer W, Mougous JD.** 2011. Type VI secretion delivers bacteriolytic effectors to target cells. *Nature* 475:343–347.

33. **Russell AB, LeRoux M, Hathazi K, Agnello DM, Ishikawa T, Wiggins PA, Wai SN, Mougous JD.** 2013. Diverse type VI secretion phospholipases are functionally plastic antibacterial effectors. *Nature* 496:508–512.

34. **Ma LS, Hachani A, Lin JS, Filloux A, Lai EM.** 2014. *Agrobacterium tumefaciens* deploys a superfamily of type VI secretion DNase effectors as weapons for interbacterial competition in planta. *Cell Host Microbe* 16:94–104.

35. **Whitney JC, Quentin D, Sawai S, LeRoux M, Harding BN, Ledvina HE, Tran BQ, Robinson H, Goo YA, Goodlett DR, Raunser S, Mougous JD.** 2015. An interbacterial NAD (P)(+) glycohydrolase toxin requires elongation factor Tu for delivery to target cells. *Cell* 163:607–619.

36. **Pissaridou P, Allsopp LP, Wettstadt S, Howard SA, Mavridou DAI, Filloux A.** 2018. The *Pseudomonas aeruginosa* T6SS-VgrG1b spike is topped by a PAAR protein eliciting DNA damage to bacterial competitors. *Proc Natl Acad Sci U S A* 115:12519–12524.

37. **Ting SY, Bosch DE, Mangiameli SM, Radey MC, Huang S, Park YJ, Kelly KA, Filip SK, Goo YA, Eng JK, Allaire M, Veesler D, Wiggins PA, Peterson SB, Mougous JD.** 2018. Bifunctional immunity proteins protect bacteria against FtsZ-targeting ADP-ribosylating toxins. *Cell* 175:1380–1392.e14.

38. **Wang T, Si M, Song Y, Zhu W, Gao F, Wang Y, Zhang L, Zhang W, Wei G, Luo ZQ, Shen X.** 2015. Type VI secretion system transports Zn2+ to combat multiple stresses and host immunity. *PLoS Pathog* 11:e1005020.

39. **Si M, Zhao C, Burkinshaw B, Zhang B, Wei D, Wang Y, Dong TG, Shen X.** 2017. Manganese scavenging and oxidative stress response mediated by type VI secretion system in *Burkholderia thailandensis. Proc Natl Acad Sci U S A* 114:E2233–E2242.

40. **Si M, Wang Y, Zhang B, Zhao C, Kang Y, Bai H, Wei D, Zhu L, Zhang L, Dong TG, Shen X.** 2017. The type VI secretion system engages a redox-regulated dual-functional heme transporter for zinc acquisition. *Cell Rep* 20:949–959.

41. **Verster AJ, Ross BD, Radey MC, Bao Y, Goodman AL, Mougous JD, Borenstein E.** 2017. The landscape of type VI secretion across human gut microbiomes reveals its role in community composition. *Cell Host Microbe* 22:411–419.e4.

42. **Bernard CS, Brunet YR, Gueguen E, Cascales E.** 2010. Nooks and crannies in type VI secretion regulation. *J Bacteriol* 192:3850–3860.

43. **Silverman JM, Brunet YR, Cascales E, Mougous JD.** 2012. Structure and regulation of the type VI secretion system. *Annu Rev Microbiol* 66:453–472.

44. **Miyata ST, Bachmann V, Pukatzki S.** 2013. Type VI secretion system regulation as a consequence of evolutionary pressure. *J Med Microbiol* 62:663–676.

45. **LeRoux M, Peterson SB, Mougous JD.** 2015. Bacterial danger sensing. *J Mol Biol* 427:3744–3753.

46. **Ho BT, Dong TG, Mekalanos JJ.** 2014. A view to a kill: the bacterial type VI secretion system. *Cell Host Microbe* 15:9–21.

47. **Basler M.** 2015. Type VI secretion system: secretion by a contractile nanomachine. *Philos Trans R Soc Lond B Biol Sci* 370:20150021.

48. **Coulthurst S.** 2019. The type VI secretion system: a versatile bacterial weapon. *Microbiology* 165:503–515.

49. **Durand E, Nguyen VS, Zoued A, Logger L, Péhau-Arnaudet G, Aschtgen MS, Spinelli S, Desmyter A, Bardiaux B, Dujeancourt A, Roussel A, Cambillau C, Cascales E, Fronzes R.** 2015. Biogenesis and structure of a type VI secretion membrane core complex. *Nature* 523:555–560.

50. **Brunet YR, Zoued A, Boyer F, Douzi B, Cascales E.** 2015. The type VI secretion TssEFGK-VgrG phage-like baseplate is recruited to the TssJLM membrane complex *via* multiple contacts and serves as assembly platform for tail tube/sheath polymerization. *PLoS Genet* 11: e1005545.

51. **Gerc AJ, Diepold A, Trunk K, Porter M, Rickman C, Armitage JP, Stanley-Wall NR, Coulthurst SJ.** 2015. Visualization of the *Serratia* type VI secretion system reveals unprovoked attacks and dynamic assembly. *Cell Rep* 12:2131–2142.

52. **Basler M, Pilhofer M, Henderson GP, Jensen GJ, Mekalanos JJ.** 2012. Type VI secretion requires a dynamic contractile phage tail-like structure. *Nature* 483:182–186.

53. **Aschtgen MS, Gavioli M, Dessen A, Lloubès R, Cascales E.** 2010. The SciZ protein anchors the enteroaggregative *Escherichia coli* type VI secretion system to the cell wall. *Mol Microbiol* 75:886–899.

54. **Aschtgen MS, Bernard CS, De Bentzmann S, Lloubès R, Cascales E.** 2008. SciN is an outer membrane lipoprotein required for type VI

secretion in enteroaggregative *Escherichia coli*. *J Bacteriol* 190:7523–7531.

55. **Ma LS, Lin JS, Lai EM.** 2009. An IcmF family protein, ImpL$_M$, is an integral inner membrane protein interacting with ImpK$_L$, and its walker a motif is required for type VI secretion system-mediated Hcp secretion in *Agrobacterium tumefaciens*. *J Bacteriol* 191:4316–4329.

56. **Aschtgen MS, Zoued A, Lloubès R, Journet L, Cascales E.** 2012. The C-tail anchored TssL subunit, an essential protein of the enteroaggregative *Escherichia coli* Sci-1 type VI secretion system, is inserted by YidC. *Microbiologyopen* 1:71–82.

57. **Logger L, Aschtgen MS, Guérin M, Cascales E, Durand E.** 2016. Molecular dissection of the interface between the type VI secretion TssM cytoplasmic domain and the TssG baseplate component. *J Mol Biol* 428:4424–4437.

58. **Felisberto-Rodrigues C, Durand E, Aschtgen MS, Blangy S, Ortiz-Lombardia M, Douzi B, Cambillau C, Cascales E.** 2011. Towards a structural comprehension of bacterial type VI secretion systems: characterization of the TssJ-TssM complex of an *Escherichia coli* pathovar. *PLoS Pathog* 7:e1002386.

59. **Rao VA, Shepherd SM, English G, Coulthurst SJ, Hunter WN.** 2011. The structure of *Serratia marcescens* Lip, a membrane-bound component of the type VI secretion system. *Acta Crystallogr D Biol Crystallogr* 67:1065–1072.

60. **Robb CS, Assmus M, Nano FE, Boraston AB.** 2013. Structure of the T6SS lipoprotein TssJ1 from *Pseudomonas aeruginosa*. *Acta Crystallogr Sect F Struct Biol Cryst Commun* 69:607–610.

61. **Durand E, Zoued A, Spinelli S, Watson PJ, Aschtgen MS, Journet L, Cambillau C, Cascales E.** 2012. Structural characterization and oligomerization of the TssL protein, a component shared by bacterial type VI and type IVb secretion systems. *J Biol Chem* 287:14157–14168.

62. **Robb CS, Nano FE, Boraston AB.** 2012. The structure of the conserved type six secretion protein TssL (DotU) from *Francisella novicida*. *J Mol Biol* 419:277–283.

63. **Chang JH, Kim YG.** 2015. Crystal structure of the bacterial type VI secretion system component TssL from *Vibrio cholerae*. *J Microbiol* 53:32–37.

64. **Zoued A, Cassaro CJ, Durand E, Douzi B, España AP, Cambillau C, Journet L, Cascales E.** 2016. Structure-function analysis of the TssL cytoplasmic domain reveals a new interaction between the type VI secretion baseplate and membrane complexes. *J Mol Biol* 428:4413–4423.

65. **Zoued A, Duneau JP, Durand E, España AP, Journet L, Guerlesquin F, Cascales E.** 2018. Tryptophan-mediated dimerization of the TssL transmembrane anchor is required for type VI secretion system activity. *J Mol Biol* 430:987–1003.

66. **Zoued A, Durand E, Bebeacua C, Brunet YR, Douzi B, Cambillau C, Cascales E, Journet L.** 2013. TssK is a trimeric cytoplasmic protein interacting with components of both phage-like and membrane anchoring complexes of the type VI secretion system. *J Biol Chem* 288:27031–27041.

67. **Nguyen VS, Logger L, Spinelli S, Legrand P, Huyen Pham TT, Nhung Trinh TT, Cherrak Y, Zoued A, Desmyter A, Durand E, Roussel A, Kellenberger C, Cascales E, Cambillau C.** 2017. Type VI secretion TssK baseplate protein exhibits structural similarity with phage receptor-binding proteins and evolved to bind the membrane complex. *Nat Microbiol* 2:17103.

68. **Rapisarda C, Cherrak Y, Kooger R, Schmidt V, Pellarin R, Logger L, Cascales E, Pilhofer M, Durand E, Fronzes R.** 2019. *In situ* and high-resolution cryo-EM structure of a bacterial type VI secretion system membrane complex. *EMBO J* 38:e100886.

69. **Yin M, Yan Z, Li X.** 2019. Architecture of type VI secretion system membrane core complex. *Cell Res* 29:251–253.

70. **Aschtgen MS, Thomas MS, Cascales E.** 2010. Anchoring the type VI secretion system to the peptidoglycan: TssL, TagL, TagP... what else? *Virulence* 1:535–540.

71. **Santin YG, Camy CE, Zoued A, Doan T, Aschtgen MS, Cascales E.** 2019. Role and recruitment of the TagL peptidoglycan-binding protein during type VI secretion system biogenesis. *J Bacteriol* 201:e00173-19.

72. **Weber BS, Hennon SW, Wright MS, Scott NE, de Berardinis V, Foster LJ, Ayala JA, Adams MD, Feldman MF.** 2016. Genetic dissection of the type VI secretion system in *Acinetobacter* and identification of a novel peptidoglycan hydrolase, TagX, required for its biogenesis. *mBio* 7:e01253-16.

73. **Santin YG, Cascales E.** 2017. Domestication of a housekeeping transglycosylase for assembly of a type VI secretion system. *EMBO Rep* 18:138–149.

74. **Russell AB, Wexler AG, Harding BN, Whitney JC, Bohn AJ, Goo YA, Tran BQ, Barry NA, Zheng H, Peterson SB, Chou S, Gonen T, Goodlett DR, Goodman AL, Mougous JD.** 2014. A type VI secretion-related pathway in *Bacteroidetes* mediates interbacterial antagonism. *Cell Host Microbe* 16:227–236.

75. Coyne MJ, Roelofs KG, Comstock LE. 2016. Type VI secretion systems of human gut *Bacteroidales* segregate into three genetic architectures, two of which are contained on mobile genetic elements. *BMC Genomics* 17:58.

76. Nazarov S, Schneider JP, Brackmann M, Goldie KN, Stahlberg H, Basler M. 2018. Cryo-EM reconstruction of type VI secretion system baseplate and sheath distal end. *EMBO J* 37:e97103.

77. Cherrak Y, Rapisarda C, Pellarin R, Bouvier G, Bardiaux B, Allain F, Malosse C, Rey M, Chamot-Rooke J, Cascales E, Fronzes R, Durand E. 2018. Biogenesis and structure of a type VI secretion baseplate. *Nat Microbiol* 3:1404–1416.

78. English G, Byron O, Cianfanelli FR, Prescott AR, Coulthurst SJ. 2014. Biochemical analysis of TssK, a core component of the bacterial type VI secretion system, reveals distinct oligomeric states of TssK and identifies a TssK-TssFG subcomplex. *Biochem J* 461:291–304.

79. Taylor NM, Prokhorov NS, Guerrero-Ferreira RC, Shneider MM, Browning C, Goldie KN, Stahlberg H, Leiman PG. 2016. Structure of the T4 baseplate and its function in triggering sheath contraction. *Nature* 533:346–352.

80. Büttner CR, Wu Y, Maxwell KL, Davidson AR. 2016. Baseplate assembly of phage Mu: defining the conserved core components of contractile-tailed phages and related bacterial systems. *Proc Natl Acad Sci U S A* 113:10174–10179.

81. Kostyuchenko VA, Leiman PG, Chipman PR, Kanamaru S, van Raaij MJ, Arisaka F, Mesyanzhinov VV, Rossmann MG. 2003. Three-dimensional structure of bacteriophage T4 baseplate. *Nat Struct Biol* 10:688–693.

82. Planamente S, Salih O, Manoli E, Albesa-Jové D, Freemont PS, Filloux A. 2016. TssA forms a gp6-like ring attached to the type VI secretion sheath. *EMBO J* 35:1613–1627.

83. Brunet YR, Espinosa L, Harchouni S, Mignot T, Cascales E. 2013. Imaging type VI secretion-mediated bacterial killing. *Cell Rep* 3:36–41.

84. Ballister ER, Lai AH, Zuckermann RN, Cheng Y, Mougous JD. 2008. In vitro self-assembly of tailorable nanotubes from a simple protein building block. *Proc Natl Acad Sci U S A* 105:3733–3738.

85. Leiman PG, Basler M, Ramagopal UA, Bonanno JB, Sauder JM, Pukatzki S, Burley SK, Almo SC, Mekalanos JJ. 2009. Type VI secretion apparatus and phage tail-associated protein complexes share a common evolutionary origin. *Proc Natl Acad Sci U S A* 106:4154–4159.

86. Brunet YR, Hénin J, Celia H, Cascales E. 2014. Type VI secretion and bacteriophage tail tubes share a common assembly pathway. *EMBO Rep* 15:315–321.

87. Mougous JD, Cuff ME, Raunser S, Shen A, Zhou M, Gifford CA, Goodman AL, Joachimiak G, Ordoñez CL, Lory S, Walz T, Joachimiak A, Mekalanos JJ. 2006. A virulence locus of *Pseudomonas aeruginosa* encodes a protein secretion apparatus. *Science* 312:1526–1530.

88. Douzi B, Spinelli S, Blangy S, Roussel A, Durand E, Brunet YR, Cascales E, Cambillau C. 2014. Crystal structure and self-interaction of the type VI secretion tail-tube protein from enteroaggregative *Escherichia coli*. *PLoS One* 9:e86918.

89. Renault MG, Zamarreno Beas J, Douzi B, Chabalier M, Zoued A, Brunet YR, Cambillau C, Journet L, Cascales E. 2018. The gp27-like hub of VgrG serves as adaptor to promote Hcp tube assembly. *J Mol Biol* 430(18 Part B):3143–3156.

90. Shneider MM, Buth SA, Ho BT, Basler M, Mekalanos JJ, Leiman PG. 2013. PAAR-repeat proteins sharpen and diversify the type VI secretion system spike. *Nature* 500:350–353.

91. Leiman PG, Shneider MM. 2012. Contractile tail machines of bacteriophages. *Adv Exp Med Biol* 726:93–114.

92. Spínola-Amilibia M, Davó-Siguero I, Ruiz FM, Santillana E, Medrano FJ, Romero A. 2016. The structure of VgrG1 from *Pseudomonas aeruginosa*, the needle tip of the bacterial type VI secretion system. *Acta Crystallogr D Struct Biol* 72:22–33.

93. Uchida K, Leiman PG, Arisaka F, Kanamaru S. 2014. Structure and properties of the C-terminal β-helical domain of VgrG protein from *Escherichia coli* O157. *J Biochem* 155:173–182.

94. Bönemann G, Pietrosiuk A, Diemand A, Zentgraf H, Mogk A. 2009. Remodelling of VipA/VipB tubules by ClpV-mediated threading is crucial for type VI protein secretion. *EMBO J* 28:315–325.

95. Lossi NS, Dajani R, Freemont P, Filloux A. 2011. Structure-function analysis of HsiF, a gp25-like component of the type VI secretion system, in *Pseudomonas aeruginosa*. *Microbiology* 157:3292–3305.

96. Bröms JE, Ishikawa T, Wai SN, Sjöstedt A. 2013. A functional VipA-VipB interaction is required for the type VI secretion system activity of *Vibrio cholerae* O1 strain A1552. *BMC Microbiol* 13:96.

97. Zhang XY, Brunet YR, Logger L, Douzi B, Cambillau C, Journet L, Cascales E. 2013.

Dissection of the TssB-TssC interface during type VI secretion sheath complex formation. *PLoS One* **8**:e81074.

98. **Kube S, Kapitein N, Zimniak T, Herzog F, Mogk A, Wendler P.** 2014. Structure of the VipA/B type VI secretion complex suggests a contraction-state-specific recycling mechanism. *Cell Rep* **8**:20–30.

99. **Kudryashev M, Wang RY, Brackmann M, Scherer S, Maier T, Baker D, DiMaio F, Stahlberg H, Egelman EH, Basler M.** 2015. Structure of the type VI secretion system contractile sheath. *Cell* **160**:952–962.

100. **Wang J, Brackmann M, Castaño-Díez D, Kudryashev M, Goldie KN, Maier T, Stahlberg H, Basler M.** 2017. Cryo-EM structure of the extended type VI secretion system sheath-tube complex. *Nat Microbiol* **2**:1507–1512.

101. **Brackmann M, Wang J, Basler M.** 2018. Type VI secretion system sheath inter-subunit interactions modulate its contraction. *EMBO Rep* **19**:225–233.

102. **Kapitein N, Bönemann G, Pietrosiuk A, Seyffer F, Hausser I, Locker JK, Mogk A.** 2013. ClpV recycles VipA/VipB tubules and prevents non-productive tubule formation to ensure efficient type VI protein secretion. *Mol Microbiol* **87**:1013–1028.

103. **Zoued A, Durand E, Brunet YR, Spinelli S, Douzi B, Guzzo M, Flaugnatti N, Legrand P, Journet L, Fronzes R, Mignot T, Cambillau C, Cascales E.** 2016. Priming and polymerization of a bacterial contractile tail structure. *Nature* **531**:59–63.

104. **Vettiger A, Winter J, Lin L, Basler M.** 2017. The type VI secretion system sheath assembles at the end distal from the membrane anchor. *Nat Commun* **8**:16088.

105. **Zoued A, Durand E, Santin YG, Journet L, Roussel A, Cambillau C, Cascales E.** 2017. TssA: the cap protein of the type VI secretion system tail. *Bioessays* **39**:10.

106. **Santin YG, Doan T, Lebrun R, Espinosa L, Journet L, Cascales E.** 2018. *In vivo* TssA proximity labelling during type VI secretion biogenesis reveals TagA as a protein that stops and holds the sheath. *Nat Microbiol* **3**:1304–1313.

107. **Dix SR, Owen HJ, Sun R, Ahmad A, Shastri S, Spiewak HL, Mosby DJ, Harris MJ, Batters SL, Brooker TA, Tzokov SB, Sedelnikova SE, Baker PJ, Bullough PA, Rice DW, Thomas MS.** 2018. Structural insights into the function of type VI secretion system TssA subunits. *Nat Commun* **9**:4765.

108. **Pietrosiuk A, Lenherr ED, Falk S, Bönemann G, Kopp J, Zentgraf H, Sinning I, Mogk A.** 2011. Molecular basis for the unique role of the AAA+ chaperone ClpV in type VI protein secretion. *J Biol Chem* **286**:30010–30021.

109. **Douzi B, Brunet YR, Spinelli S, Lensi V, Legrand P, Blangy S, Kumar A, Journet L, Cascales E, Cambillau C.** 2016. Structure and specificity of the type VI secretion system ClpV-TssC interaction in enteroaggregative *Escherichia coli*. *Sci Rep* **6**:34405.

110. **Unterweger D, Kostiuk B, Pukatzki S.** 2017. Adaptor proteins of type VI secretion system effectors. *Trends Microbiol* **25**:8–10.

111. **Ma J, Pan Z, Huang J, Sun M, Lu C, Yao H.** 2017. The Hcp proteins fused with diverse extended-toxin domains represent a novel pattern of antibacterial effectors in type VI secretion systems. *Virulence* **8**:1189–1202.

112. **Brooks TM, Unterweger D, Bachmann V, Kostiuk B, Pukatzki S.** 2013. Lytic activity of the *Vibrio cholerae* type VI secretion toxin VgrG-3 is inhibited by the antitoxin TsaB. *J Biol Chem* **288**:7618–7625.

113. **Toesca IJ, French CT, Miller JF.** 2014. The type VI secretion system spike protein VgrG5 mediates membrane fusion during intercellular spread by *pseudomallei* group *Burkholderia* species. *Infect Immun* **82**:1436–1444.

114. **Silverman JM, Agnello DM, Zheng H, Andrews BT, Li M, Catalano CE, Gonen T, Mougous JD.** 2013. Haemolysin coregulated protein is an exported receptor and chaperone of type VI secretion substrates. *Mol Cell* **51**:584–593.

115. **Whitney JC, Beck CM, Goo YA, Russell AB, Harding BN, De Leon JA, Cunningham DA, Tran BQ, Low DA, Goodlett DR, Hayes CS, Mougous JD.** 2014. Genetically distinct pathways guide effector export through the type VI secretion system. *Mol Microbiol* **92**:529–542.

116. **Unterweger D, Miyata ST, Bachmann V, Brooks TM, Mullins T, Kostiuk B, Provenzano D, Pukatzki S.** 2014. The *Vibrio cholerae* type VI secretion system employs diverse effector modules for intraspecific competition. *Nat Commun* **5**:3549.

117. **Liang X, Moore R, Wilton M, Wong MJ, Lam L, Dong TG.** 2015. Identification of divergent type VI secretion effectors using a conserved chaperone domain. *Proc Natl Acad Sci U S A* **112**:9106–9111.

118. **Unterweger D, Kostiuk B, Ötjengerdes R, Wilton A, Diaz-Satizabal L, Pukatzki S.** 2015. Chimeric adaptor proteins translocate diverse type VI secretion system effectors in *Vibrio cholerae*. *EMBO J* **34**:2198–2210.

119. **Alcoforado Diniz J, Coulthurst SJ.** 2015. Intraspecies competition in *Serratia marcescens* is mediated by type VI-secreted Rhs

effectors and a conserved effector-associated accessory protein. *J Bacteriol* **197**:2350–2360.

120. **Flaugnatti N, Le TT, Canaan S, Aschtgen MS, Nguyen VS, Blangy S, Kellenberger C, Roussel A, Cambillau C, Cascales E, Journet L.** 2016. A phospholipase A1 antibacterial type VI secretion effector interacts directly with the C-terminal domain of the VgrG spike protein for delivery. *Mol Microbiol* **99**:1099–1118.

121. **Bondage DD, Lin JS, Ma LS, Kuo CH, Lai EM.** 2016. VgrG C terminus confers the type VI effector transport specificity and is required for binding with PAAR and adaptor-effector complex. *Proc Natl Acad Sci U S A* **113**:E3931–E3940.

122. **Cianfanelli FR, Alcoforado Diniz J, Guo M, De Cesare V, Trost M, Coulthurst SJ.** 2016. VgrG and PAAR proteins define distinct versions of a functional type VI secretion system. *PLoS Pathog* **12**:e1005735.

123. **Ma J, Sun M, Dong W, Pan Z, Lu C, Yao H.** 2017. PAAR-Rhs proteins harbor various C-terminal toxins to diversify the antibacterial pathways of type VI secretion systems. *Environ Microbiol* **19**:345–360.

124. **Quentin D, Ahmad S, Shanthamoorthy P, Mougous JD, Whitney JC, Raunser S.** 2018. Mechanism of loading and translocation of type VI secretion system effector Tse6. *Nat Microbiol* **3**:1142–1152.

125. **Burkinshaw BJ, Liang X, Wong M, Le ANH, Lam L, Dong TG.** 2018. A type VI secretion system effector delivery mechanism dependent on PAAR and a chaperone-co-chaperone complex. *Nat Microbiol* **3**:632–640.

126. **Hachani A, Allsopp LP, Oduko Y, Filloux A.** 2014. The VgrG proteins are "à la carte" delivery systems for bacterial type VI effectors. *J Biol Chem* **289**:17872–17884.

Type VI Secretion Systems and the Gut Microbiota

27

MICHAEL J. COYNE[1] and LAURIE E. COMSTOCK[1]

Type VI secretion systems (T6SSs) were first identified and characterized for pathogenic bacteria of the proteobacterial phylum (1, 2). The discovery in 2010 that these secretion systems can target and intoxicate not only eukaryotic cells but also other bacteria (3) revealed that some T6SSs help bacteria compete with other bacteria in a community setting. Indeed, many proteobacterial symbionts, including the plant symbiont *Pseudomonas putida*, the bumble bee gut symbiont *Snodgrassella alvi*, and the squid symbiont *Vibrio fischeri*, all have T6SSs that provide a competitive advantage in their natural ecosystems (4–6). An early *in silico* analysis using clusters of orthologous groups (COGs) models of proteobacterial T6SS proteins against primary sequence databases suggested that T6SSs are largely confined to proteobacterial species, which are minor members of some human-associated microbial communities such as the gut microbiota (7, 8).

In the dense microbial ecosystem of the human colon, contact-dependent mechanisms of antagonism, such as a T6SS, should be an effective means of thwarting competitors. In this microbial ecosystem, the predominant Gram-negative bacteria are of the phylum *Bacteroidetes*, comprising approximately half of the colonic bacteria in many people and vastly outnumbering commensal proteobacteria. Most human gut *Bacteroidetes* species are contained

[1]Division of Infectious Diseases, Brigham and Women's Hospital, Harvard Medical School, Boston, MA 02115
Protein Secretion in Bacteria
Edited by Maria Sandkvist, Eric Cascales, and Peter J. Christie
© 2019 American Society for Microbiology, Washington, DC
doi:10.1128/microbiolspec.PSIB-0009-2018

within the order *Bacteroidales*, which includes several different families, each with one predominant genus in the human gut: *Bacteroides* (family *Bacteroidaceae*), the *Parabacteroides* (family *Tannerellaceae*), *Prevotella* (family *Prevotellaceae*), and *Alistipes* (family *Rickenellaceae*), as well as families with more minor representatives. Humans are colonized at high density with numerous *Bacteroidales* species simultaneously (9–11), and the abundant species in one individual may not be the same as in another (9, 10). T6SSs were not discovered in species of the phylum *Bacteroidetes* until recently (12–14), mostly due to the lack of primary sequence or profile sequence similarity of the 13 core proteobacterial T6SS proteins with *Bacteroidetes* proteins. The use of profile-profile analyses revealed that *Bacteroidales* species of the human gut are a rich source of diverse T6SSs (15). As structural and mechanistic properties of T6SSs are discussed in chapter 26, here we focus on unique properties of the T6SSs of the human gut *Bacteroidales*, the ecological and functional properties of these T6SSs that are known to date, and the ecological advantages conferred by the T6SSs of human gut proteobacterial strains *in vivo*.

GENETIC CHARACTERISTICS AND PROPERTIES OF THE T6SS OF GUT *BACTEROIDALES*

An analysis of 205 human gut *Bacteroidales* genomes including 35 different species of four genera, *Bacteroides*, *Parabacteroides*, *Prevotella*, and *Alistipes*, revealed the presence of 130 T6SS loci in 115 of these strains, with 15 strains containing two different T6SS loci (15). These T6SS loci were found in the genomes of 19 different species of *Bacteroides*, *Parabacteroides*, and *Prevotella* but were not found in any of the nine *Alistipes* genomes analyzed. A notable feature is that these T6SSs segregated into three very distinct genetic architectures (GA), termed GA1, GA2, and GA3 (15) (Fig. 1A). Within a given

GA, the majority of the T6SS genes are of high DNA identity, interspersed with small regions that are variable (Fig. 1A). The variable regions of all three T6SS GAs contain genes encoding effector and immunity proteins, some of which are similar to those previously described for other organisms (15, 16). Most of the GA3 T6SS effector and immunity proteins are unlike those previously described and function by mechanisms currently unknown. On the basis of comparison of the divergent regions within a GA, there are predicted to be at least 30 different variable regions in the 48 GA1 loci analyzed, 21 in the 9 GA2 regions analyzed, and 17 in the 56 GA3 regions analyzed, with each divergent region likely encoding at least one effector immunity pair. Therefore, the T6SSs of gut *Bacteroidales* encode numerous distinct toxins, many of which operate via unknown mechanisms.

An analysis of the predicted proteins produced by these loci showed that four of the conserved T6SS proteins of proteobacterial species are missing in gut *Bacteroidales* T6SSs: TssA, TssJ, TssL, and TssM. TssJ, -L, and -M are components of the transmembrane complex (17), and TssA binds this complex and likely recruits the baseplate assemblage and coordinates tail tube and sheath biogenesis (18, 19). Instead of encoding these proteins, all three GAs of *Bacteroidales* T6SS loci encode four conserved proteins not present in proteobacterial T6SS loci, namely, TssO to TssR (13, 15). These likely serve as functional orthologs of the proteobacterial proteins comprising the transmembrane complex.

All three T6SS GAs encode multiple TssD (Hcp) needle proteins, with GA2 and GA3 loci encoding six and five distinct TssD proteins, respectively. In the GA2 loci, one of the six TssD proteins has a C-terminal extension, likely conferring toxin activity on this protein (Fig. 1A). In the GA3 locus, the main structural TssD protein was identified among the five TssD proteins (Fig. 1A) (20). The function of the accessory TssD proteins

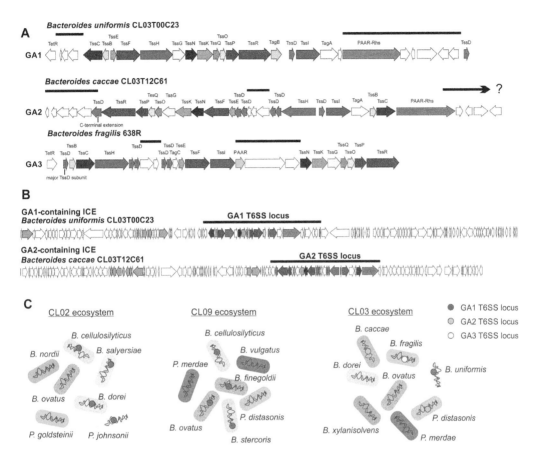

FIGURE 1 (A) Open reading frame (ORF) maps of one representative locus of each of the three genetic architectures (GA) of T6SS loci of gut *Bacteroidales*. T6SS loci of GA1 and GA2 are present in diverse *Bacteroidales* species, whereas GA3 T6SS loci are confined to *B. fragilis*. T6SS loci of a given GA are extremely similar to each other except for the divergent regions noted by lines above the genes, which encode known or putative effector and immunity proteins. The major TssD protein of GA3 is noted, as is the TssD protein of the GA2 loci that have C-terminal extensions likely conferring toxin activity. The ends of the GA1 and GA2 loci have not been precisely determined. **(B)** ORF maps of ICE containing GA1 and GA2 T6SS loci of two *Bacteroides* species. The T6SS loci are designated by a line above the map. Genes involved in conjugative transfer (*tra* genes) are colored green (15). **(C)** The abundant fecal gut *Bacteroidales* from three different healthy humans (CL02, CL09, and CL03) were analyzed for the presence of T6SSs. Seven *Bacteroidales* species were isolated and sequenced from subject CL02 and from subject CL09. Four of the seven species harbor nearly identical GA1 T6SSs loci within a subject, demonstrating transfer of these ICE between these strains in their gut (12, 15). In contrast, of the eight species isolated and sequenced from human subject CL03, two contain GA2 T6SS loci, albeit with different divergent regions. Therefore, these GA2 ICE were not transferred between these species. In addition, one species contains a GA1 T6SS locus and the *B. fragilis* strain from this individual contains a GA3 T6SS locus (15). Red, green, and yellow dots represent the GA1, GA2, and GA3 T6SS loci.

may be to bind effector proteins and incorporate them into the needle structure, as in both the GA2 and GA3 loci, four of the *tssD* genes flank the effector and immunity gene variable regions (Fig. 1A).

An important feature of the three different GAs is their distribution among gut *Bacteroidales* species. GA1 and GA2 T6SS loci are present in numerous species of *Bacteroides*, *Parabacteroides*, and *Prevotella*, whereas GA3

T6SSs are present exclusively in *Bacteroides fragilis*. The wide distribution of GA1 and GA2 in at least three distinct families of *Bacteroidales* is due to their presence on integrative and conjugative elements (ICE) (12, 15) (Fig. 1B). ICE containing GA1 T6SSs are in the range of 120 kb and ICE containing distinct GA1 T6SSs are extremely similar to each other. Excluding the T6SS divergent regions, the DNAs of ICE containing distinct GA1 T6SS loci are approximately 95% similar over their entire lengths. Although very dissimilar to the GA1 ICE, GA2-containing ICE are on the order of 100 kb and share approximately 75% to 99% DNA identity over their lengths with other GA2-containing ICE. As ICE are chromosomal self-transmissible mobile elements (21, 22), they have the ability to move between species. In fact, the GA1-containing ICE have been found to be transferred extensively between *Bacteroidales* species that are coresident in the human intestine (12, 15) (Fig. 1C). Examination of hundreds of genomes of gut *Bacteroidales* has indicated that GA2 T6SSs are not present in a strain that has either a GA1 or GA3 T6SS locus. However, *B. fragilis* strains can have both GA1 and GA3 T6SS loci in their genome. The reason for the apparent lack of GA2 T6SS loci in the same chromosome with GA1 or GA3 T6SS loci is currently unknown. *Bacteroidales* species present in the gut of an individual can, however, collectively contain all three different T6SS GAs (15) (Fig. 1C).

ECOLOGICAL CONSEQUENCES OF GUT BACTEROIDALES T6SS ANTAGONISM

The fitness benefits of the T6SSs of the gut *Bacteroidales* are still incompletely understood. The best studied are the GA3 T6SSs of *B. fragilis*. GA3 T6SSs were found in approximately 86% of *B. fragilis* strains based on genome or metagenome analyses including hundreds of strains (15, 23), making them widely distributed in the species. The GA3 T6SSs of two different *B. fragilis* strains

were found to antagonize all gut *Bacteroidales* species tested, including *Bacteroides* and *Parabacteroides* species, but not the one *Prevotella copri* strain analyzed or other *B. fragilis* strains with the same GA3 T6SS region or immunity genes (20, 24). In addition, no killing was evident against any gut proteobacterial species analyzed (20), suggesting specificity for *Bacteroidales*. The TssD needle protein of the *B. fragilis* 638R GA3 T6SS is present in the supernatant of actively growing cells *in vitro* as well as in the feces of monoassociated gnotobiotic mice (20), suggesting that this antagonistic system is constitutively synthesized and firing, rather than only responding to specific external signals or threats. The number of GA3 T6SS transmission events in a human gut colonized at typical levels with *B. fragilis* was predicted to be on the order of 6×10^{10} to 10^{11} per day (24).

Effector and immunity proteins were identified in three different GA3 T6SS loci (20, 24, 25). These toxic effector proteins are not similar to other known proteins and therefore intoxicate by as-yet-unidentified mechanisms. The toxicity of two of these effectors requires an added N-terminal periplasmic targeting sequence when they are produced inside a sensitive cell from an inducible promoter (26), suggesting that they may need to be localized to the periplasm for toxicity. In a few strains, genes encoding immunity proteins to GA3 effectors were found outside of the T6SS regions, in some cases in strains that did not have a GA3 T6SS locus (24). These immunity genes were found to confer protection from attack by a *B. fragilis* strain synthesizing the T6SS effectors to which the immunity proteins are directed, suggesting that acquisition of these immunity genes confers an advantage on an organism by protecting it from GA3 T6SS-mediated antagonism (24). A recent study analyzing human gut metagenomic data revealed arrays of orphan immunity genes, which were termed acquired interbacterial defense gene clusters (27). These orphan immunity islands reside on predicted mobile elements and include immunity genes likely derived from *B. fragilis* GA3 T6SS loci and disseminated by lateral transfer.

Although the GA3 T6SSs are very effective at antagonizing *Bacteroidales* species *in vitro*, the effects *in vivo* are more variable. As *B. fragilis* is known to coexist in the human gut with numerous other *Bacteroidales* species that are susceptible to its GA3 T6SS, the spatial organization of the microbiota likely dictates the effectiveness of this weapon. In a gnotobiotic mouse competitive colonization model, an isogenic T6SS⁺ wild-type strain outcompetes an isogenic mutant strain lacking the T6SS effector and immunity genes (20, 24). As isogenic strains should share the same spatial and nutritional niche and therefore should make frequent contacts, a strong antagonistic effect is expected. The ability of GA3 T6SSs of *B. fragilis* to antagonize other wild-type nonisogenic *B. fragilis* strains *in vivo* was also demonstrated (24, 25). However, the effectiveness of the GA3 T6SS was found to be lower when analyzing antagonism of different *Bacteroides* species (Fig. 2B). A significant effect was observed with a *Bacteroides vulgatus* strain (24) but not with *Bacteroides thetaiotaomicron* (20, 24). It may be that in this model, in which the mice are provided ample nutrients and likely utilize different nutrients in the gut (28, 29), *B. fragilis* and *B. thetaiotaomicron* would make infrequent contacts (Fig. 2B).

Analysis of metagenomic data sets of human gut samples revealed a link between the presence of *Bacteroidales* T6SSs and the composition of the microbiota, especially with regard to the GA3 T6SSs of *B. fragilis* (23). The presence of GA3 T6SS genes correlated significantly with an abundance of *Bacteroides* and a decrease in specific *Firmicutes*. Moreover, the microbiota of the developing infant gut is significantly more likely to contain GA3 T6SSs than that of adults. Also, *B. fragilis* strain replacement in the infant microbiota is more pronounced than in the adult microbiota, which appears to be dominated by a single strain of *B. fragilis*. These findings suggest that competition among *B. fragilis* strains for dominance is fiercest early in life and that the ultimate microbiota composition may be influenced by GA3 T6SSs (23).

The functions of the GA1 and GA2 T6SSs have not yet been elucidated. There has been no demonstration that these T6SSs target bacteria as do the GA3 T6SSs. Many of the identifiable effectors are predicted nucleases or other nucleic acid-targeting enzymes (15) that could function in bacteria, archaea, or eukaryotes. The fact that the ICE containing these T6SSs are shared between different coresident species of *Bacteroidales* in the human gut suggests that they likely do not have a *Bacteroidales* target and may instead provide defense against a common enemy. It is also possible that these T6SSs may allow for nutrient acquisition, or protection from environmental stressors, as has been shown for T6SSs in other organisms (30, 31). The prevalence and transfer of these systems among human gut *Bacteroidales* species make them intriguing secretion systems for continued analysis.

EFFECTS OF PROTEOBACTERIAL T6SSS IN THE MAMMALIAN GUT

Although less abundant in the healthy human gut microbiota than *Bacteroidales*, *E. coli* gut symbionts play a crucial role in colonization resistance against enteric pathogens of the proteobacterial phylum. Enteric pathogens such as *Vibrio cholerae*, *Shigella sonnei*, and *Salmonella enterica* have all been shown to utilize T6SSs to overcome colonization resistance by antagonizing resident gut *E. coli* (32–36) (Fig. 2A). What is less studied is whether symbiotic gut *E. coli* have T6SSs that function in colonization resistance against enteric pathogens. Antagonism by the diffusible microcin toxin produced by *E. coli* Nissle was shown to function as a colonization barrier to enteric pathogens (37), providing precedent

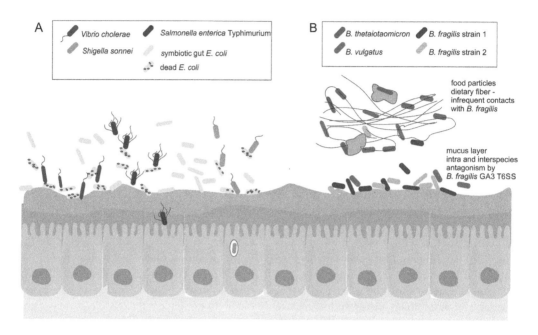

FIGURE 2 T6SS-mediated antagonism in the mammalian gut. **(A)** Three different proteobacterial enteric pathogens, *Vibrio cholerae*, *Salmonella enterica* Typhimurium, and *Shigella sonnei*, use T6SSs to target resident gut *E. coli* to overcome colonization resistance and cause disease in animal models (32–36). In the case of *V. cholerae*, the lysed *E. coli* organisms initiate innate immune responses that upregulate virulence factors and increase dissemination (32). **(B)** *Bacteroides fragilis* GA3 T6SS antagonize nearly all gut *Bacteroidales* species *in vitro*. *In vivo*, strong antagonistic effects are seen between two distinct *B. fragilis* strains likely due to their localization at the mucosal surface, where they will make frequent contacts. This intraspecies antagonism may lead to the dominance of one strain. *B. vulgatus* was also significantly antagonized by a *B. fragilis* GA3 T6SS, possibly due to overlapping nutritional niches. In contrast, a significant antagonistic effect by the GA3 T6SS of *B. fragilis* was not observed when this organism was coinoculated with *B. thetaiotaomicron*. These varied effects may be due to the substrate preferences of these species, which may spatially segregate them under normal dietary conditions.

for such an effect. Queries of a set of 1,267 human gut metagenomes consolidated into the "three cohorts gene catalog (3CGC)" (38) for matches to Pfam models identifying the four conserved proteobacterial T6SS proteins absent in *Bacteroidales* T6SSs (TssA, -J, -L, and -M) revealed that 174 of these metagenomes encoded proteins with motifs that met or exceeded the gathering threshold of all four of the proteobacterial models used (M. J. Coyne and L. E. Comstock, unpublished data). These findings suggest that resident gut *E. coli* strains likely harbor T6SSs, the *in vivo* effects of which remain to be determined.

CONCLUDING REMARKS

The discovery of T6SS loci and their prevalence in diverse human gut *Bacteroidales* species has revealed that the composition of the human gut microbiota is likely significantly influenced by these secretion systems. We still know very little about these secretion systems, including the targets of the GA1 and GA2 T6SSs, the advantage of transfer of these regions to coresident *Bacteroidales* species, and the mechanisms of action of many of the toxic effectors. In addition, it will be interesting to determine the prevalence of T6SSs in human gut *E. coli* strains and whether these antagonistic systems contribute to their ability to affect colonization by enteric pathogens.

ACKNOWLEDGMENTS

The Comstock lab is funded to study T6SSs of gut *Bacteroidales* by Public Health Service grant R01AI120633 from the NIH/National Institute of Allergy and Infectious Diseases.

CITATION

Coyne MJ, Comstock LE. 2019. Type VI secretion systems and the gut microbiota. Microbiol Spectrum 7(2):PSIB-0009-2018.

REFERENCES

1. **Mougous JD, Cuff ME, Raunser S, Shen A, Zhou M, Gifford CA, Goodman AL, Joachimiak G, Ordoñez CL, Lory S, Walz T, Joachimiak A, Mekalanos JJ.** 2006. A virulence locus of *Pseudomonas aeruginosa* encodes a protein secretion apparatus. *Science* **312:**1526–1530.

2. **Pukatzki S, Ma AT, Sturtevant D, Krastins B, Sarracino D, Nelson WC, Heidelberg JF, Mekalanos JJ.** 2006. Identification of a conserved bacterial protein secretion system in *Vibrio cholerae* using the *Dictyostelium* host model system. *Proc Natl Acad Sci U S A* **103:**1528–1533.

3. **Hood RD, Singh P, Hsu F, Güvener T, Carl MA, Trinidad RR, Silverman JM, Ohlson BB, Hicks KG, Plemel RL, Li M, Schwarz S, Wang WY, Merz AJ, Goodlett DR, Mougous JD.** 2010. A type VI secretion system of *Pseudomonas aeruginosa* targets a toxin to bacteria. *Cell Host Microbe* **7:**25–37.

4. **Bernal P, Allsopp LP, Filloux A, Llamas MA.** 2017. The *Pseudomonas putida* T6SS is a plant warden against phytopathogens. *ISME J* **11:**972–987.

5. **Steele MI, Kwong WK, Whiteley M, Moran NA.** 2017. Diversification of type VI secretion system toxins reveals ancient antagonism among bee gut microbes. *mBio* **8:**e1630-17.

6. **Speare L, Cecere AG, Guckes KR, Smith S, Wollenberg MS, Mandel MJ, Miyashiro T, Septer AN.** 2018. Bacterial symbionts use a type VI secretion system to eliminate competitors in their natural host. *Proc Natl Acad Sci U S A* **115:**E8528–E8537.

7. **Human Microbiome Project Consortium.** 2012. Structure, function and diversity of the healthy human microbiome. *Nature* **486:**207–214.

8. **Grice EA, Segre JA.** 2012. The human microbiome: our second genome. *Annu Rev Genomics Hum Genet* **13:**151–170.

9. **Hayashi H, Sakamoto M, Benno Y.** 2002. Phylogenetic analysis of the human gut microbiota using 16S rDNA clone libraries and strictly anaerobic culture-based methods. *Microbiol Immunol* **46:**535–548.

10. **Eckburg PB, Bik EM, Bernstein CN, Purdom E, Dethlefsen L, Sargent M, Gill SR, Nelson KE, Relman DA.** 2005. Diversity of the human intestinal microbial flora. *Science* **308:**1635–1638.

11. **Zitomersky NL, Coyne MJ, Comstock LE.** 2011. Longitudinal analysis of the prevalence, maintenance, and IgA response to species of the order *Bacteroidales* in the human gut. *Infect Immun* **79:**2012–2020.

12. **Coyne MJ, Zitomersky NL, McGuire AM, Earl AM, Comstock LE.** 2014. Evidence of extensive DNA transfer between *Bacteroidales* species within the human gut. *mBio* **5:**e01305-14.

13. **Russell AB, Wexler AG, Harding BN, Whitney JC, Bohn AJ, Goo YA, Tran BQ, Barry NA, Zheng H, Peterson SB, Chou S, Gonen T, Goodlett DR, Goodman AL, Mougous JD.** 2014. A type VI secretion-related pathway in *Bacteroidetes* mediates interbacterial antagonism. *Cell Host Microbe* **16:**227–236.

14. **Wilson MM, Anderson DE, Bernstein HD.** 2015. Analysis of the outer membrane proteome and secretome of *Bacteroides fragilis* reveals a multiplicity of secretion mechanisms. *PLoS One* **10:**e0117732.

15. **Coyne MJ, Roelofs KG, Comstock LE.** 2016. Type VI secretion systems of human gut *Bacteroidales* segregate into three genetic architectures, two of which are contained on mobile genetic elements. *BMC Genomics* **17:**58.

16. **Zhang D, de Souza RF, Anantharaman V, Iyer LM, Aravind L.** 2012. Polymorphic toxin systems: comprehensive characterization of trafficking modes, processing, mechanisms of action, immunity and ecology using comparative genomics. *Biol Direct* **7:**18.

17. **Durand E, Nguyen VS, Zoued A, Logger L, Péhau-Arnaudet G, Aschtgen MS, Spinelli S, Desmyter A, Bardiaux B, Dujeancourt A, Roussel A, Cambillau C, Cascales E, Fronzes R.** 2015. Biogenesis and structure of a type VI secretion membrane core complex. *Nature* **523:**555–560.

18. **Zoued A, Durand E, Brunet YR, Spinelli S, Douzi B, Guzzo M, Flaugnatti N, Legrand P,**

Journet L, Fronzes R, Mignot T, Cambillau C, Cascales E. 2016. Priming and polymerization of a bacterial contractile tail structure. *Nature* **531**:59–63.

19. Zoued A, Durand E, Santin YG, Journet L, Roussel A, Cambillau C, Cascales E. 2017. TssA: the cap protein of the type VI secretion system tail. *Bioessays* **39**:1600262.

20. Chatzidaki-Livanis M, Geva-Zatorsky N, Comstock LE. 2016. *Bacteroides fragilis* type VI secretion systems use novel effector and immunity proteins to antagonize human gut *Bacteroidales* species. *Proc Natl Acad Sci U S A* **113**:3627–3632.

21. Waters JL, Salyers AA. 2013. Regulation of CTnDOT conjugative transfer is a complex and highly coordinated series of events. *mBio* **4**:e00569-13.

22. Wozniak RA, Waldor MK. 2010. Integrative and conjugative elements: mosaic mobile genetic elements enabling dynamic lateral gene flow. *Nat Rev Microbiol* **8**:552–563.

23. Verster AJ, Ross BD, Radey MC, Bao Y, Goodman AL, Mougous JD, Borenstein E. 2017. The landscape of type VI secretion across human gut microbiomes reveals its role in community composition. *Cell Host Microbe* **22**:411–419.e4.

24. Wexler AG, Bao Y, Whitney JC, Bobay LM, Xavier JB, Schofield WB, Barry NA, Russell AB, Tran BQ, Goo YA, Goodlett DR, Ochman H, Mougous JD, Goodman AL. 2016. Human symbionts inject and neutralize antibacterial toxins to persist in the gut. *Proc Natl Acad Sci U S A* **113**:3639–3644.

25. Hecht AL, Casterline BW, Earley ZM, Goo YA, Goodlett DR, Bubeck Wardenburg J. 2016. Strain competition restricts colonization of an enteric pathogen and prevents colitis. *EMBO Rep* **17**:1281–1291.

26. Lim B, Zimmermann M, Barry NA, Goodman AL. 2017. Engineered regulatory systems modulate gene expression of human commensals in the gut. *Cell* **169**:547–558.e15.

27. Ross BD, Verster AJ, Radey MC, Schmidtke DT, Pope CE, Hoffman LR, Hajjar A, Peterson SB, Borenstein E, Mougous J. 2018. Acquired interbacterial defense systems protect against interspecies antagonism in the human gut microbiome. *bioRxiv*.

28. Pudlo NA, Urs K, Kumar SS, German JB, Mills DA, Martens EC. 2015. Symbiotic human gut bacteria with variable metabolic priorities for host mucosal glycans. *mBio* **6**:e01282-15.

29. Tuncil YE, Xiao Y, Porter NT, Reuhs BL, Martens EC, Hamaker BR. 2017. Reciprocal prioritization to dietary glycans by gut bacteria in a competitive environment promotes stable coexistence. *mBio* **8**:01068-17.

30. Si M, Zhao C, Burkinshaw B, Zhang B, Wei D, Wang Y, Dong TG, Shen X. 2017. Manganese scavenging and oxidative stress response mediated by type VI secretion system in *Burkholderia thailandensis*. *Proc Natl Acad Sci U S A* **114**:E2233–E2242.

31. Lin J, Zhang W, Cheng J, Yang X, Zhu K, Wang Y, Wei G, Qian PY, Luo ZQ, Shen X. 2017. A *Pseudomonas* T6SS effector recruits PQS-containing outer membrane vesicles for iron acquisition. *Nat Commun* **8**:14888.

32. Zhao W, Caro F, Robins W, Mekalanos JJ. 2018. Antagonism toward the intestinal microbiota and its effect on *Vibrio cholerae* virulence. *Science* **359**:210–213.

33. Fast D, Kostiuk B, Foley E, Pukatzki S. 2018. Commensal pathogen competition impacts host viability. *Proc Natl Acad Sci U S A* **115**:7099–7104.

34. Anderson MC, Vonaesch P, Saffarian A, Marteyn BS, Sansonetti PJ. 2017. *Shigella sonnei* encodes a functional T6SS used for interbacterial competition and niche occupancy. *Cell Host Microbe* **21**:769–776.e3.

35. Sana TG, Flaugnatti N, Lugo KA, Lam LH, Jacobson A, Baylot V, Durand E, Journet L, Cascales E, Monack DM. 2016. *Salmonella* Typhimurium utilizes a T6SS-mediated antibacterial weapon to establish in the host gut. *Proc Natl Acad Sci U S A* **113**:E5044–E5051.

36. Pezoa D, Blondel CJ, Silva CA, Yang HJ, Andrews-Polymenis H, Santiviago CA, Contreras I. 2014. Only one of the two type VI secretion systems encoded in the *Salmonella enterica* serotype Dublin genome is involved in colonization of the avian and murine hosts. *Vet Res (Faisalabad)* **45**:2.

37. Sassone-Corsi M, Nuccio SP, Liu H, Hernandez D, Vu CT, Takahashi AA, Edwards RA, Raffatellu M. 2016. Microcins mediate competition among *Enterobacteriaceae* in the inflamed gut. *Nature* **540**:280–283.

38. Li J, Jia H, Cai X, Zhong H, Feng Q, Sunagawa S, Arumugam M, Kultima JR, Prifti E, Nielsen T, Juncker AS, Manichanh C, Chen B, Zhang W, Levenez F, Wang J, Xu X, Xiao L, Liang S, Zhang D, Zhang Z, Chen W, Zhao H, Al-Aama JY, Edris S, Yang H, Wang J, Hansen T, Nielsen HB, Brunak S, Kristiansen K, Guarner F, Pedersen O, Doré J, Ehrlich SD, MetaHIT Consortium, Bork P, Wang J. 2014. An integrated catalog of reference genes in the human gut microbiome. *Nat Biotechnol* **32**:834–841.

ESX/Type VII Secretion Systems—An Important Way Out for Mycobacterial Proteins

28

FARZAM VAZIRI[1,2,3] and ROLAND BROSCH[1]

INTRODUCTION

The different bacterial species within the tree of life (1) possess a range of secretion systems, which play important roles in the transport of proteins across the various types of bacterial cell envelopes. Classically, Gram staining was used for differentiating Gram-positive and Gram-negative bacteria, but classifications on cell envelope architecture might come closer to the biological reality, and thus, bacteria may also be differentiated according to their cell envelopes into diderm-lipopolysaccharide (archetypal Gram-negative), monoderm (archetypal Gram-positive), and diderm-mycolate (archetypal acid-fast) bacteria (2). For Gram-negative bacteria a range of at least eight different secretion systems has been described (types I to VI, VIII, and IX) (3–5). While in monoderm bacteria secretion and export are synonymous, in diderm bacteria the secretion is completed only upon translocation of the substrates across the outer membrane (2). The here-reviewed mycobacterial ESAT-6 secretion (ESX) systems (6, 7), which were also named type VII secretion (T7S) systems (8), represent a particular

[1]Institut Pasteur, Unit for Integrated Mycobacterial Pathogenomics, UMR3525 CNRS, 75015 Paris, France
[2]Department of Mycobacteriology and Pulmonary Research, Pasteur Institute of Iran, 13164 Tehran, Iran
[3]Microbiology Research Center, Pasteur Institute of Iran, 13164 Tehran, Iran

Protein Secretion in Bacteria
Edited by Maria Sandkvist, Eric Cascales, and Peter J. Christie
© 2019 American Society for Microbiology, Washington, DC
doi:10.1128/microbiolspec.PSIB-0029-2019

class of protein export and/or secretion systems, for which at present only the inner-membrane translocation machinery has been explored in more detail (9, 10), whereas it remains unknown how ESX/T7S-exported proteins get transported through the mycobacterial outer membrane into the extracellular environment (11). Indeed, one of the remarkable characteristics of mycobacteria is their complex cell envelope, which is shared to some extent with other members of the *Corynebacterineae*, a suborder of the phylum *Actinobacteria* (1, 12, 13). Mycobacteria are surrounded by a diderm cell envelope, consisting of an inner membrane, a peptidoglycan layer, an arabinogalactan layer, an outer membrane, named mycomembrane, which is composed of covalently linked mycolic acids and extractable lipids, and a capsule (14, 15). This unusual cell envelope requires complex secretion systems for the export/secretion of proteins, such as those of the SecA and twin-arginine translocation pathways, as well as the specialized ESX/T7S systems (7, 8, 16), which were first discovered almost 20 years ago during *in silico* analyses of the genome sequence and the proteome of *Mycobacterium tuberculosis* H37Rv (17, 18). Moreover, T7S-like systems that share some core components of mycobacterial ESX/T7S systems exist in various genera of the phylum *Firmicutes*, representing many classical Gram-positive bacterial species (19), which, however, are not the subject of the current review.

   M. tuberculosis possesses five ESX/T7S systems (ESX-1 to ESX-5) (7, 8, 11, 16). All five ESX systems share several common features: the presence of small secreted proteins (of about 100 amino acids) with a conserved Trp-X-Gly (WXG) motif (20), an FtsK-SpoIIIE ATPase, several transmembrane proteins, and a subtilisin-like mycosin (MycP) (11, 16) (Fig. 1). These systems, encoded in different sections of the mycobacterial chromosome, seem to have evolved by gene duplication and diversification from simpler systems that were shuffled around in

different actinobacterial and mycobacterial species, often mediated by plasmids encoding ESX/T7S elements as well as elements of type IV secretion systems (21–23). ESX/T7S systems play an important role in the biology of *M. tuberculosis*, as well as in the interactions *M. tuberculosis* has with its host. Indeed, a number of secreted effectors, including EsxA (ESAT-6), EsxB (CFP-10), and ESX-1 secretion-associated proteins (Esp), such as EspA or EspC, as well as proteins that carry the characteristic N-terminal motifs Pro-Glu (PE) and Pro-Pro-Glu (PPE), have been suggested to intervene in host cellular and immune signaling pathways (11, 24, 25).

   Here we focus on recent updates on the ESX/T7S systems of mycobacteria and summarize new findings on their structure, function, and role in host-pathogen interactions and briefly touch on their significance in translational research.

RECENT INSIGHTS INTO THE STRUCTURAL AND FUNCTIONAL CHARACTERISTICS OF ESX/T7S NANOMACHINES

Five ESX systems are encoded in the genome of *M. tuberculosis* (16, 18), and this number is the highest found in mycobacteria so far; other mycobacterial species show fewer systems (e.g., *Mycobacterium marinum* shows four systems and *Mycobacterium abscessus* shows three systems) (21, 22). While ESX-4, ESX-3, and ESX-1 are present in most fast-growing and slow-growing mycobacteria, ESX-2 and ESX-5 systems are found only in selected slow-growing mycobacteria and thus represent the most recently evolved systems (21, 22). For ESX-2, currently not much is known on its putative function. In contrast, for ESX-1, ESX-3, and ESX-5, recent research has determined that they all contribute to virulence of *M. tuberculosis*, although the insights into the exact molecular functions often remain vague; also, because many studies have been undertaken with different

FIGURE 1 Genetic organization of the ESX loci. Shown is a schematic representation of the approximative genomic sites of the ESX-1 to ESX-5 clusters in the *M. tuberculosis* H37Rv genome. Gene nomenclature and gene color scheme were adapted from reference 16.

mycobacterial species (*M. tuberculosis, M. marinum, M. abscessus,* and/or *Mycobacterium smegmatis*), which may show some species-specific differences (reviewed in references 11 and 26 to 28). ESX-3 is important for metal homeostasis, pathogenicity, and immunogenicity (29–31). ESX-5 was suggested to be crucial for nutrient uptake and for the export of members of the PE and PPE protein families (9, 32, 33). These two large protein families have expanded during mycobacterial evolution (34, 35), and they include representative proteins that are associated with ESX/T7S systems and others with highly repetitive sequence motifs that are exported by ESX-5 and impact virulence and immunogenicity (32, 36–39). The ESX-5 nanomachine, which is integrated in the inner mycobacterial membrane, is composed of four proteins, namely, EccB5, EccC5, EccD5, and EccE5 (9, 10), that are organized in a hexameric complex, as recently determined by cryo-electron microscopy and single-particle analysis (10). This organization differs substantially from those of secretion systems of Gram-negative bac-

teria (10) (Fig. 2). Among the Ecc proteins (ESX conserved components), the structural and functional roles of EccC (FtsK/SpoIIIE ATPase) have been studied in more detail with the thermophilic actinobacterium *Thermomonospora curvata* (40). A certain flexibility of the cytosolic domains of EccC in interaction with effectors was suggested (10, 40), which is different from the cognate ATPase in type IV secretion systems (10).

Another conserved ESX component in mycobacterial ESX systems is the serine protease MycP, although this protein is not directly integrated in the EccBCDE complex (10, 41–43). Moreover, different proteins, such as EspA, EspC, EspD, and EccA1, may be essential to contributing to secretion and stabilizing the core ESX/T7S complex in the case of ESX-1 (44–47). How these related effectors are explicitly recognized and targeted towards the specific system in a single mycobacterial species with different ESX/T7S systems can be a matter of debate. Recently, it has become clear that some of the

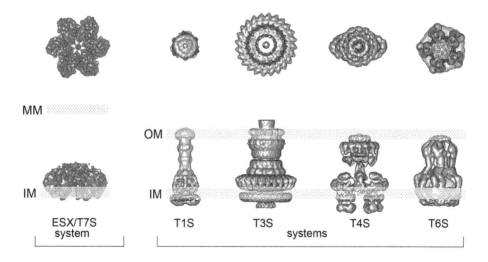

FIGURE 2 Representation of top and side views of the ESX/T7S system based on recent structural data generated by cryo-electron microscopy and single-particle analysis on an ESX-5 system from *Mycobacterium xenopi*, in comparison to selected examples of secretion systems from Gram-negative bacteria. The positions of the inner membrane (IM), outer membrane (OM), and mycomembrane (MM) are indicated. Adapted from reference 10, with permission.

conserved ESX components could potentially exhibit chaperone-like activity (e.g., EspG or EccA) (48). It was suggested that besides their chaperone activity, these proteins are also involved in determining the secretion system specificity. Indeed, by substituting the binding domain of EspG, the ESX-1-dependent substrate can be rerouted to the ESX-5 system (49). In addition, it was suggested that EspL can have a role as a chaperone and is essential for ESX-1-dependent virulence (50). Therefore, scrutinizing the role of chaperones will certainly help to provide a better understanding of the ESX/T7S functions and mechanisms.

It is also intriguing that certain ESX/T7S systems may have a dual function. For example, the ESX-1 system, which is present in fast- and slow-growing mycobacterial species, is required for distributive conjugal transfer (DCT) of chromosomal DNA from donor into recipient strains of *M. smegmatis* (51, 52). This procedure apparently also involves the ESX-4 system of *M. smegmatis* (53). Moreover, it was shown that SigM, an extracytoplasmic function σ factor, is an activator of ESX-4 expression and necessary for DCT in the recipient strain of *M. smegmatis* (54). Intriguingly, experimental strain-to-strain transfer of chromosomal DNA was also observed in selected *Mycobacterium canettii* strains (55), representing a group of rare tubercle bacilli that are thought to resemble the ancestor of *M. tuberculosis* and have been isolated mainly from tuberculosis patients in the region of the Horn of Africa (East Africa) (56, 57). In contrast to *M. canettii* strains, DNA transfer between *M. tuberculosis* strains was not observed despite numerous trials (55), emphasizing the clonal population structure of *M. tuberculosis* strains (58, 59). While it is predicted that the interstrain DNA transfer between *M. canettii* strains might also involve an ESX-1 system in the recipient strain, in analogy to the situation in *M. smegmatis*, experimental confirmation for this hypothesis has not yet been reported.

ESX SYSTEMS IN HOST-PATHOGEN INTERACTIONS

The potential dual function of the ESX-1 system is best visible by the fact that in slow-growing mycobacteria, in contrast to fast-growing mycobacteria, the ESX-1 system is also involved in the pathogenic potential of the strains. It has been speculated that this phenotype might be associated with horizontal gene transfer of a putative genomic island harboring the ESX-1-associated *espACD* locus (60). *M. tuberculosis* and *M. marinum* mutants with deletion of ESX-1 are attenuated in their respective hosts (61–64), which is in line with the attenuation of "natural" ESX-1 deletion mutants, such as the *Mycobacterium bovis* BCG (bacillus Calmette-Guérin) vaccine, which has lost ESX-1 functions due to the deletion of the region of difference RD1 (65). The ESX-1 system was shown to be involved in bacterial phagosome-to-cytosol transition of *M. tuberculosis* and host cell death (66–68), an important cell biological process that has numerous consequences for the host cell, such as induction of the cGAS/STING/TBK1/IRF-3/type I interferon signalling axis and NLRP3 inflammasome activation (69–75). However, ESX-1 is not the only factor involved in the process; it has been shown that besides ESX-1, the mycobacterial virulence lipids phthiocerol dimycocerosates also contribute to phagosomal rupture (76–78). Moreover, recent studies have also demonstrated that the endosomal sorting complex required for the transport III (ESCRT-III) system promotes the repair of small perforations in the endolysosomal membrane (79). Intriguingly, certain ESX-3-secreted effectors can block ESCRT-dependent receptor trafficking to the lysosome (80). It was shown that effectors of ESX systems differentially respond to the ESCRT endomembrane damage response. In an ESX-1-dependent manner, the ESCRT machinery is recruited to phagosomes, while ESX-3 effectors (EsxG-EsxH) antagonize the host damage response by blocking the recruitment of HRS, ESCRT-III, and GAL3 (81) (Fig. 3).

ESX-4 is one of the least well-characterized ESX systems, although it is considered the most ancestral *esx* locus in mycobacteria (21, 82). The ESX-4 loci usually lack *pe/ppe* and *espG* genes, which may be involved in host-pathogen interactions (34), as well as the *eccE* gene. However, the ESX-4 locus of *M. abscessus* is different from that of other species, as it does contain EccE4 (21). In a recent study, by using an *M. abscessus*

genome-scale *Himar mariner* transposon library, it was shown that an intact ESX-4 system is needed for full virulence in this fast-growing mycobacterium and emerging human pathogen, whereby the ESX-4 function in infection was associated with phago-somal rupture and transition of bacteria to the cytosol of amoebae and human macro-phages (83). As *M. abscessus* does not pos-sess an ESX-1 system, in this particular case, ESX-4 might be considered a surrogate of ESX-1.

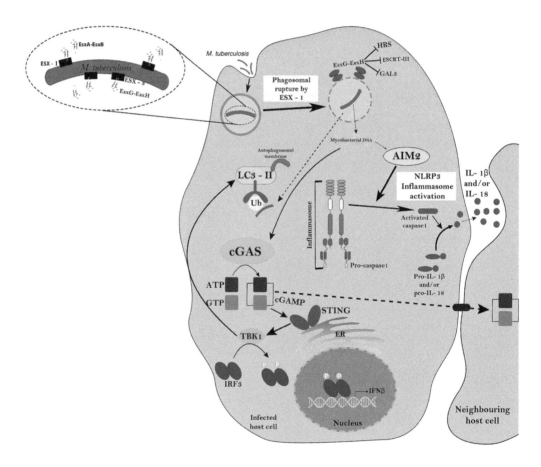

FIGURE 3 Interplay of ESX-1 and ESX-3 in host-pathogen interactions. ESX-1 is essential for the bacterial phagosome-to-cytosol transition by involving a cGAS/STING/TBK1/IRF-3/type I interferon signalling axis and AIM2 and NLRP3 inflammasome activities. In an ESX-1-dependent manner, the ESCRT machinery is recruited to phagosomes, while ESX-3 effectors (EsxG-EsxH) antagonize the host damage response by blocking the recruitment of HRS, ESCRT-III, and GAL3. The scheme is adapted from reference 11, with some additions from reference 81, with permission.

Because of extensive sequence similarities and immune cross-reactions among Esx and PE/PPE proteins secreted by the ESX/T7S systems, investigation of the secretion and regulation of these effectors is challenging. Recently, a technology termed multiplexed analysis of substrate secretion by transduced T cell hybridomas (MASSTT) was developed to explore the intra-host cell secretion profiles of various mycobacterial strains via fluorescence-mediated detection of specific *M. tuberculosis* major histocompatibility complex class II (MHC-II) epitopes by highly discriminative T cell hybridomas (84). This method thus allows investigators to follow the intracellular secretion profiles of selected mycobacterial ESX proteins, such as EsxA or EspC, as well as other secreted proteins, such as the members of the Ag85 complex. The secretion of the latter proteins (e.g., Ag85B) is regulated by the PhoP/PhoR two-component regulatory system and the small RNA Mcr7 (85). Interestingly, strains of different phylogenetic lineages showed distinct secretion levels of Ag85B proteins in a preliminary set of *M. tuberculosis* strains by the MASSTT assay (84), information that needs to be confirmed with a larger strain collection.

It is clear from the few examples mentioned here that ESX systems have a strong impact on mycobacterial host-pathogen interaction, although more work is needed to elucidate the various molecular mechanisms by which the effects are generated. New insights into these phenomena are also of interest for translational implications, as is shown by the example of attenuated whole-cell vaccines against tuberculosis. Loss of ESX-1 is one of the main reasons for the attenuation of the BCG vaccine (65), while ESX-1 effectors are important antigens in immune responses (6). Due to the absence of ESX-1, BCG does not gain access to the host cell cytosol and thus lacks the induction of certain immune signaling pathways (74, 86). Several recombinant BCG vaccine candidates have been constructed to overcome these limitations of BCG (75, 87–

90). Alternatively, rationally attenuated *M. tuberculosis* vaccine candidates may also secrete particular ESX antigens that are absent from BCG (36, 91–93) and thereby may induce improved protection.

CONCLUDING COMMENTS AND PERSPECTIVES

In summary, we have presented here a few examples showing that mycobacterial ESX/T7S systems represent dynamic molecular machines which play important roles in various aspects of the biology of mycobacteria and the interaction with their hosts. Advances in structural biology together with the use of new approaches (e.g., MASSTT) will be very helpful for better understanding the functional details that are linked to their biological activities and for exploiting this knowledge for improved intervention strategies against tuberculosis.

ACKNOWLEDGMENTS

We are grateful to our many colleagues who were involved in the original work of the subjects reviewed in this article.

Research in the authors' laboratories is in part supported by the European Union's Horizon 2020 Research and Innovation Program (grant 643381 TBVAC2020), the Agence National de Recherche (ANR-10-LABX-62-IBEID, ANR-16-CE15-0003, and ANR-16-CE35-0009), the Fondation pour la Recherche Médicale (DEQ20130326471), and the Institut Pasteur. The research internship of F.V. at the Institut Pasteur in Paris was enabled and supported by the Pasteur International Network Programme and Campus France (grant 936638E).

Citation

Vaziri F, Brosch R. 2019. ESX/Type VII secretion systems—an important way out for mycobacterial proteins. Microbiol Spectrum 7(4):PSIB-0029-2019.

REFERENCES

1. Hug LA, Baker BJ, Anantharaman K, Brown CT, Probst AJ, Castelle CJ, Butterfield CN, Hernsdorf AW, Amano Y, Ise K, Suzuki Y, Dudek N, Relman DA, Finstad KM, Amundson R, Thomas BC, Banfield JF. 2016. A new view of the tree of life. *Nat Microbiol* **1:**16048.
2. Chagnot C, Zorgani MA, Astruc T, Desvaux M. 2013. Proteinaceous determinants of surface colonization in bacteria: bacterial adhesion and biofilm formation from a protein secretion perspective. *Front Microbiol* **4:**303.
3. Gerlach RG, Hensel M. 2007. Protein secretion systems and adhesins: the molecular armory of Gram-negative pathogens. *Int J Med Microbiol* **297:**401–415.
4. Green ER, Mecsas J. 2016. Bacterial secretion systems: an overview. *Microbiol Spectr* **4:** VMBF-0012-2015.
5. Veith PD, Glew MD, Gorasia DG, Reynolds EC. 2017. Type IX secretion: the generation of bacterial cell surface coatings involved in virulence, gliding motility and the degradation of complex biopolymers. *Mol Microbiol* **106:**35–53.
6. Sørensen AL, Nagai S, Houen G, Andersen P, Andersen AB. 1995. Purification and characterization of a low-molecular-mass T-cell antigen secreted by *Mycobacterium tuberculosis*. *Infect Immun* **63:**1710–1717.
7. Brodin P, Rosenkrands I, Andersen P, Cole ST, Brosch R. 2004. ESAT-6 proteins: protective antigens and virulence factors? *Trends Microbiol* **12:**500–508.
8. Abdallah AM, Gey van Pittius NC, Champion PA, Cox J, Luirink J, Vandenbroucke-Grauls CM, Appelmelk BJ, Bitter W. 2007. Type VII secretion—mycobacteria show the way. *Nat Rev Microbiol* **5:**883–891.
9. Houben EN, Bestebroer J, Ummels R, Wilson L, Piersma SR, Jiménez CR, Ottenhoff TH, Luirink J, Bitter W. 2012. Composition of the type VII secretion system membrane complex. *Mol Microbiol* **86:**472–484.
10. Beckham KS, Ciccarelli L, Bunduc CM, Mertens HD, Ummels R, Lugmayr W, Mayr J, Rettel M, Savitski MM, Svergun DI, Bitter W, Wilmanns M, Marlovits TC, Parret AH, Houben EN. 2017. Structure of the mycobacterial ESX-5 type VII secretion system mem-

brane complex by single-particle analysis. *Nat Microbiol* **2:**17047.
11. Gröschel MI, Sayes F, Simeone R, Majlessi L, Brosch R. 2016. ESX secretion systems: mycobacterial evolution to counter host immunity. *Nat Rev Microbiol* **14:**677–691.
12. Zuber B, Chami M, Houssin C, Dubochet J, Griffiths G, Daffé M. 2008. Direct visualization of the outer membrane of mycobacteria and corynebacteria in their native state. *J Bacteriol* **190:**5672–5680.
13. Kaur D, Guerin ME, Skovierová H, Brennan PJ, Jackson M. 2009. Chapter 2: biogenesis of the cell wall and other glycoconjugates of *Mycobacterium tuberculosis*. *Adv Appl Microbiol* **69:**23–78.
14. Daffé M. 2015. The cell envelope of tubercle bacilli. *Tuberculosis (Edinb)* **95**(Suppl 1):S155–S158.
15. Touchette MH, Seeliger JC. 2017. Transport of outer membrane lipids in mycobacteria. *Biochim Biophys Acta Mol Cell Biol Lipids* **1862:**1340–1354.
16. Bitter W, Houben EN, Bottai D, Brodin P, Brown EJ, Cox JS, Derbyshire K, Fortune SM, Gao LY, Liu J, Gey van Pittius NC, Pym AS, Rubin EJ, Sherman DR, Cole ST, Brosch R. 2009. Systematic genetic nomenclature for type VII secretion systems. *PLoS Pathog* **5:**e1000507.
17. Cole ST, Brosch R, Parkhill J, Garnier T, Churcher C, Harris D, Gordon SV, Eiglmeier K, Gas S, Barry CE III, Tekaia F, Badcock K, Basham D, Brown D, Chillingworth T, Connor R, Davies R, Devlin K, Feltwell T, Gentles S, Hamlin N, Holroyd S, Hornsby T, Jagels K, Krogh A, McLean J, Moule S, Murphy L, Oliver K, Osborne J, Quail MA, Rajandream MA, Rogers J, Rutter S, Seeger K, Skelton J, Squares R, Squares S, Sulston JE, Taylor K, Whitehead S, Barrell BG. 1998. Deciphering the biology of *Mycobacterium tuberculosis* from the complete genome sequence. *Nature* **393:**537–544.
18. Tekaia F, Gordon SV, Garnier T, Brosch R, Barrell BG, Cole ST. 1999. Analysis of the proteome of *Mycobacterium tuberculosis* in silico. *Tuber Lung Dis* **79:**329–342.
19. Unnikrishnan M, Constantinidou C, Palmer T, Pallen MJ. 2017. The enigmatic Esx proteins: looking beyond mycobacteria. *Trends Microbiol* **25:**192–204.
20. Pallen MJ. 2002. The ESAT-6/WXG100 super-family—and a new Gram-positive secretion system? *Trends Microbiol* **10:**209–212.
21. Dumas E, Christina Boritsch E, Vandenbogaert M, Rodríguez de la Vega RC, Thiberge JM,

Caro V, Gaillard JL, Heym B, Girard-Misguich F, Brosch R, Sapriel G. 2016. Mycobacterial pangenome analysis suggests important role of plasmids in the radiation of type VII secretion systems. *Genome Biol Evol* **8**:387–402.

22. Newton-Foot M, Warren RM, Sampson SL, van Helden PD, Gey van Pittius NC. 2016. The plasmid-mediated evolution of the mycobacterial ESX (type VII) secretion systems. *BMC Evol Biol* **16**:62.

23. Ummels R, Abdallah AM, Kuiper V, Aâjoud A, Sparrius M, Naeem R, Spaink HP, van Soolingen D, Pain A, Bitter W. 2014. Identification of a novel conjugative plasmid in mycobacteria that requires both type IV and type VII secretion. *mBio* **5**:e01744-14.

24. Stoop EJ, Bitter W, van der Sar AM. 2012. Tubercle bacilli rely on a type VII army for pathogenicity. *Trends Microbiol* **20**:477–484.

25. Queval CJ, Brosch R, Simeone R. 2017. The macrophage: a disputed fortress in the battle against *Mycobacterium tuberculosis*. *Front Microbiol* **8**:2284.

26. Majlessi L, Prados-Rosales R, Casadevall A, Brosch R. 2015. Release of mycobacterial antigens. *Immunol Rev* **264**:25–45.

27. Ates LS, Houben EN, Bitter W. 2016. Type VII secretion: a highly versatile secretion system. *Microbiol Spectr* **4**:VMBF-0011-2015.

28. Madacki J, Mas Fiol G, Brosch R. 2019. Update on the virulence factors of the obligate pathogen *Mycobacterium tuberculosis* and related tuberculosis-causing mycobacteria. *Infect Genet Evol* **72**:67–77.

29. Serafini A, Boldrin F, Palù G, Manganelli R. 2009. Characterization of a *Mycobacterium tuberculosis* ESX-3 conditional mutant: essentiality and rescue by iron and zinc. *J Bacteriol* **191**:6340–6344.

30. Siegrist MS, Steigedal M, Ahmad R, Mehra A, Dragset MS, Schuster BM, Philips JA, Carr SA, Rubin EJ. 2014. Mycobacterial Esx-3 requires multiple components for iron acquisition. *mBio* **5**:e01073-14.

31. Tufariello JM, Chapman JR, Kerantzas CA, Wong KW, Vilchèze C, Jones CM, Cole LE, Tinaztepe E, Thompson V, Fenyö D, Niederweis M, Ueberheide B, Philips JA, Jacobs WR Jr. 2016. Separable roles for *Mycobacterium tuberculosis* ESX-3 effectors in iron acquisition and virulence. *Proc Natl Acad Sci U S A* **113**:E348–E357.

32. Bottai D, Di Luca M, Majlessi L, Frigui W, Simeone R, Sayes F, Bitter W, Brennan MJ, Leclerc C, Batoni G, Campa M, Brosch R, Esin S. 2012. Disruption of the ESX-5 system of *Mycobacterium tuberculosis* causes loss of PPE protein secretion, reduction of cell wall integrity and strong attenuation. *Mol Microbiol* **83**:1195–1209.

33. Ates LS, Ummels R, Commandeur S, van de Weerd R, Sparrius M, Weerdenburg E, Alber M, Kalscheuer R, Piersma SR, Abdallah AM, Abd El Ghany M, Abdel-Haleem AM, Pain A, Jiménez CR, Bitter W, Houben EN. 2015. Essential role of the ESX-5 secretion system in outer membrane permeability of pathogenic mycobacteria. *PLoS Genet* **11**:e1005190.

34. Gey van Pittius NC, Sampson SL, Lee H, Kim Y, van Helden PD, Warren RM. 2006. Evolution and expansion of the *Mycobacterium tuberculosis* PE and PPE multigene families and their association with the duplication of the ESAT-6 (esx) gene cluster regions. *BMC Evol Biol* **6**:95.

35. Bottai D, Brosch R. 2009. Mycobacterial PE, PPE and ESX clusters: novel insights into the secretion of these most unusual protein families. *Mol Microbiol* **73**:325–328.

36. Sayes F, Sun L, Di Luca M, Simeone R, Degaiffier N, Fiette L, Esin S, Brosch R, Bottai D, Leclerc C, Majlessi L. 2012. Strong immunogenicity and cross-reactivity of *Mycobacterium tuberculosis* ESX-5 type VII secretion: encoded PE-PPE proteins predicts vaccine potential. *Cell Host Microbe* **11**:352–363.

37. Fishbein S, van Wyk N, Warren RM, Sampson SL. 2015. Phylogeny to function: PE/PPE protein evolution and impact on *Mycobacterium tuberculosis* pathogenicity. *Mol Microbiol* **96**:901–916.

38. Ates LS, Dippenaar A, Ummels R, Piersma SR, van der Woude AD, van der Kuij K, Le Chevalier F, Mata-Espinosa D, Barrios-Payán J, Marquina-Castillo B, Guapillo C, Jiménez CR, Pain A, Houben ENG, Warren RM, Brosch R, Hernández-Pando R, Bitter W. 2018. Mutations in ppe38 block PE_PGRS secretion and increase virulence of *Mycobacterium tuberculosis*. *Nat Microbiol* **3**:181–188.

39. Ates LS, Dippenaar A, Sayes F, Pawlik A, Bouchier C, Ma L, Warren RM, Sougakoff W, Majlessi L, van Heijst JWJ, Brossier F, Brosch R. 2018. Unexpected genomic and phenotypic diversity of *Mycobacterium africanum* lineage 5 affects drug resistance, protein secretion, and immunogenicity. *Genome Biol Evol* **10**:1858–1874.

40. Rosenberg OS, Dovala D, Li X, Connolly L, Bendebury A, Finer-Moore J, Holton J, Cheng Y, Stroud RM, Cox JS. 2015. Substrates control multimerization and activation of the multi-domain ATPase motor of type VII secretion. *Cell* **161**:501–512.

41. **van Winden VJC, Damen MPM, Ummels R, Bitter W, Houben ENG.** 2019. Protease domain and transmembrane domain of the type VII secretion mycosin protease determine system-specific functioning in mycobacteria. *J Biol Chem* **294:**4806–4814.

42. **Bosserman RE, Champion PA.** 2017. Esx systems and the mycobacterial cell envelope: what's the connection? *J Bacteriol* **199:**e00131-17.

43. **van Winden VJ, Ummels R, Piersma SR, Jiménez CR, Korotkov KV, Bitter W, Houben EN.** 2016. Mycosins are required for the stabilization of the ESX-1 and ESX-5 type VII secretion membrane complexes. *mBio* **7:**01471-16.

44. **Fortune SM, Jaeger A, Sarracino DA, Chase MR, Sassetti CM, Sherman DR, Bloom BR, Rubin EJ.** 2005. Mutually dependent secretion of proteins required for mycobacterial virulence. *Proc Natl Acad Sci U S A* **102:**10676–10681.

45. **MacGurn JA, Raghavan S, Stanley SA, Cox JS.** 2005. A non-RD1 gene cluster is required for Snm secretion in *Mycobacterium tuberculosis*. *Mol Microbiol* **57:**1653–1663.

46. **Chen JM.** 2016. Mycosins of the mycobacterial type VII ESX secretion system: the glue that holds the party together. *mBio* **7:**02062-16.

47. **Lou Y, Rybniker J, Sala C, Cole ST.** 2017. EspC forms a filamentous structure in the cell envelope of *Mycobacterium tuberculosis* and impacts ESX-1 secretion. *Mol Microbiol* **103:**26–38.

48. **Phan TH, Houben ENG.** 2018. Bacterial secretion chaperones: the mycobacterial type VII case. *FEMS Microbiol Lett* **365:**fny197.

49. **Phan TH, Ummels R, Bitter W, Houben EN.** 2017. Identification of a substrate domain that determines system specificity in mycobacterial type VII secretion systems. *Sci Rep* **7:**42704.

50. **Sala C, Odermatt NT, Soler-Arnedo P, Gülen MF, von Schultz S, Benjak A, Cole ST.** 2018. EspL is essential for virulence and stabilizes EspE, EspF and EspH levels in *Mycobacterium tuberculosis*. *PLoS Pathog* **14:**e1007491.

51. **Coros A, Callahan B, Battaglioli E, Derbyshire KM.** 2008. The specialized secretory apparatus ESX-1 is essential for DNA transfer in *Mycobacterium smegmatis*. *Mol Microbiol* **69:**794–808.

52. **Derbyshire KM, Gray TA.** 2014. Distributive conjugal transfer: new insights into horizontal gene transfer and genetic exchange in mycobacteria. *Microbiol Spectr* **2:**MGM2-0022-2013.

53. **Gray TA, Clark RR, Boucher N, Lapierre P, Smith C, Derbyshire KM.** 2016. Intercellular communication and conjugation are mediated by ESX secretion systems in mycobacteria. *Science* **354:**347–350.

54. **Clark RR, Judd J, Lasek-Nesselquist E, Montgomery SA, Hoffmann JG, Derbyshire KM, Gray TA.** 2018. Direct cell-cell contact activates SigM to express the ESX-4 secretion system in *Mycobacterium smegmatis*. *Proc Natl Acad Sci U S A* **115:**E6595–E6603.

55. **Boritsch EC, Khanna V, Pawlik A, Honoré N, Navas VH, Ma L, Bouchier C, Seemann T, Supply P, Stinear TP, Brosch R.** 2016. Key experimental evidence of chromosomal DNA transfer among selected tuberculosis-causing mycobacteria. *Proc Natl Acad Sci U S A* **113:**9876–9881.

56. **Supply P, Marceau M, Mangenot S, Roche D, Rouanet C, Khanna V, Majlessi L, Criscuolo A, Tap J, Pawlik A, Fiette L, Orgeur M, Fabre M, Parmentier C, Frigui W, Simeone R, Boritsch EC, Debrie AS, Willery E, Walker D, Quail MA, Ma L, Bouchier C, Salvignol G, Sayes F, Cascioferro A, Seemann T, Barbe V, Locht C, Gutierrez MC, Leclerc C, Bentley SD, Stinear TP, Brisse S, Médigue C, Parkhill J, Cruveiller S, Brosch R.** 2013. Genomic analysis of smooth tubercle bacilli provides insights into ancestry and pathoadaptation of *Mycobacterium tuberculosis*. *Nat Genet* **45:**172–179.

57. **Boritsch EC, Frigui W, Cascioferro A, Malaga W, Etienne G, Laval F, Pawlik A, Le Chevalier F, Orgeur M, Ma L, Bouchier C, Stinear TP, Supply P, Majlessi L, Daffé M, Guilhot C, Brosch R.** 2016. *pks5*-recombination-mediated surface remodelling in *Mycobacterium tuberculosis* emergence. *Nat Microbiol* **1:**15019.

58. **Godfroid M, Dagan T, Kupczok A.** 2018. Recombination signal in *Mycobacterium tuberculosis* stems from reference-guided assemblies and alignment artefacts. *Genome Biol Evol* **10:**1920–1926.

59. **Orgeur M, Brosch R.** 2018. Evolution of virulence in the *Mycobacterium tuberculosis* complex. *Curr Opin Microbiol* **41:**68–75.

60. **Ates LS, Brosch R.** 2017. Discovery of the type VII ESX-1 secretion needle? *Mol Microbiol* **103:**7–12.

61. **Lewis KN, Liao R, Guinn KM, Hickey MJ, Smith S, Behr MA, Sherman DR.** 2003. Deletion of RD1 from *Mycobacterium tuberculosis* mimics bacille Calmette-Guérin attenuation. *J Infect Dis* **187:**117–123.

62. **Hsu T, Hingley-Wilson SM, Chen B, Chen M, Dai AZ, Morin PM, Marks CB, Padiyar J, Goulding C, Gingery M, Eisenberg D, Russell RG, Derrick SC, Collins FM, Morris SL, King CH, Jacobs WR Jr.** 2003. The primary mechanism of attenuation of bacillus Calmette-Guerin is a loss of secreted lytic function required for invasion of lung interstitial tissue. *Proc Natl Acad Sci U S A* **100:**12420–12425.

63. **Stanley SA, Raghavan S, Hwang WW, Cox JS.** 2003. Acute infection and macrophage subversion by *Mycobacterium tuberculosis* require a specialized secretion system. *Proc Natl Acad Sci U S A* **100:**13001–13006.

64. **Gao LY, Guo S, McLaughlin B, Morisaki H, Engel JN, Brown EJ.** 2004. A mycobacterial virulence gene cluster extending RD1 is required for cytolysis, bacterial spreading and ESAT-6 secretion. *Mol Microbiol* **53:**1677–1693.

65. **Pym AS, Brodin P, Brosch R, Huerre M, Cole ST.** 2002. Loss of RD1 contributed to the attenuation of the live tuberculosis vaccines *Mycobacterium bovis* BCG and *Mycobacterium microti*. *Mol Microbiol* **46:**709–717.

66. **van der Wel N, Hava D, Houben D, Fluitsma D, van Zon M, Pierson J, Brenner M, Peters PJ.** 2007. *M. tuberculosis* and *M. leprae* translocate from the phagolysosome to the cytosol in myeloid cells. *Cell* **129:**1287–1298.

67. **Simeone R, Bobard A, Lippmann J, Bitter W, Majlessi L, Brosch R, Enninga J.** 2012. Phagosomal rupture by *Mycobacterium tuberculosis* results in toxicity and host cell death. *PLoS Pathog* **8:**e1002507.

68. **Aguilo JI, Alonso H, Uranga S, Marinova D, Arbués A, de Martino A, Anel A, Monzon M, Badiola J, Pardo J, Brosch R, Martin C.** 2013. ESX-1-induced apoptosis is involved in cell-to-cell spread of *Mycobacterium tuberculosis*. *Cell Microbiol* **15:**1994–2005.

69. **Wong KW, Jacobs WR Jr.** 2011. Critical role for NLRP3 in necrotic death triggered by *Mycobacterium tuberculosis*. *Cell Microbiol* **13:**1371–1384.

70. **Wassermann R, Gulen MF, Sala C, Perin SG, Lou Y, Rybniker J, Schmid-Burgk JL, Schmidt T, Hornung V, Cole ST, Ablasser A.** 2015. *Mycobacterium tuberculosis* differentially activates cGAS- and inflammasome-dependent intracellular immune responses through ESX-1. *Cell Host Microbe* **17:**799–810.

71. **Watson RO, Bell SL, MacDuff DA, Kimmey JM, Diner EJ, Olivas J, Vance RE, Stallings CL, Virgin HW, Cox JS.** 2015. The cytosolic sensor cGAS detects *Mycobacterium tuberculosis* DNA to induce type I interferons and activate autophagy. *Cell Host Microbe* **17:**811–819.

72. **Collins AC, Cai H, Li T, Franco LH, Li XD, Nair VR, Scharn CR, Stamm CE, Levine B, Chen ZJ, Shiloh MU.** 2015. Cyclic GMP-AMP synthase is an innate immune DNA sensor for *Mycobacterium tuberculosis*. *Cell Host Microbe* **17:**820–828.

73. **Majlessi L, Brosch R.** 2015. *Mycobacterium tuberculosis* meets the cytosol: the role of cGAS in anti-mycobacterial immunity. *Cell Host Microbe* **17:**733–735.

74. **Kupz A, Zedler U, Stäber M, Perdomo C, Dorhoi A, Brosch R, Kaufmann SH.** 2016. ESAT-6-dependent cytosolic pattern recognition drives noncognate tuberculosis control in vivo. *J Clin Invest* **126:**2109–2122.

75. **Gröschel MI, Sayes F, Shin SJ, Frigui W, Pawlik A, Orgeur M, Canetti R, Honoré N, Simeone R, van der Werf TS, Bitter W, Cho SN, Majlessi L, Brosch R.** 2017. Recombinant BCG expressing ESX-1 of *Mycobacterium marinum* combines low virulence with cytosolic immune signaling and improved TB protection. *Cell Rep* **18:**2752–2765.

76. **Augenstreich J, Arbues A, Simeone R, Haanappel E, Wegener A, Sayes F, Le Chevalier F, Chalut C, Malaga W, Guilhot C, Brosch R, Astarie-Dequeker C.** 2017. ESX-1 and phthiocerol dimycocerosates of *Mycobacterium tuberculosis* act in concert to cause phagosomal rupture and host cell apoptosis. *Cell Microbiol* **19:**e12726.

77. **Quigley J, Hughitt VK, Velikovsky CA, Mariuzza RA, El-Sayed NM, Briken V.** 2017. The cell wall lipid PDIM contributes to phagosomal escape and host cell exit of *Mycobacterium tuberculosis*. *mBio* **8:**e00148-17.

78. **Barczak AK, Avraham R, Singh S, Luo SS, Zhang WR, Bray MA, Hinman AE, Thompson M, Nietupski RM, Golas A, Montgomery P, Fitzgerald M, Smith RS, White DW, Tischler AD, Carpenter AE, Hung DT.** 2017. Systematic, multiparametric analysis of *Mycobacterium tuberculosis* intracellular infection offers insight into coordinated virulence. *PLoS Pathog* **13:**e1006363.

79. **Skowyra ML, Schlesinger PH, Naismith TV, Hanson PI.** 2018. Triggered recruitment of ESCRT machinery promotes endolysosomal repair. *Science* **360:**eaar5-78.

80. **Mehra A, Zahra A, Thompson V, Sirisaengtaksin N, Wells A, Porto M, Köster S, Penberthy K, Kubota Y, Dricot A, Rogan D, Vidal M, Hill DE, Bean AJ, Philips JA.** 2013. *Mycobacterium tuberculosis* type VII secreted effector EsxH targets host ESCRT to impair trafficking. *PLoS Pathog* **9:**e1003734.

81. **Mittal E, Skowyra ML, Uwase G, Tinaztepe E, Mehra A, Köster S, Hanson PI, Philips JA.** 2018. *Mycobacterium tuberculosis* type VII secretion system effectors differentially impact the ESCRT endomembrane damage response. *mBio* **9:**01765-18.

82. **Houben EN, Korotkov KV, Bitter W.** 2014. Take five—type VII secretion systems of mycobacteria. *Biochim Biophys Acta* **1843:**1707–1716.

83. **Laencina L, Dubois V, Le Moigne V, Viljoen A, Majlessi L, Pritchard J, Bernut A, Piel L,**

Roux AL, Gaillard JL, Lombard B, Loew D, Rubin EJ, Brosch R, Kremer L, Herrmann JL, Girard-Misguich F. 2018. Identification of genes required for *Mycobacterium abscessus* growth in vivo with a prominent role of the ESX-4 locus. *Proc Natl Acad Sci U S A* **115**: E1002–E1011.

84. Sayes F, Blanc C, Ates LS, Deboosere N, Orgeur M, Le Chevalier F, Gröschel MI, Frigui W, Song OR, Lo-Man R, Brossier F, Sougakoff W, Bottai D, Brodin P, Charneau P, Brosch R, Majlessi L. 2018. Multiplexed quantitation of intraphagocyte *Mycobacterium tuberculosis* secreted protein effectors. *Cell Rep* **23**:1072–1084.

85. Solans L, Gonzalo-Asensio J, Sala C, Benjak A, Uplekar S, Rougemont J, Guilhot C, Malaga W, Martín C, Cole ST. 2014. The PhoP-dependent ncRNA Mcr7 modulates the TAT secretion system in *Mycobacterium tuberculosis*. *PLoS Pathog* **10**:e1004183.

86. Simeone R, Sayes F, Song O, Gröschel MI, Brodin P, Brosch R, Majlessi L. 2015. Cytosolic access of *Mycobacterium tuberculosis*: critical impact of phagosomal acidification control and demonstration of occurrence in vivo. *PLoS Pathog* **11**:e1004650.

87. Pym AS, Brodin P, Majlessi L, Brosch R, Demangel C, Williams A, Griffiths KE, Marchal G, Leclerc C, Cole ST. 2003. Recombinant BCG exporting ESAT-6 confers enhanced protection against tuberculosis. *Nat Med* **9**:533–539.

88. Bottai D, Frigui W, Clark S, Rayner E, Zelmer A, Andreu N, de Jonge MI, Bancroft GJ, Williams A, Brodin P, Brosch R. 2015. Increased protective efficacy of recombinant BCG strains expressing virulence-neutral proteins of the ESX-1 secretion system. *Vaccine* **33**:2710–2718.

89. Gengenbacher M, Nieuwenhuizen N, Vogelzang A, Liu H, Kaiser P, Schuerer S, Lazar D, Wagner I, Mollenkopf HJ, Kaufmann SH. 2016. Deletion of *nuoG* from the vaccine candidate *Mycobacterium bovis* BCG ΔureC::hly improves protection against tuberculosis. *mBio* **7**:e00679-16.

90. Ates LS, Sayes F, Frigui W, Ummels R, Damen MPM, Bottai D, Behr MA, van Heijst JWJ, Bitter W, Majlessi L, Brosch R. 2018. RD5-mediated lack of PE_PGRS and PPE-MPTR export in BCG vaccine strains results in strong reduction of antigenic repertoire but little impact on protection. *PLoS Pathog* **14**:e1007139.

91. Aguilo N, Gonzalo-Asensio J, Alvarez-Arguedas S, Marinova D, Gomez AB, Uranga S, Spallek R, Singh M, Audran R, Spertini F, Martin C. 2017. Reactogenicity to major tuberculosis antigens absent in BCG is linked to improved protection against *Mycobacterium tuberculosis*. *Nat Commun* **8**:16085.

92. Sayes F, Pawlik A, Frigui W, Gröschel MI, Crommelynck S, Fayolle C, Cia F, Bancroft GJ, Bottai D, Leclerc C, Brosch R, Majlessi L. 2016. CD4+ T cells recognizing PE/PPE antigens directly or via cross reactivity are protective against pulmonary *Mycobacterium tuberculosis* infection. *PLoS Pathog* **12**:e1005770.

93. Marinova D, Gonzalo-Asensio J, Aguilo N, Martin C. 2017. MTBVAC from discovery to clinical trials in tuberculosis-endemic countries. *Expert Rev Vaccines* **16**:565–576.

Bacteroidetes Gliding Motility and the Type IX Secretion System

29

MARK J. MCBRIDE[1]

INTRODUCTION

Members of the phylum *Bacteroidetes* have many unique features, including novel machinery for protein secretion and gliding motility (1–3). Most members secrete proteins across the outer membrane (OM) using the type IX protein secretion system (T9SS), which is confined to this phylum. Many also crawl rapidly over surfaces by gliding motility. For these gliding bacteria, the motility machinery and T9SS appear to be intertwined. Here we explore gliding motility, the T9SS, and the connections between them.

GLIDING MOTILITY

Bacteroidetes gliding motility was observed 100 years ago (4), but the mechanism of movement was unknown until recently. These bacteria crawl rapidly over surfaces without the aid of flagella or pili. Instead, motility involves the rapid movement of cell surface adhesins (5, 6). Gliding also occurs in bacteria that belong to other phyla (myxobacteria, mycoplasmas,

[1]Department of Biological Sciences, University of Wisconsin-Milwaukee, Milwaukee, WI 53201
Protein Secretion in Bacteria
Edited by Maria Sandkvist, Eric Cascales, and Peter J. Christie
© 2019 American Society for Microbiology, Washington, DC
doi:10.1128/microbiolspec.PSIB-0002-2018

cyanobacteria, and others), but these have their own unique motility machineries (1, 7–10). Gliding of *Flavobacterium johnsoniae* is typical of members of the *Bacteroidetes* (Fig. 1A). The long slender cells (0.4 μm by 5 to 10 μm) move over surfaces at speeds of

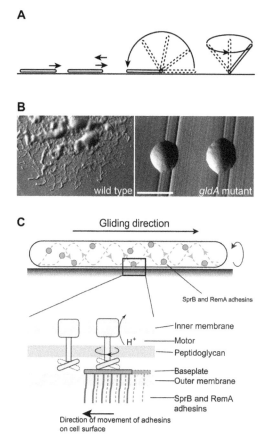

FIGURE 1 Gliding of *F. johnsoniae* cells. (A) Characteristic movements of cells. (B) Spreading colonies formed by wild-type cells and nonspreading colonies formed by cells of a nonmotile *gldA* mutant. Bar corresponds to 1 mm. (C) Model of *F. johnsoniae* gliding. Gld proteins in the cell envelope form the PMF-powered rotary motors that are attached to the cell wall and propel adhesins, such as SprB and RemA, along looped helical tracks on the cell surface. The action of the motors on adhesins that are attached to the substratum results in forward movement and rotation of the cell. Two rotary motors are shown. Rotation of one motor propels a baseplate carrying SprB and RemA adhesins and delivers it to the next motor. Modified from reference 13.

approximately 2 μm/s. Cells glide following their long axes and may reverse direction, with the head becoming the tail. Cells on wet surfaces often flip or pivot and may also attach by one pole and rotate at frequencies of about 2 revolutions per s (11, 12). The proton motive force (PMF) powers movement, and uncouplers that dissipate it block gliding (5, 12). Movement of cells on agar often results in thin spreading colonies (Fig. 1B).

Genetic analyses revealed *F. johnsoniae* Gld proteins that are essential for gliding and Spr proteins whose absence results in severe but incomplete motility defects (13). Most of the Gld and Spr proteins are novel, complicating prediction of function. The 669-kDa, repetitive cell surface protein SprB was the first *F. johnsoniae* motility protein to be assigned a function. SprB is an adhesin that is required for cell movement over agar, and *sprB* mutants thus form nonspreading colonies (6). SprB filaments extend from the cell surface. Antibodies against SprB were used to label live cells, revealing that SprB is propelled rapidly from pole to pole following an apparent closed helical loop (5, 6). Cells lacking SprB retain the ability to move on some surfaces other than agar because of the presence of additional motility adhesins, such as RemA (14). The multiple motility adhesins allow cells to move over diverse surfaces. PMF-dependent rotary motors embedded in the cell envelope are thought to propel the motility adhesins along a helical track (13, 15–17). When the adhesins attach to the substratum, the motors pushing against them result in rotation of and forward movement of the cell in a screw-like manner (Fig. 1C). The cytoplasmic membrane (CM) proteins GldL and GldM (Fig. 2B) are candidates for the motor proteins that harvest the PMF (5, 16, 18, 19). The only other known motility proteins that span the CM are components of the GldA-GldF-GldG ATP-binding cassette (ABC) transporter. This is unlikely to be the gliding motor because the ABC transporter is presumably powered by ATP rather than PMF and because several members of the *Bacteroidetes*

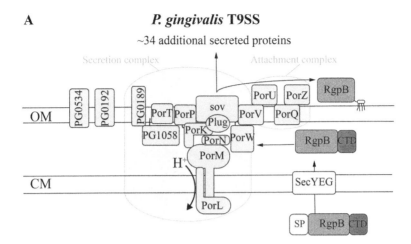

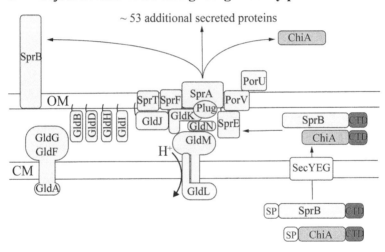

FIGURE 2 T9SS and gliding motility proteins. Proteins in blue are associated with the T9SS, and proteins in green are motility proteins that are not directly associated with the T9SS. Orthologous T9SS proteins between panels A and B are shown in the same relative positions, color, and shapes. *F. johnsoniae* GldK, GldL, GldM, GldN, SprA, SprE, and SprT correspond to *P. gingivalis* PorK, PorL, PorM, PorN, Sov, PorW, and PorT, respectively. Black lines are lipid tails on lipoproteins. Proteins secreted by the T9SS have predicted N-terminal signal peptides (yellow) that target them to the Sec system for export across the cytoplasmic membrane (CM) and C-terminal domains (red) that target them to the T9SS for secretion across the outer membrane (OM). Proteins are not drawn to scale, and stoichiometry of components is not illustrated. (A) *P. gingivalis* T9SS proteins. Where protein names were not available, locus tags (from *P. gingivalis* strain W83) were used. The gingipain protease RgpB is shown covalently attached to the outer membrane acidic lipopolysaccharide (A-LPS). Secretion complex and attachment complex are indicated by the large and small blue barrels, respectively. (B) *F. johnsoniae* T9SS and gliding motility proteins. SprB is a motility adhesin that is propelled by some of the other proteins shown. SprF is required for secretion of SprB but not for secretion of other proteins. SprF and nine other *F. johnsoniae* proteins are related to *P. gingivalis* PorP. *F. johnsoniae* PorV is required for secretion of ChiA and many other proteins, but not for secretion of SprB.

that lack this transporter exhibit gliding motility (1, 20).

Analyses of many nonmotile *gld* mutants revealed surprising phenotypes. The mutants failed to digest chitin or to adhere to surfaces and they were resistant to bacteriophages (18, 19, 21). The reason for these phenotypes is now known. *gldK*, *gldL*, *gldM*, and *gldN* encode core components of the T9SS that secretes SprB and RemA to the cell surface (19, 22, 23). The unexpected phenotypes of these *gld* mutants were caused by failure to secrete adhesins, phage receptors, a chitinase, and dozens of other proteins.

DISCOVERY OF THE T9SS

The first hint that the *Bacteroidetes* might have a novel protein secretion system came from studies of the nonmotile human pathogen *Porphyromonas gingivalis*. *P. gingivalis* secretes virulence factors, such as gingipain proteases, that are important in periodontal disease (24, 25). The secreted proteins typically remain attached to the *P. gingivalis* cell surface. A conserved carboxy-terminal domain (CTD) of secreted proteins was suggested to be linked to secretion and surface attachment (26–28). In 2005, genetic experiments revealed the first component of the secretion system, the OM protein PorT (29). Two years later, studies on *F. johnsoniae* motility and *P. gingivalis* secretion converged with the publication of papers on the *F. johnsoniae* OM motility protein SprA (30) and the orthologous *P. gingivalis* protein involved in secretion of virulence factors, Sov (31). Another OM protein involved in secretion, PorV, was recognized soon after (32). Many of the remaining T9SS proteins were discovered by comparative genome analyses that revealed genes that co-occur with the secretion gene *porT* (23). Targeted mutagenesis of these identified 10 *P. gingivalis* genes that were required for optimal secretion. Among these were orthologs of six *F. johnsoniae* gliding motility genes (*gldK*, *gldL*, *gldM*,

gldN, *sprA*, and *sprE*, named *porK*, *porL*, *porM*, *porN*, *sov*, and *porW* in *P. gingivalis*) and *porP*, *porQ*, and *porU*. Mutations in any of these *P. gingivalis* genes resulted in defects ranging from partial to complete loss of secretion. Analysis of *F. johnsoniae* strains with mutations in the corresponding genes revealed similar secretion defects (19, 22, 23), except that *porQ* and *porU* mutants were largely competent for secretion (33, 34). The secretion system was initially referred to as the Por secretion system and was later named the T9SS (1, 35). T9SSs are found in most *Bacteroidetes* but are absent in most members of the genus *Bacteroides* (1, 36), which instead rely on other secretion systems (37). Some bacteria that have T9SSs also use other secretion systems (1, 38).

SECRETED PROTEINS

Proteins secreted by T9SSs have cleavable N-terminal signal peptides that are thought to target them to the Sec system for export across the CM. They also typically have conserved 70- to 100-amino-acid CTDs that are required for secretion across the OM and are often cleaved during this process (26, 27, 29, 34, 39–43). Attachment of a T9SS CTD to a foreign protein results in its efficient secretion (34, 40, 41). The structures of two CTDs have been determined (44, 45). Both have an Ig-like fold with seven β-strands forming a sandwich-like structure that is thought to interact with components of the T9SS. The conserved CTD sequences allow predictions of secreted proteins from genomic data. Most T9SSs are predicted to secrete dozens of proteins, and some secrete many more. The cellulolytic bacterium *Cytophaga hutchinsonii* is predicted to use its T9SS to secrete at least 147 proteins (46), many of which have been verified by proteomic analyses (43), and *Fluviicola taffensis* is predicted to secrete 230 proteins (36). Given the large number of proteins secreted, it is not surprising that T9SSs are important for many processes in environmental and host-associated bacteria, including virulence (23, 47, 48), polymer degradation (40, 46), adhesion (19, 22), S-layer formation (49, 50), motility (19, 22, 46, 47, 51), and biofilm formation (49–51).

Many T9SS-secreted proteins are large. *F. johnsoniae* SprB, for example, is 669 kDa. Some proteins are secreted in soluble form, and others become attached to the cell surface. Most proteins secreted by *P. gingivalis* are of the latter type and are cell surface associated as a result of covalent bonding to an acidic form of lipopolysaccharide (A-LPS) (52). This modification occurs after transit of the OM and removal of the CTD (53).

COMPONENTS AND STRUCTURE OF THE T9SS

At least 17 proteins are thought to have roles in T9SS-mediated secretion (Fig. 2; Table 1). Several regulatory proteins, not discussed here, have also been identified (23, 54, 55). *P. gingivalis* T9SS proteins that are essential or nearly essential for secretion include PorK, PorL, PorM, PorN, Sov, PorW, PorT, PorP, PorV, and PG1058 (23, 31, 56–58). The *F. johnsoniae* orthologs of the first 7 of these (GldK, GldL, GldM, GldN, SprA, SprE, and SprT) are also required for secretion (19, 22, 23), but the situation is more complicated for PorP, PorV, and PG1058. *P. gingivalis* has a single PorP, whereas *F. johnsoniae* has 10 *porP*-like genes (59). Requirement of these genes for *F. johnsoniae* secretion has likely been masked by redundancy. One *F. johnsoniae* PorP-like protein, SprF, is required for T9SS-mediated secretion of SprB, but it is not required for secretion of other proteins examined (59). Redundancy may also occur for the five *F. johnsoniae* proteins that are related to PG1058. The genes encoding these PG1058-like proteins are each adjacent to *sprF*-like genes. These have not yet been examined for roles in secretion. *P. gingivalis* PorV is required for secretion

(32), whereas *F. johnsoniae* PorV is required only for secretion of some proteins that are targeted to its T9SS (34). PG0189 interacts with PorK and PorN and thus may be involved in secretion (60), but mutants lacking PG0189 have apparently not been examined.

The *P. gingivalis* proteins described above all localize to the cell envelope and may constitute the core of the secretion complex (Fig. 2A). Only PorL and PorM span the CM, and they may be involved in energy transduction to power secretion, similar to the proposed role of GldL and GldM in gliding motility (5, 18, 19, 61). PorL/GldL proteins have a completely conserved glutamate predicted to be buried in the membrane that might facilitate transit of protons to energize both processes (19, 61).

Recent studies of protein-protein interactions, protein complexes, and protein structures have added greatly to our understanding of the T9SS secretion complex (60–63). PorL/PorM and PorK/PorN complexes were isolated, and additional protein-protein interactions were identified (Table 1), suggesting a potential envelope-spanning complex comprised of PorK, PorL, PorM, PorN, PorP, and PG0189. Stoichiometry was suggested to be $PorL_3/PorM_2/PorN_2/PorK_2$, with perhaps 3 or 4 copies of this basic structure forming the core of the secretion machine. Crystal structures of dimers of the periplasmic domains of PorM and GldM were recently reported (62). Despite only 22% amino acid identity, the proteins formed remarkably similar four-domain structures predicted to span most of the periplasm. In GldM the domains were linear, whereas in PorM there was a bend between domains 2 and 3. The PorM and GldM structures may represent different dynamic states associated with protein translocation. Transition between these states could involve PorL/GldL-mediated PMF-driven energy transduction. The very large (267-kDa) SprA protein appears to form the OM pore of the secretion system (64). Cryo-electron microscopic analysis revealed that a single SprA protein forms a 36-strand transmembrane β-barrel with an

TABLE 1 T9SS components[a]

P. gingivalis protein name or locus tag[b]	F. johnsoniae protein name or locus tag	Localization	Predicted role in secretion/motility	Essential for F. johnsoniae gliding?	Interaction partner	Notes	Reference(s)
Secretion complex							
PorL	GldL	CM	Energizes secretion/motility	Essential	PorM	Forms complex with PorK, PorM, and PorN	18, 19, 23, 61
PorM	GldM	CM	Energizes secretion/motility	Essential	PorK, PorL, PorN, PorP	Forms complex with PorK, PorL, and PorN	18, 19, 23, 61, 62
PorK	GldK	OM lipoprotein	Component of periplasmic channel	Essential	PorM, PorN, PorP, PG0189	Forms complex with PorL, PorM, and PorN	18, 19, 23, 60, 61
PorN	GldN and GldO	P	Component of periplasmic channel	Essential	PorK, PorM, PG0189	Forms complex with PorK, PorL, and PorM	18, 19, 22, 23, 60, 61
Sov	SprA	OM	OM pore	Nearly essential	PorV, plug		19, 30, 31, 58, 64
PorT	SprT	OM	Unknown	Nearly essential	ND		23, 29, 71
PorP	SprF and 9 other PorP-like proteins[c]	OM	Unknown	ND	PorK, PorM		23, 46, 59, 61
PorW	SprE	OM lipoprotein	Unknown	Nearly essential	ND		23, 72
PG1058 (PGN_1296)	Fjoh_1647,[d] Fjoh_2275, Fjoh_3950, Fjoh_3973, Fjoh_4540	OM lipoprotein	Anchors T9SS to peptidoglycan	ND	ND	OmpA C-terminal peptidoglycan-binding domain	57
PG0189 (PGN_0297)	Fjoh_1692	OM	Unknown	ND	PorK, PorN		60
PG2092 (PGN_0144)	Fjoh_1759, plug	P/OM	Forms plug on periplasmic side of SprA pore. Prevents nonspecific leakage of periplasmic contents.	ND	SprA		64

	Fjoh homolog	Localization	Function			Complex	References
Attachment complex							
PorV[e]	PorV	OM	Shuttles secreted proteins from secretion complex to attachment complex	Not needed for gliding	PorU, SprA	Forms complex with PorU, PorQ, and PorZ	32, 34, 56, 64, 73, 74
PorU	PorU	Cell surface	CTD cleavage; attachment of substrates to A-LPS	Not needed for gliding	PorV	Forms complex with PorV, PorQ, and PorZ	39, 56, 73
PorQ	Fjoh_2755	OM	Unknown	Not needed for gliding		Forms complex with PorU, PorV, and PorZ	23, 56
PorZ	Fjoh_0707	Cell surface	Assembly of PorU on cell surface; secretion and/or modification of secreted proteins	ND		Forms complex with PorU, PorV, and PorQ	45, 56
Other							
PG0534 (PGN_1437)	Fjoh_0118	OM	Secretion and/or modification of secreted proteins	ND			65
Omp17; PG0192 (PGN_0300)	Fjoh_0599[f], Fjoh_1000, Fjoh_1688, Fjoh_1689	OM	PorU maturation or stabilization; post-secretion processing	ND		OmpH-like	66

[a] Abbreviations: CM, cytoplasmic membrane; OM, outer membrane; P, periplasm; ND, not determined.
[b] Locus tags for *P. gingivalis* strains W83 (PG) and ATCC33277 (PGN_) are listed, with the PGN_ locus tags in parentheses.
[c] PorP-like proteins recognized by assignment to Tigrfam "type IX secretion system membrane protein PorP/SprF family" TIGR03519.
[d] PG1058-like proteins recognized by the presence of domains corresponding to pfam00691 (OmpA_C terminal domain), pfam07676 (WD40-like Beta propeller repeat), and pfam13620 (carboxypeptidase regulatory-like domain).
[e] PorV may be a component of the secretion complex and the attachment complex.
[f] Proteins related to PG0192 (PGN_0300) recognized by the presence of domains corresponding to pfam03938 (OmpH-like).

internal pore of approximately 70 Å in diameter, which should allow transit of folded proteins. The SprA pore appears to be alternately occluded on the periplasmic side by the Plug protein (Fjoh_1759) and on the external side by PorV. This may prevent nonspecific leakage of periplasmic contents. PorV is thought to escort proteins from the lumen of the SprA channel to the outside of the cell.

P. gingivalis PorU, PorQ, and PorZ appear to form a complex with PorV that modifies secreted proteins and attaches them to the cell surface (56). PorU is the peptidase that removes CTDs after secretion (39). It may also covalently attach the newly exposed C termini of the secreted proteins to A-LPS via a "sortase-like" mechanism (53). Cells of *porU* mutants secrete proteins, but these retain their CTDs and fail to attach to the cell surface. PorZ is required for proper localization and stability of PorU, and thus, *porZ* mutants behave similarly to *porU* mutants (45). PorV is thought to function as the shuttle that delivers proteins from SprA and the secretion complex to the attachment complex (56). Cells lacking PG0534 and PGN_0300 (PG0192) have phenotypes that may indicate that they also function in modification of secreted proteins (65, 66).

RELATIONSHIP BETWEEN THE T9SS AND GLIDING MOTILITY

The *F. johnsoniae* T9SS and gliding motility machines appear to be intertwined, since many mutations disrupt gliding and secretion (67). This is reminiscent of the bacterial flagellum, which has a type III secretion system involved in flagellar assembly at its core (68, 69). GldL and GldM have been suggested to be part of the PMF-driven rotary gliding motor, and they are also thought to energize secretion (5, 18, 19, 61). Other core components of the T9SS (GldK, GldN, SprA, SprE, and SprT) are also essential for gliding, suggesting the possibility that a transmembrane complex of these T9SS proteins may be central to gliding and secretion. Loss of some other motility proteins, GldA, GldB, GldD, GldF, GldG, GldH, GldI, and GldJ (green in Fig. 2B), also results in defects in motility and secretion (67). The secretion defects were unexpected, since these proteins are not associated with the *P. gingivalis* T9SS. The reason for the loss of secretion became clear when it was discovered that the *F. johnsoniae* motility protein GldJ is required to stabilize the T9SS protein GldK (67). It was already known that GldA, GldB, GldD, GldF, GldG, GldH, and GldI are needed to stabilize GldJ (70). Apparently, the absence of any of these proteins results in loss of GldJ, which results in loss of GldK and thus in defects in protein secretion. Truncated forms of GldJ were identified that are nonfunctional for motility but that stabilize GldK and thus allow secretion (67). This partially untangles gliding from secretion, but if current suggestions that GldL and GldM are motor proteins for secretion and motility are correct, it may be difficult to completely separate the two processes.

CONCLUSIONS

The novel machines described above, involved in protein secretion and cell movement, have remarkable properties. The T9SS efficiently secretes huge proteins across the OM, and the gliding motility machinery rapidly propels some of these along the cell surface. Rapid progress has been made in our understanding of gliding and secretion, but many mysteries remain regarding the functioning of the machines responsible for these processes.

ACKNOWLEDGMENTS

This work was supported by grant MCB-1516990 from the National Science Foundation.

CITATION

McBride MJ. 2019. *Bacteroidetes* gliding motility and the type IX secretion system. Microbiol Spectrum 7(1):PSIB-0002-2018.

REFERENCES

1. **McBride MJ, Zhu Y.** 2013. Gliding motility and Por secretion system genes are widespread among members of the phylum *Bacteroidetes*. *J Bacteriol* **195**:270–278.

2. **Lasica AM, Ksiazek M, Madej M, Potempa J.** 2017. The type IX secretion system (T9SS): highlights and recent insights into its structure and function. *Front Cell Infect Microbiol* **7**:215.

3. **Veith PD, Glew MD, Gorasia DG, Reynolds EC.** 2017. Type IX secretion: the generation of bacterial cell surface coatings involved in virulence, gliding motility and the degradation of complex biopolymers. *Mol Microbiol* **106**: 35–53.

4. **Hutchinson HB, Clayton J.** 1919. On the decomposition of cellulose by an aerobic organism (*Spirochaeta cytophaga*, n. sp.). *J Agric Sci* **9**:143–173.

5. **Nakane D, Sato K, Wada H, McBride MJ, Nakayama K.** 2013. Helical flow of surface protein required for bacterial gliding motility. *Proc Natl Acad Sci U S A* **110**:11145–11150.

6. **Nelson SS, Bollampalli S, McBride MJ.** 2008. SprB is a cell surface component of the *Flavobacterium johnsoniae* gliding motility machinery. *J Bacteriol* **190**:2851–2857.

7. **Jarrell KF, McBride MJ.** 2008. The surprisingly diverse ways that prokaryotes move. *Nat Rev Microbiol* **6**:466–476.

8. **Miyata M.** 2010. Unique centipede mechanism of *Mycoplasma* gliding. *Annu Rev Microbiol* **64**:519–537.

9. **Nan B, McBride MJ, Chen J, Zusman DR, Oster G.** 2014. Bacteria that glide with helical tracks. *Curr Biol* **24**:R169–R173.

10. **Cho YW, Gonzales A, Harwood TV, Huynh J, Hwang Y, Park JS, Trieu AQ, Italia P, Pallipuram VK, Risser DD.** 2017. Dynamic localization of HmpF regulates type IV pilus activity and directional motility in the filamentous cyanobacterium *Nostoc punctiforme*. *Mol Microbiol* **106**:252–265.

11. **Lapidus IR, Berg HC.** 1982. Gliding motility of *Cytophaga* sp. strain U67. *J Bacteriol* **151**:384–398.

12. **Pate JL, Chang L-YE.** 1979. Evidence that gliding motility in prokaryotic cells is driven by rotary assemblies in the cell envelopes. *Curr Microbiol* **2**:59–64.

13. **McBride MJ, Nakane D.** 2015. *Flavobacterium* gliding motility and the type IX secretion system. *Curr Opin Microbiol* **28**:72–77.

14. **Shrivastava A, Rhodes RG, Pochiraju S, Nakane D, McBride MJ.** 2012. *Flavobac-*

terium johnsoniae RemA is a mobile cell surface lectin involved in gliding. *J Bacteriol* **194**:3678–3688.

15. **Shrivastava A, Berg HC.** 2015. Towards a model for *Flavobacterium* gliding. *Curr Opin Microbiol* **28**:93–97.

16. **Shrivastava A, Lele PP, Berg HC.** 2015. A rotary motor drives *Flavobacterium* gliding. *Curr Biol* **25**:338–341.

17. **Shrivastava A, Roland T, Berg HC.** 2016. The screw-like movement of a gliding bacterium is powered by spiral motion of cell-surface adhesins. *Biophys J* **111**:1008–1013.

18. **Braun TF, Khubbar MK, Saffarini DA, McBride MJ.** 2005. *Flavobacterium johnsoniae* gliding motility genes identified by *mariner* mutagenesis. *J Bacteriol* **187**:6943–6952.

19. **Shrivastava A, Johnston JJ, van Baaren JM, McBride MJ.** 2013. *Flavobacterium johnsoniae* GldK, GldL, GldM, and SprA are required for secretion of the cell surface gliding motility adhesins SprB and RemA. *J Bacteriol* **195**:3201–3212.

20. **Zhu Y, McBride MJ.** 2016. Comparative analysis of *Cellulophaga algicola* and *Flavobacterium johnsoniae* gliding motility. *J Bacteriol* **198**:1743–1754.

21. **Chang LE, Pate JL, Betzig RJ.** 1984. Isolation and characterization of nonspreading mutants of the gliding bacterium *Cytophaga johnsonae*. *J Bacteriol* **159**:26–35.

22. **Rhodes RG, Samarasam MN, Shrivastava A, van Baaren JM, Pochiraju S, Bollampalli S, McBride MJ.** 2010. *Flavobacterium johnsoniae* *gldN* and *gldO* are partially redundant genes required for gliding motility and surface localization of SprB. *J Bacteriol* **192**:1201–1211.

23. **Sato K, Naito M, Yukitake H, Hirakawa H, Shoji M, McBride MJ, Rhodes RG, Nakayama K.** 2010. A protein secretion system linked to bacteroidete gliding motility and pathogenesis. *Proc Natl Acad Sci U S A* **107**:276–281.

24. **Pathirana RD, O'Brien-Simpson NM, Brammar GC, Slakeski N, Reynolds EC.** 2007. Kgp and RgpB, but not RgpA, are important for *Porphyromonas gingivalis* virulence in the murine periodontitis model. *Infect Immun* **75**:1436–1442.

25. **O'Brien-Simpson NM, Paolini RA, Hoffmann B, Slakeski N, Dashper SG, Reynolds EC.** 2001. Role of RgpA, RgpB, and Kgp proteinases in virulence of *Porphyromonas gingivalis* W50 in a murine lesion model. *Infect Immun* **69**:7527–7534.

26. **Seers CA, Slakeski N, Veith PD, Nikolof T, Chen YY, Dashper SG, Reynolds EC.** 2006. The RgpB C-terminal domain has a role in attach-

ment of RgpB to the outer membrane and belongs to a novel C-terminal-domain family found in *Porphyromonas gingivalis*. *J Bacteriol* **188**:6376–6386.

27. **Nguyen KA, Travis J, Potempa J.** 2007. Does the importance of the C-terminal residues in the maturation of RgpB from *Porphyromonas gingivalis* reveal a novel mechanism for protein export in a subgroup of Gram-negative bacteria? *J Bacteriol* **189**:833–843.

28. **Veith PD, Talbo GH, Slakeski N, Dashper SG, Moore C, Paolini RA, Reynolds EC.** 2002. Major outer membrane proteins and proteolytic processing of RgpA and Kgp of *Porphyromonas gingivalis* W50. *Biochem J* **363**:105–115.

29. **Sato K, Sakai E, Veith PD, Shoji M, Kikuchi Y, Yukitake H, Ohara N, Naito M, Okamoto K, Reynolds EC, Nakayama K.** 2005. Identification of a new membrane-associated protein that influences transport/maturation of gingipains and adhesins of *Porphyromonas gingivalis*. *J Biol Chem* **280**:8668–8677.

30. **Nelson SS, Glocka PP, Agarwal S, Grimm DP, McBride MJ.** 2007. *Flavobacterium johnsoniae* SprA is a cell surface protein involved in gliding motility. *J Bacteriol* **189**:7145–7150.

31. **Saiki K, Konishi K.** 2007. Identification of a *Porphyromonas gingivalis* novel protein Sov required for the secretion of gingipains. *Microbiol Immunol* **51**:483–491.

32. **Ishiguro I, Saiki K, Konishi K.** 2009. PG27 is a novel membrane protein essential for a *Porphyromonas gingivalis* protease secretion system. *FEMS Microbiol Lett* **292**:261–267.

33. **Shrivastava A.** 2013. *Cell surface adhesins, exopolysaccharides and the Por (type IX) secretion system of* Flavobacterium johnsoniae. *PhD dissertation.* University of Wisconsin, Milwaukee, Milwaukee, WI.

34. **Kharade SS, McBride MJ.** 2015. *Flavobacterium johnsoniae* PorV is required for secretion of a subset of proteins targeted to the type IX secretion system. *J Bacteriol* **197**:147–158.

35. **Sato K, Yukitake H, Narita Y, Shoji M, Naito M, Nakayama K.** 2013. Identification of *Porphyromonas gingivalis* proteins secreted by the Por secretion system. *FEMS Microbiol Lett* **338**:68–76.

36. **Kulkarni SS, Zhu Y, Brendel CJ, McBride MJ.** 2017. Diverse C-terminal sequences involved in *Flavobacterium johnsoniae* protein secretion. *J Bacteriol* **199**:e00884-16.

37. **Wilson MM, Anderson DE, Bernstein HD.** 2015. Analysis of the outer membrane proteome and secretome of *Bacteroides fragilis* reveals a multiplicity of secretion mechanisms. *PLoS One* **10**:e0117732.

38. **Russell AB, Wexler AG, Harding BN, Whitney JC, Bohn AJ, Goo YA, Tran BQ, Barry NA, Zheng H, Peterson SB, Chou S, Gonen T, Goodlett DR, Goodman AL, Mougous JD.** 2014. A type VI secretion-related pathway in Bacteroidetes mediates interbacterial antagonism. *Cell Host Microbe* **16**:227–236.

39. **Glew MD, Veith PD, Peng B, Chen YY, Gorasia DG, Yang Q, Slakeski N, Chen D, Moore C, Crawford S, Reynolds EC.** 2012. PG0026 is the C-terminal signal peptidase of a novel secretion system of *Porphyromonas gingivalis*. *J Biol Chem* **287**:24605–24617.

40. **Kharade SS, McBride MJ.** 2014. *Flavobacterium johnsoniae* chitinase ChiA is required for chitin utilization and is secreted by the type IX secretion system. *J Bacteriol* **196**:961–970.

41. **Shoji M, Sato K, Yukitake H, Kondo Y, Narita Y, Kadowaki T, Naito M, Nakayama K.** 2011. Por secretion system-dependent secretion and glycosylation of *Porphyromonas gingivalis* hemin-binding protein 35. *PLoS One* **6**:e21372.

42. **Slakeski N, Seers CA, Ng K, Moore C, Cleal SM, Veith PD, Lo AW, Reynolds EC.** 2011. C-terminal domain residues important for secretion and attachment of RgpB in *Porphyromonas gingivalis*. *J Bacteriol* **193**:132–142.

43. **Veith PD, Nor Muhammad NA, Dashper SG, Likić VA, Gorasia DG, Chen D, Byrne SJ, Catmull DV, Reynolds EC.** 2013. Protein substrates of a novel secretion system are numerous in the *Bacteroidetes* phylum and have in common a cleavable C-terminal secretion signal, extensive post-translational modification, and cell-surface attachment. *J Proteome Res* **12**:4449–4461.

44. **de Diego I, Ksiazek M, Mizgalska D, Koneru L, Golik P, Szmigielski B, Nowak M, Nowakowska Z, Potempa B, Houston JA, Enghild JJ, Thøgersen IB, Gao J, Kwan AH, Trewhella J, Dubin G, Gomis-Rüth FX, Nguyen KA, Potempa J.** 2016. The outer-membrane export signal of *Porphyromonas gingivalis* type IX secretion system (T9SS) is a conserved C-terminal β-sandwich domain. *Sci Rep* **6**:23123.

45. **Lasica AM, Goulas T, Mizgalska D, Zhou X, de Diego I, Ksiazek M, Madej M, Guo Y, Guevara T, Nowak M, Potempa B, Goel A, Sztukowska M, Prabhakar AT, Bzowska M, Widziolek M, Thøgersen IB, Enghild JJ, Simonian M, Kulczyk AW, Nguyen KA, Potempa J, Gomis-Rüth FX.** 2016. Structural and functional probing of PorZ, an essential bacterial surface component of the type-IX secretion system of human oral-microbiomic *Porphyromonas gingivalis*. *Sci Rep* **6**:37708.

46. **Zhu Y, McBride MJ.** 2014. Deletion of the *Cytophaga hutchinsonii* type IX secretion system gene *sprP* results in defects in gliding motility and cellulose utilization. *Appl Microbiol Biotechnol* **98:**763–775.

47. **Li N, Zhu Y, LaFrentz BR, Evenhuis JP, Hunnicutt DW, Conrad RA, Barbier P, Gullstrand CW, Roets JE, Powers JL, Kulkarni SS, Erbes DH, García JC, Nie P, McBride MJ.** 2017. The type IX secretion system is required for virulence of the fish pathogen *Flavobacterium columnare. Appl Environ Microbiol* **83:**e01769-17.

48. **Guo Y, Hu D, Guo J, Wang T, Xiao Y, Wang X, Li S, Liu M, Li Z, Bi D, Zhou Z.** 2017. *Riemerella anatipestifer* type IX secretion system is required for virulence and gelatinase secretion. *Front Microbiol* **8:**2553.

49. **Narita Y, Sato K, Yukitake H, Shoji M, Nakane D, Nagano K, Yoshimura F, Naito M, Nakayama K.** 2014. Lack of a surface layer in *Tannerella forsythia* mutants deficient in the type IX secretion system. *Microbiology* **160:** 2295–2303.

50. **Tomek MB, Neumann L, Nimeth I, Koerdt A, Andesner P, Messner P, Mach L, Potempa JS, Schäffer C.** 2014. The S-layer proteins of *Tannerella forsythia* are secreted via a type IX secretion system that is decoupled from protein O-glycosylation. *Mol Oral Microbiol* **29:**307–320.

51. **Kita D, Shibata S, Kikuchi Y, Kokubu E, Nakayama K, Saito A, Ishihara K.** 2016. Involvement of the type IX secretion system in *Capnocytophaga ochracea* gliding motility and biofilm formation. *Appl Environ Microbiol* **82:**1756–1766.

52. **Paramonov N, Rangarajan M, Hashim A, Gallagher A, Aduse-Opoku J, Slaney JM, Hounsell E, Curtis MA.** 2005. Structural analysis of a novel anionic polysaccharide from *Porphyromonas gingivalis* strain W50 related to Arg-gingipain glycans. *Mol Microbiol* **58:**847–863.

53. **Gorasia DG, Veith PD, Chen D, Seers CA, Mitchell HA, Chen YY, Glew MD, Dashper SG, Reynolds EC.** 2015. *Porphyromonas gingivalis* type IX secretion substrates are cleaved and modified by a sortase-like mechanism. *PLoS Pathog* **11:**e1005152.

54. **Kadowaki T, Yukitake H, Naito M, Sato K, Kikuchi Y, Kondo Y, Shoji M, Nakayama K.** 2016. A two-component system regulates gene expression of the type IX secretion component proteins via an ECF sigma factor. *Sci Rep* **6:**23288.

55. **Vincent MS, Durand E, Cascales E.** 2016. The PorX response regulator of the *Porphyromonas gingivalis* PorXY two-component system does not directly regulate the type IX secretion genes but binds the PorL subunit. *Front Cell Infect Microbiol* **6:**96.

56. **Glew MD, Veith PD, Chen D, Gorasia DG, Peng B, Reynolds EC.** 2017. PorV is an outer membrane shuttle protein for the type IX secretion system. *Sci Rep* **7:**8790.

57. **Heath JE, Seers CA, Veith PD, Butler CA, Nor Muhammad NA, Chen YY, Slakeski N, Peng B, Zhang L, Dashper SG, Cross KJ, Cleal SM, Moore C, Reynolds EC.** 2016. PG1058 is a novel multidomain protein component of the bacterial type IX secretion system. *PLoS One* **11:**e0164313.

58. **Saiki K, Konishi K.** 2010. The role of Sov protein in the secretion of gingipain protease virulence factors of *Porphyromonas gingivalis. FEMS Microbiol Lett* **302:**166–174.

59. **Rhodes RG, Nelson SS, Pochiraju S, McBride MJ.** 2011. *Flavobacterium johnsoniae sprB* is part of an operon spanning the additional gliding motility genes *sprC, sprD,* and *sprF. J Bacteriol* **193:**599–610.

60. **Gorasia DG, Veith PD, Hanssen EG, Glew MD, Sato K, Yukitake H, Nakayama K, Reynolds EC.** 2016. Structural insights into the PorK and PorN components of the *Porphyromonas gingivalis* type IX secretion system. *PLoS Pathog* **12:**e1005820.

61. **Vincent MS, Canestrari MJ, Leone P, Stathopulos J, Ize B, Zoued A, Cambillau C, Kellenberger C, Roussel A, Cascales E.** 2017. Characterization of the *Porphyromonas gingivalis* type IX secretion trans-envelope PorKLMNP core complex. *J Biol Chem* **292:**3252–3261.

62. **Leone P, Roche J, Vincent MS, Tran QH, Desmyter A, Cascales E, Kellenberger C, Cambillau C, Roussel A.** 2018. Type IX secretion system PorM and gliding machinery GldM form arches spanning the periplasmic space. *Nat Commun* **9:**429.

63. **Glew MD, Veith PD, Chen D, Seers CA, Chen YY, Reynolds EC.** 2014. Blue native-PAGE analysis of membrane protein complexes in *Porphyromonas gingivalis. J Proteomics* **110:**72–92.

64. **Lauber F, Deme JC, Lea SM, Berks BC.** 2018. Type 9 secretion system structures reveal a new protein transport mechanism. *Nature* **564:**77–82.

65. **Saiki K, Konishi K.** 2010. Identification of a novel *Porphyromonas gingivalis* outer membrane protein, PG534, required for the production of active gingipains. *FEMS Microbiol Lett* **310:**168–174.

66. **Taguchi Y, Sato K, Yukitake H, Inoue T, Nakayama M, Naito M, Kondo Y, Kano K, Hoshino T, Nakayama K, Takashiba S, Ohara**

N. 2015. Involvement of an Skp-like protein, PGN_0300, in the type IX secretion system of *Porphyromonas gingivalis*. *Infect Immun* **84:** 230–240.

67. **Johnston JJ, Shrivastava A, McBride MJ.** 2017. Untangling *Flavobacterium johnsoniae* gliding motility and protein secretion. *J Bacteriol* **200:**e00362-17.

68. **Minamino T, Namba K.** 2008. Distinct roles of the FliI ATPase and proton motive force in bacterial flagellar protein export. *Nature* **451:** 485–488.

69. **Paul K, Erhardt M, Hirano T, Blair DF, Hughes KT.** 2008. Energy source of flagellar type III secretion. *Nature* **451:**489–492.

70. **Braun TF, McBride MJ.** 2005. *Flavobacterium johnsoniae* GldJ is a lipoprotein that is required for gliding motility. *J Bacteriol* **187:**2628–2637.

71. **Nguyen KA, Zylicz J, Szczesny P, Sroka A, Hunter N, Potempa J.** 2009. Verification of a

topology model of PorT as an integral outer-membrane protein in *Porphyromonas gingivalis*. *Microbiology* **155:**328–337.

72. **Rhodes RG, Samarasam MN, Van Groll EJ, McBride MJ.** 2011. Mutations in *Flavobacterium johnsoniae sprE* result in defects in gliding motility and protein secretion. *J Bacteriol* **193:**5322–5327.

73. **Saiki K, Konishi K.** 2014. *Porphyromonas gingivalis* C-terminal signal peptidase PG0026 and HagA interact with outer membrane protein PG27/LptO. *Mol Oral Microbiol* **29:**32–44.

74. **Chen YY, Peng B, Yang Q, Glew MD, Veith PD, Cross KJ, Goldie KN, Chen D, O'Brien-Simpson N, Dashper SG, Reynolds EC.** 2011. The outer membrane protein LptO is essential for the O-deacylation of LPS and the co-ordinated secretion and attachment of A-LPS and CTD proteins in *Porphyromonas gingivalis*. *Mol Microbiol* **79:**1380–1401.

Similarities and Differences between Colicin and Filamentous Phage Uptake by Bacterial Cells

30

DENIS DUCHÉ[1] and LAETITIA HOUOT[1]

INTRODUCTION

The cell envelope of Gram-negative bacteria, such as *Escherichia coli*, is characterized by the presence of two membranes, the inner (IM) and outer (OM) membranes, separated by the periplasm and a thin layer of peptidoglycan (PG). This envelope is a formidable barrier against a myriad of harmful compounds, while simultaneously allowing the entry of nutrients necessary for cell survival. However, this barrier, like the Maginot Line in France during the Second World War, is not completely impenetrable, and exogenous particles, including some toxins and viruses, can pierce it.

Colicins are plasmid-encoded toxins (40 to 80 kDa), produced by *E. coli* under conditions of stress, that allow the efficient killing of related bacteria competing for space and resources. The toxicity of colicins relies on various modes of action, including IM depolarization, the inhibition of PG synthesis, and the degradation of DNA or RNA of the target cell. Colicins can be classified into two groups, A and B, depending on whether they use the Tol or Ton protein complexes to cross the OM and translocate through the periplasm of their host (for a review, see reference 1).

[1]Laboratoire d'Ingénierie des Systèmes Macromoléculaires, UMR7255, Institut de Microbiologie de la Méditerranée, Aix-Marseille Université—CNRS, 13402 Marseille, France

Protein Secretion in Bacteria
Edited by Maria Sandkvist, Eric Cascales, and Peter J. Christie
© 2019 American Society for Microbiology, Washington, DC
doi:10.1128/ecosalplus.ESP-0030-2018

Filamentous bacteriophages of the *Inoviridae* family are elongated viruses (~7 by ~800 to 2,000 nm) that must pierce the whole envelope to deliver their nucleic acid into the cytoplasm of the target cell to complete their life cycle. The virion particle consists of an assembly of the major coat protein pVIII, capped with the adsorption protein pIII and minor coat proteins (pVI, pVII, and pIX), encapsulating the circular single-stranded viral DNA. Among inoviruses, Ff coliphages have garnered particular scientific attention because of their extensive application in genetic engineering and phage display technology (reviewed in reference 2), along with the vibriophage CTX, which converts *Vibrio cholerae* to pathogenicity (3).

While colicins and phages are very different particles, their modes of uptake across the bacterial envelope share both structural and functional similarities. They both use a multistep process that relies on the parasitism of structures produced by sensitive bacteria. Group A colicins and filamentous phages are first specifically recruited via an interaction with a bacterial surface receptor (reception step), followed by crossing of the OM and transport through the periplasm (translocation step). This step requires interactions with one or several proteins of the Tol system, a conserved macromolecular motor of the cell (4). This review describes the similarities and differences observed between the group A colicin and filamentous phage models of adsorption and translocation into their host and the role of cellular energy in these processes.

COLICINS AND THE FILAMENTOUS PHAGE ADSORPTION PROTEIN pIII SHOW SIMILAR STRUCTURAL ORGANIZATIONS

Like many toxins, colicins are organized into three structural domains that perform specific functions (Fig. 1A); the central domain (receptor [R] domain) binds a specific bacterial OM protein, the N-terminal domain (translocation [T] domain) is required for colicin translocation across the OM and interacts with the Tol system, and the C-terminal domain encodes the toxic activity (5–10). The crystallographic structures of some colicins have highlighted two important features: long antiparallel coiled-coil α-helices within the R domain that separate this domain from the T and C domains and a disordered region within the T domain (11–14). Importantly, nuclease colicins are produced in a complex with an immunity protein, preventing their lethal action in the producing cell (Fig. 1B).

Filamentous phage infection is driven by three to five copies of the minor coat protein pIII (also called G3P), located at the tip of the particle (2, 15). As for colicins, phage pIII protein is organized into three functional domains, separated by two flexible regions (16, 17) (Fig. 1A). The central (N2) domain is responsible for adsorption to a pilus, whereas the N-terminal (N1) domain interacts with the Tol system. The pIII C-terminal domain anchors pIII to the phage particle and is required for DNA injection into the host. The structure of pIII-N1 and pIII-N1-N2 has been solved for the Ff and CTX phages by crystallography and nuclear magnetic resonance studies (Fig. 1B) (16, 18–22).

BINDING TO THE PRIMARY RECEPTOR AND CROSSING OF THE OM

The initial cellular reception of most group A colicins requires a high-affinity OM protein receptor, such as BtuB for colicins A and E1 to E9 (Table 1) (23–25). The structure of the colicin E3 and E2 R domain (135 residues) bound to BtuB has been solved (26, 27). In the current model of colicin uptake, the elongated helical coiled-coil R domain binds BtuB, with the T and C domains extending at a 45° angle out over the OM. Then the intrinsically disordered region of the T domain, adjacent to the OM, searches for a more abundant second receptor (called the translocator), in

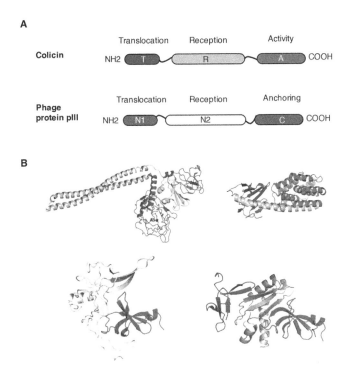

FIGURE 1 Colicin and phage minor coat protein organization and crystal structures. (A) Schematic representation highlighting the similar general organization of colicin and phage pIII proteins for translocation (T or N1 domain), reception (R or N2 domain), and activity or anchoring (A or C domain). (B) Structures of full-length colicin E3 (top left; PDB code 1JCH) bound to its immunity protein (in green), full-length colicin N (top right; PDB code 1A87), and M13 phage protein pIII-N1 and -N2 domains (bottom left; PDB code 1G3P) and superposition (bottom right) of *E. coli* TolAIII domain (gray) interacting with the colicin A T domain on its convex side (cocrystal; PDB code 3QDR) and interacting with G3P-N1 on its concave side (cocrystal PDB code 1TOL). The color code used for each protein domain is the same for panels A and B.

this case OmpF, and crosses the OM through one of the OmpF pores, which are wide enough to accommodate unfolded peptides. This model is based on numerous experiments, such as occlusion of the OmpF channel in planar lipid bilayers by colicin T domains and isolation of an intact complex of colicin with BtuB and OmpF, confirmed by crystallization of colicin T-domain peptides within the OmpF pore (28–32). The resulting colicin-BtuB-OmpF complex is called the colicin OM translocon (Fig. 2A).

Filamentous phages have been reported to use host-specific type IV pili or fimbriae that extend from the cell surface as extracellular receptors (Table 1). Adsorption of the phage to the pilus is dependent on pIII-N2, the central domain of the capsid protein pIII (17, 33). For Ff phages, N2 binding to the *E. coli* F pilus leads to conformational changes in pIII that unmask the TolA binding site on N1, allowing phage infection to proceed (34) (Fig. 2B). How these phages cross the OM to reach the periplasmic space is unknown, but they may follow pilus retraction through the pilus secretin that spans the OM (35–37). Interestingly, the reception step is not strictly required for phage infection, as phages can bypass the pilus to infect cells at low frequencies (17, 38, 39). In the absence of the F pilus receptor, the addition of $CaCl_2$ to the medium has been reported to enhance phage infection, possibly by neutralizing negative charges at the cell surface (38).

TABLE 1 Host proteins required for reception and translocation of filamentous phages and group A colicins[a]

Name	Receptor	Translocator	Uptake proteins	Cytotoxicity/host	Reference
Group A colicins					
ColA	BtuB	OmpF	TolA, -B, -Q, -R	Pore forming	
ColE2, E7, E8, E9	BtuB	OmpF	TolA, -B, -Q, -R	DNase	
ColE3, E4, E6	BtuB	OmpF	TolA, -B, -Q, -R	rRNase	
ColE1	BtuB	TolC	TolA, -Q	Pore forming	
ColN	LPS	OmpF	TolA, -Q, -R	Pore forming	117
ColE5	BtuB	OmpF	TolA, -B, -Q, -R	tRNase	
ColK	Tsx	OmpF/ OmpA	TolA, -B, -Q, -R	Pore forming	118
ColU	OmpA	OmpF	TolA, -B, -Q, -R	Pore forming	119
Filamentous phages					
Ff (Fd, f1, M13)	F pilus		TolA, -Q, -R	*E. coli*	
IKE	N pilus		TolA, -Q, -R	*E. coli*	120
IF1	I pilus		ND	*E. coli*	
CTX	TCP pilus		TolA, -Q, -R	*V. cholerae*	
VGJ	MshA pilus		ND	*V. cholerae*	121
Pf1	Type IV PAK pilus		ND	*Pseudomonas aeruginosa*	122
Pf3	RP4 pilus		n.d.	*P. aeruginosa*	122

[a]Only references that do not appear in the text are cited in this table. ND, not determined.

After the binding step, which allows specific targeting of the bacterial host, both group A colicins and filamentous phages parasitize a *trans*-envelope protein complex, the Tol system, to cross the periplasm. The Tol system components required for this step can vary depending on the group A colicin or phage involved (Table 1).

THE Tol MACROCOMPLEX: A CONSERVED MOLECULAR MOTOR OF THE BACTERIAL ENVELOPE

The Tol system is highly conserved among Gram-negative bacteria (40). It consists of five proteins, TolQ, TolR, TolA, TolB, and Pal, which form a complex in the cell envelope. TolQRA are IM proteins that interact via their transmembrane (TM) segments to form an IM complex (41–44). TolQ is a polytopic protein with three TM segments, whereas TolA and TolR are bitopic proteins with a large periplasmic domain (43, 45–48). TolB is a periplasmic protein composed of two subdomains, called D1 and D2, in which D2

forms a six-bladed β-propeller (49, 50). Finally, Pal is a lipoprotein tethered to the OM via its N-terminal acylated residue, and its C-terminal domain is free to interact with the PG layer and TolB (51–55). Thus, TolB-Pal forms the OM complex of the Tol system and TolB competes with the PG to bind Pal (56). The two Tol subcomplexes are transiently connected via the C-terminal domain of TolA (called TolAIII), which interacts with both TolB-D1 and Pal (57–60). The TolQ-TolR complex uses the proton motive force (PMF) to induce conformational changes in the periplasmic domain of TolA, promoting the TolA-Pal interaction (4, 41, 61). Thus, the Tol system works as a molecular motor, using the PMF to form a link between PG and the IM and OM.

The Tol system has been reported to be essential in many bacterial species, whereas it is dispensable in the *E. coli* K-12 strain (62–66). In *E. coli*, deletion of any of the *tol* genes causes a pleiotropic phenotype. Indeed, *tol* mutants are highly sensitive to detergents and some antibiotics, they release periplasmic proteins into the extracellular medium, and they

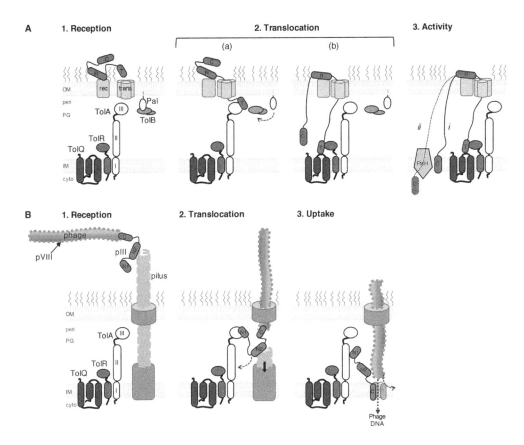

FIGURE 2 Models of group A colicin (A) and Tol-dependent filamentous phage import (B) into sensitive *E. coli* cells. (A) In stage 1, colicin binds to the OM receptor by its central domain (26, 27). In stage 2(a), the disordered N-terminal segment of the T domain translocates through the OM β-barrel and interacts with a free periplasmic TolB or dissociates TolB from Pal (28–31, 77, 79, 81). In stage 2(b), the N-terminal segment interacts with other Tol proteins (82–85, 95). At this stage, the immunity protein of nuclease colicins is released (108, 109). Then the unfolded C-terminal domain is thought to cross the OM through the interface between OmpF and the lipid bilayers (112) or directly through the OmpF porin (28). In stage 3, for pore-forming colicins (*i*) the C-terminal domain inserts spontaneously into the IM and forms voltage-gated channels that depolarize and kill the target bacteria (for a review, see reference 113). For nuclease colicins (*ii*), the C-terminal domain is cleaved by FtsH (114, 115), an essential ATP-dependent IM protease, and spontaneously crosses the IM (116) or uses FtsH for its transfer (115). (B) In stage 1, the phage minor coat protein pIII-N2 domain binds to the tip of an F pilus protruding from the cell surface (33). In stage 2, pilus retraction pulls the phage into the cell periplasm, possibly through the pilus secretin pore. Once there, the phage pIII-N1 domain interacts with the globular domain of TolA (TolAIII) (86, 87). In *E. coli*, a direct interaction between TolAII and phage pIII-N2 has been reported (dashed arrow) (88). The PMF-dependent TolQR motor may trigger conformational changes of TolA that bring the phage particle in close contact with the IM. The phage uncapping process during the uptake stage (stage 3) is speculative. In the model, pIII oligomerizes to form a channel in the IM of the host through its C-ter domain (pIII-C). Then diffusion of the phage pVIII major coat protein in the IM leads to disassembly of the capsid, releasing the internal pressure of the structure. This force is thought to drive phage DNA injection through the IM pIII-C channel (102, 103). The phage is composed of three to five copies of pIII, but only one copy has been represented, and other minor virion coat proteins have been omitted for simplicity. OM, outer membrane; IM, inner membrane; PG, peptidoglycan; peri, periplasm; cyto, cytoplasm; rec, receptor; trans, translocator.

form numerous OM vesicles (67–71). These phenotypes all suggest a potential role for the Tol system in the maintenance of OM integrity. This role can be partially explained by the involvement of the Tol complex in OM lipid homeostasis (72, 73). Finally, the Tol system has been described to be involved in the late stage of the cell division process, more precisely in OM invagination during cell division (62, 74).

THE Tol SYSTEM SERVES AS A VERSATILE IMPORT MACHINERY FOR PARASITES AND TOXINS

Various *tol* mutants have been reported to show resistance to both group A colicins and filamentous phages. Indeed, direct or indirect interactions of these particles with components of the Tol complex have been described. When they reach the periplasmic space, most group A colicins first recruit the β-propeller domain of TolB through their N-terminal T domains (50, 58, 75–77). Conversely, TolB is not required for phage uptake (39, 78). The affinity of ColE9 for TolB has been shown to be strong enough to competitively displace TolB from Pal (77, 79). Structural data have shown that in the presence of Pal, the N-terminal 12 residues of TolB, encompassing the TolA binding domain, are ordered and sequestered to the TolB surface, whereas these residues are disordered and accessible to TolA binding in the presence of the ColE9 T domain (79, 80). Thus, in the presence of ColE9, TolB binds to TolAIII and the toxin can continue its translocation. Although this model can be applied to the other nuclease colicins, some results suggest that the mechanism for the recruitment of TolB by the pore-forming colicin A may be different. As ColA is unable to competitively displace TolB in the TolB-Pal complex, the model suggests that the toxin binds to free periplasmic TolB and then to TolAIII, for which it has a higher affinity (81) (Fig. 2A).

All group A colicins and filamentous bacteriophages studied to date require TolA for their transit into the periplasm (Table 1). Direct interaction between the TolAIII domain and the colicin A, E1, N, and K T domains has been shown by numerous *in vivo* and *in vitro* experiments (82–85). Similarly, the pIII protein at the tip of the phage particle is responsible for TolA binding (19, 22, 86, 87). The phage pIII-N1 and pIII-N2 domains have been reported to bind the *E. coli* TolAIII and TolAII domains, respectively (88). No direct interaction has yet been observed between TolA and enzymatic colicins, even though *tolA* mutants are tolerant to these toxins.

Several studies indicate that TolA has diverse binding sites, which enable multiple interactions with other partners. First, point mutations or deletions in *tolA* differentially affect colicin import; some remain active, whereas others lose their ability to kill mutated strains (59, 89). Second, structural data have shown that the colicin A T domain interacts with the convex site of TolAIII, whereas the phage Ff and IF1 pIII-N1 domain and colicin N T domain bind to the concave side of TolAIII (18, 21, 90–93) (Fig. 1B). In contrast to coliphage Ff, the vibriophage CTX pIII-N1 domain binds to the convex site of *V. cholerae* TolAIII (22).

The role of the TolQ and TolR proteins in both phage and colicin uptake is less clear. Although TolQ and TolR are necessary but not essential for phage infection (39, 94), both are absolutely required for the translocation of group A colicins (excepted ColE1). However, only a few studies have reported a direct interaction between colicins and either TolQ or TolR. The ColA and ColE3 T domain has been shown to interact with TolR by cross-linking experiments (95), and the interaction between the ColK T domain and TolQ or TolR has been revealed by coimmunoprecipitation (85). The final steps concerning the toxic effect of colicins and phage DNA injection are not discussed here but are briefly summarized in Fig. 2.

IMPORTANCE OF PROTEIN STRUCTURAL FLEXIBILITY IN UPTAKE PATHWAYS

An interesting aspect of the current models of colicin uptake is that during the process, colicins T domains are disordered and they fold into an ordered structure upon binding to a Tol protein. This disorder-to-order transition could explain the ability of colicins to bind several target partners and progress through the periplasm.

Another important aspect in both colicin and phage translocation is the large-scale structural modification of the molecules during the process. First, studies have shown that colicins A and E2 remain bound to their receptor and the Tol machinery, whereas their C and T domains are translocated (96, 97). Second, *in vivo* and *in vitro* experiments have shown that colicin A unfolds during its import into the cells (98, 99). This unfolding either is triggered by binding to the first ColE1 receptor, which is then propagated to the distal ends of the T and C domains of the toxin (26, 100), or occurs after recruitment of the second receptor and formation of the colicin OM translocon for ColE9 (101). Similarly, the extensive structural modifications of coliphage pIII protein during initial binding to the F pilus, followed by interaction with TolA in the periplasm, have been proposed to lead to the insertion of pIII-C into the IM of the host and the formation of a multimeric channel that allows phage DNA injection into the bacterial cytoplasm, concomitant with disassembly of the capsid by diffusion of the major coat protein pVIII in the membrane (102, 103).

ENERGETICS OF COLICIN AND FILAMENTOUS PHAGE UPTAKE

Despite the large number of publications on colicin and phage import, only a few studies have focused on the energetic aspect of this process. As discussed above, the Tol system uses the IM PMF, by means of the TolQR motor, to power mechanisms involved in the maintenance of OM integrity and cell constriction. However, initial studies suggested that the Tol-dependent import of two pore-forming colicins, ColA and ColE1, is energy independent (104, 105). These results were later confirmed by point mutations introduced into the TolQR motor that affect the physiological function of the Tol system without preventing the killing action of pore-forming colicins. These mutations were called discriminative mutations (106, 107). Energy requirements for the import of nuclease colicins appear to be different. It is clear that release of the immunity protein from nuclease colicins in the external medium requires functional import machinery (108, 109). Intriguingly, the release of the ColE9 immunity protein was shown to be energy dependent (109), whereas strains harboring discriminative TolQR mutations remain susceptible to ColE2 (107). Based on these divergent results, the question of the energy requirement for the translocation of group A colicins is still open. For coliphages, experiments using arsenate or protonophores have led to the hypothesis that phage uptake may be dependent on both ATP and PMF, possibly for pilus retraction and phage particle transit through the periplasm, respectively (110, 111).

In summary, it is fascinating to observe that small toxins, such as colicins, and complex viral structures, such as filamentous phages, are shaped so as to use similar molecular mechanisms to pierce the envelope of their target cells. These include a sequential docking mechanism, the strict requirement for the hub protein TolA, and large-scale structural modifications of the proteins during their transit. Although some key steps in these processes have been well characterized, numerous questions are still unanswered concerning the dynamics, sequence of events, and role of host cell energy in the uptake of these nanostructures.

ACKNOWLEDGMENTS

We thank Roland Lloubes, Laure Journet, and Jean Peuplu for careful reading of the manuscript.

Work in the laboratory is supported by the Centre National de la Recherche Scientifique (CNRS), the Aix-Marseille Université, and the Agence Nationale de la Recherche (MEMOX, ANR-18-CE11-0027).

CITATION

EcoSal Plus 2019; doi:10.1128/ecosalplus.ESP-0030-2018.

REFERENCES

1. **Cascales E, Buchanan SK, Duché D, Kleanthous C, Lloubès R, Postle K, Riley M, Slatin S, Cavard D.** 2007. Colicin biology. *Microbiol Mol Biol Rev* **71:**158–229.

2. **Rakonjac J, Bennett NJ, Spagnuolo J, Gagic D, Russel M.** 2011. Filamentous bacteriophage: biology, phage display and nanotechnology applications. *Curr Issues Mol Biol* **13:**51–76.

3. **Waldor MK, Mekalanos JJ.** 1996. Lysogenic conversion by a filamentous phage encoding cholera toxin. *Science* **272:**1910–1914.

4. **Cascales E, Lloubès R, Sturgis JN.** 2001. The TolQ-TolR proteins energize TolA and share homologies with the flagellar motor proteins MotA-MotB. *Mol Microbiol* **42:**795–807.

5. **Baty D, Frenette M, Lloubès R, Géli V, Howard SP, Pattus F, Lazdunski C.** 1988. Functional domains of colicin A. *Mol Microbiol* **2:**807–811.

6. **Dankert JR, Uratani Y, Grabau C, Cramer WA, Hermodson M.** 1982. On a domain structure of colicin E1. A COOH-terminal peptide fragment active in membrane depolarization. *J Biol Chem* **257:**3857–3863.

7. **Escuyer V, Mock M.** 1987. DNA sequence analysis of three missense mutations affecting colicin E3 bactericidal activity. *Mol Microbiol* **1:**82–85.

8. **Martinez MC, Lazdunski C, Pattus F.** 1983. Isolation, molecular and functional properties of the C-terminal domain of colicin A. *EMBO J* **2:**1501–1507.

9. **Ohno-Iwashita Y, Imahori K.** 1980. Assignment of the functional loci in colicin E2 and E3 molecules by the characterization of their proteolytic fragments. *Biochemistry* **19:**652–659.

10. **Ohno-Iwashita Y, Imahori K.** 1982. Assignment of the functional loci in the colicin E1 molecule by characterization of its proteolytic fragments. *J Biol Chem* **257:**6446–6451.

11. **Wiener M, Freymann D, Ghosh P, Stroud RM.** 1997. Crystal structure of colicin Ia. *Nature* **385:**461–464.

12. **Vetter IR, Parker MW, Tucker AD, Lakey JH, Pattus F, Tsernoglou D.** 1998. Crystal structure of a colicin N fragment suggests a model for toxicity. *Structure* **6:**863–874.

13. **Soelaiman S, Jakes K, Wu N, Li C, Shoham M.** 2001. Crystal structure of colicin E3: implications for cell entry and ribosome inactivation. *Mol Cell* **8:**1053–1062.

14. **Deprez C, Blanchard L, Guerlesquin F, Gavioli M, Simorre JP, Lazdunski C, Marion D, Lloubès R.** 2002. Macromolecular import into *Escherichia coli*: the TolA C-terminal domain changes conformation when interacting with the colicin A toxin. *Biochemistry* **41:**2589–2598.

15. **Grant RA, Lin TC, Konigsberg W, Webster RE.** 1981. Structure of the filamentous bacteriophage fl. Location of the A, C, and D minor coat proteins. *J Biol Chem* **256:**539–546.

16. **Holliger P, Riechmann L.** 1997. A conserved infection pathway for filamentous bacteriophages is suggested by the structure of the membrane penetration domain of the minor coat protein g3p from phage fd. *Structure* **5:**265–275.

17. **Heilpern AJ, Waldor MK.** 2003. pIIICTX, a predicted CTXphi minor coat protein, can expand the host range of coliphage fd to include *Vibrio cholerae*. *J Bacteriol* **185:**1037–1044.

18. **Lubkowski J, Hennecke F, Plückthun A, Wlodawer A.** 1999. Filamentous phage infection: crystal structure of g3p in complex with its coreceptor, the C-terminal domain of TolA. *Structure* **7:**711–722.

19. **Lubkowski J, Hennecke F, Plückthun A, Wlodawer A.** 1998. The structural basis of phage display elucidated by the crystal structure of the N-terminal domains of g3p. *Nat Struct Biol* **5:**140–147.

20. **Holliger P, Riechmann L, Williams RL.** 1999. Crystal structure of the two N-terminal domains of g3p from filamentous phage fd at 1.9 A: evidence for conformational lability. *J Mol Biol* **288:**649–657.

21. **Lorenz SH, Jakob RP, Weininger U, Balbach J, Dobbek H, Schmid FX.** 2011. The filamentous phages fd and IF1 use different mechanisms to infect *Escherichia coli*. *J Mol Biol* **405:**989–1003.

22. **Ford CG, Kolappan S, Phan HT, Waldor MK, Winther-Larsen HC, Craig L.** 2012. Crystal structures of a CTXphi pIII domain unbound and in complex with a *Vibrio cholerae* TolA domain reveal novel interaction interfaces. *J Biol Chem* **287:**36258–36272.

23. **Di Masi DR, White JC, Schnaitman CA, Bradbeer C.** 1973. Transport of vitamin B12 in *Escherichia coli*: common receptor sites for vitamin B12 and the E colicins on the outer membrane of the cell envelope. *J Bacteriol* **115:**506–513.

24. **Cavard D, Lazdunski C.** 1981. Involvement of BtuB and OmpF proteinsin binding and uptake of colicin A. *FEMS Microbiol Lett* **12:**311–316.

25. **Chai T, Wu V, Foulds J.** 1982. Colicin A receptor: role of two *Escherichia coli* outer membrane proteins (OmpF protein and *btuB* gene product) and lipopolysaccharide. *J Bacteriol* **151:**983–988.

26. **Kurisu G, Zakharov SD, Zhalnina MV, Bano S, Eroukova VY, Rokitskaya TI, Antonenko YN, Wiener MC, Cramer WA.** 2003. The structure of BtuB with bound colicin E3 R-domain implies a translocon. *Nat Struct Biol* **10:**948–954.

27. **Sharma O, Yamashita E, Zhalnina MV, Zakharov SD, Datsenko KA, Wanner BL, Cramer WA.** 2007. Structure of the complex of the colicin E2 R-domain and its BtuB receptor. The outer membrane colicin translocon. *J Biol Chem* **282:**23163–23170.

28. **Zakharov SD, Eroukova VY, Rokitskaya TI, Zhalnina MV, Sharma O, Loll PJ, Zgurskaya HI, Antonenko YN, Cramer WA.** 2004. Colicin occlusion of OmpF and TolC channels: outer membrane translocons for colicin import. *Biophys J* **87:**3901–3911.

29. **Yamashita E, Zhalnina MV, Zakharov SD, Sharma O, Cramer WA.** 2008. Crystal structures of the OmpF porin: function in a colicin translocon. *EMBO J* **27:**2171–2180.

30. **Housden NG, Wojdyla JA, Korczynska J, Grishkovskaya I, Kirkpatrick N, Brzozowski AM, Kleanthous C.** 2010. Directed epitope delivery across the *Escherichia coli* outer membrane through the porin OmpF. *Proc Natl Acad Sci U S A* **107:**21412–21417.

31. **Housden NG, Hopper JT, Lukoyanova N, Rodriguez-Larrea D, Wojdyla JA, Klein A, Kaminska R, Bayley H, Saibil HR, Robinson CV, Kleanthous C.** 2013. Intrinsically disordered protein threads through the bacterial outer-membrane porin OmpF. *Science* **340:**1570–1574.

32. **Housden NG, Rassam P, Lee S, Samsudin F, Kaminska R, Sharp C, Goult JD, Francis ML,**

Khalid S, Bayley H, Kleanthous C.** 2018. Directional porin binding of intrinsically disordered protein sequences promotes colicin epitope display in the bacterial periplasm. *Biochemistry* **57:**4374–4381.

33. **Deng LW, Perham RN.** 2002. Delineating the site of interaction on the pIII protein of filamentous bacteriophage fd with the F-pilus of *Escherichia coli*. *J Mol Biol* **319:**603–614.

34. **Martin A, Schmid FX.** 2003. A proline switch controls folding and domain interactions in the gene-3-protein of the filamentous phage fd. *J Mol Biol* **331:**1131–1140.

35. **Jacobson A.** 1972. Role of F pili in the penetration of bacteriophage fl. *J Virol* **10:** 835–843.

36. **Gao Y, Hauke CA, Marles JM, Taylor RK.** 2016. Effects of tcpB mutations on biogenesis and function of TCP, the type IVb pilus of *Vibrio cholerae*. *J Bacteriol* **198:**2818–2828.

37. **Ng D, Harn T, Altindal T, Kolappan S, Marles JM, Lala R, Spielman I, Gao Y, Hauke CA, Kovacikova G, Verjee Z, Taylor RK, Biais N, Craig L.** 2016. The *Vibrio cholerae* minor pilin TcpB initiates assembly and retraction of the toxin-coregulated pilus. *PLoS Pathog* **12:**e1006109.

38. **Russel M, Whirlow H, Sun TP, Webster RE.** 1988. Low-frequency infection of F- bacteria by transducing particles of filamentous bacteriophages. *J Bacteriol* **170:**5312–5316.

39. **Heilpern AJ, Waldor MK.** 2000. CTXphi infection of *Vibrio cholerae* requires the *tolQRA* gene products. *J Bacteriol* **182:**1739–1747.

40. **Sturgis JN.** 2001. Organisation and evolution of the *tol-pal* gene cluster. *J Mol Microbiol Biotechnol* **3:**113–122.

41. **Derouiche R, Bénédetti H, Lazzaroni JC, Lazdunski C, Lloubès R.** 1995. Protein complex within *Escherichia coli* inner membrane. TolA N-terminal domain interacts with TolQ and TolR proteins. *J Biol Chem* **270:**11078–11084.

42. **Germon P, Clavel T, Vianney A, Portalier R, Lazzaroni JC.** 1998. Mutational analysis of the *Escherichia coli* K-12 TolA N-terminal region and characterization of its TolQ-interacting domain by genetic suppression. *J Bacteriol* **180:**6433–6439.

43. **Journet L, Rigal A, Lazdunski C, Bénédetti H.** 1999. Role of TolR N-terminal, central, and C-terminal domains in dimerization and interaction with TolA and TolQ. *J Bacteriol* **181:**4476–4484.

44. **Lazzaroni JC, Vianney A, Popot JL, Bénédetti H, Samatey F, Lazdunski C, Portalier R, Géli V.**

1995. Transmembrane alpha-helix interactions are required for the functional assembly of the *Escherichia coli* Tol complex. *J Mol Biol* **246**:1–7.

45. **Bourdineaud JP, Howard SP, Lazdunski C.** 1989. Localization and assembly into the *Escherichia coli* envelope of a protein required for entry of colicin A. *J Bacteriol* **171**:2458–2465.

46. **Kampfenkel K, Braun V.** 1993. Membrane topologies of the TolQ and TolR proteins of *Escherichia coli*: inactivation of TolQ by a missense mutation in the proposed first transmembrane segment. *J Bacteriol* **175**:4485–4491.

47. **Vianney A, Lewin TM, Beyer WF Jr, Lazzaroni JC, Portalier R, Webster RE.** 1994. Membrane topology and mutational analysis of the TolQ protein of *Escherichia coli* required for the uptake of macromolecules and cell envelope integrity. *J Bacteriol* **176**:822–829.

48. **Levengood SK, Beyer WF Jr, Webster RE.** 1991. TolA: a membrane protein involved in colicin uptake contains an extended helical region. *Proc Natl Acad Sci U S A* **88**:5939–5943.

49. **Abergel C, Bouveret E, Claverie JM, Brown K, Rigal A, Lazdunski C, Bénédetti H.** 1999. Structure of the *Escherichia coli* TolB protein determined by MAD methods at 1.95 A resolution. *Structure* **7**:1291–1300.

50. **Carr S, Penfold CN, Bamford V, James R, Hemmings AM.** 2000. The structure of TolB, an essential component of the tol-dependent translocation system, and its protein-protein interaction with the translocation domain of colicin E9. *Structure* **8**:57–66.

51. **Lazzaroni JC, Portalier R.** 1992. The *excC* gene of *Escherichia coli* K-12 required for cell envelope integrity encodes the peptidoglycan-associated lipoprotein (PAL). *Mol Microbiol* **6**:735–742.

52. **Bouveret E, Bénédetti H, Rigal A, Loret E, Lazdunski C.** 1999. In vitro characterization of peptidoglycan-associated lipoprotein (PAL)-peptidoglycan and PAL-TolB interactions. *J Bacteriol* **181**:6306–6311.

53. **Ray MC, Germon P, Vianney A, Portalier R, Lazzaroni JC.** 2000. Identification by genetic suppression of *Escherichia coli* TolB residues important for TolB-Pal interaction. *J Bacteriol* **182**:821–824.

54. **Cascales E, Bernadac A, Gavioli M, Lazzaroni JC, Lloubes R.** 2002. Pal lipoprotein of *Escherichia coli* plays a major role in outer membrane integrity. *J Bacteriol* **184**:754–759.

55. **Parsons LM, Lin F, Orban J.** 2006. Peptidoglycan recognition by Pal, an outer membrane lipoprotein. *Biochemistry* **45**:2122–2128.

56. **Bouveret E, Derouiche R, Rigal A, Lloubès R, Lazdunski C, Bénédetti H.** 1995. Peptidoglycan-associated lipoprotein-TolB interaction. A possible key to explaining the formation of contact sites between the inner and outer membranes of *Escherichia coli*. *J Biol Chem* **270**:11071–11077.

57. **Cascales E, Gavioli M, Sturgis JN, Lloubès R.** 2000. Proton motive force drives the interaction of the inner membrane TolA and outer membrane Pal proteins in *Escherichia coli*. *Mol Microbiol* **38**:904–915.

58. **Walburger A, Lazdunski C, Corda Y.** 2002. The Tol/Pal system function requires an interaction between the C-terminal domain of TolA and the N-terminal domain of TolB. *Mol Microbiol* **44**:695–708.

59. **Dubuisson JF, Vianney A, Lazzaroni JC.** 2002. Mutational analysis of the TolA C-terminal domain of *Escherichia coli* and genetic evidence for an interaction between TolA and TolB. *J Bacteriol* **184**:4620–4625.

60. **Cascales E, Lloubès R.** 2004. Deletion analyses of the peptidoglycan-associated lipoprotein Pal reveals three independent binding sequences including a TolA box. *Mol Microbiol* **51**:873–885.

61. **Germon P, Ray MC, Vianney A, Lazzaroni JC.** 2001. Energy-dependent conformational change in the TolA protein of *Escherichia coli* involves its N-terminal domain, TolQ, and TolR. *J Bacteriol* **183**:4110–4114.

62. **Yeh YC, Comolli LR, Downing KH, Shapiro L, McAdams HH.** 2010. The *Caulobacter* Tol-Pal complex is essential for outer membrane integrity and the positioning of a polar localization factor. *J Bacteriol* **192**:4847–4858.

63. **Houot L, Navarro R, Nouailler M, Duché D, Guerlesquin F, Lloubes R.** 2017. Electrostatic interactions between the CTX phage minor coat protein and the bacterial host receptor TolA drive the pathogenic conversion of *Vibrio cholerae*. *J Biol Chem* **292**:13584–13598. (Erratum, *J Biol Chem* **293**:7263, 2018.)

64. **Gaspar JA, Thomas JA, Marolda CL, Valvano MA.** 2000. Surface expression of O-specific lipopolysaccharide in *Escherichia coli* requires the function of the TolA protein. *Mol Microbiol* **38**:262–275.

65. **Dennis JJ, Lafontaine ER, Sokol PA.** 1996. Identification and characterization of the *tolQRA* genes of *Pseudomonas aeruginosa*. *J Bacteriol* **178**:7059–7068.

66. **Meury J, Devilliers G.** 1999. Impairment of cell division in *tolA* mutants of *Escherichia coli* at low and high medium osmolarities. *Biol Cell* **91**:67–75.

67. **Bernstein A, Rolfe B, Onodera K.** 1972. Pleiotropic properties and genetic organiza-

tion of the *tolA,B* locus of *Escherichia coli* K-12. *J Bacteriol* **112:**74–83.

68. **Bernadac A, Gavioli M, Lazzaroni JC, Raina S, Lloubès R.** 1998. *Escherichia coli tol-pal* mutants form outer membrane vesicles. *J Bacteriol* **180:**4872–4878.

69. **Fognini-Lefebvre N, Lazzaroni JC, Portalier R.** 1987. *tolA, tolB* and *excC*, three cistrons involved in the control of pleiotropic release of periplasmic proteins by *Escherichia coli* K12. *Mol Gen Genet* **209:**391–395.

70. **Lazzaroni JC, Portalier RC.** 1981. Genetic and biochemical characterization of periplasmic-leaky mutants of *Escherichia coli* K-12. *J Bacteriol* **145:**1351–1358.

71. **Prouty AM, Van Velkinburgh JC, Gunn JS.** 2002. *Salmonella enterica* serovar Typhimurium resistance to bile: identification and characterization of the *tolQRA* cluster. *J Bacteriol* **184:**1270–1276.

72. **Shrivastava R, Jiang X, Chng SS.** 2017. Outer membrane lipid homeostasis via retrograde phospholipid transport in *Escherichia coli. Mol Microbiol* **106:**395–408.

73. **Masilamani R, Cian MB, Dalebroux ZD.** 2018. Salmonella Tol-Pal reduces outer membrane glycerophospholipid levels for envelope homeostasis and survival during bacteremia. *Infect Immun* **86:**e00173-18.

74. **Gerding MA, Ogata Y, Pecora ND, Niki H, de Boer PA.** 2007. The trans-envelope Tol-Pal complex is part of the cell division machinery and required for proper outer-membrane invagination during cell constriction in *E. coli. Mol Microbiol* **63:**1008–1025.

75. **Bouveret E, Rigal A, Lazdunski C, Bénédetti H.** 1997. The N-terminal domain of colicin E3 interacts with TolB which is involved in the colicin translocation step. *Mol Microbiol* **23:**909–920.

76. **Bouveret E, Rigal A, Lazdunski C, Bénédetti H.** 1998. Distinct regions of the colicin A translocation domain are involved in the interaction with TolA and TolB proteins upon import into *Escherichia coli. Mol Microbiol* **27:**143–157.

77. **Loftus SR, Walker D, Maté MJ, Bonsor DA, James R, Moore GR, Kleanthous C.** 2006. Competitive recruitment of the periplasmic translocation portal TolB by a natively disordered domain of colicin E9. *Proc Natl Acad Sci U S A* **103:**12353–12358.

78. **Sun TP, Webster RE.** 1986. fii, a bacterial locus required for filamentous phage infection and its relation to colicin-tolerant tolA and tolB. *J Bacteriol* **165:**107–115.

79. **Bonsor DA, Grishkovskaya I, Dodson EJ, Kleanthous C.** 2007. Molecular mimicry enables competitive recruitment by a natively disordered protein. *J Am Chem Soc* **129:**4800–4807.

80. **Bonsor DA, Hecht O, Vankemmelbeke M, Sharma A, Krachler AM, Housden NG, Lilly KJ, James R, Moore GR, Kleanthous C.** 2009. Allosteric beta-propeller signalling in TolB and its manipulation by translocating colicins. *EMBO J* **28:**2846–2857.

81. **Zhang Y, Li C, Vankemmelbeke MN, Bardelang P, Paoli M, Penfold CN, James R.** 2010. The crystal structure of the TolB box of colicin A in complex with TolB reveals important differences in the recruitment of the common TolB translocation portal used by group A colicins. *Mol Microbiol* **75:**623–636.

82. **Bénédetti H, Lazdunski C, Lloubès R.** 1991. Protein import into *Escherichia coli*: colicins A and E1 interact with a component of their translocation system. *EMBO J* **10:**1989–1995.

83. **Derouiche R, Zeder-Lutz G, Bénédetti H, Gavioli M, Rigal A, Lazdunski C, Lloubès R.** 1997. Binding of colicins A and E1 to purified TolA domains. *Microbiology* **143:**3185–3192.

84. **Raggett EM, Bainbridge G, Evans LJ, Cooper A, Lakey JH.** 1998. Discovery of critical Tol A-binding residues in the bactericidal toxin colicin N: a biophysical approach. *Mol Microbiol* **28:**1335–1343.

85. **Barnéoud-Arnoulet A, Gavioli M, Lloubès R, Cascales E.** 2010. Interaction of the colicin K bactericidal toxin with components of its import machinery in the periplasm of *Escherichia coli. J Bacteriol* **192:**5934–5942.

86. **Click EM, Webster RE.** 1997. Filamentous phage infection: required interactions with the TolA protein. *J Bacteriol* **179:**6464–6471.

87. **Riechmann L, Holliger P.** 1997. The C-terminal domain of TolA is the coreceptor for filamentous phage infection of *E. coli. Cell* **90:**351–360.

88. **Karlsson F, Borrebaeck CA, Nilsson N, Malmborg-Hager AC.** 2003. The mechanism of bacterial infection by filamentous phages involves molecular interactions between TolA and phage protein 3 domains. *J Bacteriol* **185:**2628–2634.

89. **Schendel SL, Click EM, Webster RE, Cramer WA.** 1997. The TolA protein interacts with colicin E1 differently than with other group A colicins. *J Bacteriol* **179:**3683–3690.

90. **Anderluh G, Gökçe I, Lakey JH.** 2004. A natively unfolded toxin domain uses its receptor as a folding template. *J Biol Chem* **279:**22002–22009.

91. **Li C, Zhang Y, Vankemmelbeke M, Hecht O, Aleanizy FS, Macdonald C, Moore GR, James R, Penfold CN.** 2012. Structural evi-

dence that colicin A protein binds to a novel binding site of TolA protein in *Escherichia coli* periplasm. *J Biol Chem* **287**:19048–19057.

92. **Hecht O, Ridley H, Lakey JH, Moore GR.** 2009. A common interaction for the entry of colicin N and filamentous phage into *Escherichia coli. J Mol Biol* **388**:880–893.

93. **Ridley H, Lakey JH.** 2015. Antibacterial toxin colicin N and phage protein G3p compete with TolB for a binding site on TolA. *Microbiology* **161**:503–515.

94. **Sun TP, Webster RE.** 1987. Nucleotide sequence of a gene cluster involved in entry of E colicins and single-stranded DNA of infecting filamentous bacteriophages into *Escherichia coli. J Bacteriol* **169**:2667–2674.

95. **Journet L, Bouveret E, Rigal A, Lloubes R, Lazdunski C, Bénédetti H.** 2001. Import of colicins across the outer membrane of *Escherichia coli* involves multiple protein interactions in the periplasm. *Mol Microbiol* **42**:331–344.

96. **Duché D, Letellier L, Géli V, Bénédetti H, Baty D.** 1995. Quantification of group A colicin import sites. *J Bacteriol* **177**:4935–4939.

97. **Duché D.** 2007. Colicin E2 is still in contact with its receptor and import machinery when its nuclease domain enters the cytoplasm. *J Bacteriol* **189**:4217–4222.

98. **Bénédetti H, Lloubès R, Lazdunski C, Letellier L.** 1992. Colicin A unfolds during its translocation in *Escherichia coli cells* and spans the whole cell envelope when its pore has formed. *EMBO J* **11**:441–447.

99. **Duché D, Baty D, Chartier M, Letellier L.** 1994. Unfolding of colicin A during its translocation through the *Escherichia coli* envelope as demonstrated by disulfide bond engineering. *J Biol Chem* **269**:24820–24825.

100. **Griko YV, Zakharov SD, Cramer WA.** 2000. Structural stability and domain organization of colicin E1. *J Mol Biol* **302**:941–953.

101. **Housden NG, Loftus SR, Moore GR, James R, Kleanthous C.** 2005. Cell entry mechanism of enzymatic bacterial colicins: porin recruitment and the thermodynamics of receptor binding. *Proc Natl Acad Sci U S A* **102**:13849–13854.

102. **Bennett NJ, Rakonjac J.** 2006. Unlocking of the filamentous bacteriophage virion during infection is mediated by the C domain of pIII. *J Mol Biol* **356**:266–273.

103. **Glaser-Wuttke G, Keppner J, Rasched I.** 1989. Pore-forming properties of the adsorption protein of filamentous phage fd. *Biochim Biophys Acta* **985**:239–247.

104. **Braun V, Frenz J, Hantke K, Schaller K.** 1980. Penetration of colicin M into cells of *Escherichia coli. J Bacteriol* **142**:162–168.

105. **Bourdineaud JP, Boulanger P, Lazdunski C, Letellier L.** 1990. In vivo properties of colicin A: channel activity is voltage dependent but translocation may be voltage independent. *Proc Natl Acad Sci U S A* **87**:1037–1041.

106. **Goemaere EL, Cascales E, Lloubès R.** 2007. Mutational analyses define helix organization and key residues of a bacterial membrane energy-transducing complex. *J Mol Biol* **366**:1424–1436.

107. **Lloubès R, Goemaere E, Zhang X, Cascales E, Duché D.** 2012. Energetics of colicin import revealed by genetic cross-complementation between the Tol and Ton systems. *Biochem Soc Trans* **40**:1480–1485.

108. **Duché D, Frenkian A, Prima V, Lloubès R.** 2006. Release of immunity protein requires functional endonuclease colicin import machinery. *J Bacteriol* **188**:8593–8600.

109. **Vankemmelbeke M, Zhang Y, Moore GR, Kleanthous C, Penfold CN, James R.** 2009. Energy-dependent immunity protein release during tol-dependent nuclease colicin translocation. *J Biol Chem* **284**:18932–18941.

110. **Yamamoto M, Kanegasaki S, Yoshikawa M.** 1981. Role of membrane potential and ATP in complex formation between *Escherichia coli* male cells and filamentous phage fd. *J Gen Microbiol* **123**:343–349.

111. **Häse CC.** 2003. Ion motive force dependence of protease secretion and phage transduction in *Vibrio cholerae* and *Pseudomonas aeruginosa. FEMS Microbiol Lett* **227**:65–71.

112. **Baboolal TG, Conroy MJ, Gill K, Ridley H, Visudtiphole V, Bullough PA, Lakey JH.** 2008. Colicin N binds to the periphery of its receptor and translocator, outer membrane protein F. *Structure* **16**:371–379.

113. **Lakey JH, Slatin SL.** 2001. Pore-forming colicins and their relatives. *Curr Top Microbiol Immunol* **257**:131–161.

114. **Walker D, Mosbahi K, Vankemmelbeke M, James R, Kleanthous C.** 2007. The role of electrostatics in colicin nuclease domain translocation into bacterial cells. *J Biol Chem* **282**:31389–31397.

115. **Chauleau M, Mora L, Serba J, de Zamaroczy M.** 2011. FtsH-dependent processing of RNase colicins D and E3 means that only the cytotoxic domains are imported into the cytoplasm. *J Biol Chem* **286**:29397–29407.

116. **Mosbahi K, Lemaître C, Keeble AH, Mobasheri H, Morel B, James R, Moore GR, Lea EJ, Kleanthous C.** 2002. The cytotoxic domain of colicin E9 is a channel-forming endonuclease. *Nat Struct Biol* **9**:476–484.

117. **Johnson CL, Ridley H, Marchetti R, Silipo A, Griffin DC, Crawford L, Bonev B, Molinaro**

A, Lakey JH. 2014. The antibacterial toxin colicin N binds to the inner core of lipopolysaccharide and close to its translocator protein. *Mol Microbiol* **92**:440–452.

118. Bradley DE, Howard SP. 1992. A new colicin that adsorbs to outer-membrane protein Tsx but is dependent on the tonB instead of the tolQ membrane transport system. *J Gen Microbiol* **138**:2721–2724.

119. Smajs D, Pilsl H, Braun V. 1997. Colicin U, a novel colicin produced by *Shigella boydii*. *J Bacteriol* **179**:4919–4928.

120. Jakob RP, Geitner AJ, Weininger U, Balbach J, Dobbek H, Schmid FX. 2012. Structural and energetic basis of infection by the filamentous bacteriophage IKe. *Mol Microbiol* **84**:1124–1138.

121. Campos J, Martínez E, Suzarte E, Rodríguez BL, Marrero K, Silva Y, Ledón T, del Sol R, Fando R. 2003. VGJ phi, a novel filamentous phage of *Vibrio cholerae*, integrates into the same chromosomal site as CTX phi. *J Bacteriol* **185**:5685–5696.

122. Holland SJ, Sanz C, Perham RN. 2006. Identification and specificity of pilus adsorption proteins of filamentous bacteriophages infecting *Pseudomonas aeruginosa*. *Virology* **345**:540–548.

A Hybrid Secretion System Facilitates Bacterial Sporulation: A Structural Perspective

31

NATALIE ZEYTUNI[1] and NATALIE C.J. STRYNADKA[1]

SECRETION AND SPORULATION

Bacteria utilize sophisticated nanomachines to transport proteins, small molecules, and DNA across membranes to the extracellular environment. These transport machineries, also known as secretion systems, are involved in various cellular functions, such as adhesion to surfaces or host cells, cell-cell communication, motility (flagella), virulence effector protein secretion, and, notably, bacterial pathogenesis (1–5). Several of the identified protein secretion systems comprise large complexes that localize and assemble in and around the bacterial membrane(s), forming specialized channels through which the selected substrate(s) is actively delivered (6–9). Although exhibiting significant diversity in structure, substrate, and function, the dedicated type II, III, IV, and IV-pilus secretion systems (T2SS, T3SS, T4SS, and T4PS, respectively) in didermic Gram-negative bacteria each transport a specific subset of proteins to the extracellular milieu via passage through large stacked ring-shaped channels that span the inner membrane (IM) and outer membrane (OM).

Recently, a novel variant of these secretion systems has been proposed to play a central role during bacterial sporulation (10). Sporulation is an

[1]Department of Biochemistry and Molecular Biology and the Center for Blood Research, University of British Columbia, Vancouver, British Columbia, Canada V6T 1Z4

Protein Secretion in Bacteria
Edited by Maria Sandkvist, Eric Cascales, and Peter J. Christie
© 2019 American Society for Microbiology, Washington, DC
doi:10.1128/microbiolspec.PSIB-0013-2018

ancient developmental process, observed most typically in Gram-positive bacteria from the *Firmicutes* phylum but also observed in some Gram-negative species (*Myxococcus xanthus* [11], for example), that allows starving cells to differentiate into metabolically dormant spores that can survive extreme environmental conditions. The sporulation process involves multiple steps, each with corresponding morphological changes that are governed by intercellular signaling pathways through the activation of cell-specific sigma factors that control gene expression (Fig. 1). Early in sporulation, an asymmetric septum divides

the rod-shaped bacterium into two cells: a smaller "forespore," which becomes the spore, and a larger "mother cell" that contributes to the development of the forespore but ultimately dies. After asymmetric division is complete, σ^F is activated in the forespore and then signals for σ^E activation in the mother cell. σ^E expression triggers the start of the engulfment step, in which the mother cell membrane migrates around the forespore in a phagocytic-like process, resulting in the forespore being engulfed as a double-membraned protoplast within the mother cell, ending with σ^G activation in the forespore. Later on, σ^K is

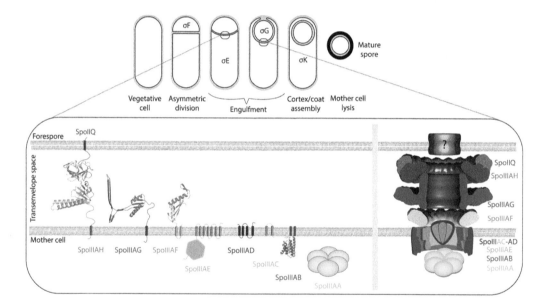

FIGURE 1 Schematic representation of the sporulation process and the active sporulation channel architecture model. (Top) Morphological changes mediated by cell-specific sigma factors that regulate gene expression in *Bacillus subtilis*. (Bottom) Sporulation channel assembly and function during the engulfment stage involve the expression of nine core component proteins forming a channel that crosses the mother cell membrane, the transenvelope space, and the forespore membrane. (Left) Monomeric topology and known structures of the essential core proteins. (Right) Schematic illustration of the suggested model of the assembled SpoIIIA-IIQ channel. Based on the similarities of the individual components to proteins from other bacterial secretion systems, it is predicted that the core components oligomerize into ring-like structures that are stacked to form this sporulation-specialized secretion system. In this model, the stacked rings of SpoIIIAF, SpoIIIAG, and SpoIIIAH-SpoIIQ form the main conduit in the transenvelope space connecting the mother cell and the forespore. SpoIIIAC and SpoIIIAD form a simplified version of an export apparatus through the mother cell membrane. SpoIIIAE utilizes the proton motif force for substrate transportation and also to mediate the interaction with the SpoIIIAA ATPase and its possible docking platform formed by oligmerized (?) SpoIIIAB. Any additional pore-forming protein(s) required at the forespore membrane has yet to be identified.

activated in the mother cell and supports spore maturation, mother cell lysis, and mature spore release to the environment.

During engulfment, at least nine proteins assemble into a channel apparatus that spans the two opposing membranes separating the mother cell and forespore (10). Eight of these proteins (SpoIIIAA to -AH) are encoded in a single operon (*spoIIIA*) expressed in the mother cell under the control of σ^E (12, 13) and a ninth protein, SpoIIQ, is produced in the forespore under the control of σ^F (14). Mutants lacking any of these channel genes display collapsed forespores that are unable to carry out gene expression either dependent or independent of the late-acting σ^G (14–18). Collectively, these observations suggest that the channel transports one or more substrates, yet to be identified, that support forespore physiology and the capacity to carry out macromolecular synthesis.

THE INDIVIDUAL CHANNEL COMPONENTS

Although many questions remain regarding the SpoIIIA-IIQ sporulation channel, recent studies have helped to bring into focus its evolutionary origins and some of its structural features (19). Intriguingly, the emerging picture is of a novel hybrid secretion system with proteins that share common elements with those of diverse bacterial secretion systems.

SpoIIIAA

Homology searches reveal that SpoIIIAA resembles AAA+ superfamily ATPases of the T2SS, the related T4PS, and T4SS (18 to 20% sequence similarity) (17, 20). In these systems the secretion ATPase is found closely associated with the IM complex and likely adopts a distinct hexameric structure (21, 22) (Fig. 1). The predicted SpoIIIAA hexamer is proposed to utilize energy from ATP hydrolysis to drive substrate export (17).

SpoIIIAB

SpoIIIAB is predicted to be a bitopic membrane protein with anchoring transmembrane helices (TMHs) at the N and C termini and an intervening soluble domain (Fig. 1) (23, 24). The X-ray crystallographic structure of the SpoIIIAB soluble domain (residues 27 to 153) adopts a six-helix bundle similar to that of soluble regions of the polytopic GspF/PilC membrane proteins of the T2SS and evolutionarily related T4PS (22 to 28% sequence similarity) (25–28) (Fig. 2A). The latter variants are localized to the IM platform of the T2SS/T4PS, although their specific function is unknown. Additional structural similarity was found between SpoIIIAB and the C-subunit protein from the bacterial V-ATPase complex (3.4-Å root mean square deviation [RMSD] over 108 Cα atoms) that serves as a "socket" attaching the cytosolic V_1 central stalk subunits to the membrane-bound V_O domain (29). It was hypothesized that by analogy, SpoIIIAB could oligomerize and be positioned to serve as a structural link between the membrane-bound protein components and other soluble components of the SpoIIIA-IIQ channel, such as the SpoIIIAA ATPase (28).

SpoIIIAC and SpoIIIAD

SpoIIIAC and SpoIIIAD are predicted to be small, polytopic membrane proteins with two and four TMHs, respectively (Fig. 1). Although they share no detectable similarity with any protein of known function, their size, number, and orientation of transmembrane segments do resemble components of the flagellar and T3SS proteins, namely, FliQ/SpaQ (SpoIIIAC) and FliP/SpaP (SpoIIIAD), of the IM export apparatus proposed to play a central role in substrate selection and chronological secretion. Recently the near-atomic-resolution, cryo-electron microscopic (cryo-EM) structure of the *Salmonella enterica* serovar Typhimurium flagellar IM export apparatus revealed that none of the three integral membrane

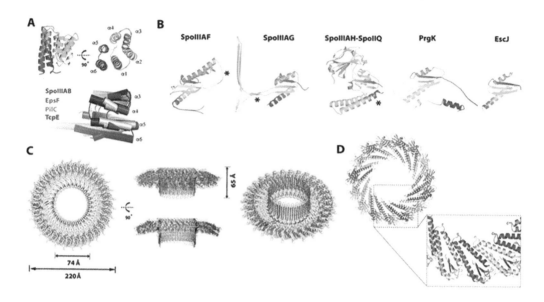

FIGURE 2 **Structures of the core components of the sporulation channel share similar structural motifs with homologs from other bacterial secretion systems. (A) (Top) The SpoIIIAB soluble domain adopts a six-helix bundle fold with both N and C termini in close proximity and facing the mother cell membrane. The molecule is shown in two views, related by a 90° rotation. (Bottom) SpoIIIAB shares a fold similar to that of homologous proteins from the T2SS and T4PS. Shown is a structural overlay of SpoIIIAB with EpsF, TcpE (both from *Vibrio cholerae*), and PilC (*Thermus thermophilus*) proteins in blue, wheat, green, and pink, respectively (PDB codes 6BS9, 3C1Q, 2WHN, and 4HHX, respectively). Two regions of structural variation are seen in the helix 6 angle and the increasing dimensions of helices 4 and 5 and the loop connecting them. (B) SpoIIIAF, SpoIIIAG, and the SpoIIIAH-SpoIIQ heterodimer contain an RBM fold similar to that of the T3SS basal body proteins, EscJ (*Escherichia coli*) and PrgK (*Salmonella Tryphimurium*) (PDB codes 6DCS, 5WC3, 3UZ0, 1YJ7, and 3J6D, respectively). All five structures are displayed in cartoon representation and rainbow color scheme and for clarity are individually shown in identical orientations originating from structural superposition. SpoIIIAF is presented as an overlay of the two monomers seen in the crystal structure, with the region of alternate conformation associated with regulation marked with an asterisk. SpoIIIAG adopts the canonical RBM fold, with a large insertion of the β-triangle motif marked with an asterisk. An SpoIIIAH additional N-terminal helix is marked with an asterisk. (C) Cryo-EM structure of the SpoIIIAG soluble domain 30-meric ring. A three-dimensional reconstruction and atomic model are shown in top side, cropped, and tilted views. The SpoIIIAG ring structure is colored according to distinctive ring elements: RBM in cyan, planar β-ring in green, and vertical β-ring in pink, with the single protomer in red. (D) SpoIIIAH-SpoIIQ representative computational modeled ring, here in C15 symmetry with zoomed-in view of the predicted interaction region between the RBMs of SpoIIIAH. Ring model coordinates were obtained from Meisner et al. (46).**

proteins FliP, FliQ, and FliR (5:4:1 stoichiometry, respectively) adopts the canonical integral membrane protein topologies previously predicted by primary sequence analysis. Instead, the cryo-EM structure showed that the intimately associating helix-turn-helix structural elements common to all 3 flagellar proteins mediate a soluble, coiled-coil-like export apparatus complex formation that sits atop

rather than directly embedded in the IM (30). Further, in the same study, the FliR monomer was shown to be a structural fusion of FliP and FliQ partners in the export apparatus assembly, suggesting that in primitive secretion systems two proteins might be functionally sufficient for export apparatus assembly. Based on the overall secondary-structure similarities, one can therefore speculate that

SpoIIIAC and SpoIIIAD may also form a homologous helix-turn-helix, coiled-coil export apparatus in the sporulation channel and play a central role in substrate recognition and secretion (Fig. 1).

SpoIIIAE

Primary sequence analysis suggests that SpoIIIAE is a multipass polytopic membrane protein with 7 TMHs and an N-terminal Sec-type signal peptide that is followed by a small (~75-residue) mother cell cytoplasmic domain (Fig. 1). Mutations of the SpoIIIAE signal peptide arrest sporulation following the engulfment stage and thereby prevent activation of σ^G (31). SpoIIIAE was shown to interact with SpoIIIJ, a membrane protein translocase (YidC homolog), promoting correct localization and topology in the membrane by the Sec system (15, 31). Bioinformatic analyses suggest that SpoIIIAE is similar to the permease domain of the ATP-binding cassette transporters of the type I secretion system and to electrochemical-potential-driven transporters involved in the shuttle of various drugs and other proteins across membranes (18 to 24% sequence similarity) (32). Furthermore, the SpoIIIAE N-terminal domain shares remote similarity to SD3 (see below) of InvA/FlhA proteins from the T3SS/flagella (~8% sequence similarity). These are highly conserved, polytopic IM proteins which form an additional major cytosol-facing component of the export apparatus (described in part above—SpaPQRS and InvA or FliPQR and FlhAB in T3SS and flagellar nomenclature, respectively) and are associated with secretion regulation. InvA/FlhA have been shown to oligomerize into a non-americ ring with each protomer containing an N-terminal integral membrane domain of 7 or 8 TMHs that has been hypothesized to employ proton motive force to promote protein export, as well as a ring-forming C-terminal cytoplasmic domain (33). The latter domain is thought to comprise a central component of the cytoplasmic, substrate "docking platform" of the apparatus and was shown to interact with members of the FlhB superfamily, secretion substrates in complex with their chaperones, and the conserved ATPase and its regulators (33–36). The monomeric structure of the InvA/FlhA cytosolic domain contains four subdomains, SD1 to SD4, with SD3 shaping the inner pore surface and participating in ring-stabilizing interactions (35, 37). It is possible that the SpoIIIAE N-terminal domain and following transmembrane domain, although swapped in position along the polypeptide chain relative to InvA/FlhA, may share function and oligomerization propensity similar to that of InvA/FlhA in the assembled SpoIIIA-IIQ channel.

SpoIIIAF

SpoIIIAF is a bitopic membrane protein anchored to the mother cell membrane by two N-terminal TMHs followed by a larger soluble domain (residues 60 to 206) (Fig. 1). The latter was shown to oligomerize into homomeric rings through a canonical ring-building motif (RBM) fold (residues 85 to 206) as determined by X-ray crystallography (Fig. 2B) (38). The RBM fold is defined by a growing group of characterized small mixed α/β modular domain structures that pack into oligomeric rings of large assemblies. RBMs characterized thus far share a superficial common architecture divided into two groups based on secondary-structure topology: an αββαβ fold distributed in the IM basal body proteins of the T3SS (PrgK/EscJ and PrgH from the T3SS and FliF from the flagella) (39) and a βαββα fold predominantly found in the OM secretins common to the T2SS, T3SS, and T4PS (40). While the SpoIIIAF core RBM fold shares significant similarity to the T3SS PrgK/EscJ variants (Cα RMSD < 2 Å), it also contains two unique features: an extended N-terminal helix, associated with multimerization and possible interaction with the mother cell membrane, and an 11-residue insertion within a loop region observed to adopt two distinct conformations. The ability of the same primary sequence to adopt different secondary-structure conformations is associated with protein regulation, suggesting dual structural and regulatory roles for the SpoIIIAF RBM (38).

SpoIIIAG

SpoIIIAG contains a short N-terminal mother cell-cytosolic region followed by a single TMH and a large soluble domain, shown to face the intermembrane region (Fig. 1) (41). A near-atomic-resolution single-particle cryo-EM structure of the SpoIIIAG soluble domain revealed that the monomeric form contains a long disordered region (33 residues) followed by two structural motifs: an RBM and a novel β-triangle motif insertion (Fig. 2B and C) (42). Interestingly, SpoIIIAG was found to self-assemble into a large and stable 30-mer complex comprised of three distinctive circular structures: an inner 60-strand vertical β-barrel and a 60-strand planar β-ring, both formed by the β-triangle motif repeats, and an external ring formed by the RBMs (42). Overall, the SpoIIIAG complex shares striking similarity with two major components of the T3SS/flagellar basal body: the IM rings and the secretin. The IM components of these three systems all display a high oligomerization number (30/24/26-mers of SpoIIIAG/PrgK/FliF from SpoIIIA-IIQ/T3SS/flagella, respectively), common repetitive RBM interaction interfaces (Cα RMSD < 2 Å), and large outer ring diameters (~22/19/24 nm for SpoIIIAG/PrgK/FliF, respectively) (6, 42, 43). The SpoIIIAG complex and secretins found in the T2SS and T3SS also share distinct architectures of giant β-barrels comprised of 60 β-strands that are vertical to the membrane axis. These secretin megastructures are formed by a completely unique 15-mer repeat of a four-stranded β-sandwich that assembles the central domain of the massive double-layered β-barrel (6, 44). The outer repeating four-stranded β-sheet from each monomeric sandwich forms an overall 60-stranded anti-parallel β-barrel that constitutes the outer wall, while the inner sheet forms a vertical anti-parallel barrel that serves as the inner "periplasmic gate." Notably, a low-resolution cryo-EM structure of the R-domain for FliF revealed a complex with a similar overall shape and with an almost identical vertical dimension of ~6.5 to 7 nm and similar inner channel diameters of ~7.5 nm in SpoIIIAG (and PrgK) and ~9 nm in FliF (6, 42, 43). Although the high-resolution FliF complex structure is currently unavailable, prior model-building bioinformatic predictions have proposed that FliF is comprised of an RBM with a β-strand insertion similar to that of SpoIIIAG and therefore potentially a similar β-barrel structure (42).

SpoIIIAH and SpoIIQ

Individual monomeric structures and dimerization

SpoIIIAH is anchored to the mother cell membrane by a single N-terminal TMH followed by a large soluble domain that faces the intermembrane space (45, 46) (Fig. 1). The SpoIIIAH soluble domain structure (residues 32 to 218) was determined to a 2.26-Å resolution using X-ray crystallography in complex with SpoIIQ (residues 43 to 283). The SpoIIIAH structure was shown to contain a long disordered region (~65 residues) and a globular domain that adopts the canonical RBM fold with an additional helix packed against the first helix and β-sheet of the motif (45, 46) (Fig. 2B). The binding partner, SpoIIQ, is anchored to the forespore membrane by a single N-terminal TMH followed by a large soluble domain that also faces the intermembrane space (Fig. 1). The SpoIIQ soluble domain contains a long disordered region (~30 residues) and a globular domain that adopts a LytM-metalloendopeptidase fold (~20% sequence similarity). Despite its fold, SpoIIQ is not an active enzyme, as the typical catalytic and zinc-coordinating histidine residues at the active site have been evolutionarily lost. This stable heterodimer is formed through tight pairing of the third β-strands of both proteins, resulting in the formation of a continuous intermolecular five-stranded β-sheet accompanied by other auxiliary stabilizing interactions (45, 46) (Fig. 2B).

Ring formation modeling of the heterodimer

Based on the structural similarity of SpoIIIAH to the IM ring-forming proteins of the T3SS and the flagella (PrgK/EscJ and FliF, respectively), it has been speculated that SpoIIIAH in complex with SpoIIQ can also oligomerize into a circular complex, and several planar ring-shaped models have been proposed with high-order symmetries imposed (42, 45, 46) (Fig. 1 and 2D). These SpoIIIAH-SpoIIQ ring models present outer surfaces that are largely electronegative and therefore unlikely to be directly abutting the inherently anionic bacterial membranes due to charge repulsion; this suggests that additional proteins are likely required to form a continuous conduit between the mother cell and the forespore membranes.

SpoIIIAH-SpoIIQ heterodimeric complex as a protein interaction hub

In addition to their direct role as key members of the SpoIIIA-IIQ apparatus (47, 48), both proteins are at the center of a protein interaction network that may have implications for both mother cell and forespore (49). Knockouts of either SpoIIIAH or SpoIIQ result in severe cellular defects that include mislocalization of SpoIID and SpoIIP involved in engulfment (50), inactivation of σ^K and mislocalization of the SpoIVFA complex (48, 50), altered spore coat formation by CotE (51), altered regulation of σ^F and σ^G by SpoIIE (52), accumulation of dipicolinic acid in the forespore by SpoVV (53), and spore germination by GerM (49). It is yet to be determined if these defects are the result of direct interaction of these proteins with the core SpoIIIAH-SpoIIQ complex or rather indirect effects on downstream developmental processes.

OVERALL ARCHITECTURE AND HIERARCHY OF THE SpoIIIA-IIQ CHANNEL MODEL

To date, a full biochemical and structural elucidation of the sporulation secretion system has yet to be deduced, but significant efforts have been made to progressively assemble this puzzle piece by piece. Coimmunoprecipitation studies have demonstrated that SpoIIIAB, SpoIIIAD, SpoIIIAE, SpoIIIAF, and SpoIIIAG reside in a single complex in vivo (17). Microscopy experiments have indicated that SpoIIQ localization to the forespore membrane depends on the interaction with SpoIIIAH and GerM proteins through the thinning septal cell wall, an interaction that facilitates SpoIIIAG localization to the mother cell septal membrane (49). Taking these facts into consideration and by analogy to the architecture of other secretion systems that rely on hierarchical stacking of oligomerized rings for assembly, a composite sketch of the SpoIIIA-IIQ channel can be proposed (Fig. 1), as follows.

1. At the mother cell cytosol, there is an interaction between the hexameric SpoIIIAA ATPase and the possible docking platform of SpoIIIAB (oligomerization number unknown) (28).
2. A connecting link between the ATPase docking platform and the mother cell membrane export apparatus can be mediated by SpoIIIAE soluble and transmembrane domains.
3. A simplified export apparatus of only two components, SpoIIIAC and SpoIIIAD, is embedded at the mother cell membrane.
4. At the intermembrane space, a model of three stacked RBM-containing protein rings (SpoIIIAF, SpoIIIAG, and SpoIIIAH in complex with SpoIIQ) that are closely associated, with the length of the flexible polypeptide linkers connecting the TMH anchors to large C-terminal domains, is suggested. In this stacked ring model, SpoIIIAF is proposed to have a regulatory role via its

observed alternate structural conformations, SpoIIIAG acts as the primary stable scaffold for the complex, and the SpoIIIAH-SpoIIQ ring acts as the mediator between the mother cell and forespore membranes (38, 42, 45, 46).

5. Additional accessory proteins could provide mechanical support and/or be involved in signal transduction, such as the soluble GerM proteins that have been shown to interact with SpoIIQ and SpoIIIAG *in vivo* (49).

6. Based on the highly electronegative charge of the modeled SpoIIIAH-SpoIIQ rings and the length of the disordered linker connecting the soluble domain to the forespore membrane, it is postulated that SpoIIQ is unlikely to serve as the forespore membrane translocon but rather an additional, yet to be identified, protein(s) is required to allow cargo transport.

7. For SpoIIIA-IIQ complex disassembly, cleavage of SpoIIQ by the SpoIVB forespore-produced protease serves as a signal for the end of the engulfment stage (54, 55).

HINTS FOR SpoIIIA-IIQ CHANNEL SUBSTRATE(S)

The striking similarity between the SpoIIIAG assembled 30-mer to the IM and OM components of the T2SS, T3SS, and T4PS (6, 40, 44) suggests that the SpoIIIA-IIQ complex may also support the transport of large macromolecules rather than simply small-molecule nutrients as previously supposed (18). One possibility is that the SpoIIIAG pore, like the T3SS basal body equivalent, provides a hyperstable shell or "locknut" for an additional inner channel that, in turn, modulates the final pore size for smaller substrates (in the case of the T3SS, the IM export apparatus channel described above and an extended hollow polymerized needle that serves as a direct conduit for partially unfolded virulence

effector protein substrates). Alternatively, as in the case of the T4PS and T4SS, the substrates themselves may be large polymers (pilus or DNA). Given the likelihood of large-macromolecule passage suggested by the dimensions of the SpoIIIAG conduit, additional components would be required to maintain and regulate unidirectional transport of specific substrates from the mother cell to the forespore while at the same time preventing potentially deleterious uptake as seen in other well-characterized translocation machineries.

CONCLUDING REMARKS

The SpoIIIA-IIQ sporulation channel represents a hybrid, secretion-like machinery, involving homologous proteins from different bacterial secretion systems that support forespore physiology, including the capacity to carry out macromolecular synthesis. The remarkable homology to other secretion systems accompanied by the available component structures represents major pieces of the forespore development puzzle and sets a solid foundation for better understanding of the sporulation channel machinery. Notably, RBM-containing components of the channel have broadened our understanding the motif plasticity, which supports diverse functions in large assemblies with only subtle variations in sequence, extensions, or insertions. Still, many questions remain unanswered and can fuel future research, including the following questions. What is the transported substrate(s)? What are the additional components required, especially on the forespore side? Does the sporulation channel represent an ancient ancestor to all secretion systems, or has it evolved from multiple parts of different secretion systems?

ACKNOWLEDGMENTS

We thank Liam Worrall, Bronwyn Lyons, and Dorothy Majewski for fruitful discussions and technical support.

This work was supported by a Banting fellowship to N.Z. from the Canadian Institutes of Health Research (CIHR), operating grants from CIHR and the Howard Hughes International Senior Scholar program to N.C.J.S. N.C.J.S. is a Tier I Canada Research Chair in Antibiotic Discovery.

CITATION

Zeytuni N, Strynadka NCJ. 2019. A hybrid secretion system facilitates bacterial sporulation: a structural perspective. Microbiol Spectrum 7(1):PSIB-0013-2018.

REFERENCES

1. **Kostakioti M, Newman CL, Thanassi DG, Stathopoulos C.** 2005. Mechanisms of protein export across the bacterial outer membrane. *J Bacteriol* **187:**4306–4314.

2. **Costa TRD, Felisberto-Rodrigues C, Meir A, Prevost MS, Redzej A, Trokter M, Waksman G.** 2015. Secretion systems in Gram-negative bacteria: structural and mechanistic insights. *Nat Rev Microbiol* **13:**343–359.

3. **Korotkov KV, Sandkvist M, Hol WGJ.** 2012. The type II secretion system: biogenesis, molecular architecture and mechanism. *Nat Rev Microbiol* **10:**336–351.

4. **Christie PJ, Whitaker N, González-Rivera C.** 2014. Mechanism and structure of the bacterial type IV secretion systems. *Biochim Biophys Acta* **1843:**1578–1591.

5. **Burkinshaw BJ, Strynadka NCJ.** 2014. Assembly and structure of the T3SS. *Biochim Biophys Acta* **1843:**1648–1693.

6. **Worrall LJ, Hong C, Vuckovic M, Deng W, Bergeron JR, Majewski DD, Huang RK, Spreter T, Finlay BB, Yu Z, Strynadka NC.** 2016. Near-atomic-resolution cryo-EM analysis of the Salmonella T3S injectisome basal body. *Nature* **540:**597–601.

7. **Durand E, Nguyen VS, Zoued A, Logger L, Péhau-Arnaudet G, Aschtgen MS, Spinelli S, Desmyter A, Bardiaux B, Dujeancourt A, Roussel A, Cambillau C, Cascales E, Fronzes R.** 2015. Biogenesis and structure of a type VI secretion membrane core complex. *Nature* **523:**555–560.

8. **Reichow SL, Korotkov KV, Hol WGJ, Gonen T.** 2010. Structure of the cholera toxin secretion channel in its closed state. *Nat Struct Mol Biol* **17:**1226–1232.

9. **Low HH, Gubellini F, Rivera-Calzada A, Braun N, Connery S, Dujeancourt A, Lu F, Redzej A, Fronzes R, Orlova EV, Waksman G.** 2014. Structure of a type IV secretion system. *Nature* **508:**550–553.

10. **Crawshaw AD, Serrano M, Stanley WA, Henriques AO, Salgado PS.** 2014. A mother cell-to-forespore channel: current understanding and future challenges. *FEMS Microbiol Lett* **358:**129–136.

11. **Rosenbluh A, Rosenberg E.** 1989. Sporulation of *Myxococcus xanthus* in liquid shake flask cultures. *J Bacteriol* **171:**4521–4524.

12. **Guillot C, Moran CP Jr.** 2007. Essential internal promoter in the *spoIIIA* locus of *Bacillus subtilis. J Bacteriol* **189:**7181–7189.

13. **Illing N, Errington J.** 1991. The spoIIIA operon of *Bacillus subtilis* defines a new temporal class of mother-cell-specific sporulation genes under the control of the sigma E form of RNA polymerase. *Mol Microbiol* **5:**1927–1940.

14. **Londoño-Vallejo J-A, Fréhel C, Stragier P.** 1997. SpoIIQ, a forespore-expressed gene required for engulfment in *Bacillus subtilis. Mol Microbiol* **24:**29–39.

15. **Camp AH, Losick R.** 2008. A novel pathway of intercellular signalling in *Bacillus subtilis* involves a protein with similarity to a component of type III secretion channels. *Mol Microbiol* **69:**402–417.

16. **Kellner EM, Decatur A, Moran CP Jr.** 1996. Two-stage regulation of an anti-sigma factor determines developmental fate during bacterial endospore formation. *Mol Microbiol* **21:**913–924.

17. **Doan T, Morlot C, Meisner J, Serrano M, Henriques AO, Moran CP Jr, Rudner DZ.** 2009. Novel secretion apparatus maintains spore integrity and developmental gene expression in *Bacillus subtilis. PLoS Genet* **5:**e1000566.

18. **Camp AH, Losick R.** 2009. A feeding tube model for activation of a cell-specific transcription factor during sporulation in *Bacillus subtilis. Genes Dev* **23:**1014–1024.

19. **Morlot C, Rodrigues CDA.** 2018. The new kid on the block: a specialized secretion system during bacterial sporulation. *Trends Microbiol* **26:**663–676.

20. **Planet PJ, Kachlany SC, DeSalle R, Figurski DH.** 2001. Phylogeny of genes for secretion NTPases: identification of the widespread tadA subfamily and development of a diagnostic key for gene classification. *Proc Natl Acad Sci U S A* **98:**2503–2508.

21. **Lu C, Turley S, Marionni ST, Park YJ, Lee KK, Patrick M, Shah R, Sandkvist M, Bush MF, Hol WG.** 2013. Hexamers of the type II secretion ATPase GspE from *Vibrio cholerae*

with increased ATPase activity. *Structure* **21**:1707–1717.

22. **Mancl JM, Black WP, Robinson H, Yang Z, Schubot FD.** 2016. Crystal structure of a type IV pilus assembly ATPase: insights into the molecular mechanism of PilB from *Thermus thermophilus. Structure* **24**:1886–1897.

23. **Söding J, Biegert A, Lupas AN.** 2005. The HHpred interactive server for protein homology detection and structure prediction. *Nucleic Acids Res* **33**(Suppl 2):W244–W248.

24. **Slabinski L, Jaroszewski L, Rychlewski L, Wilson IA, Lesley SA, Godzik A.** 2007. XtalPred: a web server for prediction of protein crystallizability. *Bioinformatics* **23**:3403–3405.

25. **Py B, Loiseau L, Barras F.** 2001. An inner membrane platform in the type II secretion machinery of Gram-negative bacteria. *EMBO Rep* **2**:244–248.

26. **Abendroth J, Mitchell DD, Korotkov KV, Johnson TL, Kreger A, Sandkvist M, Hol WG.** 2009. The three-dimensional structure of the cytoplasmic domains of EpsF from the type 2 secretion system of *Vibrio cholerae. J Struct Biol* **166**:303–315.

27. **Karuppiah V, Hassan D, Saleem M, Derrick JP.** 2010. Structure and oligomerization of the PilC type IV pilus biogenesis protein from *Thermus thermophilus. Proteins* **78**:2049–2057.

28. **Zeytuni N, Flanagan KA, Worrall LJ, Massoni SC, Camp AH, Strynadka NCJ.** 2018. Structural characterization of SpoIIIAB sporulation-essential protein in *Bacillus subtilis. J Struct Biol* **202**:105–112.

29. **Iwata M, Imamura H, Stambouli E, Ikeda C, Tamakoshi M, Nagata K, Makyio H, Hankamer B, Barber J, Yoshida M, Yokoyama K, Iwata S.** 2004. Crystal structure of a central stalk subunit C and reversible association/dissociation of vacuole-type ATPase. *Proc Natl Acad Sci U S A* **101**:59–64.

30. **Kuhlen L, Abrusci P, Johnson S, Gault J, Deme J, Caesar J, Dietsche T, Mebrhatu MT, Ganief T, Macek B, Wagner S, Robinson CV, Lea SM.** 2018. Structure of the core of the type III secretion system export apparatus. *Nat Struct Mol Biol* **25**:583–590.

31. **Serrano M, Vieira F, Moran CP Jr, Henriques AO.** 2008. Processing of a membrane protein required for cell-to-cell signaling during endospore formation in *Bacillus subtilis. J Bacteriol* **190**:7786–7796.

32. **Green ER, Mecsas J.** 2016. Bacterial secretion systems: an overview. *Microbiol Spectr* **4**:215–239.

33. **Minamino T, Morimoto YV, Hara N, Namba K.** 2011. An energy transduction mechanism used in bacterial flagellar type III protein export. *Nat Commun* **2**:475.

34. **Minamino T.** 2014. Protein export through the bacterial flagellar type III export pathway. *Biochim Biophys Acta* **1843**:1642–1648.

35. **Abrusci P, Vergara-Irigaray M, Johnson S, Beeby MD, Hendrixson DR, Roversi P, Friede ME, Deane JE, Jensen GJ, Tang CM, Lea SM.** 2013. Architecture of the major component of the type III secretion system export apparatus. *Nat Struct Mol Biol* **20**:99–104.

36. **Hu B, Lara-Tejero M, Kong Q, Galán JE, Liu J.** 2017. In situ molecular architecture of the *Salmonella* type III secretion machine. *Cell* **168**:1065–1074.e10.

37. **Worrall LJ, Vuckovic M, Strynadka NCJ.** 2010. Crystal structure of the C-terminal domain of the *Salmonella* type III secretion system export apparatus protein InvA. *Protein Sci* **19**:1091–1096.

38. **Zeytuni N, Flanagan KA, Worrall LJ, Massoni SC, Camp AH, Strynadka NCJ.** 2018. Structural and biochemical characterization of SpoIIIAF, a component of a sporulation-essential channel in *Bacillus subtilis. J Struct Biol* **204**:1–8.

39. **Bergeron JRC, Worrall LJ, Sgourakis NG, DiMaio F, Pfuetzner RA, Felise HB, Vuckovic M, Yu AC, Miller SI, Baker D, Strynadka NC.** 2013. A refined model of the prototypical *Salmonella* SPI-1 T3SS basal body reveals the molecular basis for its assembly. *PLoS Pathog* **9**:e1003307.

40. **Korotkov KV, Gonen T, Hol WGJ.** 2011. Secretins: dynamic channels for protein transport across membranes. *Trends Biochem Sci* **36**:433–443.

41. **Rodrigues CDA, Henry X, Neumann E, Kurauskas V, Bellard L, Fichou Y, Schanda P, Schoehn G, Rudner DZ, Morlot C.** 2016. A ring-shaped conduit connects the mother cell and forespore during sporulation in *Bacillus subtilis. Proc Natl Acad Sci U S A* **113**:11585–11590.

42. **Zeytuni N, Hong C, Flanagan KA, Worrall LJ, Theiltges KA, Vuckovic M, Huang RK, Massoni SC, Camp AH, Yu Z, Strynadka NC.** 2017. Near-atomic resolution cryoelectron microscopy structure of the 30-fold homooligomeric SpoIIIAG channel essential to spore formation in *Bacillus subtilis. Proc Natl Acad Sci U S A* **114**:E7073–E7081.

43. **Suzuki H, Yonekura K, Namba K.** 2004. Structure of the rotor of the bacterial flagellar motor revealed by electron cryomicroscopy and single-particle image analysis. *J Mol Biol* **337**:105–113.

44. **Yan Z, Yin M, Xu D, Zhu Y, Li X.** 2017. Structural insights into the secretin transloca-

tion channel in the type II secretion system. *Nat Struct Mol Biol* **24:**177–183.

45. **Levdikov VM, Blagova EV, McFeat A, Fogg MJ, Wilson KS, Wilkinson AJ.** 2012. Structure of components of an intercellular channel complex in sporulating *Bacillus subtilis. Proc Natl Acad Sci U S A* **109:**5441–5445.

46. **Meisner J, Maehigashi T, André I, Dunham CM, Moran CP Jr.** 2012. Structure of the basal components of a bacterial transporter. *Proc Natl Acad Sci U S A* **109:**5446–5451.

47. **Blaylock B, Jiang X, Rubio A, Moran CP Jr, Pogliano K.** 2004. Zipper-like interaction between proteins in adjacent daughter cells mediates protein localization. *Genes Dev* **18:**2916–2928.

48. **Doan T, Marquis KA, Rudner DZ.** 2005. Subcellular localization of a sporulation membrane protein is achieved through a network of interactions along and across the septum. *Mol Microbiol* **55:**1767–1781.

49. **Rodrigues CDA, Ramírez-Guadiana FH, Meeske AJ, Wang X, Rudner DZ.** 2016. GerM is required to assemble the basal platform of the SpoIIIA-SpoIIQ transenvelope complex during sporulation in *Bacillus subtilis. Mol Microbiol* **102:**260–273.

50. **Aung S, Shum J, Abanes-De Mello A, Broder DH, Fredlund-Gutierrez J, Chiba S, Pogliano K.** 2007. Dual localization pathways for the engulfment proteins during *Bacillus subtilis* sporulation. *Mol Microbiol* **65:**1534–1546.

51. **McKenney PT, Eichenberger P.** 2012. Dynamics of spore coat morphogenesis in *Bacillus subtilis. Mol Microbiol* **83:**245–260.

52. **Flanagan KA, Comber JD, Mearls E, Fenton C, Wang Erickson AF, Camp AH.** 2016. A membrane-embedded amino acid couples the SpoIIQ channel protein to anti-sigma factor transcriptional repression during *Bacillus subtilis* sporulation. *J Bacteriol* **198:**1451–1463.

53. **Ramírez-Guadiana FH, Meeske AJ, Rodrigues CDA, Barajas-Ornelas RDC, Kruse AC, Rudner DZ.** 2017. A two-step transport pathway allows the mother cell to nurture the developing spore in *Bacillus subtilis. PLoS Genet* **13:**e1007015.

54. **Chiba S, Coleman K, Pogliano K.** 2007. Impact of membrane fusion and proteolysis on SpoIIQ dynamics and interaction with SpoIIIAH. *J Biol Chem* **282:**2576–2586.

55. **Jiang X, Rubio A, Chiba S, Pogliano K.** 2005. Engulfment-regulated proteolysis of SpoIIQ: evidence that dual checkpoints control σ activity. *Mol Microbiol* **58:**102–115.

Index